Hunting Down Social Darwinism

Hunting Down Social Darwinism

Will This Canard Go Extinct?

Stuart K. Hayashi

LEXINGTON BOOKS
Lanham • Boulder • New York • London

Published by Lexington Books
An imprint of The Rowman & Littlefield Publishing Group, Inc.
4501 Forbes Boulevard, Suite 200, Lanham, Maryland 20706
www.rowman.com

Unit A, Whitacre Mews, 26-34 Stannary Street, London SE11 4AB

British Library Cataloguing in Publication Information Available

Library of Congress Cataloging-in-Publication Data Available

ISBN 978-0-7391-8670-1 (cloth : alk. paper)
ISBN 978-0-7391-8671-8 (electronic)

∞ ™ The paper used in this publication meets the minimum requirements of American National Standard for Information Sciences Permanence of Paper for Printed Library Materials, ANSI/NISO Z39.48-1992.

Printed in the United States of America

Contents

Acknowledgments

I wish to thank Robert C. Bannister, author of *Social Darwinism: Science and Myth in Anglo-American Social Thought* and one of the earliest and most prominent scholars to challenge the prevailing interpretations about social Darwinism and Herbert Spencer which Richard Hofstadter established in *Social Darwinism in American Thought*. In the process of writing this work, I came across an odd quotation ascribed to Herbert Spencer, and I suspected that this was a misattribution. Then unable to trace the source of that mysterious quotation, I contacted Dr. Bannister to ask him if he was aware of the quotation's origin. As explained in chapter 2, Dr. Bannister proved immensely helpful in this regard.

Preface

To Oppose the Welfare State Is to Be a Bigot?

I type these words during a phase in which the United States and Europe face a fiscal crisis. Following decades of spending tax money to support welfare programs not merely for the poor but also for the middle class, the USA and Europe find themselves confronting the great likelihood that their welfare systems cannot be sustained indefinitely. Some commentators of the free-market persuasion, recognizing the dangers of national governments digging themselves deep into debt through their welfare programs, advise that these governments exercise greater fiscal discipline, reducing their spending and allowing for their economies to recover and prosper through economic deregulation and liberalization. Such free-market suggestions come from a sound basis, but the partisans of the entrenched welfare-state mechanisms stubbornly deny the need to abolish or so much as shrink the pervasive influence of these welfare systems. In response to any call for the freeing of enterprise and disempowerment of the welfare state, the partisans of the welfare state howl that any proposal to discontinue welfare payments—however gradually they are phased out—amounts to a genuflection to the discredited doctrines of social Darwinism.

Social Darwinism, we are told, is a doctrine that right-wing bigots invented in the nineteenth century. In this age, Charles Darwin laid forth his theory of natural selection, pointing out that various organisms of the same species competed with one another for limited resources in an effort to survive. On account of mutations, some individual organisms developed traits unlike those of other members of the same species. Individuals possessing the traits that proved most advantageous in adapting to the environment would survive and procreate. To subsequent generations, these survivors would transmit the very same traits that had helped them adapt and flourish. Conversely, the competitors of the same species possessing the less advantageous properties would be unable to survive long enough to reproduce. They would perish and thereby prove unable to transmit their disadvantageous traits to other generations.

Here, the advocates of the welfare state—which I also call the *regulatory-entitlement state*—proclaim that nineteenth-century social Darwinists sought inappropriately to apply Darwin's theory to the study of human society, particularly in the area of government policies pertaining to political economy. The welfare state's partisans charge that the social Darwinists, most prominently Great Britain's Herbert Spencer and the USA's William Graham Sumner, posited a theory to this effect:

Competition in the free-market capitalist economy recreates the competition in the jungle that Darwin described, as well it should. Should someone ascend from penury to wealth and prestige in the market economy, it is on account of that person possessing superior biological traits. Such a person is obligated to have children, in whom those same superior biological traits shall manifest. Contrariwise, if someone remains in penury her entire life, her failure to ascend to wealth is to be attributed to her possessing low-quality inborn traits. The best outcome is for such a low-quality person to die in the absence of bearing any children, for any children she raises shall likely inherit her inferior biological traits and continue to retard the progress of the human race. As private charity and tax-funded welfare enable the inferior poor to survive and support their children, private charity and tax-funded welfare enfeeble Western society. What is best is to institute a strict laissez-faire capitalist policy, cutting off tax-funded welfare and letting the inferior classes die off as they ought.

The welfare-state proponents identify Spencer and Sumner as having originated that cold-hearted prescription, and they then proclaim that anyone who favors a reduction in tax-funded welfare must favor such a reduction, on some psychological level, on the same bigoted basis as the one ascribed to Spencer and Sumner. In short, it is said, to resent the welfare system is to be a social Darwinist—a right-wing bigot whose policy recommendations, if implemented, would result in poor people dying when their lives would otherwise be saved. Social Darwinism, then, rationalized classism, elitism, and able-ism. And it gets worse, add the proponents of the welfare state, because the evils of the right-wingers' social Darwinist doctrine go farther. Seeking to continue the misapplication of Darwin's theory to the rationalization of right-wing policy proposals, claim the welfare statists, the social Darwinists developed the pseudoscientific ideology of eugenics. Eugenics proclaimed that a person's genes were what primarily determined his or her character and behavior. As mental illness was ascertained as genetically heritable, eugenicists deemed epileptics and sufferers of bipolar disorder to be defective, just as social Darwinists had stigmatized the poor. In turn, social Darwinists urged for the oppression of the mentally ill. That is why social Darwinists supported the policy of state governments to mandate highly invasive medical procedures that sterilized epileptics and manic depressives, against their consent. Moreover, as sexual dimorphism renders inborn biological differences between men and women, social Darwinists cited biological theory to rationalize Victorian gender roles. Consistent with the assumptions about the inferiority of the poor and the mentally ill, social Darwinists judged women to be congenitally biologically inferior to men and worthy of nothing but domination by men.

Likewise, the social Darwinists noticed that, in general, two people of different "races" were more genetically dissimilar than were two people of the same "race." Assuming Caucasians to be the most genetically superlative race, social Darwinists decided to expand their bigoted conclusions about the poor and the mentally ill to the nonwhite races. Following in their stigmatization of paupers and depressives as persons too weak to deserve to live, they likewise stigmatized nonwhites as weaklings who ought to be subjugated or exterminated. Hence, Spencer's and Sumner's social Darwinism rationalized European conquest of indigenous peoples in Africa, Asia, and the Americas. In total, social Darwinism misused biological science to legitimize the status quo of white, male, right-wing capitalist patriarchy to engender classism, elitism, able-ism, sexism, racism, colonialism, and fascism. When these conservative right-wing *ism*'s were integrated into a single political philosophy, the result was Nazism. Applying social Darwinism, the Nazis decreed Jews to be the congenitally inferior race of underdogs, and therefore sought to squelch them. Insofar as we account for such considerations, say the critics of laissez faire, it follows that if a free-marketer takes his own latent social Darwinist premises to their logically necessary conclusion, the free-marketer's philosophy leads to Naziism. And if the free-marketer does not go that far, adds the welfare proponent, the fact remains that any implementation of the free-marketer's suggestion for cutting welfare will result in poor people perishing. In short, goes the welfare proponent's argument, someone who demands constraints on tax-funded welfare payments is a right-wing bigot at best and a Nazi at worst.

As an example of a partisan of the regulatory-entitlement state denouncing free-marketers as social Darwinists, I present a speech that Barack Obama provided in 2006 in a commencement address at Knox College. Then-Senator Obama announced:

> At the end of the Civil War, when farmers and their families began moving into the cities to work in the big factories that were sprouting up all across America, we had to decide: Do we do nothing and allow captains of industry and robber barons to run roughshod over the economy and workers by competing to see who can pay the lowest wages at the worst working conditions? Or we do try to make the system work by setting up basic rules for the market, instituting the first public schools, busting up monopolies...?
>
> We chose to act, and we rose together. . . .
>
> Today, at the beginning of this young century, we have to decide again. . . . Once again, there are those who believe...that the best idea is to give everyone one big refund on their government...in the form of tax breaks,...and encourage everyone to use their share to go buy their own health care, their own retirement plan, their own child care, their own education, and so on.
>
> In Washington, they call this the Ownership Society. But in our past there has been another term for it—Social Darwinism—every man or woman for him- or herself. It's a tempting idea, because it doesn't require much thought or ingenuity. It allows us to say [to] those whose health care [costs] or tuition may rise faster than they can afford—tough luck. It allows us to say to the . . . workers who lost their job—life isn't fair. It lets us say to the child who was born into poverty—pull yourself up by your bootstraps. And it is especially tempting because each of us believes we will always be the winner in life's lottery, that we're the ones who will be the next Donald Trump . . .

Such an idea, continued then-Senator Obama, "is a problem. It won't work. It ignores our history." Then, upon proclaiming that support for the free market is social Darwinism, then-Senator Obama promoted, as a wonderful alternative, his version of intrusive government control and social collectivism.[1]

This common denunciation of free-market advocates as social Darwinists—a code for *classist, elitist, sexist, racist, right-wing bigots*—is of concern to me, as it contains implications for the work you are currently reading. This work, *Hunting Down Social Darwinism*, is the final installment of the trilogy *The Nature of Liberty*, which provides a secular, philosophical defense of free enterprise against the regulatory-entitlement state.

Book One of Three

To review, the first installment of the trilogy was *The Freedom of Peaceful Action: On the Origin of Individual Rights* (*FOPA*). This first book was divided into two parts. Part I argued that there can be no case for free enterprise except a rational one. To begin, there is no sort of truth except that which is ascertained through the procedures of inductive, observational reason. On that foundation, this work defends the veracity of inductive cogitation against the criticisms of the philosophers David Hume and Immanuel Kant. Having defended the proposition that truth is found exclusively through inductive reasoning, I proceed to argue that the principle applies to moral truth—ethical truth can be determined by no means other than inductive reason. This means that ethical truth cannot be ascertained through religious revelation or through inborn, emotional intuition. In proposing that ethical principles must derive from observational reason, I challenge David Hume's dictum that ethical principles cannot properly be discovered from the facts. Once establishing that ethics originates from facts, I cite an argument from the twentieth-century Russian-born author Ayn Rand, whose philosophy—called Objectivism—bears considerable influence over this entire trilogy. Ethics is about deciphering which general principles are best for guiding one's actions, and the first question to ask of ethics is, "Why should I care about ethics to any degree?" The rational answer is that it is in one's long-range self-interest to care about ethics. Such self-interest does not consist in indulging every whim—doing so would result in many terribly uncomfortable and dangerous consequences—but in attaining a long-range general happiness and well-being. That well-being is what such ancient Athenians as Aristotle identified as *eudemonia, eudaemonia,* or *eudaimonia,* which translates to "good spirits." Contrary, then, to mainstream admonitions that acting in ethical principle precludes being primarily concerned with one's own self-interest, it happens that following one's long-range, eudemonia-seeking self-interest serves as the most logical foundation for the discipline of ethics. Indeed, self-interest also serves as the basis for caring about philosophy and reason. As observational rationality serves as the sole means of discerning truth, it follows that the lone sensible purpose in exercising rationality is that doing so is consonant with the long-range pursuit of eudemonia. Exercising the capacity to care for other people—as opposed to harming those other people or being indifferent to their plights—is also consistent with serving one's long-range self-interest. It is therefore ethically proper to care for other people, but—despite the claims of most philosophers—the ethical properness of caring about others arrives not from some unspecified duty or commandment, but from recognition of the manner in which one directly and indirectly benefits in the long run from exercising such care. Book One also explains that such ethical principles are absolute, and that they are not "absolute" in the manner that philosophical writings have conventionally meant by *absolute.* That is, for a philosophic principle to be properly absolute means that it remains consistently applicable within its own properly specified context. The notion that a principle cannot be absolute unless it applies in any and every context, including contexts not yet conceived, is the notion that I call that of *platonic absolutes,* called such because of its derivation from the philosophy of Plato, though Kant popularized the idea as well. By contrast, I argue that a principle that is properly recognized as absolute is a *contextual absolute*—a principle that is consistently applicable within a specific context, and which can be applied to a separate context only insofar as that separate context is similar enough to the original context in the pertinent respects. Contextually absolute principles remain consistently applicable in the hard sciences, and they remain consistently applicable in the area of human interaction.

And that is where we come to the central issue of the trilogy—the proper ethical policy that must guide a government. In Part II of Book One, I explain that ethical policy, which I call the Rule of Peace. The Rule of Peace states that it is immoral to initiate the use of force against another sapient party's person or private property when that other sapient party is peaceful and innocent. The initiation of the use of force refers to such actions as bodily injury, murder, rape, poisoning, theft, extortion, abduction, vandalism, property damage, contract breach, fraud, and intellectual property infringement. Such initiations of the use of force are called *spoliation.* Should Brad initiate the use of force against Constance, the immorality of that action derives not merely from the direct harm that Brad imposes upon Constance, but from the adverse consequences such actions will likely impose upon Brad himself in the long run. As Constance initiates the use of force upon no one else, she is to be deemed *peaceful* and *innocent,* and to initiate the use of force upon her is evil. Once Brad initiates the use of force upon Constance, though, Constance's ethical right to long-range self-interest does justify that she retaliate—that either she herself exercises retaliatory force against Brad or that she seeks out a third party to help violently defend her against Brad. That is where the role of proper government emerges.

A government refers to a specific party in a society. That party is a government when it fulfills the following conditions.

1. That party ultimately holds the final say over who may or may not exercise violence within that society.
2. When that party exercises force against other parties in that society, the majority of that society's members at least tacitly acquiesce to that party's authority. Should there be a portion of that society that rebels against this dominating party's author-

> ity, that rebellion is not large enough to overpower that dominating party or defeat it.

Note that a society need not be confined to a specific geographic spot to have a government. As a group of nomads possesses a central authority that exercises violence to punish dissenters, a nomadic group has a government; it is not anarchist. Moreover, people need not be in close geographic proximity to one another to be properly considered members of the same society under the same government. The inhabitants of a small island near a continent may be considered members of the same society as the continent's inhabitants, and a central authority located on the continent may serve as the government to the inhabitants of the nearby island. In any case, a party is a society's government when a large enough number of that society's members tacitly acquiesce to that party holding a monopoly on the exercise of violence. Nongovernmental parties might be able to exercise violence on their own accord, but not in the absence of the government's permission. Should any party exercise any violence in defiance of the government's authority, that party is *criminal* and the government exercises violence itself to stop or punish those who inflicted the violence the government had not authorized.

Throughout history there have been different forms of government, such as feudalism and communism. Yet a government is just—by the standards of the rational ethics we have advocated—insofar as it abides by the Rule of Peace. That entails that the government acts in no capacity other than to exercise violence to stop or punish the initiation of the use of force. A government that does nothing but act to punish or stop the initiation of the use of force is called a "night watchman state," as it behaves as a night watchman who does not intervene unless it is to fight violent wrongdoing. As every government action is violent, a government itself initiates the use of force on any occasion on which it does something other than violently deter or penalize spoliation. Suppose, for instance, that Perry and Jordana, being consenting adults, peaceably negotiate their own contract. Jordana agrees to be Perry's maid, and Perry will pay her two U.S. dollars per hour. The government then declares that the agreement between Perry and Jordana is illegal, based on their agreement violating minimum-wage statutes. The government informs Perry that it will fine him 5,000 dollars and, should he fail to cough up the fine, it will imprison him. This would be a case of the government itself threatening to initiate the use of force upon Perry. Perry was peaceful with Jordana, and yet the State threatens to send armed men after to Perry to manhandle him if he does not relinquish his peaceably acquired 5,000 dollars to the State. That sort of government regulation, then, is spoliation, and not the action of a night watchman state. It is instead a manifestation of what I identify as *governism*. *Governism* is present when at least one of the following conditions is present.

1. The belief is professed that the government is morally obligated to engage in compulsory taxation and to take on activities other than those to which a night watchman state is limited.
2. The government actually does engage in compulsory taxation or take on some other activity that is forbidden to a night watchman state.

Governism and the Rule of Peace stand in contrast to one another. A system of society that abides by the Rule of Peace—the system of the night watchman state—goes by various names. Other names include *freedom, liberty, the free market, free enterprise, laissez faire, consensualism, voluntarism, volitionism, autonomism, deregulation, privatization, Manchesterism,* and *capitalism*. *Capitalism* here must be qualified. Too often, the opponents of the night watchman state attempt to conflate any and every commercial transaction with the system of capitalism. Should Max kidnap Akuns and sell Akuns into slavery, the sale of Akuns as a slave would be a commercial exchange. Likewise, if Max hires Lucky to murder Max's wife, Max's payment to Lucky for that grisly service would also be a commercial exchange. Equating capitalism with commercial exchange, the welfare-state proponent says that slave traders and hit men practice capitalism, which goes to show that capitalism is not morally superior to governism. Yet that argument arrives through misunderstanding. Capitalist free enterprise entails that each person be free to enterprise peaceably. That means each person must maintain possession over his or her own most important capital asset—his or her own life. When Max pays for the murder of his wife and sells Akuns into slavery, those actions deprive Max's wife and Akuns of their own respective freedom to enterprise, and thereby rob Max's wife and Akuns of that capital asset which is their own bodily autonomy. Therefore, commercial exchanges that involve spoliation do not count as pure capitalism.

Other terms for a society that follows the Rule of Peace are *liberalization* and *liberalism*, and someone who advocates such a society is sometimes called a *liberal*. These terms must be qualified as well. Throughout the early 1800s, supporters of the night watchman state were known as *liberals*, as they supported *liberty*, the word from which *liberal* derived. In the late nineteenth-century England, though, some opponents of the night watchman state, such as Leonard Hobhouse, began to tout their own governism as a new, better liberalism. In short, British governists co-opted the word *liberalism*, and so too did the USA's governists by the 1930s. As I write this trilogy, it happens that when an American refers to *liberalism*, it is usually the case that he or she means not the Rule of Peace that I favor, but the Rule's opposite—governism. Except in cases where I specify otherwise, though, almost any time this book uses *liberal* or *liberalism*, it will refer respectively to free-marketers and capitalism. And, as shall be explicated throughout the book you presently read, *social Darwinism* is not among the synonyms for the night watchman state.

A night watchman state is about stopping spoliation—stopping the initiation of the use of force. To initiate the use of force is to violate individual rights. These individual rights are what I christen *Lockean rights*—the rights to life, liberty, and private property, as identified by John Locke in the 1680s. As Locke constructed much of the foundation for the philosophy of the night watchman state, the theory and practice of it can also be dubbed *Lockeanism*.

To be sure, no society—not even that of the United States—is, or ever has been, perfectly Lockean. Every society has had some freedom of enterprise in some places but intrusive government controls in others. The term for this unstable mixture of freedom and controls is *mixed economy*. Although every society is a mixed economy, some are more governist than others. The most extreme manifestations of governism include Naziism and communism. Absolute monarchy was another extreme version of governism. Less extreme, but still daunting, are socialism, Progressivism, and the regulatory-entitlement state, which, as we shall learn throughout this work, are frequently but incorrectly held up as morally exemplary foils to social Darwinism, eugenics, and Naziism.

Also significant to the matter is republicanism. A republic is a representative government in which citizens exercise the ability to vote for public officials. On account of a republic maintaining citizen suffrage, from the twentieth century onward political scientists have usually equivocated republicanism with democracy, saying that a republic is a specific type of democracy. That equivocation, though, proves misleading. Democracy, in its undiluted form, allows for citizens to vote on anything and everything—they can go as far as voting on whether a philosopher should be killed by the State, simply because that philosopher annoyed people. Starting in the twentieth century, political scientists pointed out that in a republic, not every measure is put to a vote but that, rather, citizens delegate voting authority to specific representatives to vote on what legislation should or should not be implemented. But such political scientists omit mention of another important distinction. It is that a republic is governed by a charter which spells out important principles and procedures that the State must implement. A republic considers those principles and procedures to be so important that the republic stipulates that citizens either cannot vote down such procedures, or cannot simply amend those procedures through a majority vote, instead requiring a supermajority. Consistent with the principles of liberalism, a liberal republic prioritizes the individual's autonomy over any attempt by the majority to spoliate the individual for some ostensive collective benefit. Were the police or prosecutors of a government to be able to prosecute people with impunity, there is a great likelihood that they could falsely accuse innocent people of crimes and then inflict violence upon them as punishment. Such a government, then, would be spoliating innocent people. It is to guard against such violations that a liberal republic institutes the rule of law and Due Process. As I argued in Book One, though, a regulatory-entitlement state can maintain due process while continuing to spoliate people wrongfully. Imagine, for instance, that a government instituted a law forbidding anyone from keeping a feline in her home. Angela peaceably keeps a cat in her house, harming no one else. That government could go through the necessary due-process procedures to prove, beyond any shadow of a doubt, that Angela indeed committed the infraction for which the authorities had charged her. If Angela receives the necessary due process, the government will still spoliate her by punishing her for her peaceable action. Therefore, a regulatory-entitlement state can maintain republican Due Process in the absence of the pure freedom of a night watchman state. Despite that fact, Due Process is necessary to guard against violent punishment of the innocent and falsely accused, and therefore the freedom of the night watchman state cannot persist, in the long run, in the absence of Due Process.

Throughout part II of Book One, I explained the principles governing a night watchman state and how, according to the standards of rational ethics, any system of government other than the night watchman state proves a moral failure. I went as far as explaining that, in the long run, a perfectly free society would finance its night watchman state's police, military, and courts through consensual contracts and not through compulsory taxation. I also went into detail to refute the notion, popularized by Thomas Hobbes and Jean-Jacques Rousseau, which states that, simply by taking part in a community, a person tacitly consents to some Social Contract that justifies compulsory taxation and every other State encroachment upon the individual's peaceful autonomy.

Book Two of Three

Book Two of the trilogy—*Life in the Market Ecosystem* (*LIME*)—is divided into three parts, and in it I explain what evolutionary theory has to say on this issue. Despite such evolutionary psychologists as Edward O. Wilson and Charles Lumsden opposing the night watchman state, I contend that when one takes Wilson's and Lumsden's well-argued theory of gene-culture co-evolution to its final conclusion, one finds that free enterprise is the social system most conducive for every individual's prosperity and therefore, by the standards of rational ethics, the morally supreme social system. Part I explains the basics of how evolutionary theory applies. The evolutionary psychologists' gene-culture co-evolution theory recognizes that whether or not a person practices a specific custom is not genetically predetermined; it is not as if a person's genes cause him or her to observe that custom. Likewise, should someone practice a specific custom, his or her skill in doing so is not necessarily genetically predetermined. Rather, when someone practices a custom, it is on account of that person taking the initiative to invent it him- or herself—an act of free will—or of having adopted that custom upon learning it from others. In any case, continue the evolutionary psychologists, although no custom is genetically inborn, it does happen that whether or not a person practices a specific custom can help influence the likelihood of that person surviving long enough to have children and raise those same children into adulthood. Therefore, whether or not a person practices a specific custom does help determine whether or not that person transmits his or her genes to the next generation. And, add the evolutionary psychologists, given that parents usually teach those same customs to their own children, it follows that when practicing a custom helps a man survive and raise his own children, it is not merely that the custom helps transmit the man's genes across generations; it also happens that, insofar as the custom helps the man survive and reproduce, that custom helps transmit itself across generations. A custom, then, counts as an evolutionary adaptation.

Furthermore, when several customs are grouped together, those customs constitute a culture. Among such sets of cultural customs are political systems—that is, capitalism versus governism. And although no society is purely capitalist, it is the case that some societies are freer than others. You can compare two societies that are similar ethnically, which have a similar racial background and which involve people speaking the same language, but in which one society is freer than the other. You can compare the progress of less-free Hungary against freer Austria; less-free mainland China against freer Taiwan; less-free East Germany against freer West Germany; less-free North Korea against freer South Korea; and less-free Cuba against freer Puerto Rico. There was a point where the people of both less-free Hungary and freer Austria started out in roughly the same place—extreme poverty—but in which people in the freer society ultimately gained a higher standard of living than those in the more governist society. That trend applies in each of the comparisons. You can also examine the southern and northern halves of the town of Nogales, which is divided by the U.S.-Mexican border. The people of both halves are of similar ancestry and speak Spanish, yet, throughout the decades, those of the freer U.S. half prospered more significantly than those of the less-free Mexican half.

These results illustrate that the survival of the fittest does apply to social science, though not in the manner in which social Darwinists were purported to believe. A social Darwinist, we are expected to think, would proclaim that if the people of North Korea and South Korea started in roughly the same poverty, and if the South Koreans ultimately grew more successful than the North Koreans, it was on account of the South Koreans being of a genetically superior "race" than the North Koreans. But the races and genetics are, in the pertinent respects, basically the same. The survival of the fittest applies in a different respect. Herbert Spencer coined *survival of the fittest*, and what he meant by that was that when a group of organisms survive and thrive in a particular environment, it is on account of an adequate match between the organisms and their environment. *Survival of the fittest* refers to the survival of organisms that *fit* in with their surroundings. When organisms are unable to survive and prosper in a specific environment, it indicates a mismatch between those organisms and that environment. Should organisms be unable to survive in a particular environment, it is not necessarily because the environment is perfect whereas the organisms are unworthy of it. It may be the case that the organisms are just fine and that the environment itself is inadequate. Note that the political system of a human society is not merely a human-created adaptation, but that it is also a human-created *environment*. When humans thrive in a capitalist society and suffer under a communist one, it is not on account of the people in the capitalist society being genetically or racially superior to those of the communist society. Rather, in the pertinent respects, the genes of the people in the capitalist society are roughly the same in quality as the genes of those in the communist society. Prosperity under capitalism is the survival of the fittest on account of how the peaceable freedom that capitalism affords happens to be the sort of environment which is most fit for humans. By contrast, governism produces an environment that stifles the ability of people to support their own well-being. Governism, then, is an environment unfit for humans—and the more severely governism is applied, the less prosperous are those humans in comparison to how they would fare under capitalism.

Just as a single custom counts as an evolutionary adaptation, so too are cultures and political systems a sort of evolutionary adaptation. The principle demonstrates that amidst every possible social system humans can adopt, capitalism is the evolutionary adaptation that has been most conducive to helping human beings survive and thrive. I call this the Survival of the Freest. Part I of Book Two devotes itself to explaining that principle. Part II then moves on to a common rejoinder advanced by environmentalists. Numerous environmentalists proclaim that although capitalism has benefited Western civilization in the past, the West prospers on borrowed time, as capitalism depletes a fixed quantity of nonrenewable natural resources and dangerously alters the Earth's climate. Therefore, conclude these environmentalists, any commercial community inevitably self-destructs in the long run, which means our society must become more governist if it is to avert disaster. I spent part II explicating how capitalist principles—applied most consistently—are what best redress these supposed weaknesses of capitalism. The final part of Book Two—part III—addresses the governists' proclamation that laissez-faire capitalism is immoral on account of how capitalism leaves people free to refrain from charity for the poor. The governists, predictably, pronounce that it is the duty of the State to take money from the rich and provide for the poor. I rebutted that argument as well.

Yet the governist argument utilizes the idea that we discussed at the start of this preface—that to leave people free to refrain from charity, as opposed to forcing them to give money to the poor, is social Darwinism. Because my argument in Book Two applies evolutionary theory to a defense of capitalism, my argument also leaves critics an opportunity to accuse it of attempting to revitalize social Darwinism. I therefore intend to devote this book—the final installment of our trilogy—to exploring the extent to which it is fair or unfair to accuse a conviction of philosophy or science of promulgating social Darwinism.

To wit, Book Three—the present volume—is about this expression. This is a search for—a stalking of—the meaning of that notorious phrase. We are hunting down *social Darwinism*.

Book Three of Three—The Present Volume

Book Three is broken into three Parts, the first two of which take the form of a historiography. In part I, we shall explore the history of the expression *social Darwinism*. We shall examine whether anyone in the nineteenth century called himself a social Darwinist, and whether Herbert Spencer and William Graham Sumner really did argue that capitalist principles require that poor people be left to die and unable to raise kids. Then, given that the partisans of the regulatory-entitlement state finger Spencer and Sumner as the two persons to inspire the eugenicists' pseudoscientific rationalizations for racism, we shall investigate the extent to which the free-market writings of Spencer and Sumner did or did not influence the eugenics movement. Because such governist Progressives as Theodore Roosevelt, Louis D. Brandeis, and Lester Frank Ward, as well as socialists in the vein of George Bernard Shaw and Beatrice Webb, opposed the laissez-capitalism of Spencer and Sumner, such Progressives and socialists are usually venerated as the arch-opponents of social Darwinism and, by extension, the arch-opponents of eugenics. For that reason, we will probe into whether these Progressives and socialists themselves supported eugenicist legislation and "scientific racism," which would actually leave them vulnerable to the charge of social Darwinism themselves, insofar as that label means anything. Were we to discover that it was Progressives, socialists, and leftists themselves who championed eugenics and politically mandated racial discrimination, with Spencer and Sumner and other free-marketers being the ones to denounce such measures, it would call into question the accusation that bigotry and racism are furthered by either free-market advocacy or the application of biological theories to the social sciences.

Of special attention is the nature of the Nazi regime. As they argue that the free-market advocacy of Spencer and Sumner inevitably devolved into eugenicist legislation, the supporters of the welfare state extend that allegation farther to implicate Spencer's and Sumner's free-market advocacy as the inspiration for the eugenicist and genocidal policies of the Third Reich. Numerous supporters of the welfare state assert the following accusations:

1. The regimes of Benito Mussolini and the Nazis alike put fascism into practice.
2. Capitalism, taken to its logical conclusion, is fascism.
3. Therefore, accepting *1* and *2* above, capitalism ultimately changes itself into Naz-ism.
4. Big business brought Mussolini and Hitler to political prominence, and as big business represents capitalism, capitalism itself brought Mussolini and Hitler to prominence.
5. Capitalism, fascism, and Naziism are all politically right-wing.
6. Spencer and Sumner applied evolutionary theory to argue in favor of capitalism.
7. Nazis applied evolutionary theory to argue for their racial bigotry.
8. Accepting *5* as correct, it follows that Spencer and Sumner applied evolutionary theory to advance a right-wing conclusion, and that Nazis applied evolutionary theory to advance a right-wing conclusion. Given that Spencer, Sumner, and Nazis were uniformly right-wing, it is reasonable to infer that the applications of evolutionary theory by Spencer and Sumner were what inspired the right-wing applications of evolutionary theory by the Nazis.
9. The Third Reich, being extremely right-wing, possessed no features in common with socialism, Progressivism, the regulatory-entitlement state, or any ideas or policies associated with the political Left.

Part II shall query into whether any or all of the above propositions—*1* through *9*—are accurate, and to what degree they are accurate or inaccurate.

Part III of this work summarizes some final lessons for this trilogy. If Spencer and Sumner did not cite evolutionary theory in order to rationalize indifference or cruelty toward the poor, then the reader may inquire, "In what manner did Spencer and Sumner really attempt to apply evolutionary theory to their study of social science in general and free enterprise in particular?" In the final part, chapter 11 shall examine the specifics of how Spencer and Sumner sought to apply their knowledge of biology and evolution to their convictions about human society. This chapter shall also examine the extent to which the free-market economists, evolutionary psychologists, and Complexity Theorists of the late twentieth- and early twenty-first centuries have been influenced by theories pioneered by Spencer and Sumner. I shall also come to conclusions about the degree to which such free-market economists, evolutionary psychologists, and Complexity Theorists have been fair or unfair in their forceful disavowals and repudiations of Spencer and Sumner even as they seemingly benefit from the work that Spencer and Sumner did before them. Chapters 12 and 13 then revisit some topics that Book One explored. Considering that the entire trilogy has argued against government overreach, a number of purported defenders of capitalism—so-called libertarians—propose that the best method of guarding against government overreach is for Western society to adopt anarchy wholesale. Chapter 12 will consider whether anarchy is a viable alternative either to governism or the night watchman state. This chapter refers back to the discussion from Book One about the fact that before anyone accused of spoliation can be justly punished for such spoliation, that person must receive Due Process if that society is to maintain freedom.

Chapter 13 comes full circle, returning to the question that opened Book One: *Can liberty be properly justified on a philosophic basis other than inductive reason?* I address whether libel, slander, and defamation are permissible in a truly free society, and also whether a libertarian truly supports liberty if he believes that liberty does not require a philosophic justification based on inductive reasoning or an ethical code founded on self-interest. Following chapter 13 is a conclusion that summarizes the findings of the trilogy.

As I mentioned in the preface of Book One, I write in a time of social upheaval. The Western realm is at a crossroads. We can persist in condoning the manner in which the State has thwarted free enterprise, for fear of being labeled social Darwinists if we reject this unsustainable system of coercive taxation and mounting national debt. On the converse, we can consider whether the accusation of social Darwinism is but a canard. Then, no longer permitting such an epithet to intimidate us, we can stand firm in the conviction that free enterprise is just and necessary, not to be killed off by those who claim to know of the evils that social Darwinism brings.

NOTES

1. Barack Obama, Knox College commencement address, June 4, 2005, republished as Obama 2006, 36–40.

I

Stalking Social Darwinism

ONE

Did Nineteenth-Century Capitalists Want the Poor to Die?

To Support Laissez Faire Is to Murder the Poor?

The frequently repeated accusation is as follows. In the nineteenth century, the free-market advocates Herbert Spencer and William Graham Sumner argued for an unregulated economy and against the welfare state. They supposedly likened competition in the marketplace to struggle in the jungle. According to this interpretation, persons who rose to wealth or maintained their wealth were the fittest on account of their inborn biological makeup, on account of their genes. Due to their beneficent physiology, these plutocrats earned every opportunity to thrive and have children. Through their children, these plutocrats shall transmit, to future generations, the biological traits that had advantaged them throughout their lives. By contrast, persons who remained in penury were unfit and deserved to die. We are told that this is the reason why Spencer and Sumner opposed tax-funded welfare—if tax-funded welfare kept poor people alive, financially supporting these stragglers wasted resources and held back everyone else, particularly the much-better rich people. Supposedly, this misapplication of Darwinian theories rationalizes right-wing public policy. We are told that this theory has the name *social Darwinism*. Worse, this misapplied theory was extended to justify other forms of bigotry. Ultimately, social Darwinists employed their rationalizations to advance every conceivable right-wing *-ism*: capitalism, classism, elitism, able-ism, scientism, racism, sexism, imperialism, militarism, and fascism. Each of these *–isms* ultimately came together to form one final *–ism*: Nazism. By implication, this fable instructs us, free-market advocacy indirectly brought about World War Two and the Holocaust, and it follows that anyone in the present who wishes to reduce tax-funded welfare spending is a bigot as the Nazis were. Therefore, any and every opponent of tax-funded welfare spending ought to be shunned as a bigot, and particular opprobrium must be cast not merely upon Spencer and Sumner, but also upon Ayn Rand, whom the political Left has additionally castigated as a social Darwinist.

Henceforth in this tome, I shall, for the most part, dispense with placing the expression social Darwinism in scare quotation marks. Bear in mind that, in almost any instance wherein you come across my usage of the expression, it should be read as if there are scare quotation marks.

In Book Two I mentioned the monograph *Social Darwinism in American Thought*, authored by Columbia University historian Richard Hofstadter (1916–1970). I will henceforth abbreviate this work as *SDAT* (pronouncing it "ESS-dât"), alternating that abbreviation with *American Thought*. Originally written in 1938 as the author's Ph.D. dissertation,[1] Hofstadter had this published as a popular work in 1944. It is this book, more than any other, which is culpable for the pervasive belief among the university-educated that twenty-first century support for a laissez faire economy can be traced to a nineteenth-century ideology known as social Darwinism.

SDAT quotes a 1905 prediction by business-bashing[2] politician William Jennings Bryan. Bryan proclaims that Charles Darwin's discoveries will "weaken the cause of democracy and strengthen class pride and the power of wealth." Afterward Hofstadter comments, "Here, as in other matters, Bryan had sound intuitions. . . ."[3] The Columbia historian concludes as much for a very specific reason. His entire book is a purported survey of how various individuals—every one of whom the author designates "conservative"—have exploited Darwin's discoveries for their own nefarious agendas. The two individuals whom the Columbia historian most famously indicts on this count are Herbert Spencer and William Graham Sumner. Subsequent to the publication of *SDAT*, governists on the political Left have had *carte blanche* on denouncing Sumner and Spencer on the most pernicious of terms, repeatedly going as far as putting words into their mouths. The "doctrines of Herbert Spencer and William Graham Sumner," writes Yale University historian Gabriel Kolko, were those of a "conservative Social Darwinism . . ." Conservative Darwinism, hisses Kolko, conjured up "a justification of the existing distribution of economic power and laissez faire."[4] As of this writing, the disrepute of Spencer and Sumner as laissez-faire social Darwinists—who wanted the underprivileged to commit suicide—is a weapon that left-wing governists effectively wield to ruin the public image of free-market proponents *per se*. Here is another example of the infamy of Spencer and Sumner being exploited to hurt any and every free-enterpriser. In 1982 *The New Republic* magazine rebuked President Ronald Reagan as a social Darwinist on account of his pro-capitalism rhetoric and his desire for cuts in the personal income tax.[5] A more recent incident was in April 2012, when Democrats and Republicans were arguing over the size of the federal budget, particularly with respect to spending on welfare programs. President Barack Obama characterized Republican Congressional efforts to limit federal spending as "thinly-veiled social Darwinism."[6]

One of the implicit targets of Obama's slur was U.S. Rep. Paul Ryan, then-chair of the U.S. House of Representatives Budget Committee. Ryan had proposed an alternate budget intending to slow the growth of federal spending. In response, Georgetown University theologian Thomas J. Reese publicly denounced Ryan's proposed cuts to welfare spending as being inimical toward Ryan's self-professed Catholic faith. "I am afraid," Reese proclaimed, "that Chairman Ryan's budget reflects the values of his favorite philosopher Ayn Rand rather than the gospel of Jesus Christ. Survival of the fittest may be okay for Social Darwinists but not for followers of the gospel of compassion and love."[7]

Comparably, Griffiths University humanities professor Hiram Caton takes offense at Thomas Jefferson's espousal of Lockean political philosophy. In particular is Jefferson's approving inference that a "natural aristocracy" will develop in a market economy. On that basis, Dr. Caton curses the Declaration of Independence's principal author as a practitioner of "social Darwinism a century before Darwin . . ."[8] Nor is the antislavery campaigner Frederick Douglass immune to this putdown. Douglass proclaimed that every person "is the rightful owner of his own body."[9] On account of such rhetoric, Yale University history professor David Blight derides Douglass for extolling "a laissez-faire individualism that echoed the reigning Social Darwinism of the day."[10] Professor David's inaccurate pronouncement blights this debate. And in a style similar to Blight's and Caton's, governist journalist Paulina Borsook criticizes social Darwinism as the "philosophy of that-which-governs-best-governs-least" and a "love" for "laissez-faire free-market economics . . ."[11]

The simplistic syllogism can be summarized as:

1. Spencer and Sumner were free-market advocates.
2. Spencer and Sumner were social Darwinists who wanted the poor to die.
3. Therefore, every free-market advocate is a social Darwinist who wants the poor to die.

This syllogism is implicit in the denigration of Nobel Prize-winning economist Friedrich August von Hayek and other free-marketers in an April 2004 op-ed by advertising-executive-turned-journalist Robert M. Rees. Rees's op-ed first states that Spencer and Sumner "worked to establish Social Darwinism and the survival-of-the-fittest doctrine upon which Libertarian thought. . . rests its case against government pursuit of social justice." This Libertarian dogma is none other than "the Social Darwinist view that any individual who can't get along should be ignored . . ." Rees imputes into free-marketers the belief that "the suffering of the poor and oppressed" is merely "the weeding out of the unfit." Rees then pegs Spencer and Sumner as Professor Hayek's "intellectual precursors" and intones, "To understand them is to understand Hayek."[12] Here, the syllogism goes:

1. Spencer and Sumner championed laissez-faire economics, and eagerly awaited the collective demise of the needy.
2. F. A. Hayek also championed laissez-faire economics.
3. Therefore, Hayek likewise eagerly awaited the collective demise of the needy.

Yes, Rees associates Spencer and Sumner with Hayek, hoping that the unsavory reputations of the first two men can tarnish the third's. So, too, does John Kenneth Galbraith invoke Spencer's name to cast a pall upon anyone who expresses some objection to welfare programs. "The poor. . . in the Spencerian view were the weaklings; their euthanasia was nature's way of improving the species." Galbraith curses that the "voice of Herbert Spencer is also still heard in powerful resistance to the more generally protective role of the state." He further promulgates that William Graham Sumner wanted the USA divided into two classes—"the rich and self-reliant and below them the ragged fringe. There could be a Darwinian selection of individuals, a Darwinian euthanasia of the fringe . . ."[13]

To my sorrow, Galbraith's misrepresentations of free-market advocates strongly influence the popular culture, gaining control over the opinions of non-academicians. Take, for instance, ostensibly educational videos put out by internet celebrity John Green (b. 1977). Most famous for having written the best-selling novel-to-motion-picture *The Fault in Our Stars*, Green also puts out left-wing propaganda videos on YouTube in praise of Keynesian fiscal stimulus and socialized medicine. He also puts out a putatively apolitical "educational" series titled *Crash Course in History*, which is shown in high schools. Not surprisingly, his *Crash Course* videos repeat the usual clichés about the evilness of capitalism and industrialization and of the need for Progressivism. One of Green's videos smears Spencer as an apologist who equated "corporations" with "people," and who said the poor deserve to get "poorer" on account of their having an "inherent flaw" in their genes. Failing to examine the context behind it, Green has his video exhibit a newspaper editorial cartoon of Herbert Spencer being labeled with the banner "Science" and being depicted as having the body of a lion. This lion-bodied Spencer is muzzled, as Spencer's critics are attempting to silence him. Green has this cartoon labeled with the caption "persecution complex" and implies that the cartoon was drawn by a thin-skinned supporter of consensualist economics. Green would have us believe that this editorial cartoonist, having sided with Spencer, resents anyone who questions Spencer's free-market advocacy.[14] In fact, the cartoon is in reference to a different controversy. It refers to when William Graham Sumner intended to assign one of Spencer's books on biological evolution for his sociology class. As Yale was founded as a religious institution, the college's president, Noah Porter (1811–1892), censured Sumner for assigning that book, as Spencer's theories contradicted a literal interpretation of Genesis. Sumner stood up to Porter, telling him that he would indeed teach the book, and that if Porter prioritized Biblical doctrine above scientific evidence, Porter would have to go ahead and terminate Sumner's professorship. In the end, Sumner remained at Yale and refrained from assigning the work, saying that the controversy undermined its usefulness as a textbook. The matter resulted in increased press coverage for Spencer's writings in the USA, though, and Porter inadvertently drew more interest to the man's philosophy. The editorial cartoon that Green showcases was in reference to that controversy—of religious authorities trying to suppress discussion of Spencer's ideas concerning biological evolution. Professors who are aware of the context happen to side with Sumner and Spencer in this case, calling it a victory in the history of the academic-freedom movement.[15] *American Thought* itself tells this story.[16] But, on account of his yielding to assumptions and failure to consider evidence, John Green derides a cartoon in defense of academic freedom as an apologia for corporations.

On Thursday, November 10, 2011, one letter to the editor in the *Honolulu Star-Advertiser* delivered a smorgasbord of the capitalism-haters' favorite clichés: "While some of our American middle class still believe in trickle-down economics, the upper-class believes in the sucking-up theory, in which more and more wealth and political power are placed in the hands of fewer and fewer individuals and large corporations. A crude and sometimes cruel social Darwinism, the survival of the economic fittest, has become a part of the radical right's agenda and ideology. . . " To complete the triteness of the letter, its writer included a dig against the title of one of Ayn Rand's best books—"Whatever happened to the fundamental belief that we are our brothers' and sisters' keepers? Selfishness has become a primary virtue."[17]

The *Miami Herald* is far more direct in its attack, ridiculing free-market advocates for "their Ayn Rand-modeled social Darwinism and adherence to libertarian precepts like completely deregulating business, and stopping the government from providing Medicare, Social Security or regulating food, drugs or civil rights."[18]

A similar commentary appears in the Friday, May 11, 2012, Henderson, Kentucky's *Gleaner*. Laboring under the prejudice that every last free-marketer belongs to the Religious Right, the letter writer sarcastically snips, "After listening to all of today's right-wing political leaders' adoration of the renowned Russian atheist, Ayn Rand ('Atlas Shrugged'), I realize why they hold the anti-Christian opinion that love of self and greed is the ultimate basis for a virtuous and productive life." The letter-writer then goes on to provide a completely un-sourced, fabricated quotation from Ayn Rand: "Rand states that, 'Anything other than putting your own desires above all others is immoral.' She and her Republican followers reject the compassion and mercy of Christ's teachings (Luke 10, Good Samaritan). They believe all ethical philosophy from other religions are also contrary to the natural rule of 'Social Darwinism' (survival of only the economically fittest is justified)."[19] Despite the Religious Right's objection to natural-selection theory, though, some prominent figures of the political Left recognized Darwin as their enemy, and juxtaposed the image of Darwin against the moral supremacy of the political Left. In a 1934 fresco in Mexico City, which he titled "Man at the Crossroads," the Marxian painter Diego Rivera (1886–1957) made known which historical figures he regarded as good and which as evil. On the right of the painting he portrayed historical personalities he admired—Karl Marx, Vladimir Ilyich Lenin, and Leon Trotsky—whereas, on the left, in opposition to them, was Charles Darwin, whose theory Rivera surmised to be a rationalization for capitalist competition.[20]

Although he did not employ the iconic expression *social Darwinism*—more on that expression's origin in chapter 2—John Maynard Keynes did invoke the image that free enterprise is Darwinian savagery. "The parallelism between economic *laissez-faire* and Darwinianism," he wrote ". . . is now seen, as Herbert Spencer was foremost to recognise, to be close indeed." And Lord Keynes did not approve. He laments that, insofar as free-market economists influenced public policy, "Socialistic interference" in the market had been judged "not merely inexpedient, but impious, as calculated to retard the onward movement of the mighty process by which we ourselves had risen . . . out of the primeval slime of the Ocean." Lord Keynes rejected the idea that the complex adaptive system of natural selection is comparable to the complex adaptive system of the market.[21]

More recently, in 2003 the political commentator David Sirota and real-estate investor Minnie Lush propounded, "Proponents of *laissez-faire* promulgated Darwin's theory of survival of the fittest as the most equitable and efficient manner for distributing U.S. resources—until they went bankrupt. Then, those proponents cried out to the government to save them." Unsurprisingly, Lush and Sirota provide no examples of any free-market proponent seeking a taxpayer-funded bailout. They nevertheless go on, "The social reforms of the 1930s were the result of these pleas for government intervention."[22]

The left-wing billionaire George Soros—himself often accused of ruthlessness as a businessman—propounds, "The laissez-faire argument against income redistribution invokes the doctrine of the survival of the fittest. . . . This social Darwinism is based on an outmoded theory of evolution. . . . I mention this because social Darwinism is one of the misconceptions driving human affairs. . . . The main point I want to make is that cooperation is as much a part of the system as competition, and the slogan 'survival of the fittest' distorts this fact."[23] It is Soros distorting facts when he presumes cooperation and free-market economics to be mutually exclusive. Every free-market advocate recognizes that a buyer and seller cooperate with each other when they agree to a trade. Likewise, every free-market advocate recognizes that an employer and employee freely cooperate with one another, just as a team of employees cooperates. These forms of cooperation occur because each participant recognizes it is in her long-term self-interest to do so. Separate firms in the same industry compete for customers, but every link in a single supply chain cooperates. A supplier cooperates with a manufacturer. A manufacturer cooperates with its distributors and wholesalers. A wholesaler cooperates with a retailer. Further, when firms in the same industry compete for the same customers, cooperation is involved. Suppose Rolf and I are in the same industry, and we each wish to have Yasmine as our customer. We each try different prices and offer different features to our wares to lure Yasmine. Should Yasmine choose Rolf's business over mine, Yasmine and Rolf will cooperate with one another. Conversely, should Yasmine choose me over Rolf, Yasmine will cooperate with me. Therefore, in a competition among firms for customers and clients, the prize for winning the competition is that the winner gets to cooperate with the customer. Therefore, competition in the free market results in multiple parties cooperating.[24] Free-market advocates are unanimous in recognizing that such cooperation exists, based on mutual self-interest. The anti-capitalists' false dichotomy between self-interest and cooperation is therefore lame—quite frankly, unfit for rational discourse. Anyhow, from his own verbiage, Soros concludes, "The main enemy of the open society, I believe, is no longer the communist but the capitalist threat."[25]

Unequivocally, the falsehood that every free-marketer is a social Darwinist continues to endure. Consequently, almost any invocation of that misconception will catch modern free-marketers on the defensive. Upon being called social Darwinists, most free-marketers behave in a predictable fashion. The free-marketers first vociferously protest that their arguments for Manchesterism are entirely unlike those of Spencer and Sumner. Then they uphold their own moral superiority by proclaiming their own hatred for Spencer and Sumner for giving free-market economics a bad name. One example of this phenomenon can be found with Brink Lindsey—the former vice-president of research at the libertarian Cato Institute.

In his book *The Age of Abundance*, Lindsey trashes Sumner's meticulous defenses of capitalism as corrupt "rationalizations" that did nothing but "provide comfort for the comfortable . . ."[26] And it is conspicuous that Lindsey himself would accuse a fellow free-marketer of social Darwinism. In a prior book of his, *Against the Dead Hand*, Lindsey took after Spencer and Sumner by likening the market to a biological ecosystem. *Against the Dead Hand* proclaims that private enterprise functions according to the same principles of "the evolutionary process of natural selection. The market system accelerates the evolution of useful new ideas in a two-step process: First, it increases the number of 'mutations' by decentralizing investment decisions." That is, says Lindsey, a purely socialist state confiscates funds from everyone and places these funds into a common pool of capital. Then the central government is the sole party to decide which enterprises any of this capital is to be invested in. By contrast, the capitalist ecosystem permits anyone who is both willing and able to invest his own capital in whichever enterprise he judges best. Budding entrepreneurs compete with one another to draft the best business plan. They try to win the favor of potential investors and persuade them to commit their capital to the endeavor. Secondly, the market ecosystem "then applies" the "mutations" to

> the ruthless selection pressures of profit and loss. . . . Entrepreneurs who successfully develop and apply new good ideas are rewarded with profits. Those profits perform vital signaling functions: They encourage the original entrepreneur to expand operations, while at the same time they lure in new competitors to the market. In other words, profits are the signals that lead to the propagation of good ideas throughout the economy by attracting additional resources that will be devoted to applying those ideas. Meanwhile, entrepreneurs whose ideas fail are stuck with losses. Losses, likewise, act as signals: They drive entrepreneurs to contract their operations or else fold completely. Accordingly, losses are the means for reducing the resources devoted to less successful ideas. The profit-and-loss system thus creates feedback loops that constantly push the rearrangement of resources to concentrate them on applying the best ideas for creating value.[27]

Disappointingly, as a contrast to Sumner's take-no-prisoners style of argumentation, Lindsey has his books deliver a milquetoast "defense" of capitalism that does not explain the virtues of a radically unfettered market. Lindsey does not deign to defend Lockean rights explicitly. Instead *Abundance* argues that modern Democrats should acquiesce to some modest deregulation. This is on the grounds that, throughout the twentieth century, market forces have furthered the enfranchisement of women and racial minorities and the preservation of endangered species. Lindsey is right on this count, though there is no wisdom in the denigration of Sumner.

No less than the eminent Ludwig von Mises was naïve in proclaiming his disagreement with social Darwinism, unaware that the term was coined for the purpose of making a straw man of such free-marketers as himself.[28] In that tradition, a 1981 *Wall Street Journal* editorial defended President Reagan's tax cuts and deregulation policies by assuring readers that they were not "just another version of late-nineteenth-century robber-baron social Darwinism."[29]

And the social Darwinism appellation has nastier applications. Take, for instance, American Enterprise Institute president and Syracuse University political scientist Arthur C. Brooks—the same Arthur Brooks whose findings about the links to happiness and income I cited in Book Two.

Who Else Is a Social Darwinist?

Arthur Brooks employs a tactic similar to John Kenneth Galbraith's. Much akin to Georgetown theologian Thomas Reese and the *Miami Herald* op-ed I quoted, Brooks exploits Spencer's ruined public image in order to soil that of Ayn Rand and other historical figures he dislikes. Brooks snivels that "one extreme line of reasoning" that proliferated from the late 1800s to the early 1900s, argued that low-income men

> were flawed by their very nature. By giving aid and comfort to the underperformers in society, charity promotes the worst in society by artificially helping the weak to survive and bring forth new generations of underperformers.
>
> The intellectual roots of this kind of thinking can be traced to the nineteenth-century British philosopher Herbert Spencer, the father of so-called "Social Darwinism". . . Adherents of Social Darwinism believed that observed inequalities between people represented a natural process by which the "fit" (successful people) rose to the top of social and economic hierarchies, and the "unfit" (the poor and needy) were weeded out. Since political and social institutions (such as welfare and charity) cannot permanently alter the process, the thinking went, these institutions created a disservice to society by delaying the inevitable population decline of the unfit. . . .
>
> This sort of philosophy may sound radically right wing to modern ears, reminiscent of Ayn Rand (who said that "suffering is a not a claim check, and its relief is not the goal of existence") and her like.[30]

The evolutionary psychology-promoting primatologist Frans de Waal, dubs Herbert Spencer a social Darwinist and proclaims that Ayn Rand follows Spencer's social Darwinist tradition.[31] But the potshots from Brooks and Waal miss their targets. Examine the ostensive implication behind Brooks' partial quotation of Rand that "suffering is a not a claim check" and that "its relief is not the goal of existence . . ." Brooks quotes Rand in such a manner to sour the reader's opinion of her. If Rand holding such views is evidence of her being a bad person, one should pose the following questions to Dr. Brooks. "Well, then, should I believe that other people's suffering *is* a claim check upon me? Do you mean that the relief of everyone else's suffering *is* the goal of my existence?" Anyone who answers in the affirmative is invited to provide a logical explanation of why everyone should believe that, complete with empirical substantiation. We have yet to see this happen. More often, conservatives of Brooks's sort attempt to evade that consideration by citing David Hume's Is-Ought dictum. I mean the dictum that facts have nothing to do with morals. Those who cite Hume's dictum, though, are advised to remember our dissection of this dictum from Book One.

To accuse Rand of categorically opposing private charity is to ascribe to her sensibilities that she did not express. Read her actual commentary on the matter. "There is nothing wrong in helping other people, if and when they are worthy of the help and you can afford to help them. . . . What I am fighting is the idea that charity is a moral duty and a primary virtue."[32] In *Atlas Shrugged*, Rand has protagonist John Galt state, "Do you ask if it's ever proper to help another man? No—if he claims it as his right or as a moral duty that you owe him. Yes—if such is your own desire based on your own selfish pleasure in the value of his person and his struggle. . . . If you choose to help a man who suffers, do it only on the ground of his virtues, of his fight to recover, of his rational record, or of the fact that he suffers unjustly; then your action is still a trade, and his virtue is the payment for your help."[33] It would behoove us to dispense with the presupposition that Rand preached against aiding the misfortunate when she counseled against self-sacrifice and altruism. Remember that altruism is defined as the dogma that one carries a duty to forfeit one's higher priorities for the benefit of one's lower priorities. As Rand spelled it out,

> altruism gauges a man's virtue by the degree to which he surrenders, renounces or betrays his values (since help to a stranger or an enemy is regarded as more virtuous, less "selfish," than help to those one loves). . . .
> Concern for the welfare of those one loves is a rational part of one's own selfish interests. If a man who is passionately in love with his wife spends a fortune to cure her of a dangerous illness, it would be absurd to claim that he does it as a "sacrifice" for *her* sake, not his own, and that it makes no difference to *him*, personally and selfishly, whether she lives or dies. . . .
> But suppose he let her die in order to spend his money on saving the lives of ten other women, none of whom meant anything to him—as the ethics of altruism would require. *That* would be a sacrifice. [Emphases Rand's.]

In contrast to mainstream society's fetishism over self-renunciation, Rand taught, "your highest moral purpose is the achievement of your own happiness, your money is yours, use it to save your wife, *that* is your moral right . . ."[34] (emphasis hers).

Rand stated something wise that is in implicit rebuttal to John Kenneth Galbraith's asseveration that the right to one's own life and fortune are merely "the unspoken excuse . . . for passing the beggar with the outstretched hand."[35] In this statement, Rand spots Galbraith's and Brooks's question-begging (pun intended) about the inherent evilness of being uncharitable. You may recount this admonition of hers from Book Two: "Do not hide behind such superficialities as whether you should or should not give a dime to a beggar. That is not the issue. The issue is whether you *do* or do *not* have the right to exist *without* giving him that dime. The issue is whether you must keep buying your life, dime by dime, from any beggar who might choose to approach you. The issue is whether the needs of others is the first mortgage on your life . . ." [36] (emphases hers). Journalist Robert W. Tracinski paraphrased it more succinctly: "The issue is not whether you give a dime to a beggar. The issue is whether you have a right to exist if you don't."[37] If social Darwinism has the same meaning as brutality, then consider which option is more socially Darwinian.

1. Miles abstains from contributing his inheritance to the economically misfortunate.
2. The government threatens violence against Miles if he does not shell out a specific percentage of his income to the destitute.

Clergyman-turned-Yale-sociologist William Graham Sumner has the same answer that I do. For the State to spoliate Miles is far crueler.

There are very few documented cases of a prominent opponent of the welfare state seriously suggesting that the indigent ought to die. One such case is a series of letters that Baltimore, Maryland journalist H. L. Mencken wrote in debate with a socialist. Mencken quite unwisely assumed that somehow socialism would be better than capitalism at producing wealth and abundance and lowering the mortality rate. Proceeding from the bizarre Malthusian streak we witnessed from Book Two, Mencken thought that a reduction in the death rate would be calamitous, as this would exacerbate overpopulation. Hence Mencken really did argue that capitalism should be chosen over socialism on the basis that the former did more than the latter to weed out the weaker members of the species. The Baltimore journalist thinks it foolish that "the concerted efforts to put an end" to the Darwinian "struggle for existence will, for a time at least, reduce the death-rate among what are now the lowest orders toward that of what is now the highest, and that this reduction will quickly swell the population of the world." Such a denial of "the law of natural selection" shall precipitate the fiercest of repercussions.[38] He recognizes no benefit in welfare, as "the state's efforts to keep England's loafers and incompetents from starving to death has certainly not transformed them into efficient men . . ." Then Mencken exhibits racist sentiment by bewailing black suffrage—"The possession of the franchise did not make the American negro a civilized man, though every one knows that the franchise is an important part of every civilized man's heritage." Acting in accordance with the left-winger's image of an anti-welfare, eugenicist social Darwinian, Mencken cites both welfare and black suffrage as case studies in the obstruction of "the operation of the law of natural selection in the lower orders." To Mencken, the best that could be stated about early-twentieth-century America's mostly-commercial system was that, insofar as it is maintained, "the law of natural selection" could be said to be "aiding the man-made laws of artificial selection." But under socialism, Mencken sniffed, "the unfit would survive."[39] That is, the unfit would survive until overpopulation led to resource depletion and a war-of-all-against-all among men of every class.[40] Mencken's fallacious line of argumentation is disappointing, particularly given that he had publicly praised both Spencer and Sumner on numerous occasions.[41]

More recently, a similar argument was expressed by James Fallon (b. 1947), a neuroscientist at the University of California at Irvine. The national news media covered Fallon in 2010 when he took an fMRI of his own brain and discovered traits in his own brain starkly similar to the traits of brains of convicts professionally diagnosed as psychopaths. Fallon claims that what his brain has in common with those of diagnosed psychopaths is that there are low levels of activity in his ventromedial prefrontal cortex, to which blood must flow for affective empathy to be felt. Fallon argues, though, that his having been raised in a loving and caring household, as opposed to an abusive and devaluing one, has prevented him from growing up to be the violent felon he otherwise would have become. Still, Fallon continues, the trait he shares with the more dangerous, more criminalistic, and spoliative psychopaths is that he is bereft of the capacity for genuine affective empathy for others. That Fallon proudly proclaims himself both a libertarian and a non-empathetic borderline-psychopath poses a dilemma for the reputation of free-market advocates. It is easy for someone to observe Fallon's example and cite it as further evidence that to support free-market economics, one must be a psychopath devoid of affective empathy. Fallon worsens the dilemma by opening his mouth to declare, "As a Libertarian. . . . I don't think we should spend every dime we have to save one child. . . . If one person croaks tomorrow for the sake of society, it's too bad, but I don't care. . . . If the system weeds out weak or lazy individuals, fine."[42]

But the remarks of Mencken and Fallon do not eradicate the fact that the line of argumentation put forth by Mencken and Fallon cannot be found in the works of either Spencer or Sumner. To impute Mencken's sentiments into those of either Spencer or Sumner is therefore to fault someone for sentiments he has not conveyed.

It chagrins me that contemporary critics insist on besmirching Ayn Rand, Spencer, and Sumner, equating them with somebody who wishes to sacrifice other people for one's own ostensive benefit. The Rand-haters appear to be confusing Rand with the French philosopher-pornographer Donatien-Alphonse-François de Sade (1740–1814), who indeed advocated that one should spoliate others for one's own supposed profit. To quote the Marquis de Sade, "Get into your head . . . that what fools call *humaneness* is nothing but a weakness born of fear and egoism . . ."[43] (emphasis Sade's). When Sade accuses humanitarians of *egoism*, he means that they are too arrogant to admit to themselves that his cynical worldview, which condones spoliation, is the correct one. Sade cannot help but join the altruist philosophers in presuming that *egoism* must be a pejorative, a judgment of opprobrium.

In his erotic fiction *Juliette*, Sade proclaims through one of his characters that Mother Nature "has given us . . . weak individuals to be our slaves: they are her gift to us, a sacrifice. . . the strong man may hence use the weak as he sees fit; may he not aid them in some instances? No; for if he does; he acts contrary to Nature's will." Insofar as a strong man derives pleasure from exploiting a weaker person, "he behaves as Nature's friend" in withholding mercy; "but . . . if . . . he aids the abject, raises the lowly to a level of parity with himself . . . , then he necessarily disrupts the natural order and perverts the natural law: whence it results that pity, far from being a virtue, becomes a real vice . . ." As far as the marquis ascertains, the idea of charity "could have been dreamt up only by some feeble individual; for it to have occurred to one of the mighty, in need of nothing, would not have been natural: to bind the weak to his will, he already had what the task required: his strength . . ." The belief in mercy was "invented by some puny wretch, and it is founded upon arguments quite as futile as would be this one addressed by the lamb to the wolf: *You mustn't eat me, I am four-footed too*." It has not occurred to the marquis that an individual whom many people might have once misperceived as weak—he may be handicapped—may later prove to be strong in contexts they had not priorly anticipated, such as his being an inventive genius.

Nonetheless, Sade continues that he hates that the "religion of that wily little sneak Jesus—feeble, sickly, persecuted, singularly desirous to outmaneuver the tyrants of the day—can bully the tyrants into granting clemency." "Here we see Christianity in the role of the weaker party; Christianity represents the weak and must speak and sound like them. . . . But that he who is neither weak nor Christian subject himself to such restrictions, voluntarily entangle himself in this mythical snarl of brotherly relationships which without benefiting him in the least deprive him enormously—it's unthinkable; and from these arguments we must conclude" that the idea of mercy was initially "proposed by the weak . . ." and that ". . . its existence is frivolous and that we must not under any circumstances submit to it."[44]

University of the Arts professor Camille Paglia observes Sade's low opinion of *Homo sapiens* in general.[45] She brings her readers' attention to this statement of his: "What is man? And what difference is there between him and other plants, between him and all the other animals of the world? None, obviously."[46] In contrast to Rand, Sade holds no reverence for the rational faculty. Sade's rhetoric astonishingly matches the straw-man social-Darwinism rhetoric that governists have falsely imputed to the free-market advocates they aim to defame. But any exhortation that the strong subjugate the weak is not Randian self-interest in any respect. Nor does Sade's philosophy match that of Spencer and Sumner.

Basically, *Hunting Down Social Darwinism* will address the vagaries of the social Darwinism label in greater depth than did Book Two. In this chapter, I shall explore the notion that Spencer and Sumner opposed private charity. Most of the rest of this part will address the additionally scandalous idea that the pro-capitalism theories of Spencer and Sumner inspired a social Darwinism movement that gave rise to eugenics legislation and Naziism. The record evinces that the laissez-faire philosophy, especially when explicated by Spencer and Sumner, is the opposite of everything that animated eugenics legislation and Naziism. Further, the early twentieth-century's socialists and Progressives—the putative detractors of such so-called social Darwinists such as Spencer and Sumner—were the true brains behind eugenics legislation. Finally, I propose that Naziism was not the intellectual progeny of Spencer-Sumner ideology but instead a governist effort to combat it.

Are Millionaires the Product of Natural Selection?

Here, one may ask me why, if Sumner was not an advocate of dog-eat-dog spoliation or social Darwinism, did he judge self-made industrialists to be at the peak of the evolutionary pyramid? Governists quote Sumner saying, "The millionaires are a product of natural selection. . . ." Among the governists quoting this expression are John Kenneth Galbraith, Robert M. Rees, Charles Derber, and Philip Taft Award-winning financial journalist and labor historian Steve Fraser (b. 1945).[47] When these critics quote Sumner in this manner, they do so to give the appearance that Spencer approved whenever a business executive resorted to violence, thievery, backstabbing, or blackmail in the amassing of his fortune.

In refutation to those who defame Sumner and Spencer, I shall cite the work of Swarthmore College historian Robert C. Bannister. His book is *Social Darwinism: Science and Myth in Anglo-American Social Thought*. From this moment forward I shall abbreviate its title as *SD-SMAAST* and pronounce it "S. D. smast." Alternately, I shall allude to this same work by the first three words of its subtitle, *Science and Myth*. Anyhow, in this volume Bannister puts to rest any such misperception that Sumner condoned spoliation. *SD-SMAAST* proves that Sumner "deplored" any "ill-gotten wealth" and "the entire 'get-rich-quick' mentality . . ." Spoliation yielded nothing except "a picture of society quite unacceptable to him."[48] Sumner identified spoliation as an assault on what he called civil liberty. He identified spoliation as such whether inflicted by a private corporation or the government.[49]

Robert Rees huffs that if Sumner is right about millionaires being the creation of natural selection, then it implies that "the rest of us are worker bees, destined by natural selection to a life of unquestioning hard labor."[50] Charles Derber's book *Corporation Nation* quotes that same passage of Sumner's and then affixes it to this one: ". . . if we do not like the survival of the fittest, we have only one possible alternative, and that is the survival of the unfittest. The former is the law of civilization; the latter is the law of anti-civilization."[51] The Derber book then adds tartly that Sumner "argued that workers were nature's losers and deserved to be treated as such."[52] New York University historian Kim Phillips-Fein employs the infamous quotation as she fumes that following "the Civil War, the rhetoric of market liberty became increasingly associated with rigid refusals to permit the government to take any actions to regulate the harsh tempo of industrialization." Phillips-Fein quotes the Sumner passage about millionaires, and then mistranslates it as meaning that "those who survived the rigors of the market were those best fitted for wealth and power, and those who failed should be left to the gutters."[53] Adding to my horror and disappointment, Brink Lindsey of the Cato Institute likewise quotes Sumner's statement about "anti-civilization" for the express purpose of damning Sumner. Lindsey condemns Sumner as "the leading light of the American 'social Darwinist' school . . ."[54]

When Derber's *Corporation Nation* continues Sumner's quotation with ". . . if we do not like the survival of the fittest,"[55] it quotes a completely different essay from the one wherein Sumner opines that millionaires are "the product of natural selection." Derber's book splices together two discrete Sumner quotations from separate sources to form a hybrid quotation. The part of the Sumner quotation about millionaires is from an article Sumner had published in the April-June 1902 issue of the *Independent* titled "The Concentration of Wealth." Yet the part about the survival of the fittest comes from the transcript of an address that Sumner gave to the Free Trade Club in 1879. Curiously, Derber cites the 1902 *Independent* article whereas the 1879 Free Trade Club speech goes unmentioned. The second source of the hybrid quotation is not provided. Given that, *Corporation Nation* has committed a gruesome distortion. It places together two different quotations of Sumner's that were made over thirty-two years apart from one another. This is done as if both quotations were from the same source.

Moreover, when Sumner spoke of "fitness" in the 1879 talk that Derber quotes, Sumner was referring to the survival of the fittest businesses not the fittest persons. As we observed in Book Two, a business's "death" does not kill the business's owners, managers, or employees. Examine Sumner's "millionaires" remark in its full context. Sumner was writing about technological innovations that provide an important service that was once provided exclusively by manual laborers. The power loom's output exceeded that of a hand weaver by such a magnitude, that the introduction of the power loom in mills had cost many hand weavers their jobs. Such an innovation

> displaces the acquired skill of the men who formerly performed the service . . .
> It is not surprising, under such a state of things, that some people should lose their heads and begin to doubt the economic doctrines which have been most thoroughly established. . . . Many of them are frightened at liberty, especially under the form of competition, which they elevate into a bugbear ["source, real or imaginary, of needless fright or fear," according to *Dictionary.Com*]. . . . They do not perceive, furthermore, that **if we do not like the survival of the fittest, we have only one possible alternative, and that is the survival of the unfittest. The former is the law of civilization; the latter is the law of anti-civilization**. . . .
> One of the chief characteristics of the great improvements in industry . . . is that they bring about new distributions of population. If machinery displaces laborers engaged in manufactures, these laborers are driven to small shopkeeping, if they have little capital; or to agricultural labor, if they have no capital [boldface mine].[56]

Note that Sumner is not saying the hand weavers are unfit and should die. He is saying that their outdated technological methods are going extinct, being replaced by more advanced technologies that provide greater satisfaction to consumers. That is not a long-term tragedy for the members of old industries that are supplanted by new, more efficient ones. Book Two provides statistical figures demonstrating that the majority of employees dislocated from obsolete industries acquire new occupations, and they reap the benefits from the new technologies that displaced them.

What Sumner Meant

As stated before, Sumner abhorred spoliation committed by anyone—especially a businessman—as a violation of "civil liberty." But I can tell you what conditions met Sumner's approval. He was convinced that, inasmuch as Manchesterite principles were implemented in a society, the amount of money an entrepreneur earned in it was directly commensurate with what his customers or clients paid him willingly in exchange for the goods or services the entrepreneur availed to them. Recall the lesson on entrepreneurship from Book Two in this trilogy. Now we can study the words of Sumner's that Derber, Galbraith, and Rees quoted about millionaires and natural selection, but do so with the words in an accurate context. Employing the locution *franchise* as a synonym for product or service, Sumner explains,

> **No man can acquire a million [dollars] without helping a million men to increase their little fortunes all the way down through all the social grades. . . . The millionaires are a product of natural selection**, acting on the whole body of men to pick out those who can meet the requirement of certain work to be done. In this respect they are just like the great statesmen, or scientific men, or military men. . . . Let one of them make a mistake and see how quickly the concentration [of wealth in the mistaken millionaire's hands] gives way to dispersion. . . . They get high wages and live in luxury, but the bargain is a good one for society. There is the intensest competition for their place and occupation. This assures us that all who are competent for this function will be employed in it, so that the cost of it will be reduced to the lowest terms; and furthermore that the competitors will study the proper conduct to be observed in their occupation. [boldface mine][57]

Sumner ends the essay at exactly this point. He utters no disparaging comments about blue-collar workers or those who endure financial deprivation. This belies Rees's assertion that Sumner hailed millionaires in order to imply condescendingly that blue-collar employees were "worker bees." Putting aside the governist misrepresentations of his philosophy, Sumner's inferences are correct. To the extent that a businessperson satisfies consumer demand, he is well-adapted to a free-market habitat. In lieu of conceding this truth, progressive journalist Susan Jacoby simply quotes from this same Sumner passage as she sneers, "In his repeated *arguments against taxing the rich*—'no man can acquire a million without helping a million men to increase their little fortune'—Sumner advocated what would now be called trickle-down economics." (emphasis added)[58] Observe something. In the passage that Susan Jacoby quoted—the very same one that I placed in a block quotation—Sumner was not talking about taxes on the rich.

Sadly, there is no ceasefire by intellectual detractors who take Sumner out of context. Evolutionary-psychology promoter David Berreby verbally flays Sumner for establishing himself as "one of his country's nastiest social Darwinists" with his "rabid laissez-faire writings." Berreby detests Sumner's "sermonlike essays and books" that "declared that nothing should be done to interfere with those at the top of society (the 'captains of industry'), nor should anyone waste time on its losers. 'A drunkard in the gutter,' he wrote, 'is just where he ought to be, according to the fitness and tendency of things. Nature has set up on him the process of decline and dissolution by which she removes things which have survived their usefulness.'"[59] That passage of Sumner's arrives via his book *What Social Classes Owe to Each Other*. Berreby quotes it to convey that Sumner wanted homeless winos to hurry up and expire. That way, the allegation goes, their bodies would cease to waste so much space. Actually, Sumner is contending that there should be no laws proscribing consenting adults from imbibing alcohol or other drugs. Sumner wants legality for almost any consensual activity that the general public may judge to be detrimental to the health of the participants. In Sumner's estimate, a consenting adult should have the Lockean political prerogative of running his life as he chooses, absent of busybodies overruling his choices. Sumner further intimates that if the government polices an alcoholic's vices, the government will have to extort money from honest, responsible taxpaying adults to save the would-be addict, ostensibly, from himself. Sumner ascertains that this unfairly penalizes the honest, responsible taxpayer who takes the precaution to limit his own liquor intake.

The Yale sociologist calls this taxpayer "the Forgotten Man." Whenever the populace clamors for the government to spend money on performing some function beyond the strictures of the night watchman state, it forgets about this honest taxpayer. People forget the taxpayer who is put under duress in order to pay for the government action. Finally, Sumner recommends that if you want to save someone from his own addiction, there is a much better method than simply threatening him with legal imprisonment or mandatory therapy. That would be allowing him to face fully the dire repercussions of his own choices. Let a junkie hit rock bottom. At that juncture, it is incumbent upon the junkie to contemplate a crucial choice. He can continue down this miserable road. Alternatively, he might pull himself together and try to better his own circumstances. In the end no one can implement this decision for the addict but the addict himself. With that in mind, we can read Sumner's words in context. When we come to the passage that Berreby quotes, it shall be in italics. Sumner opposes any government regulation that would protect an individual from his own unhealthy-but-peaceable choices. Such a regulation blocks a man from facing the logically inevitable

> penalty of his vice. . . . *A drunkard in the gutter is just where he ought to be, according to the fitness and tendency of things. Nature has set up on him the process of decline and dissolution by which she removes things which have survived their usefulness. . . .*
>
> Now, we never can annihilate a penalty. We can only divert it from the head of the man who has incurred it to the heads of others who have not incurred it. A vast amount of "social reform" consists in just this operation. The consequence is that those who have gone astray, being relieved from Nature's fierce discipline, go on to worse, and that there is a constantly heavier burden for the others to bear. . . . The industrious and sober

> workman, who is mulcted of a percentage of his day's wages to pay the policeman, is the one who bears the penalty. But he is the Forgotten Man. He passes by and is never noticed, because he has behaved himself, fulfilled his contracts, and asked for nothing.[60]

I have said it in Book Two, and I will say it again: there is no good excuse to deprive the Forgotten Man of his most basic survival need—the liberty to act peaceably on his own judgment, absent of the immediate threat of spoliation—for the Progressive purpose of safeguarding the ostensive longevity of the wino. To leave people be in a night watchman state does not hinder any individual from devoting himself to consensual philanthropy that may help treat the addictions of winos. Recall my quotations of Sumner from Book Two that exhibit Sumner's endorsement of private charity. Yet governists retort that if Sumner did not intend to imply that low-income persons should just *off* themselves, such an outcome was still the desire of his idol, Herbert Spencer.

Interestingly, critics of Spencer sometimes mention Spencer's worry, in the late nineteenth century, that the more-liberal Western realm would be menaced by a particularly aggressive strain of socialism.[61] Spencer referred to this particularly aggressive strain as "the coming slavery."[62] In 1903, the famed progressive economist Richard T. Ely—more about him in chapter 4—dismissed Spencer's worry as groundless.[63] Some writers of the late twentieth and early twenty-first centuries evidently believe that this prediction and warning of Spencer's is evidence of him being a crank.[64] That is odd. As a matter of fact, the history of Europe and Asia during the twentieth century indeed proved Spencer correct. Communism indeed overtook Eastern Europe and Asia, and menaced the more-liberal Western Europe and the United States. As we recognized in Book One, communism indeed rendered the citizens under it the slaves of the red menace. That Spencer predicted, in the late 1800s, the West would be threatened in the next century by a socialist form of "slavery," actually showed how prescient Spencer's judgment was. Such a correct prediction is not the mark of a crank but of someone who understood the topic upon which he elaborated.

Did Spencer Want to Proscribe Private Charity?

Yet it is easy for anti-capitalists to caricature Spencer's opposition to welfare laws as a rationalization for his selfish refusal to part with any of his money. His work as a civil engineer on British railroads—not to mention the patents on his inventions[65]—brought him to enviable fortune. Northern Illinois University scholar James Gettier Kennedy (b. 1932) disclosed that when Spencer died, his estate was appraised at what would be valued at 400,000 U.S. dollars in 1978 money.[66] The inflation-adjusted equivalent for the year 2013 would be over 1.4 million U.S. dollars.[67] Additionally, the net worth of Spencer's estate placed him within the wealthiest 1 percent of the Englishmen of his period.[68]

Governist authors Edwin Black, Susan Jacoby, Frans de Waal, and Allan Chase do much more to foster the image that Spencer resented the institution of private charity. They do so by quoting Spencer's remark that persons who "are sufficiently complete to live" are the ones who "*do* live, and it is well they should live." Conversely, those who are "not sufficiently complete to live" simply "die, and it is best they should die"[69] (emphasis Spencer's). These authors grumble that this is Spencer's opinion of those in poverty. As we shall discover, such a framing of Spencer's words is misleading. And to make Spencer look worse yet, Black, Chase, Waal, and Richard Hofstadter go on to quote Spencer saying, "The whole effort of nature is to get rid of such"—such poor persons, if we are to believe these three quoters' evaluation of what Spencer meant—"to clear the world of them, and make room for better."[70]

Charles Derber piles on the confusion. He takes the Spencer statements I quoted above, and mangles them when he puts them together: "Herbert Spencer, one of England's leading Darwinist thinkers a century ago, summed up the Gilded Age view of safety nets and social contracts: 'The whole effort of nature is to get rid of such'"—in brackets, Derber inserts "the poor," and then he continues the Spencer quotation—"to clear the world of them, and make room for better. . . . it is best that they die.'"[71] Further, Hofstadter repeats Spencer verbatim, "He who loses his life because of his stupidity, vice or idleness is in the same class as the victims of weak viscera or malformed limbs. Under nature's laws all alike are put on trial."[72] John Kenneth Galbraith continues the repetition of Spencer's words—"Partly by weeding out those of lowest development, and partly by subjecting those who remain to the never-ceasing discipline of experience, nature secures the growth of a race who shall both understand the conditions of existence, and be able to act up to them. It is impossible in any degree to suspend this discipline."[73]

At this juncture Allan Chase snipes, "Small wonder the humanitarian Darwin had such profound contempt for Spencer's preachments."[74] But that does not appear to be Charles Darwin's ultimate assessment when one reads his own words about Spencer. Darwin wrote in one letter that he "enjoyed" meeting Spencer "very much." More than that, Darwin was in awe of Spencer, saying of him, "I could bear, and rather enjoy feeling that he was twice as ingenious and clever as myself, but when I feel that he is about a dozen times my superior, . . . I feel aggrieved."[75] The sixth edition of *The Origin of Species* comments upon a case for biological evolution that Spencer presented in his 1852 essay "The Development Hypothesis." Here Darwin praises Spencer's essay for articulating itself "with remarkable skill and force."[76] In 1858, soon upon reading this essay, Darwin wrote to Spencer, "Your remarks on the general argument of the so-called development theory seem to me admirable. I am . . . preparing an Abstract of a larger work on the changes of species"—*The Origin of Species*—"but I treat the subject simply as a naturalist, and not from a general point of view, otherwise, in my opinion, your argument could not have been improved on, and might have been quoted by me with great advantage."[77] In another letter, this one from 1860 and addressed to a prominent geologist, Darwin hailed Spencer as one "who puts, to my mind, the philosophy of the argument" for biological evolution "better than anyone else. . ."[78] Then Darwin wrote in yet another letter that he thought that Spencer might be "by far the greatest living philosopher in England; perhaps equal to any that have lived."[79] Upon finishing his read of Spencer's essay on his phrase *the survival of the fittest*, Darwin wrote to Spencer that ". . . I cannot resist the wish to express my unbounded admiration of your article. . . . Everyone with eyes to see and ears to hear . . . ought to bow their knee to you, and I for one do."[80]

To some extent, I do think Spencer sounds a bit sympathetic to social Darwinism in one respect. Spencer openly wonders about whether it is entirely bad to refrain from extending the lives of individuals stricken with crippling heritable afflictions. He expresses his compunction about encouraging those with debilitating maladies, such as Type 1 diabetes—for which there was still no treatment years after Spencer's death—to live long enough to bear children. Should those with inheritable diseases procreate, they will thereupon bequeath their painful conditions to future generations. That, Spencer warns, "increases" people's "suffering," rather than diminishes it. Such treatment "favours the multiplication of those worst fitted for existence, and, by consequence, hinders the multiplication of those best fitted for existence—leaving, as it does, less room for them. It tends to fill the world with those to whom life will bring most pain, and tends to keep out of it those to whom life will bring most pleasure." Yet that conclusion of Spencer's is not drawn from a general hostility toward the poor or sick. Instead Spencer takes a somewhat euthanistic stance, suggesting that sometimes it may be preferable to permit a suffering man to die with dignity, forgoing the extension of his agony.

The previous comments of Spencer's from which I quoted are in chapter 28, section 4, of Spencer's 1851 treatise *Social Statics*. I am not surprised that Jacoby, Black, Chase,

Galbraith, and Hofstadter avoid delving into an in-depth examination of those quotations in their actual nuanced meaning. In the section of *Social Statics* that these critics cite, Spencer was not referring to the financially downtrodden; he was specifically discussing persons suffering from debilitating heritable diseases and of what can be done about their plight. He said that it is the law of nature that people more genetically prone to healthiness survive long enough to pass on their genes, whereas those genetically prone to unhealthiness are less likely to do so. When he said that it was "best" that those "not complete to live . . . should die," he was putting into his own words the cold, detached manner that one may take when studying the matter from afar. Spencer then goes on to confess his own personal feelings about it, and his evaluations bear little resemblance to the hideous straw man that Jacoby, Black, Waal, and Hofstadter set up before their readers. On second thought, they not only patch together a straw man of Spencer; they burn it in effigy.

Jacoby, Black, Waal, and Hofstadter omitted any mention of something important. I recommend that one peer at the paragraph following the one about persons "incomplete to live." This is the exact same paragraph wherein Spencer cautions against "the multiplication" of people afflicted with "the worst" heritable diseases. In this paragraph, Spencer exhorts the reader to provide some private charity. He opines that insofar as the combined horror of both debilitating disease and extreme penury "is mitigated by the spontaneous sympathy of men for each other, it is proper that it should be mitigated: albeit there is unquestionably harm done when sympathy is shown, without any regard to ultimate results. But the drawbacks hence arising are nothing like commensurate with the benefits otherwise conferred."[81]

Incidentally, watch out for something in this exact book of Spencer's—the same one from which Jacoby, Chase, Black, Galbraith, Waal, and Hofstadter quote in order to intimate that Spencer is a social Darwinist. In its twenty-fifth chapter, *Statics* explains quite clearly that Spencer likes the idea of voluntary philanthropy and wishes there were more of it: ". . . charity is in its nature essentially civilizing. The emotion accompanying every generous act adds an atom to the fabric of the ideal man. As no cruel thing can be done without character being thrust a degree back towards barbarism, so no kind thing can be done without character being moved a degree forward towards perfection." Spencer's detractors may retort that, in that very same chapter 25, Spencer erupts into a tirade against the poor. It sounds as though Spencer is lamenting that the indigent possess comparatively little moral fiber. Then Spencer goes on to share his concerns about those who claim to help the poor. He worries that both private philanthropists and supporters of government-imposed welfare "advocate an interference which not only stops the purifying process, but even increases the vitiation," i.e., bad behavior, "absolutely encourages the multiplication of the reckless and incompetent by offering them an unfailing provision. . . . And thus, in their eagerness to prevent the really salutary sufferings that surround us, these sigh-wise and groan-foolish people bequeath to posterity a continually increasing curse." But, two paragraphs later, Spencer reaffirms once again his support for consensual almsgiving.

> At first sight these considerations seem conclusive against *all* relief to the poor—voluntary as well as [governmentally] compulsory; and it is no doubt true that they imply a condemnation of whatever private charity enables the recipients to elude the necessities of our social existence. With this condemnation, however, no rational man will quarrel. That careless squandering of pence [money] which has fostered into perfection a system of organized begging—which has made skilful mendicancy [i.e., panhandling] more profitable than ordinary manual labour . . . cannot but be disapproved by every one. Now it is only against this injudicious charity that the foregoing argument tells. To that charity which may be described as helping men to help themselves, it makes no objection—

> countenances it rather. And in helping men to help themselves, there remains abundant scope for the exercise of a people's sympathies. Accidents will still supply victims on whom generosity may be legitimately expended. Men thrown upon their backs by unforeseen events, men who have failed for want of knowledge inaccessible to them, men ruined by the dishonesty of others, and men in whom hope long delayed has made the heart sick, may, with advantage to all parties, be assisted [emphases his].[82]

Recall that Spencer, unlike myself, believes in David Hume's Fact-Value Divide. We disproved the validity of that Divide in Book One. The Fact-Value Divide presumes that there is no connection between descriptive facts and value-driven prescriptive moral actions. One side of Spencer—what he regarded as his descriptive-science biologist side—contemplated whether the collective of Western society might indeed be better served if particular biologically disadvantaged people perished in the absence of any opportunity to rear children. On the other hand, Spencer simultaneously judged sympathy and compassion to be fully natural consequences of biological evolution. Moreover, he considered a compassionate, sympathetic, philanthropic person to be more biologically advanced that someone who cared nothing about the hardships of others. Upon that reasoning, Spencer repeatedly stated that he thought it made sense for someone's desire to aid disadvantaged persons to win out over any possible benefits that the societal collective may attain from having disadvantaged persons removed from the gene pool.

Sometimes some critics quote out-of-context passages from other Spencer works. In a book that defends the right of individual parents to enhance their own children through genetic engineering—we will explore that idea in chapter 10—Meridian Institute fellow Walter Truett Anderson (b. 1933) attempts to distinguish his benign pro-evolutionist view from what he judges to be the nastiness of Spencer's. Predictably crediting Allan Chase for tipping him off to the quotation, W. T. Anderson pronounces Spencer a "fierce Darwinist" and resents that Spencer "dedicated a large part of his life to fighting against any measures that might lead to what he called 'the artificial preservation of those least able to take care of themselves.'" That quotation arrives from Spencer's *Study of Sociology*, and Anderson cites it as he falsely accuses Spencer and T. Robert Malthus of opposing smallpox vaccination. Anderson presumes that Spencer and Malthus wished to block access to vaccination on account of their desire that the lowest-income members of the population die from communicable viruses.[83] In the very same chapter of *Study of Sociology*, though—a mere two pages following the passage that W. T. Anderson quoted—Spencer mentions that he judges it laudable and predictable "that parental affection, the regard of relatives, and the spontaneous sympathy of friends and even of strangers, should mitigate the pains" suffered by the misfortunate. "Doubtless, in many cases the reactive influence of this sympathetic care . . . is morally beneficial" and balances out the disadvantages of philanthropy of which Spencer previously spoke. "It may be fully admitted that individual altruism, left to itself, will work advantageously. . ."[84]

In fact, at a time when there was still great controversy over whether every adult male should have voting privileges, Spencer delineated in no uncertain terms that the impoverished had just as much a right to suffrage as anyone else. There "is no escape from the conclusion," he said, "that the interest of the *whole* society can be secured, only by giving power into the hands of the *whole* people" (emphases his). Spencer then prepared a rejoinder to those who feared that granting suffrage to paupers would impel them to vote for their own interests at the expense of the rich. Were there some validity to that worry, Spencer proclaimed, "the evidence would still preponderate in favour of popular enfranchisement."[85] It is this freedom fighter for the lower classes whom the likes of Jacoby, Black, Hofstadter, Rees, Brooks, and other anti-individualists have smeared as somebody who wants the destitute to succumb to their own mortality.

Another critic of Spencer's, Harvard University sociologist Paul Starr (b. 1949), has much in common with the others. Starr's Pulitzer Prize-winning tome, *The Social Transformation of American Medicine*, would have us believe that Spencer thought that con men should be free to market snake oil and poisons as medicines. We are encouraged to think that this is Spencer's view: *any hapless rube foolish enough to be taken in by the hucksters' sales pitches would simply get what was coming to him. That applies even if that rube is foolish enough to purchase such toxins and die from consuming them.* Starr writes, "Social Darwinists, following the English social theorist Herbert Spencer," argued that regulation over the quality and safety of medicines could be nothing but "unwise. 'Very many of the poorer classes *are* injured by druggists' prescriptions and quack medicines,' Spencer willingly conceded. But there was nothing wrong in that; it was the penalty nature attached to ignorance. If the poor died of their own foolishness, the species would improve" (italics are Spencer's).

As with the other terrible-sounding Spencer quotations, the one that Starr repeats comes from *Social Statics*.[86] And, like the others, it is quoted out of context. Note that Starr states that Spencer wanted men to *die* as a result of their ignorance. When you gaze upon the actual source material, you find that Spencer did not take it for granted that someone would perish from imbibing bad medicine. In fact, Spencer disputed the accusation that there was some epidemic of poor men dying from drinking quack medications; he observed that the vast majority of men who had taken such medication had survived the process. He then said that the quack medications did cause people enough physical discomfort to prompt them to become more discerning of the various pharmaceuticals being sold. This learning experience further encouraged these victims to educate themselves later about the quality and effectiveness and safety of the various other substances they considered consuming. Below is more of the passage from which Starr quotes. The exact part he quotes is in boldface. I have underlined another part for emphasis.

> The most specious excuse for not extending to medical advice the principles of free-trade, is . . . that the judgment of the consumer is not a sufficient guarantee for the goodness of the commodity. . . . Ignorant people say they cannot distinguish good treatment from bad . . . Hear Mr. Wakley. Speaking of a recently-revived law relating to chemists and druggists, he says, "It must have the effect of checking . . . that frightful evil called counter practice, exercised by unqualified persons, which has so long been a disgrace to the operation of the laws relating to medicine . . . , and which, doubtless, has been attended with a dreadful sacrifice of human life." (*Lancet*, Sept. 11, 1841)
> Any one may discern through these ludicrous exaggerations much more of the [governist political] partizan than of the philanthropist. . . . [But] let it be conceded that **very many of the poorer classes *are* injured by** druggists **prescriptions and quack medicines**. The allegation having been thus, for argument's sake, admitted in full, let us now consider whether it constitutes a sufficient plea for legal interference. . . . It is impossible in any degree to suspend this discipline by stepping in between ignorance and its consequences, without, to a corresponding degree, suspending the progress. If to be ignorant were as safe as to be wise, no one would become wise. And all measures which tend to put ignorance upon a par with wisdom, inevitably check the growth of wisdom. Acts of parliament to save silly people from the evils which putting faith in empirics may entail upon them, do this, and are therefore bad. . . . it is best to let the foolish man suffer the appointed penalty of his foolishness. For the pain—he must bear it as well as he can: for the experience—he must treasure it up, and act more rationally in the future.

Obviously, a man cannot learn from a painful experience if he dies from it. Spencer is stating that a man can learn from consuming unhealthy medicines because he will live through the ordeal. And Spencer continues that when a man falls ill from quack medications, the adverse effects shall serve as "a warning" to others "as well as to himself. . . . And by multiplication of such warnings, there cannot fail to be generated in all men a caution corresponding to the danger to be shunned. Are there any who desire to facilitate the process? Let them dispel error; and, provided they do this in a legitimate way, the faster they do it the better. But to guard ignorant men against the evils of their ignorance—to divorce a cause and consequence . . ."[87] Back in 1843, Spencer stated on this same subject that he intended for someone who consumed quack medicines to learn from his previous acts of poor judgment. The "misfortunes of one," he wrote, "are lessons for thousands—that the world generally learns more by its mistakes than by its successes . . ."[88]

University of Belfast science historian Peter J. Bowler demonstrates a more careful reading and much more accurate understanding of Spencer—"He extolled the sufferings that are a consequence of failure as the best possible stimulus encouraging the individual to do better next time."[89] To wit, Spencer understood that in circumstances such as this, inductive lessons were frequently the most powerful and instructive. Thus, Starr's Pulitzer-winning book paints an inaccurate picture of Spencer's argument. But such misrepresentations of Spencer are legion. My study cannot scratch the surface when it comes to the task of eroding the plaque buildup of falsifications about Spencer sticking to the academic record. For instance, I spotted such a fabrication in a hagiography of socialist propagandist Upton Sinclair written by Idaho State University humanities dean Jon A. Yoder. In that work, Yoder assumes that he discredits Spencer by repeating this quotation from Johns Hopkins University historian John Higham (1920–2003): "in their eagerness to convert social values into biological facts, Darwinian optimists unblinkingly read 'the fittest' to mean 'the best.'"[90]

I shall reveal to you Spencer's actual words on the subject. Spencer was very clear on this point when one of his contemporary critics, James Martineau, delivered the very same accusation that Yoder and Higham later would. In an 1872 issue of *The Contemporary Review*, Spencer elaborated that when he wrote of survival of the fittest, he was not implying that merely the best beings or organisms survived. Nor, he continued, was he implying that any childless person or beast that died should be judged, perforce, as inferior. "Mr. Martineau," said Spencer, "speaks of the 'survivorship of the better,' as though that were the statement" of the biological principles that Spencer famously named. "But the words" that Martineau "here uses are his own, not the words of those he opposes. The law is the survival of the *fittest*. Probably, in substituting 'better' for 'fittest,' Mr. Martineau did not suppose he was changing the meaning. . . . Had he examined the facts, he would have found that the law is not the survival of the 'better' or 'stronger'. . . It is the survival of those which are constitutionally fittest to thrive under the conditions in which they are placed . . ."[91] (emphasis Spencer's). Recall the example from Book Two about peppered moths in a forest, some white and some black, trying to camouflage themselves on the bark of trees. Should the tree bark be whitish, the white moths will camouflage better than the black moths. Should a nearby factory's soot stain the tree bark as black, the black moths will be more camouflaged. Fitness in that context depends on the circumstances and not on one type of moth being inherently superior to another.

Yes, Jon A. Yoder and John Higham proclaim that when "Darwinian optimists" read of Spencer's allusions to *survival of the fittest*, they "read 'the fittest' to mean 'the best.'" Whichever "Darwinian optimists" Yoder and Higham were indicting, the criticism cannot realistically apply to Spencer. Unfortunately, rather than actually read any of Spencer's books for comprehension, some alleged free-market advocates, too, simply opt to accept at face value the libelous allegations leveled against the man. We witnessed that from Brink Lindsey from the Cato Institute. Likewise, the anti-regulation economist Mark Skousen similarly tags Spencer as a social Darwinist. Skousen accuses Spencer of professing that society's "weakest members should go to the wall" because "this would improve society by weeding out the less intelligent and less industrious."[92]

And a particularly laughable element of Edwin Black's anti-capitalist book, *War Against the Weak*, is that, when crediting Spencer with starting "social Darwinism," Black cites Robert C. Bannister's *Science and Myth*.[93] Here, Black is not alone. Arthur C. Brooks drubs Spencer for being "the father" of "Social Darwinism . . ." To Brooks, Spencer espoused that "the poor and the needy" should be "weeded out" and denied "incentives" to "reproduce" their genes. In this drubbing, one of the two sources that Arthur C. Brooks cites is Bannister's 1973 essay "William Graham Sumner's 'Social Darwinism': A Reconsideration" in volume Five of the journal *History of Political Economy*.[94] Bannister revised this very same article and had it republished it in *Science and Myth* as its fifth chapter.[95]

It is amusing that Black and Brooks both cite Bannister to validate that Spencer was a social Darwinist. I am amused, as Bannister's book actually clarifies that "the conventional portrait of Spencer" as a social Darwinist "is both inaccurate and ironic." *Science and Myth* continues, "The label social Darwinism, as eventually applied to Spencer, made good propaganda" as well as "bad history . . ." According to Bannister, "Darwin's rhetoric of struggle and survival provided" anti-capitalists not unlike Edwin Black "the perfect vocabulary to caricature the alleged inhumanity and brutality" of free industry. Moreover, Bannister further admonishes the likes of Black for invoking this caricature "to parody anyone who continued to ground policy in natural law or even science."[96] *Parody* is an appropriate description for what Brooks and Black have done to Spencer in their writings. Black and Brooks cite Bannister's research as if it proves their point when it directly contradicts their point. As observed by sociologist Howard L. Kaye in a work published by Yale University Press, "a close reading of the theories of Sumner and Spencer exonerates them from the century-old charge of social Darwinism in the strictest sense of the term. They themselves did not advocate the application of . . . 'the law of the jungle' to human society."[97] To my consternation, the parody has other victims as well. And the leftists who have invoked the social Darwinism epithet compound the injustice. Subsequent to proclaiming that the social Darwinism of Spencer and Sumner hatched from free-market economics, they say that the same social Darwinism of Spencer and Sumner sired the advocacy of eugenicist government policies. To wit, leftists allege that, on account of Spencer's and Sumner's involvement in free-market advocacy, free-market advocacy itself must be blamed for originating eugenics legislation and World War Two. We must address this allegation.

NOTES

1. G. Hodgson 2004.
2. That William Jennings Bryan had such presumptions of guilt against big business in general explains why, according to Chernow 1998, 388, he was "adored" by "socialists" and "populists."
3. Hofstadter 1959, 200.
4. Kolko 1963, 57.
5. Bannister 1988, xxx.
6. Obama 2012, Para. 45, accessed online Friday, April 6, 2012.
7. Qtd. in "Paul Ryan Challenged on Budget By Georgetown Faculty," *Huffington Post*, Tuesday, April 24, 2012, http://www.huffingtonpost.com/2012/04/24/paul-ryan-challenged-by-georgetown-faculty_n_1449437.html, accessed Friday, April 27, 2012.
8. H. Caton 1981, 106.
9. Frederick Douglass, speech, July 4, 1852, qtd. in D. W. Root 2012, accessed online Saturday, April 26, 2014.
10. David Blight, qtd. by D. W. Root 2012, accessed online Saturday, April 26, 2014.
11. Borsook 2000, 3.
12. Rees 2004, A6, accessed online Thursday, April 26, 2007. This same man previously used the "Social Darwinism" smear in Rees 2003, accessed online Thursday, April 26, 2007. Rees's 2004 attack was met with applause by retired University of Hawaii sociologist Jerome G. Manis in "Contemporary Sociology

Debunks Hayek's Beliefs," *Honolulu Advertiser*, letters to the editor, Monday, May 10, 2004, A7, http://tinyurl.com/3btzsj, accessed Thursday, April 26, 2007.

13. Galbraith 1987, 121, 243, 165.

14. John Green with the username "CrashCourse," "The Industrial Economy: Crash Course US History #23," July 25, 2013, https://www.youtube.com/watch?v=r6tRp-zRUJs, accessed Monday, June 9, 2014.

15. Bannister 1988, 98; and R. N. Smith 1997, 67.

16. Hofstadter 1959, 20-21.

17. Roman Leverenz, "Virtue of Selfishness Drives Public Policy," *Honolulu Star-Advertiser*, letters to the editor, Thursday, November 10, 2011, A18.

18. Joy-Ann Reid, "A First-Hand Look at What Motivates Paulites," *The Miami Herald*, Wednesday, December 28, 2011, http://www.miamiherald.com/2011/12/28/2564668/a-first-hand-look-at-what-motivates.html, accessed Wednesday, December 28, 2011.

19. Glenn Hall, "Letter: Conservatives All for Social Darwinism," *The Gleaner*, Friday, May 11, 2012, http://www.courierpress.com/news/2012/may/11/no-headline---gl_hall_letter_to_editor/, accessed Saturday, May 12, 2012.

20. G. Hodgson 2004.

21. Keynes 2009a, 20, 31.

22. M. Lush and D. Sirota 2003, 5th ed., 330.

23. Soros 1997, 45–58, accessed online Saturday, November 23, 2013.

24. Moataz "Michael" Philip Kadada pointed this out to me over Facebook on Saturday, November 23, 2013.

25. Soros 1997, 45–58, accessed online Saturday, November 23, 2013.

26. Lindsey 2007, 53.

27. Lindsey 2002, 52.

28. See Mises's "disagreements" with "social Darwinism" in Mises 1966 3d revised ed., 176.

29. James Ring Adams, "Supply-Side Roots of the Founding Fathers," *Wall Street Journal*, November 17, 1981, 26, qtd. by S. Blumenthal 1986, 171.

30. A. Brooks 2006, 68–69.

31. Waal 2009, 28–32.

32. Ayn Rand, interviewed by Alvin Toffler, "Ayn Rand: A Candid Conversation With the Fountainhead of Objectivism," *Playboy* vol. 11 (no. 3, March 1964), republished in Rand 2000, 10.

33. Rand, 2007 paperback, 970.

34. Rand, "The Ethics of Emergencies," in Rand 1970, paperback 51.

35. Galbraith 1987, 123.

36. Rand, "Faith and Force: The Destroyers of the Modern World," a speech delivered to Yale University on February 17, 1960; at Brooklyn College on April 4, 1960; and at Columbia University on May 5, 1960; republished in Rand 1984c, 61.

37. Tracinski 2005, accessed online Thursday, May 10, 2007.

38. Mencken 1910a, 70.

39. Mencken 1910b, 189, 206.

40. Mencken 1910a, 70–71.

41. For Mencken's praise of Herbert Spencer, see Mencken 1922c, 248; H. L. Mencken, "Criticism of Criticism of Criticism," *Smart Set*, August 1917, 142, qtd. in Mencken 1990-EE, 745; H. L. Mencken, "The Effect on the Race," *Smart Set*, November 1919, 63, qtd. in Mencken 1990-LL, 276–77 (unfortunately, this article contains some eugenicist assumptions of H. L. Mencken's); H. L. Mencken, "Heredity," *In Defense of Women*, originally published 1918, 2d revised edition in 1922, 85–86, republished as Mencken 2006-H, 129 (this passage also contains eugenicist thought); and H. L. Mencken's Introduction to *George Bernard Shaw: His Plays*, 1905, ix–xi, qtd. in Mencken 1990-I, 287 and Mencken 1990-VV, 281–82.

42. J. Fallon 2013, 164–65.

43. Sade 1965b, 360.

44. Sade 1988, 177–78.

45. C. Paglia 1990, 236.

46. Sade 1965b, 329-330.

47. This is in Derber 1998, 95; S. Fraser 2005, 274–75; Galbraith 1987, 122–23; and Rees 2004, A6.

48. Bannister 1988, 101–102.

49. See William Graham Sumner's definition of *civil liberty* in Sumner, "The Influence of Commercial Crises on Opinions About Economic Doctrines," an address to the Free Trade Club, New York City, May 15, 1879, republished as Sumner 1969-E, 61.

50. Rees 2004, A6, accessed online Friday, May 25, 2007.

51. Qtd. by Derber 1998, 95.

52. Derber 1998, 95.

53. K. Phillips-Fein 2009, 39.

54. Lindsey 2007, 53.

55. See Derber's use of the quotation in Derber 1998, 95, and then see that only one source was cited, when Derber actually spliced together two different Sumner quotations from two different speeches given over thirty-two years apart, in Derber 1998, 345–46 n. 6. That same note also cites a book by Richard Rubenstein, but nowhere does the note acknowledge that the quotation of Sumner is actually two different quotations put together.

56. Sumner, 1969-E, 52–58.

57. Sumner, "The Concentration of Wealth: Its Economic Justification," *The Independent*, April–June 1902, republished as Sumner 1969-B, 170–72.

58. S. Jacoby 2008, 71.

59. Berreby 2005, 211. Interestingly, this same David Berreby wrote to Julian L. Simon in 1991 that he would prefer seven thousand more elephants on Earth over a billion more people. This is according to J. L. Simon 1998 trade paperback, 566 n. 10, 649 n. 10.

60. Sumner 1911, ch. 9, para. 7–10, http://tinyurl.com/2gsog3, accessed Tuesday, November 6, 2007.

61. R. H. Gabriel 1956, 241.

62. Spencer 1981a.

63. R. T. Ely 2006, 57.

64. For instance, W. H. Spiegel 2002 3d ed., 400, notices Spencer's warning about intrusive government in the nineteenth century leading to a "coming slavery," but this author—William Henry Spiegel (b. 1911)—fails to notice that communism was indeed a form of slavery that emerged from the trend about which Spencer warned.

65. Schumpeter 1996, revised ed., 773 n. 5.

66. J. G. Kennedy 1978, 119.

67. This calculation was made using the Federal Reserve Bank of Minneapolis 2014, accessed online Wednesday, July 2, 2014.

68. J. G. Kennedy 1978, 153 n. 3.

69. Black 2003, 12; Allan Chase 1982, 67; and S. Jacoby 2004, 140.

70. Black 2003, 12; Allan Chase 1982, 67; Hofstadter 1959, 41; and Waal 2009, 28–29.

71. Derber 1998, 95.

72. Hofstadter 1959, 41.

73. Galbraith 1987, 122.

74. Allan Chase 1982, 68.

75. W. Irvine 1959, 166, citing Darwin 1896 vol. 2, 239. Darwin found Spencer annoying and snooty, but he was not morally outraged by Spencer's economic theories.

76. Darwin 1872, Preface, para. 17, accessed online Sunday, January 8, 2012.

77. Qtd. by R. L. Carneiro 1981, 157.

78. G. H. Smith 1981, 143 n. 6, citing an 1860 letter from Darwin to Charles Lyell in Darwin 1896 vol. 2, 84.

79. G. H. Smith 1981, 143 n. 6, quoting a letter from Darwin to E. Ray Lankester in Darwin 1896 vol. 2, 301.

80. Qtd. in G. H. Smith 1981, 143 n. 6, citing Darwin 1896 vol. 2, 344.

81. Spencer 1851, pt. 3, ch. 28, sec. 4, para. 3, 5, http://tinyurl.com/2ayk3m, accessed Tuesday, July 17, 2007

82. Spencer 1851, pt. 3, ch. 25, sec. 5, para. 1, and then sec. 6, para. 3-5, accessed online Tuesday, July 17, 2007.

83. W. T. Anderson 1996, 67, citing Allan Chase 1982, 68. The quotation is from Spencer 1886, 343.

84. Spencer 1886, 345.

85. Spencer 1851, pt. 3, ch. 20, sec. 3, para. 3, accessed online Tuesday, July 17, 2007.

86. P. Starr 1982, 105, 461 n. 70.

87. Spencer 1851, pt. 3, ch. 28, sec. 4, para. 1–3, accessed online Sunday, October 26, 2008.

88. Spencer 1981e, 249, Letter 9, para. 6, accessed online Wednesday, January 11, 2012.

89. P. J. Bowler 1996, 25.

90. Yoder 1975, 32, quoting J. Higham 1967, 135.

91. "Mr. Martineau on Evolution," *Contemporary Review*, June 1872, republished as Spencer 1891b, 379.

92. Skousen 2001, trade paperback, 212

93. Black 2003, 12, which is in chapter 2, credits Spencer with starting Social Darwinism, and then leads to chapter 2, endnote #18 in the books back on page 446. Black 2003, 446 n. 18 cites page xii of Bannister's *Social Darwinism: Science in Myth in Anglo-American Social Thought*. Looking at Bannister 1988, xii, right now, I do not see any text demonstrating Black's point that Spencer started social Darwinism.

94. The excerpts quoted are in A. Brooks 2006, 68–69. The derogatory assessment of Spencer is in chapter 3 and ends with a superscripted "27." When one goes to the back of the book, in its Endnotes section, one finds chapter 3, Endnote 27 in *ibid.*, 219 n. 27. One of the two sources cited is Bannister 1973, which Bannister revised and had republished as chapter 5 of Bannister 1988. Brooks gets the title slightly wrong. A. Brooks 2006, 219 n. 27 says the title is "William Graham Sumner's 'Social Darwinism' Reconsidered" but it is actually "William Graham Sumner's 'Social Darwinism': A Reconsideration."

95. Bannister states this very clearly in Bannister 1988, ix (the Acknowledgments page): "Chapter 5 appears in slightly different form as 'William Graham Sumner's' "Social Darwinism" (*History of Political Economy* 5 [1973], 89-109), copyright by Duke University Press, which grants permission for republication." Observe that "Social Darwinism" is in scare quotes in Bannister's title "William Graham Sumner's 'Social Darwinism.'" This is because this paper, much like the chapter it was remade into in Bannister 1988, argued that Sumner was not a social Darwinist; he was only smeared as such by many of his political opponents.

96. Bannister 1988, 55.

97. H. Kaye 1986, 33–34, qtd. by T. C. Leonard 2009, 41, accessed online Sunday, November 6, 2011.

TWO

The Conflation of Laissez Faire with Regulation-Imposed Eugenics

Two Opposing Ideologies Squeezed Together, Beneath the Social Darwinism Umbrella

It is not enough for the critics of Herbert Spencer, William Graham Sumner, and John D. Rockefeller to try to impugn them as haters of the underclass. Richard Hofstadter, Edwin Black, and others have gone as far as insinuating that Spencer and Sumner paved the way for European imperialism and Naziism. These critics have largely succeeded in perpetuating this misconception. As I write this, supposed defenders of the free market accept that belief uncritically. For instance, Matt Ridley—a British ornithologist and evolutionary-psychology promoter—is often quite reliable. It therefore disappoints me that Ridley utters a falsehood about Herbert Spencer. According to Ridley, Spencer "called" his own ideology of "laissez-faire economics" and "individualism" by the name of "social darwinism."[1] Citing Ridley, journalist Marion Roach from the National Public Radio's *All Things Considered* repeats the claim—"the term 'social Darwinism'" comes directly "from Spencer's work." Then in inadvertent hypocrisy Ms. Roach comments that this phrase is "wrongly attributed to Darwin."[2]

Scholars have yet to unearth any record of Spencer placing the words *social* and *Darwinism* together consecutively. There were indeed late-nineteenth-century pundits who applied theories about natural selection to their defense of the night watchman state. But the reader may be surprised to learn that these pundits did not actually employ the words *social Darwinism* to describe their systems of thought. A laudable source of information on this subject is Alfred Kelly, Hamilton College's Edgar B. Graves Professor of Modern European History. Quite plainly, Kelly states, "No one in the late nineteenth century called himself a social Darwinist."[3] Business management professor Geoffrey Hodgson of the University of Hertfordshire concurs: "There was no self-declared school of Social Darwinists. . . . In fact, neither Spencer nor Sumner used the term 'Social Darwinism.'"[4]

In 1877, the term appeared in a paper that Joseph Fisher published in *Transactions of the Royal Historical Society*. This paper, "A History of Landholding in Ireland," described the British's feudalistic occupation of Ireland as cruel social Darwinism.[5] Oscar Schmidt, a professor of the University of Strasburg, employed the expression *Darwinism* in connection with the social sciences in an 1879 issue of the American magazine *Popular Science*. But in this article Schmidt had criticized socialists and the pretense that evolutionary theory somehow proved socialism's veracity. Schmidt did not find it convincing when the socialists invoked "Darwinism."[6] The expression of social Darwinism appeared much more frequently in the literature of the European continent than in the U.K. or the Americas. The year of 1880 saw French thinker Emile Gautier employing *le Darwinisme sociale* in the manner that it is normally employed at the date of this writing—as a denunciation of consensualist economics. This might have been the first occasion on which *social* and *Darwinism* were placed right next to each other. By contrast, in 1910 the Russian writer Jacques Novicow fumed about *Darwinisme social* that was the antithesis of laissez faire—governist militarism.[7] In 1884 and 1890, Gabriel Tarde proved to be a rarity among academicians, for he was one of the handful of academicians from Geoffrey Hodgson's study who applied the phrase social Darwinism with approval. When he spoke of social Darwinism, though, he referred not to violent behavior or economic transactions, but simply the manner in which people learned new customs by copying behaviors they had witnessed being practiced by others.[8] None of these references targeted either Spencer or William Graham Sumner as major proponents of a social Darwinist ideology.

The term started to see more Anglophone usage around the early 1900s in the Progressive Era. There are recorded instances of it being used in 1907 by sociologist-paleobotanist Lester Frank Ward,[9] who detested the politics of Spencer and Sumner.[10] You will recall from Book One of our trilogy, *The Freedom of Peaceful Action* (*FOPA*), that Lester Ward argued that the biggest perk of having tax-funded government-run schools, complete with mandatory attendance laws, is that such schools can manipulate the citizens' children into becoming subservient to the State. At any rate, Spencer had already died in 1903 and Sumner's demise would follow seven years later, when the Anglophone version of that epithet *Darwinisme social—social Darwinism*—was still in its infancy. Geoffrey Hodgson has inquired into the history of the social Darwinism designation. He employed "large electronic databases," such as the online *Journal Storage* (*JSTOR*) to plow through "leading Anglophone academic journals" published from the nineteenth century to the present, to account for every academic reference to the term *social Darwinism*. The search was done through 203,000 articles and reviews.[11] Hodgson notes that the idea that social Darwinism was a popular philosophy among Westerners in the nineteenth century has "acquired mythological attributes, referring to a pre-1914 era" when the use of the expression social Darwinism "was assumed to be prevalent. At least as far as the Anglophone academic journals are concerned, this assumption is false." Hodgson found that usage of the phrase social Darwinism was "rare up to the 1940s."[12] From 1800 to 1915, the expression pops up but eleven times. Between 1916 and 1943, the term appears in forty-nine different articles and reviews.[13] Hodgson adds, "Neither Herbert Spencer nor William Graham Sumner were described as Social Darwinists in this early literature."[14] The epithet would not gain tremendous circulation among social science scholars—particularly with regard to how they described Spencer and Sumner—until 1944. That is the year that Richard Hofstadter brought *SDAT* into the world.[15] Subsequently, the expression *social Darwinism* could be found in a whopping 4,236 articles and reviews, out of a possible 2,475,225, between the years 1944 and 2007. Moreover, prior to 1944, a mere *two* articles from the entire database had referred to Spencer as a social Darwinist, and one of these articles was written by Hofstadter himself, 1941's "William Graham Sumner: Social Darwinist." In the entire database, Hodgson found but one paper explicitly advocating what it referred to as "social Darwinism." This was a paper of Collin Wells that the *American Journal of Sociology* published in 1907. And when Wells said that he favored social Darwinism, he simply referred to broad changes in the customs of society; not a strict adherence to laissez faire.[16]

Thus, writes Stanford University economist Thomas Leonard, "though the epithet 'social Darwinist' is today more closely associated with Sumner and Spencer than with any other writers, that association was all but nonexistent before Hofstadter published

SDAT." Incidentally, of the 4,236 articles and reviews that cite *social Darwinism* from 1944 to 2004, a bit more than *one-third* of them—that is 1,416 academic essays—also mention Spencer, Sumner, or both of them. As Leonard sums up the situation, "Before Hofstadter, Spencer and Sumner were rarely, if ever, known as social Darwinists. After Hofstadter, the two men were transformed into arch-social Darwinists." Suffice it to say, "Hofstadter in *SDAT* succeeded brilliantly in affixing the epithet 'social Darwinism' to free-market economics. . . . In the popular and scholarly mind, Spencer and Sumner *are* social Darwinism"[17] (emphasis Leonard's).

After decades of this misconception running rampant, a small portion of the academic community is wising up. The proudly governist Columbia University historian Eric Foner (b. 1943), who earned his Ph.D. under Hofstadter, admits that prior to *SDAT*'s publication, "the term social Darwinism...was used only on rare occasions..."[18] Likewise, progressive journalist Susan Jacoby shockingly admits in her book *The Age of American Unreason*, "I use the term 'social Darwinism' even though no one employed it in nineteenth century America or England. The phrase was . . . not in common usage in America, even among academics, until the publication in 1944 of Richard Hofstadter's *Social Darwinism in American Thought*."[19]

Alfred Kelly learns that the expression *social Darwinism* was barely used in German literature until 1906 and "the term" did not enter "the historians' vocabulary" until *SDAT* arrived on the scene.[20] Yet, in spite of her own admission, Jacoby displays no reservations about going on to employ the epithet to denounce nineteenth-century American and English intellectuals who never described themselves with this locution. The emotional impact of the epithet is apparently so strong that governists cannot resist exploiting it to hurt the reputations of their free-market adversaries. Indeed, observes Sogang University historian Donald C. Bellomy, the deployment of the social Darwinism tag has always been "heavily polemical, reserved for ideas with which a writer disagreed."[21]

There were indeed theorists in the nineteenth century who, under the imprimatur of scientific respectability, cited biology to rationalize racial prejudices. The anthropologist Daniel Shute declared that "the Caucasian stands at the head of the racial scale and the Negro at the bottom." Joseph Le Conte, who served as the president of the American Association for the Advancement of Science (AAAS) much agreed.[22] Yet these scientific racists did not identify themselves as social Darwinists or capitalists, nor did they cite the convictions of Spencer or Sumner as the source of their ideas.

Remember someone I argued against in Book Two—psychology professor Barry Schwartz, who, alongside Bannister, has taught at Swarthmore College. Imitating everyone else, Schwartz invokes *SDAT* to stigmatize Sumner as a social Darwinist.[23] Meanwhile, historian Arthur Alphonse Ekirch, Jr. (1915–2000), of the State University of New York at Albany, refers to *SDAT* in order to paint both Spencer and Sumner with that same unflattering stroke.[24] And *New Republic* senior editor Robert C. Wright, who bears no ill will toward Spencer himself,[25] still lists Hofstadter as his source as he demonizes free-market economists as social Darwinists.[26] Worse, the mostly-pro-capitalist Larry Schweikart and Michael Allen also do this in their *Patriot's History of the United States*, a book that otherwise contains many good points.[27]

Hofstadter introduces social Darwinism as an umbrella term. Shoved beneath this umbrella are two very different political movements that Hofstadter deplored, both of which invoked natural selection in their arguments: (1) free-market economics and (2) government-imposed eugenics. A few academicians of a political bent similar to Hofstadter's had been exercising the same tactic as Hofstadter decades prior to *American Thought*, possibly as early as 1905.[28] Nonetheless, it was Hofstadter who most famously touted the largely illusory connection between laissez-faire evolutionism and governist eugenics that currently haunts the psyche of social-studies academicians.

The agenda of government-regulated eugenics, which some modern commentators have christened *scientific racism*,[29] stated that free will does not exist. Along that line, scientific racism proposed that a person's DNA largely predetermines his personality traits and actions, including his moral character. Likewise, it argued that some lineages are better than others. This belief implies the inherent superiority of some ethnic bloodlines over others. This prompted members of the governist eugenics movement to proclaim the justness of the vast conquest of other societies by European powers and the United States. Consequently the state-decreed eugenics movement called for America's federal government to restrict the influx of immigrants. This is purportedly because newcomers from Eastern Europe, Southern Europe, and Asia were genetically inferior to native-born Americans descended from Anglo-Saxon Protestants.[30] The dogma also rationalized the practice of U.S. states coercively sterilizing persons when it was discovered that mental illness ran in these people's families. This sterilization was to preclude the mentally ill from passing on conditions such as bipolar disorder and epilepsy.[31] As is widely known in the social sciences field, government-implemented eugenics was consistently praised, preached, and praticed under the Nazi regime. This consideration should evince that members of the government-pushed eugenics movement were avowed opponents of the night watchman state. One of the leaders of the early-twentieth-century eugenics movement,[32] for instance, was the trailblazing statistician and self-described "socialist"[33] Karl Pearson (1857–1936). University of Wisconsin historian George L. Mosse writes that Pearson's brand of socialism was "opposed to individualism and the undue concentration of wealth." And "Pearson was not alone among racial biologists in advocating this kind of socialism."[34]

Ignoring those points, various scholars persist in drawing a link between Spencer and governist eugenics. They prefer to convince their students and readers of capitalism's evil by arguing that laissez-faire ideology is the springboard for racist genocide à la the Nazis'. In that tradition, the atheist website NoBeliefs.Com proclaims that Adolf Hitler did "at times express ideas...from Herbert Spencer's concept of Social Darwinism. . . Spencer's Social Darwinism tried to connect Darwin's biological theory with the field of social relations. The result of Social Darwinism resulted in many eugenics programs that began in America and were adopted by the Nazis."[35] As far as NoBeliefs.Com is concerned, to agree with Spencer's social theories is to be a proto-Nazi.

The person who started the ideology of eugenics, which was previously known as *stiripculture*,[36] was not Spencer but Sir Francis Galton (1822–1911).[37] One cannot find any so-called social Darwinist defining social Darwinism, but one can find Sir Francis's introduction of the word *eugenics* in his 1883 tract *Inquiries Into Human Faculty*.[38] Sir Francis wrote, "We greatly want a brief word to express the science of improving" human genetic "stock... The word *eugenics* would sufficiently express the idea"[39] (italics Galton's). In 1908 he further clarified that the "first object" of eugenics "is to check the birth rate of the unfit instead of allowing them to come into being. . . . the second object is the improvement of the race by furthering the productivity of the fit by early marriages and the healthful rearing of children."[40] Despite Galton having coined the expression in 1883, he delivered eugenic arguments much earlier, in a series of articles that he would publish in 1869 as the book *Hereditary Genius*.[41] Governist eugenics, Galton explained, was a state policy intending "to replace natural selection by other processes that are more merciful and not less effective."[42] Note that choice of wording. Whereas free-market economics is identified with natural selection, Galton proposes eugenics as a means "to *replace* natural selection" (emphasis added). This serves as an early indication of the eugenics movement's opposition to Herbert Spencer's decentralized, complex-adaptive-system economics.

Yet the notion that the State should control human breeding, purportedly to improve biologically driven traits in human beings, predates Galton. In *The Republic,* Plato approvingly attributes the idea to Socrates. Plato recommends that for "our rulers" to be "skillful physicians of the State," they must "bring about desirable unions"—legislatively imposed sexual couplings—"between their subjects." Healthy people must be paired with other healthy people, and likewise the unhealthy paired with the unhealthy, "and the offspring of the one must be reared, and of the other destroyed; in this way the flock will be preserved in prime." The children of the healthy are not to be raised by their own parents, incidentally, but an entire team of women. That is consistent with the idea of *allomothering* discussed by evolutionary psychologist Sarah Blaffer Hrdy. Plato adds that when a child is raised by this team of women, the child may come into contact with his or her own mother, but that the information of who is whose parent or child is to be kept a secret from everyone. "The mothers will be brought to the fold and will suckle the children; care however must be taken that none of them recognise their own offspring . . ."[43] The Nazis would carry out such a program. Under the Lebensborn program led by Heinrich Himmler (1900–1945), the State paired off the Aryan men and women it judged to be of highest quality, and the children who resulted from this endeavor were never notified of the identifies of their actual parents.[44] Such is the logical conclusion of the notion that a child should be raised by the community as a whole, not merely his or her parents.

Lester Thurow (b. 1938)—dean of MIT's management school, economic adviser to 1981 Democratic presidential nominee George McGovern,[45] and a onetime editorial board member of the *New York Times*—exhibits his own obliviousness to the history of eugenics as he poses as someone knowledgeable about it. In 1997, Thurow blithely asserted, "A nineteenth century economist, Herbert Spencer, formulated a concept he called survival-of-the-fittest capitalism . . . Spencer believed it was the duty of the economically strong to drive the economically weak into extinction. . . . *Spencer created the eugenics movement* to stop the unfit from reproducing, since this was simply the most humane way to do what the economy would do in a more brutal way (starvation) if left to itself" (emphasis added). Predictably, Thurow then raised the specter of Spencer in order to spook his own contemporary political opponents. As Thurow wrote these words, the Republican-controlled Congress proposed a *Contract With America* that would ostensibly reduce federal welfare spending. Thurow howled, "The Contract with America is very Spencerian in tone and offers a return to survival-of-the-fittest capitalism."[46] In case anyone missed it, "Spencerian in tone" is a euphemism for *evil.* Occasionally the actual word is employed. One piece in the London *Guardian* blares that Spencer was "a down-right evil man . . . whose passion for eugenics and elimination made him the daydreamer of [Nazi] things to come."[47]

Through and through, the very rationale behind *American Thought* is Hofstadter's attempt to take the dirtied postwar reputation of governist eugenics and smear it onto free-market economics. Thurow's ignorant remarks are a testament to this influence. Hofstadter's former student, Eric Foner, confesses that Hofstadter was an ardent anti-capitalist throughout the 1930s. Though published in 1944, *SDAT* was a revised version of the Ph.D. dissertation that Hofstadter completed in 1932. In this decade Hofstadter was so strident in his leftism that he joined the Communist Party in 1938. Hofstadter was clear about his impetus—"My fundamental reason for joining is that I don't like capitalism and want to get rid of it." Hofstadter left the party in disillusionment the following year,[48] but he stated that one aspect of his ideology remained unchanged—"I hate capitalism and everything that goes with it."[49] Throughout his twenties, Hofstadter increasingly distanced himself from the communist label as he remained committed to propagating a Progressive reformist agenda. In 1955 Hofstadter admitted that *SDAT*'s bias "was naturally influenced by the political and moral controversy of the New Deal era."[50]

Further, Thomas C. Leonard adds, the passage of time "did not much soften Hofstadter's hostility to free markets."[51] As late as 1964, Hofstadter was still wont to exhibit his contempt for those who publicly suggested that consensualism was preferable to governism. In the October 8 issue the *New York Review of Books* from that same year, Hofstadter lambasted then-U.S. Senator and presidential hopeful Barry Goldwater (1909–1998) over the candidate's minor, watered-down advocacy of night-watchman-state principles. The Columbia scholar fumed, "When in our history has anyone with ideas so bizarre, so archaic, so self-confounding, so remote from the basic American consensus, ever got so far" in public support? Hofstadter lamented that Barry was "within a hair's breadth of ruining one of our great and long-standing institutions," the Republican Party, which was then dominated by such left-wing governists as John D. Rockefeller, Sr.'s grandson, Nelson Aldrich Rockefeller (1908–1979).[52] These confessions on Hofstadter's part remove the obstacles to our gaining an understanding of why *SDAT* is written as it is. "Dr. Hofstadter's own sympathies," said a favorable article in the *English Historical Review*, "are not hard to divine: he is hostile to laissez-faire, to Spencerian sociology, to classical economics."[53]

There is but one area where Spencer advocated a policy similar to those prescribed by the governist eugenicists. While otherwise favoring a night watchman state, Spencer did want the government to prohibit interracial breeding—what was then known as *miscegenation*. He wrote a letter to a Japanese admirer of his that he condoned regulations that forbade sex between whites and East Asians. The Briton believed that, just as a mule, being the hybrid of horse and donkey, was condemned to infertility, it followed that whites and East Asians were so biologically dissimilar from one another that any interbreeding between the disparate groups would result in children cursed with the same defects as those found in mules.[54] It was on this one count—of supporting legislation to discourage miscegenation—where Spencer and the governist eugenicists were of the same mind. For Spencer to back such miscegenation laws was indeed misguided and based on false assumptions. Nonetheless, Spencer's agreement with the eugenicists extends no farther than that. The majority of eugenicists believed that whites were inherently superior to nonwhites. By contrast, although Spencer wanted the races kept apart, he continued to judge them as being the same in moral merit. To quote a notorious phrase, Spencer wanted the races to be separate but equal under the law.[55] As I shall demonstrate in the next chapter, in no other manner did Spencer approve of legislation that discriminated against people based on their ancestry.

Not to concede that *SDAT* is mostly a spurious screed against free-enterprisers, scholars in Hofstadter's vein can point out that Spencer had some social correspondence with historical figures that were important to the development of eugenics. Just as Spencer was a friend of Darwin's, it was Darwin's ideologically governist cousin, Galton, who founded the eugenics movement. Spencer and Galton had befriended one another.[56] In the final page of his manifesto in support of governist eugenics, paleontologist Henry Fairfield Osborn, Sr. (1857–1935), favorably quotes from Spencer. Irving Fisher (1867–1947), a pioneer in what came to be known as monetarist economics, was both a student of Sumner's and a vehement advocate of eugenics,[57] going as far as writing to Galton to seek moral support.[58] Moreover, William Graham Sumner's brother-in-law, famed Yale University football coach Walter Camp (1859–1925), held membership in the Eugenics Council of the USA (ECUSA).[59]

Was Governist Eugenics a Capitalist Conspiracy?

There has been no suspension to the effort to trace state-sponsored eugenics' origins to consensualist economics. In this mission, Edwin Black notes that the estates of various

Inventive-Period industrialists helped finance the genetic research of one of Galton's protégés and correspondents[60]: biologist cum eugenics promoter Charles Davenport (1866–1944). It turns out that Davenport's research and propaganda received funding from Alexander Graham Bell; John D. Rockefeller, Jr.; and Mary Williamson Averell Harriman (1851–1932), the widow to the railroad tycoon and corporate-turnaround expert Edward Henry Harriman (1848–1909).[61] It will be remembered from Book Two that these Inventive-Period industrialists have been widely derogated as robber barons. What is less well-known is that these same parties also donated to the pro-eugenics research of Davenport's colleague, New York Zoological Society founder Madison Grant (1865–1937).[62] Eerily, Grant received fan letters from Adolf Hitler himself.[63] George Eastman (1854–1932), the inventor of photographic film and the founder of Kodak Corporation, donated to the American Eugenics Society.[64] The same goes for John Harvey Kellogg (1852–1943), a doctor and sanitarium owner.[65] Worse yet, Kellogg attended the society conferences.[66] This man was the elder brother to Will K. Kellogg (1860–1951), who would become a multimillionaire as the founder of Kellogg's Cereal. And as *National Review* contributor Robert Zubrin is quick to point out, the pioneering American eugenicist Henry Fairfield Osborn, Sr., director of the American Museum of Natural History and employer of Madison Grant,[67] was a nephew of the banker John Pierpont Morgan, Jr. Morgan, Jr., personally attended the Second International Congress of Eugenics in 1921, which was also peopled by Charles Davenport, Mary Rumsey née Harriman, and Madison Grant.[68] Topping these other figures might be Henry Ford, Sr., being praised by name in *Mein Kampf*.[69] Prior to U.S. entry into the second World War, Ford hung a photograph of Hitler on the wall of his office.[70] Anti-capitalist professor Michael Parenti mentions Ford's anti-Semitism and Nazi sympathies as if these are traits inherent to businesspeople.[71] There, Parenti ignores the fact that a significant percentage of successful businesspeople are Jewish. He also ignores that Europeans have long mentally associated Jews with business and finance at least as early as the Middle Ages. Hitler exploited that mental association in his demagoguery against Jews. That mental association is something we will address in chapters 7 through 9.

Especially damning, it seems, is that the seed money with which Davenport established his Cold Springs Harbor Laboratory for his genetics research in 1904 came from the bank account of steel magnate Andrew Carnegie.[72] And Carnegie would donate an additional $10 million to Davenport in 1911.[73] Hofstadter is eager to disclose that Carnegie was a fan and personal friend of Herbert Spencer's. Hofstadter elaborates that, when waiting at an American dock for the large passenger liner that would sail him back to England, Spencer pointed to Carnegie and *Popular Science* magazine founder Edward L. Youmans (1821–1887) and announced to nearby reporters, "Here are my two best American friends."[74] Carnegie himself wrote in his 1889 treatise, *The Gospel of Wealth*, that although commercial competition "may sometimes be hard for the individual, it is best for the race, because it insures the survival of the fittest in every department."[75]

Furthermore, Carnegie played a large role in throwing a major party in Spencer's honor. Spencer visited the United States in November 1882 where, at the fancy New York restaurant Delmonico's, various pillars of the business community and the intelligentsia attended to pay tribute to Spencer. Among these luminaries was the trailblazing paleontologist Othniel Charles Marsh (1831–1899). Carnegie made an appearance, as did the great banker August Belmont, Sr. (1813–1890).[76] *American Thought* then goes on to quote an excerpt of Carnegie's autobiography, in which Spencer's name pops up.[77] But *American Thought* uncharacteristically declines to quote the sections of the industrialist's memoir that most juicily implicate Spencer as being the source of Carnegie's social Darwinian eugenicism. For instance, Hofstadter quotes but a fraction of this passage wherein Carnegie reminisces about his first encounters with Spencer's writings— "When I . . . was in a stage of doubt about theology, including the supernatural element, and indeed of the whole scheme of salvation through vicarious atonement and all the fabric built upon it, I came fortunately upon Darwin's and Spencer's works: 'The Data of Ethics,' 'First Principles,' 'Social Statics,' 'The Descent of Man.'" That last book was by Darwin; Spencer authored the others listed. Carnegie's passage continues, "Reaching the pages which explain how man has absorbed such mental foods as were favorable to him, retaining what was salutary, rejecting what was deleterious, I remember that light came as in a food and all was clear. Not only had I got rid of theology and the supernatural, but I had found the truth of evolution. . . . Man was not created with an instinct for his own degradation, but from the lower he had risen to the higher forms. Nor is there any conceivable end to his march to perfection." The so-called robber baron then spouts this incredible series of claims about his philosophic mentor. "Spencer was always the calm philosopher. I believe that from childhood to old age . . . he never was guilty of an immoral act or did an injustice to any human being. He was certainly one of the most conscientious men in all his doings that ever was born. Few men have wished to know another man more strongly than I know Herbert Spencer, for seldom has one been more deeply indebted than I to him and to Darwin."[78] As Black puts it, Davenport and other eugenicists "relied upon the powerful, the wealthy, and the influential to make their war against the weak a conflict fought not in public, but in the administrative and bureaucratic foxholes of America." In Black's mind, eugenics legislation "was nothing less than an alliance between biological racism and mighty American power, position, and wealth against the most vulnerable, the most marginal and the least empowered in the nation."[79]

This signifies, yes, that some wealthy individuals supported state-trumpeted eugenics. But it proves absolutely nothing unseemly about unencumbered markets. First, Yale University historian Daniel J. Kevles reveals that Mrs. Mary Williamson Harriman began to bankroll Davenport's eugenics projects primarily at the request of her daughter, Mary Rumsey *née* Harriman (1882–1934). As Kevles articulates, Mary Rumsey developed such a zeal for government-backed eugenics as a direct result of her being a "social activist with a [left-wing welfare-state] liberal bent." Mary Rumsey's brother was W. Averell Harriman (1891–1986), FDR's ambassador to the Soviet Union.[80] She was also a friend of Eleanor Roosevelt and FDR's U.S. Labor Secretary, Frances Perkins (1882–1965),[81] whom, I may add, championed the 1938 Fair Labor Standards Act. Mary Rumsey herself headed the Consumers' Advisory Board of the National Recovery Administration under the New Deal. Daniel Kevles, writes that governist eugenics struck Mary Rumsey "as a means of social improvement . . ." She convinced her mother, Mary Harriman, of the same, and this persuaded Mrs. Harriman to bankroll Davenport's work.[82]

Audit the notion that Spencer inspired Carnegie's governist eugenicism. The very magazine that Spencer once helped edit, *The Economist*, has, as recently as the twenty-first century, ominously implicated Spencer in supposedly supplying Carnegie with a ruthless attitude. In 2003, the writer Peter Krass came out with his own biography of Carnegie. That same year, *The Economist* reviewed Krass's book in its pages. The periodical praises Krass for noticing that one of the "main motivating forces" of Carnegie's career was none other than "the harsh form of Darwinism taught by Herbert Spencer, Carnegie's guru (and writer for *The Economist*). Spencer held that 'society advances where its fittest members are allowed to assert their fitness with the least hindrance, and where the least fitted are not artificially prevented from dying out.'"[83]

I have already dispelled the false accusation that Spencer hated private charity. The source of the quotation about the "least fitted" being "not artificially prevented from dying out" is unclear. When you rummage through the endnotes section of Peter Krass's Carnegie biography, you see that Krass cites two works about Herbert Spencer.[84] Krass purports to be discussing Spencer's books *Social Statics* (which is the source of my Spencer quotation about charity) and *First Principles*. Krass's citation leaves it ambiguous as to whether the quotation is from one of the Spencer-authored books, or from one of the works about Spencer. That quotation is not found in either *First Principles* or *Social Statics*.[85] Because of my inability to track down the source of the quotation, I decided to query someone who knows much more about this topic than I do. In 2013 I e-mailed Robert Bannister himself to inquire as to whether he was familiar with the passage attributed to Spencer. Dr. Bannister looked into it, and pointed out to me that the quotation is found in one of the books that Krass cites. That book is *The Social and Political Thought of Herbert Spencer*, the 1978 Ph.D. dissertation of David Wiltshire.[86] In *Social and Political Thought*, Wiltshire ascribes the quotation to Spencer and says it is from page 81 of a 1965 edition of *The Man Versus the State*, edited by the libertarian Albert Jay Nock.[87] However, the quotation is not to be found in any edition of *The Man Versus the State*. As *The Man Versus the State* is available on the World Wide Web in its entirety, both on Google Books and on the Liberty Fund's *Online Library of Liberty* website, one can run a search for the "artificially prevented" quotation in these editions. Such a search comes up with zero results. The quotation is most likely one that Wiltshire accidentally misattributed to Spencer. Nevertheless, critics quote the "artificially prevented" misquotation as evidence of Spencer's callousness. Peter Krass cites the misquotation for that purpose, as do the Ph.D. dissertation of Stephen H. LeDrew[88] and an article in the periodical that itself once employed Spencer—*The Economist*.

Despite his professed admiration for the English philosopher, Carnegie did not convey deep comprehension of Spencer's economic theories. "In reality," writes Bannister, "Carnegie's understanding of Spencer's philosophy was superficial at best . . ." The steel magnate went as far as expressing disdain for laissez faire. "I differ from my great master Herbert Spencer in regard to the duties of the state." Thus this industrialist publicly extolled protective tariffs, labor regulations, the Sherman Antitrust Act, and the Interstate Commerce Act.[89] When something goes amiss in the economy, Carnegie stated, "it always comes back to me that Government control, and that alone, will properly solve the problem . . ."[90]

In his *Gospel of Wealth*—that same manifesto where he defended competition as "the survival of the fittest"—Carnegie argued that a multimillionaire's fortune is really public property rather than his own private property. According to the Carnegie manifesto, wealthy individuals should be permitted to maintain legal control over their own estates solely because they are particularly competent at managing assets. From this perspective, one should retain control over his fortune under the condition that he fulfills a communalistic duty to donate his money to charity. According to people who agree with Carnegie, a millionaire businessman unilaterally acquires such a fortune at the expense of his consumers, as if the consumers themselves had received no remuneration from the fact that they received the goods and services they wanted from the businessman. Therefore, the millionaire has simply collected the public's money and should learn to "give back to the community," as if he had not already "given back" to his customers when he handed them the product they wanted. *The Gospel of Wealth* consequently pronounced that whether or not a millionaire deserves to live is incumbent upon whether he does the following, which Carnegie calls the millionaire's duty:

> . . . to provide moderately for the legitimate wants of those dependent upon him; and, after doing so, to consider all surplus revenues which come to him simply as trust funds, which he is called upon to administer, and strictly bound as a matter of duty to administer in the manner which, in his judgment, is best calculated to produce the most beneficial results for the community—the man of wealth thus becoming the mere trustee and agent for his poorer brethren, bringing to their service his superior wisdom, experience, and ability to administer, doing for them better than they would or could do for themselves.

The Gospel therefore assigned multimillionaires the task of

> **returning** [sic] their surplus wealth to the mass of their fellows in the forms best calculated to do them lasting good. [Remember this "return" for later; boldface added. –S.H.] . . . the millionaire will be but a trustee for the poor, intrusted for a season with a great part of the increased wealth of the community, but administering it for the community far better than it could or would have done for itself.

Hence the treatise concluded, "The man who dies rich thus dies disgraced."[91] Guess what—when Carnegie died, he still had twenty-three million dollars left over.[92]

Carnegie had numerous followers in this creed. James J. Couzens (1872–1936), the politician and Ford Motor Company vice-president who first suggested to the company's founder that he raise employee wages to five dollars per day, likewise proclaims that a millionaire does not own his fortune, but that his fortune is merely a public exchequer he holds for society.[93] Couzens told his daughter Madeleine that their family's money "doesn't belong to us. . . . It's a trust."[94] And Carnegie's words are in accord with those of a man who is called a social Darwinist as often as Carnegie himself—Adolf Hitler. In his famous first book, *Mein Kampf*, Hitler wrote that our moral estimate of a "man must be based on the manner in which he fulfills the task entrusted him by the community." His "duty" is

> to **return** to the national community with honest industry what it has given him. [Why does this sound so familiar? –S.H.] . . .
> The present era . . . sees material reward the expression of a man's worth and thereby shatters the foundation for the noblest equality that there can be. For equality does not rest and never can rest on the achievements of individuals in themselves, but it is possible in the form in which everyone fulfills his special obligations. [boldface added]. [95]

Historian James Truslow Adams concurs. To him, multimillionaires who produce large endowments are not generous, for they "merely **return**, not seldom unwisely, a part of their wealth to that society without which they could not have made it, and which too often they have plundered in the making"[96] (boldface mine).

To my dismay, Calvin Coolidge somewhat agrees—"It is recognized that in time of peace that the public may take what it may need of private property for the general welfare, . . . and the right to own property carries with it the duty of using it for the welfare of our fellow man." An industrialist "must use his property for the general good or the very right to hold private property is lost."[97]

Of course, there is a shared assumption among Carnegie, Truslow Adams, and Hitler. It is that a multimillionaire deserves to live only as far as he renounces his time and labor for the social collective. That is why Carnegie and Hitler spoke of philanthropy as a millionaire's method of "returning" wealth to the "community." They ignore that the millionaire had already given to his customers the product for which they paid him, and hence anything else he "gave" to others was actually something extra. This shared assumption of Carnegie's and Hitler's is never criticized as social Darwinism. Indeed, pro-regulation governists who frequently deploy the social Darwinism epithet largely concur with Carnegie and Hitler on this issue.

Additionally, Carnegie's *Gospel* praised the estate tax as "salutary." This same work further delivered a strange point about tax-funded libraries. Recall from Book One that Carnegie privately financed the construction of a municipality-owned library in lower-crust neighborhoods. In *Gospel*, Carnegie adds that any other millionaire who finances such library construction should ensure that this library's maintenance should not be financed through private contributions. Such maintenance, Carnegie instructs, should be financed by nothing except for taxes exacted from that neighborhood's low-income residents.[98] The Scottish-born industrialist assures his readers that this taxation will engender a strong work ethic among the lower classes.[99] *The Gospel* struck a chord with another industrialist normally derided as a social Darwinist—John D. Rockefeller, Sr. Consequently, Carnegie's homilies prompted the entire Rockefeller clan toward the very same collectivist mentality that led it to support government regulations quite readily.[100] When the Carnegie Library opened in Pittsburgh, Pennsylvania, in 1896, John D. Rockefeller, Sr., wrote to the book depository's benefactor, "I would that more men of wealth were doing what you were doing with your money; but, be assured, your example will bear fruits, and the time will come when men of wealth will more generally be willing to use it for the good of others."[101] It may appear damning for capitalism that Carnegie was good friends with one of history's most avid governist eugenics promoters, Madison Grant. Yet Madison Grant was also good friends with Sierra Club founder John Muir (1838–1914) and with protectionist labor leader Samuel Gompers (1850–1924).[102] I doubt that most governist academicians would consider this association to be damning for Muir and Gompers.

In complete contrast to Edwin Black's portrait: Carnegie, the Rockefellers, and the Harrimans became patrons of government-regulatory eugenics not because they were privy to some self-interested, plutocratic, capitalist conspiracy. Nay, it was because they were the early twentieth-century's equivalent of the present's politically correct welfare-state limousine (il)liberals.

Henry Ford was a virulent anti-Semite, but his anti-Semitism does not reflect on capitalism. In fact, a rather strange oddity is that though Ford prided himself on being a businessman, he shunned the word *capitalist*, which, like Karl Marx, he considered synonymous with *financier*.[103] The racist book that Ford paid another author to ghostwrite for him, *The International Jew*, frequently employed the word *capitalist* in a negative light. The book mainly criticizes Jews on account of the stereotype that Jews tend to be moneymen.[104] *The International Jew* repeated the same scurrilous accusations against J. Pierpont Morgan, Sr., and Jay Gould that anti-capitalists had already been leveling for decades. Then the book falsely identified both of these men as Jews, implying that being Jewish was linked to their capitalist rapacity.[105] What was ironic in this is that Morgan himself, regrettably, harbored anti-Semitic prejudices. It cannot be denied that Ford's anti-Semitism is inexcusable. With that acknowledged, it is noteworthy that Ford's personal prejudices were not reflected in his hiring practices. From the 1920s onward, there were never fewer than 3,000 Jews employed at his motor company.[106] Whereas discrimination was rampant almost everywhere else, Ford actually made a point of hiring blacks. One-tenth of Ford's 100,000-member workforce was African-American.[107] And, as author Robert Lacey notes, Ford "never paid a man more, or less, on account of his skin." Indeed, his "black employment policy was genuinely ahead of its time. Henry believed in an equal day's pay for an equal day's work . . ."[108] When Ford died, the *Journal of Negro History* said that Ford "gave Negroes along with others the chance to help themselves, and in doing so he met the highest test of being a friend . . ."[109]

The closest there came to being a real-life social-Darwinist eugenicist industrialist was Wickliffe Draper (1891–1972). A textile heir, he founded the Pioneer Fund, which from the day of its founding to that of this writing, has financed research purporting to prove the following tenets. (1) Someone's IQ number is the major predictor of whether he will be economically productive and law-abiding. Economic productivity is directly proportional to one's IQ. By contrast, one's IQ number is inversely proportional to one's likelihood of committing violent crime. (2) The extent to which one's IQ number is attributable to genetics is between 40 percent and 80 percent. Furthermore, race and ethnicity are intertwined with genetics. (3) Therefore, someone's racial and ethnic background determines one's IQ, which therefore determines one's economic productivity or criminality.[110] Generally, the work of the Pioneer Fund casts unfavorable generalizations about blacks and Latinos, and much of the data cited in the 1994 book *The Bell Curve* came from research done at the Pioneer Fund's behest.[111] As usual, it would be folly to categorize Draper as a free-marketer. Draper financed the Immigration Restrict League (IRL). It was in large part a consequence of this organization's lobbying efforts that the Immigration Restriction Act of 1924 saw passage. Moreover, just as Draper self-consciously emulated Madison Grant's governist eugenics advocacy, so too did Draper adopt Grant's environmentalism. In the tradition of Grant and Theodore Roosevelt, Draper advocated government nationalization of the wilderness and government management over wildlife preservation, irrespective of people's private property rights, and ignorant of the benefits of privatization.[112]

And though Alexander Graham Bell was a wealthy entrepreneur, the story of his involvement in the eugenics movement reveals nothing of a capitalist conspiracy against the underprivileged. Bell's interest in the scientific study of sound vibrations—which precipitated his invention of the telephone—and in eugenics came from the same source. The source was his early job as a teacher at a private school for deaf girls. One of his hearing-impaired students, Mabel Hubbard (1857–1923), would one day become his wife. Bell discerned from studying the genealogies of his students that some cases of inborn deafness might be at least partially attributable to heredity. He thereby contemplated the chance that by learning about their own genealogies and those of their prospective spouses, people could make choices in marriage partner that reduced the chances of their own children being born deaf. *Science and Myth* clarifies the extent to which the telephone inventor was interested in eugenics: "Bell observed that normally there was little prospect of individuals with similar congenital illnesses choosing one another for marriage partners. But the segregation of the deaf made their case an exception, posing the threat of a new race of deaf-mutes. . . . Moreover, he stressed he was interested only in the special case of deafness, known to be inheritable, not a broader spectrum of physiological and psychological traits." Aside from those considerations, Robert Bannister informs us, Bell was not enthused by eugenics, and he "refused to endorse coercive controls" on human reproduction.[113] Himself an immigrant, Bell grew increasingly offended by the governist eugenicists' draconian measures against immigration and their cruel sterilization policies. To quote Bell's own words, he abstained from involvement with any "attempt to interfere, by compulsory means, with the marriages of the defective and undesirable."[114] He elaborated, "We cannot [rightfully] control the marriages of men as we can the breeding of animals, and at first sight there seems to be no way of ascertaining how far human beings are susceptible of variation by selection."[115] Thus, writes Georgia State University law professor Paul A. Lombardo, "Bell would eventually leave the eugenics movement and dissociate himself" from Davenport's circle.[116]

As for George Eastman, his patronage of eugenics research was not motivated by a driving urge to further racism. Eastman was actually a major financier of education for blacks, endowing millions to Booker T. Washington's Tuskegee Institute.[117] Likewise, John D. Rockefeller, Sr., founded that historically black educational seminary, Spelman College.[118]

I cannot deny Edwin Black's observation that family dynasties founded by nineteenth-century industrialists—whom Matthew Josephson famously termed "robber barons"—financially contributed to governist eugenics research. I do doubt Black's conclusion, though, that this was a selfish capitalist conspiracy. Madison Grant's friendship with John Muir merits further comment. The same Walter Truett Anderson whose misrepresentation of Spencer I debunked in the previous chapter, remarks, "Eugenics became a wildly popular cause, as trendy in its time as environmental protection is today."[119] That comparison proves more apt than Anderson seems to realize. Edwin Black should notice that most of these exact same people financed the early twentieth-century's environmental conservation efforts. Writes Jonathan Spiro, "The Carnegie, Harriman, and Rockefeller families had all contributed to the conservation activities of Madison Grant, and now they were aiding the eugenic ambitions of Charles Benedict Davenport. The conservation and eugenics movements did not want for friends in high places in the early twentieth century."[120] Would Black and his fellow leftists conclude that this proves that environmental activism is a capitalist conspiracy as well? In the book *Conservation and the Gospel of Efficiency*, left-wing historian Samuel P. Hayes admits that the wealthy early-twentieth-century progressives who pioneered in American environmentalism were governists who lobbied for government-enforced cartelization of industry, for they held "revulsion against unrestrained competition . . ."[121] As shall be detailed in chapters 5 and 9, eugenicism and environmentalism alike grew popular in the USA and Germany on account of the very same historical figures, one of whom was Madison Grant, supporting both causes on the same basic philosophic premise.

Given the aforementioned considerations, I contest Edwin Black's insinuations (1) that laissez faire is the same as plutocracy and (2) that Herbert Spencer was some sort of apologist for a plutocracy. One of Spencer's main points of his political writings was his lamentation that the government of nineteenth-century Britain enacted regulations in order to subsidize the wealthiest classes at the expense of the lower ones. Spencer thundered, "Our legislators tax the people to a most exorbitant extent; squander the money thus wrested from the toiling artisan in the support of institutions for the benefit of the rich; . . . and when the misused subject demands of the government that it defend him in the exercise of his rights and privileges . . . what is its conduct? . . . Does it take up the cause of the poor man, and defend him against the aggressions of his rich neighbour? No! . . . Not only has our government done those things which it ought not to have done, but it has left undone those things which it ought to have done; and truly may it be said that there is no health in it."[122]

Financial Meritocracy Facilitates Racism and Sexism?

Government-practiced eugenics violates Lockeanism by definition. Recognizing this, Yale's Daniel Kevles appends his own unflattering description of Herbert Spencer's ideas with a reminder to his readers that such free-market advocates as Spencer inveighed against government regulations over human reproduction. Kevles acknowledges that such that coercion contradicts the "doctrine of laissez-faire by requiring state interference with individual liberty, and one of the most private areas of liberty at that."[123] Georgia State University law professor Paul A. Lombardo also finds it behooving, following his unfair denunciation of Spencer, to acknowledge that the advocates of forced sterilization and government-regulated eugenics virulently detested "a laissez-faire policy in a society whose charitable institutions subsidized the lives of the deficient."[124] Scholar Diane Paul cannot help but remark that governist eugenics has no logical place in a country that may be thought of as "the land of Horatio Alger."[125] Robert Bannister remarks that the very allegation that governist "eugenics was simply an extension of . . . the industrial laissez faire most eugenicists despised," simply "blurred the fact" that governist eugenics amounted to an "abandonment" of Spencer's domestic peace advocacy in favor of violent governist intrusions.[126]

Indeed, legions of writers lump Spencer and William Graham Sumner together with the governist eugenicist movement as social Darwinist just because each of these parties invoked evolution in their arguments concerning social science and politics. Those who lump these parties together place Spencer and Sumner in the same category as men of a completely dissimilar affiliation. Spencer and Sumner clashed with the governist eugenicists on:

1. Political economy (Spencer's and Sumner's free markets versus the eugenicists' regulations).

2. Western colonialism (Spencer and Sumner against, eugenicists for).
3. Compulsory sterilization (Spencer and Sumner against, eugenicists for).
4. Open immigration (Sumner supported it, eugenicists opposed).[127]

Daniel Kevles's own knowledge of the matter, though, has not stopped him from himself besmirching Spencer and Sumner. Over seventeen years after he admitted that Spencer opposed governist eugenics, Kevles nevertheless co-authored with other academicians *Inventing America: A History of the United States*. This work follows the conventions of categorizing Spencer, Sumner, and governist eugenicists in the general class of social Darwinist and of implying that governist eugenics was consensualist ideology taken to its logical conclusion. As I think that those responsible for this travesty should be publicly identified, I will name Kevles' co-authors: Pauline Maier and Merritt Roe Smith, both historians at the Massachusetts Institute of Technology; and Alexander Keyssar, a professor of history and social policy at Harvard University's John F. Kennedy School of Government. Once they have finished trashing Spencer, this brain trust announces, "The foremost American advocate of social Darwinism was William Graham Sumner. . . . A prolific writer, Sumner spent . . . decades developing and promoting the Spencerian and social Darwinist ideas that first captured his attention in the 1870s. Denouncing government intervention, socialism, protectionism, and social reform, Sumner defended economic inequality. . . . Not surprisingly, Sumner's views were widely applauded by business leaders and political conservatives." Upon establishing social Darwinism's capitalist credentials, these scholars attempt to tie social Darwinian capitalism to imperialism, as "social Darwinist notions of 'survival of the fittest' seemed readily applicable to the international arena." Hence these social Darwinists "argued that international conflict was inescapable and would surely lead to American economic, political, and cultural triumphs." This worldview therefore "lent support to efforts to implant the American way of life, and Protestant religion, in distant corners of the globe" by means of force.[128]

Besides the mutual concern that Spencer and Sumner had in common with governist eugenicists over natural selection, there is another disingenuous rationale that politically correct (PC) leftists have exploited. They exclaim that most social Darwinists—including Spencer, Sumner, and the eugenicists—engage in these two behaviors.

1. They applied theories relating to biological competition to the social sciences.
2. They provided apologias for "social inequality."

I will briefly address each of these purported characteristics, one by one. It was in 1932 that an academician first proclaimed in print that social Darwinism is to be defined as "the application of Darwinian concepts of variation and selection to social evolution." Those were the words of Harvard University sociologist Talcott Parsons (1902–1979).[129] That the free-market evolutionists and governist eugenicists alike prescribed public policy, according to their own respective interpretations of biology, cannot be denied. This practice, which Hofstadter resents, is Hofstadter's official, explicit characterization of social Darwinism. The preface of *American Thought* defines social Darwinism as "the adaptation of Darwinism and related biological concepts to social ideologies." Chapters later, *American Thought* further generalized the definition to "biologically derived social speculation." The introduction to the book's 1959 edition goes as far as saying social Darwinism encompasses "the effect of Darwin's work upon social thinking in America." In contrast to the thinking of more-modern left-wingers such as E. O. Wilson and Peter Singer, *SDAT* pronounced that knowledge about natural selection proves "utterly useless in attempting to understand society."[130] The Religious Right would not dispute Hofstadter on this count.

Now we move on to the common avowal that every brand of social Darwinism—including free-market evolutionism and governist eugenics—is a rationalization for social inequality. That is the exact term that Rice University anthropologist Sarah Blaffer Hrdy (b. 1946) enunciates when she equates free-market economics with racism—"social inequality."[131] Everywhere we find that PC equivocation about free-market individualism and racial bigotry. Within years of *SDAT*'s publication, that attitude became ubiquitous. In 1945, David F. Bowers of Princeton declared, "Spencer's biological approach to social theory provided support for the biological approach to history implicit in all racist and imperialist propaganda." Creepily, Bowers cautioned that a reading of *SDAT* laid out Spencer's "important influence on the eugenics movement of the period."[132] "It would seem a stretch," Thomas C. Leonard mused in 2009, "to claim that Herbert Spencer advocated planned state control of human breeding, but that charge emerged immediately in the wake of SDAT and can still be found today."[133]

In 2007, Binghamton University anthropologist David Sloan Wilson denounced Spencer. You may recall my mention in Book Two of D. S. Wilson's derision of Ayn Rand. On the matter of the nineteenth-century evolutionist, Wilson spews, "Herbert Spencer, an intellectual giant of Darwin's day, liked evolution because he thought it justified the inequalities of British class society. . . . Hitler liked evolution because he thought it justified the ultimate social inequality of genocide. Using evolution to justify social inequality has become known as 'social Darwinism.' "[134] According to D. Sloan Wilson's insinuation, Spencer glamorized inequality by saying that there is nothing inherently unjust about Millionaire Mike being wealthier than Thousand-aire Theo. This is with Spencer's and Sumner's proviso that Millionaire Mike acquired his affluence peacefully. Anti-capitalist intellectuals assume that that conclusion is identical to eugenicists promoting "social inequality" in the sense that eugenics regards Aryans to be congenitally superior to other races and therefore morally obliged to subjugate them. And take notice that while she starts off with condemning free-market economics as social Darwinism, Sarah Blaffer Hrdy smoothly segues into an insinuation that social Darwinism is the link that binds Lockean capitalism with racism and sexism. "Social Darwinism explicitly assumes that competition leads to 'improvement' of a species; the mechanism of improvement is the unequal survival of individuals and their offspring. Applying this theory to the human condition, social Darwinists hold that those individuals who win the competition, who survive and thrive, must necessarily be the 'best.' Social inequalities between the sexes, or between classes and races, represent the operation of natural selection and therefore should not be tampered with, since such tampering would impede the progress of the species."[135]

Observe what data Hrdy and D. S. Wilson have marginalized. To some degree, Spencer and Sumner formulated that free enterprise was good exactly because it made it possible for someone to prosper through his own individual merit by his own volition. Yes, as I stated in Book Two, Spencer disliked the term *free will* and derided it. I would not be surprised if Sumner expressed similar derision of the term. Nonetheless, at least in the sphere of market activities, they understood that a person's wealth, in the final analysis, was determined primarily by the person's own choices, and therefore the result of the person's own level of judgment and responsibility. These are the terms that Spencer himself employed in explaining why, if Mike peaceably serves consumer demand on a wider scale, and at greater efficiency, than does Theo, then Mike deserves to earn more money than Theo: "Justice . . . means preservation of the normal connexions between acts and results—the obtainment by each of as much benefit as his efforts are equivalent to—no more and no less. Living and working within the restraints imposed by one another's presence, justice requires that individuals shall severally take the consequences of their conduct, neither increased nor decreased." There is therefore great injustice, and an impediment to authentic social progress, when, through the force of violence, the State "abstracts from some men part of the advantages they have earned, and awards to other men advantages they have not earned."[136]

On the other hand, the movement for government-controlled eugenicists argued for the destruction of free enterprise on the basis that nobody possessed volition anyway, and therefore no one possessed individual responsibility.[137] Whereas we are told that laissez-faire social Darwinists are arch-advocates of individualism and selfishness, Francis Galton preached political collectivism and personal sacrifice. Sir Francis approvingly judges that in the course of natural selection, "the life of the individual is treated as of absolutely no importance, while the race is treated as everything, Nature being wholly careless of the former except as a contributor to the maintenance and evolution of the latter. . . . We must, therefore, try to render our individual aims subordinate to those which lead to the improvement of the race." Such a centrally planned, government-directed improvement of the genome ought to "be looked upon as one of the chief religious obligations. . . . It may come to be avowed as a paramount duty, to anticipate . . . natural selection, by endeavoring to breed out . . . ignoble instincts, and to breed in those which are...social." Given that Francis Galton disbelieved in the importance of personal volition, it should not be too startling that he coined that much-overrated expression I contested in Book Two—*nature versus nurture*.[138] As I wrote in Book Two, persons who proclaim that one's personal character is decided by either biology or environmental stimuli choose to elide the consideration that there may be another factor at work. That factor they presume to be meaningless is indeed individual volition, exactly what the free-market individualist esteems as important and what the determinist collectivist denies. Yes, the Nazis took it for granted that biological factors had justified the oppression of nonwhite races. And, yes, that idea was propounded by some nineteenth-century European eugenic imperialists who preceded them.

Granting this, it would not be difficult for one to draw a particular conclusion. The conclusion is that if Spencer and Sumner co-founded the governist eugenics movement, then it would not be inconceivable that they might have aired sympathies with the transatlantic slave trade. John M. Hobson, for instance, decries Spencer and his fellow social Darwinists for "legitimising to Westerners the superiority of the white race. The importation of Darwinism into social science theory was especially important . . ." It gave men the opportunity to pass off their crude racist beliefs as scientific facts. Social Darwinism "found its place alongside the emerging explicit (scientific) racist treatises . . ."[139]

Similarly, take Colin Tudge—an award-winning British science journalist and scholar at the Center for Philosophy at the London School of Economics. Tudge hurls this dubious asseveration: "Some, like Herbert Spencer, sought to extrapolate his ideas from biology into moral philosophy and politics. Thus emerged 'social Darwinism'—apparently the notion that society *ought* to be as 'red in tooth and claw' as Alfred, Lord Tennyson, supposed nature to be, and the weakest should go to the wall. The gentle and liberal Darwin, who railed against slavery in an age when it was still considered both proper and necessary . . . could hardly have approved"[140] (emphasis Tudge's). The assumption buried in Tudge's prose is that Spencer apparently *did* approve of slavery "when it was still considered both proper and necessary." In truth Spencer and Sumner both abhorred this institution. An actual reading of Spencer's corpus evinces that he was an abolitionist. As Ohio State University demographer William Petersen correctly notes, Spencer's youth at *The Nonconformist* magazine was spent campaigning against the slave trade.[141] Furthermore, Spencer unequivocally denounced chattel slavery in *Social Statics*, the same book that his enemies have cited as proof of his being a social Darwinist. *Statics* remarked upon various "schemes devised without consulting ethical principles. . . ." These are schemes wherein "benefit has been sought by going in palpable opposition to those principles—

cases in which men . . . have diverged into the by-ways of injustice . . ." Spencer finds, "The enslavement of the negroes serves for a good example." Through and through, he deplores "the reasoning of unscrupulous colonists on this matter. . . . Slavery brought in its train the multiplied curses of a diseased social state; a reign of mutual hatred and terror . . ."[142]

In this same book, Spencer curses that it was once "universally supposed that slavery was a natural and quite legitimate institution . . .; nay, indeed, a great proportion of mankind hold this opinion still. A higher social development, however, has generated in us a better faith, and we now to a considerable extent recognize the claims of humanity."[143] Spencer speculates as to how anyone could rationalize the inhumanity of chattel slavery. "Belief always bears the impress of character—is, in fact, its product. . . . Men's wishes eventually get expressed in their faiths—their real faiths, that is; not their nominal ones. Pull to pieces a man's Theory of Things, and you will find it based upon facts collected at the suggestion of his desires. A fiery passion consumes all evidences opposed to its gratification, and fusing together those that serve its purpose, casts them into weapons by which to achieve its end." In the end, Spencer concludes that "the slave-owner's assertion that negroes are not human beings . . ." are among "the strangest samples of convictions so formed."[144] For these reasons, *Social Statics* applauds some boycotts conducted by 300,000 abolitionists against produce harvested from slave plantations, boycotts performed "out of sympathy for the negroes. . . ."[145] Decades later—in 1884—Spencer continued to write bitterly about the "sins of responsible legislators seen in the long list of laws made in the interests of dominant classes—a list coming down in our own country to those under which there were long maintained slavery and the slave-trade, torturing nearly 40,000 negroes annually by close packing during a tropical voyage, and killing a large percentage of them . . ."[146]

Yes, Spencer was indeed capable of holding some condescending attitudes about nonwhite races, and often expressed such attitudes. He did believe that, to some degree, biological inheritance contributed to someone's moral character. On some level, he did think that biological heredity caused some ethnicities to be cleverer or lazier than others. That never convinced him, though, that races should be treated unequally under the law.[147] Examine this statement of his: "If you ask me what prompts me to denounce our unjust treatment of inferior races, I reply that I am prompted by a feeling which is aroused in me quite apart from any sense of duty, quite apart from any thought of Divine command, quite apart from any thought of reward or punishment here or hereafter. In part the feeling results from consciousness of the suffering inflicted, which is a painful consciousness, and in part from irritation at the breach of a law of conduct on behalf of which my sentiments are enlisted, and obedience to which I regard as needful for the welfare of humanity in general."[148] The man is indeed condescending in judging nonwhites to be *inferior races*. Nonetheless, this notion about inferiority nowhere justifies to Spencer any abrogation of the rights of nonwhites.

Upon reading Spencer's 1851 masterpiece as the U.S. Civil War raged on, American philosopher John Fiske (1842–1901) wrote to his fiancée, "No one can start from 'Social Statics' and logically deduce conclusions which shall be other than unfavorable to the [Confederate] South at present."[149]

Carl Schurz (1829–1906) fought in that war as a Union soldier. As that war went on, he spent his nights in his tent at the Chattanooga winter camp reading that same work of Spencer's by candlelight. Reminiscing about that time in subsequent decades, he stated, "It became perfectly clear in my mind that, if the people of the South had well studied and thoroughly digested that book," *Social Statics*, "there never would have been any war for the preservation of slavery." Historian Barry Werth, one of whose books was a finalist for the National Book Critics Circle Award, gets it right. Werth writes that "Schurz, like

Spencer, considered emancipation the central logic" of a free society. And drawing from Spencer, Schurz likewise concluded that the Confederation's insistence on "fighting a war to defend government-sanctioned inequality was antievolutionary: a moral dead end." In the following years, Schurz would go on to become a U.S. Senator from Missouri and then a U.S. Secretary of the Interior.[150]

Philosophically consistent with his high regard for Spencer is Schurz's distaste for Karl Marx, whom he encountered at a conference in 1848, the same year that Marx released the infamous *Communist Manifesto* he coauthored with Friedrich Engels. Of Marx, Schurz sighed, "Never have I met a man of such offensive, insupportable arrogance. . . . Everyone who disagreed with him was treated with scarcely-veiled contempt. He answered all arguments which displeased him with a biting scorn...or with a libelous questioning of their motives. I still remember the cutting, scornful tone with which he uttered—I might say 'spat'—the word 'bourgeois'; and he denounced as 'bourgeois'—that is to say an unmistakable example of the lowest moral and spiritual stagnation—everyone who dared to oppose his opinions."[151]

Carl Schurz's opposition to slavery, along with the other case studies presented, expose the flimsiness of any supposed connection that one may try to identify between Spencer and political racism. When critics allege that Spencer somehow whitewashed racial conquest, they portray an abolitionist as an apologist for human bondage. Not surprisingly, Carl Schurz gave a speech in Spencer's honor at the 1882 Delmonico's banquet,[152] where John Fiske was in attendance as well.[153] Insofar as one defines racism as the attempt to "discriminate against, or exploit members of the 'lower races,'" writes Robert L. Carneiro, an anthropologist at the American Museum of Natural History, ". . . Spencer was most certainly not a racist. . . . Indeed, with regard to the treatment of native peoples and oppressed minorities, Spencer could easily be considered a liberal."[154]

William Graham Sumner went farther than Spencer and Schurz. The Yale sociologist preached that if the United States were to incorporate other countries into its Union through military action, then those other nations' natives being of nonwhite races should not bar them from being treated with the same political rights as those of the USA's white population. He promulgated, "There are plenty of people in the United States to-day who regard negroes as human beings, perhaps, but of a different sort from white men, so that the ideas and social arrangements of white men cannot be applied to them with propriety. Others feel the same way about Indians. This attitude of mind, whenever you meet with it, is what causes tyranny and cruelty." He deduced that "the doctrine that all men are equal . . . in its absolute form, must, of course, apply to Kanakas, Malays, Tagals, and Chinese just as much to Yankees, Germans, and Irish." From this idea, Sumner deduced that it would be unethical for the U.S. to apply military force to annex parts of Asia under its rule unless it recognized these Asians as having the same rights as every white American.[155] It is noteworthy that Sumner could write such words in 1899, many decades preceding the civil rights era.

On this count I must give some qualified credit to Richard Hofstadter. This is despite his being the man who is quite possibly the one most culpable for inspiring academicians to equate both free-market evolutionism and governist eugenicism. Unlike most welfare-state advocates who follow his lead in milking the social Darwinism epithet, Hofstadter at least concedes a partial distinction between free-market evolutionism and governist eugenics. He designates the former "Darwinian individualism" and the latter "Darwinian collectivism of the nationalist or racist variety . . ."[156] The Columbia historian allowed it to slip a few times—somewhat inconspicuously—that when people advocated governist eugenics, their conclusions were based on governist collectivist premises not dissimilar from his own. Somewhat contradicting himself, though, every such admission was followed by qualifying remarks that severely downplayed the extent to which governist eugenicists were explicit members of the Progressive movement. Hofstadter also remained silent on the fact that these progressives supported eugenicist legislation on explicitly progressive grounds.

Hofstadter felt queasy about any social scientist's claim that most human social behavior could be explained primarily by genetics, as opposed to cultural conditioning. For this reason, Hofstadter felt ambivalent about the platoons of progressives who invoked eugenicist arguments to advance the very same regulatory measures that Hofstadter favored. He could forgive them their reformism, but not their biology-centered social science. By contrast, Hofstadter could forgive almost nothing about Spencer or Sumner, who were objectionable on account of their biology-based sociology and, far worse to Hofstadter, their sympathies toward capitalism.

Shamefully, here Hofstadter is much akin to the vast majority of left-wing wordsmiths who invoke *SDAT*. Once conceding the difference between Spencerian individualism and eugenicist collectivism, Hofstadter immediately states that what the laissez-faire individualists had in common with the eugenicist collectivists was that both camps were politically right-wing and "conservative." First he writes, "Conservatism and Spencer's philosophy walked hand in hand. The doctrine of selection and the biological apology for laissez faire, preached in Spencer's formal sociological writings and in a series of shorter essays, satisfied the desire of the select [that is, the wealthy few] for scientific rationale." Then he opines, "The social views of Spencer's popularizers were likewise conservative." The Columbia historian half-heartedly concedes that government-regulatory eugenicists differed "from earlier social Darwinists in that they failed to draw sweeping laissez-faire conclusions . . ." He additionally somewhat concedes that "indeed a part of their own program depended on state action . . ." Hofstadter then brazenly concludes that, in the end, the eugenicist progressives were "almost equally conservative in their general bias" as were those loathed free-enterprisers. He cannot help but promulgate that governist eugenics embodied a "fundamental conservatism." *Conservative*, in this context, must be this Columbia University historian's word for practically any political position that churns his stomach. That is, in an inept attempt to explain how fellow progressives could promote governist eugenics—which he relegates to the right wing—Hofstadter evidently writes off progressive eugenicists as crypto-rightists, and then glosses over the specific cases where most of his favorite progressives publicly endorsed eugenics. Resultantly, *American Thought* concludes that governist eugenics amounts to the same old conservative social Darwinism "decked out in a somewhat new guise." Governist eugenics was distasteful to Hofstadter, but the welfare state itself was morally supreme. When Hofstadter describes the social Darwinist ideologies of both free-marketers and governist eugenicists in his own words, he caricatures them both this way.

> Domination by the fittest is of the greatest benefit to society as a whole. In order to facilitate the process the great man must be impelled by strong motives and granted the instruments of domination. . . . The great man can exert his influence by one of the two economic means—the slave system and the capitalistic wage system, the one a system of compulsion, the other of voluntary inducement. . . . To progress, a social system must retain competition between the directors of labor, the contest for industrial domination. No matter what happens to society, the domination of the fittest great men—capitalistic competition—must be ensured. Such men are the true producers. The fundamental condition of social progress is that these leaders be obeyed by the masses. In politics, as in industry, the forms of democracy are hollow; for while executive agencies are designed to execute the will of the many, the opinions of the many are formed by the few, who manipulate them.

Hofstadter's ideological myopia accounts for his refusal to acknowledge in *SDAT* that, as we shall begin to learn in chapters 4 and 5, Hofstadter's own fellow progressives were at the forefront of government-mandated eugenics. The myopia likewise helps Hofstadter evade this conclusion: if eugenicists count as social Darwinists, then revered historical figures of the political Left's pantheon would also have to be remembered as such. Indeed, in Hofstadter's own words, "eugenics . . . has proved to be the most enduring aspect of social Darwinism." Consequently, it is quite telling that *SDAT* identifies Karl Pearson as a social Darwinist who "set the tone of eugenics . . . when he estimated the heredity accounts for nine-tenths" of a person's intellectual "capacity." What is conspicuous about this is that Hofstadter divulges not a peep about Pearson's explicitly socialist convictions.[157] It is therefore natural that left-wingers familiar with Hofstadter's works regard fascism and Spencer's Manchesterism to be equal progenitors of social Darwinism. It is also easy to comprehend the reason why those who read *SDAT* would conclude that Spencer's "conservatism" could somehow give birth to the "conservative" idea of eugenics legislation. Sadly, for decades Hofstadter's misrepresentations went unchallenged except by a tiny cadre of intellectuals such as Robert C. Bannister and David Bellomy. For the most part, notes Geoffrey Hodgson, *American Thought* has been "the seminal treatise on Social Darwinism, and it has driven the discussion of the topic ever since."[158]

The consequent misconceptions about Spencer's allegedly right-wing influence have wrought myriad negative repercussions. Among them is that Spencer and Sumner—lifelong enemies of jingoism—have been repeatedly misidentified as warmongers. They would wage no war, but I shall do so at this moment. I am at war against this misrepresentation of Spencer and Sumner.

NOTES

1. M. Ridley 1999, 288.
2. M. Roach 2005, 107–108, 223, with this part of the bibliography listing M. Ridley 1999.
3. A. Kelly 1981, 101.
4. G. Hodgson 2004.
5. T. C. Leonard 2009, 40 n. 8, accessed online Sunday, November 6, 2011.
6. O. Schmidt 1879, accessed online Thursday, October 27, 2011.
7. Bannister 1988, 258 n. 3, 4. Italian writer G. Vadalà-Papale employed the expression in 1882, Gariel de Tarde in 1884; Emile de Laveleye in 1885, Achille Lora in 1896, Gabriel Ambon in 1899. To see the exact bibliographic information on these sources, consult Bannister 1988, 257–58 n. 3.
8. G. Hodgson 2004.
9. Lester Frank Ward used the term *social Darwinism* in 1907 in the *Journal of American Sociology* (see Degler 1991, 12, 351 n. 11). However, it was Hofstadter who got it to catch on in academia, as is noted in G. Hodgson 2004 and in A. Kelly 1981, 101.
10. Ward's differences with Spencer over political economy, and his outright hostility to Sumner, can be read about in Bannister 1988, 127.
11. T. C. Leonard 2009, 40 n. 9, accessed online Sunday, November 6, 2011.
12. G. Hodgson 2004.
13. T. C. Leonard 2009, 40, accessed online Sunday, November 6, 2011.
14. Hodgson's paper is Hodgson 2004. Its abstract can be read at http://tinyurl.com/2tbg8w, accessed Saturday, May 26, 2007.
15. *SDAT* is Hofstadter 1959. That it was Hofstadter who popularized among academicians the usage of the *social Darwinism* epithet on Spencer, Sumner, and free-market advocates, is noted in Hodgson 2004. Interestingly, according to Eric Foner's Introduction to the 1992 edition of *SDAT* in Hofstadter 1992, xiii, Richard Hofstadter was but twenty-six years old when he completed *SDAT* in 1932.
16. G. Hodgson 2004.
17. T. C. Leonard 2009, 40, 45, 45 n. 24, 49, accessed online Monday, November 7, 2011.
18. Eric Foner, Introduction to 1992 edition of *SDAT*, in Hofstadter 1992, xviii.
19. S. Jacoby 2008, 61n.
20. A. Kelly 1981, 171, Ch. 6 n. 1.
21. D. C. Bellomy 1984, 2.
22. G. Hodgson 2004.
23. B. Schwartz 1986, 46.

24. Ekirch 1974, unflatteringly describes Spencer and Sumner in chapter 2 on pages 22–24, and, in the chapter 2 notes section on page 279, cites *SDAT*, calling it "the best account of its subject"
25. R. Wright 1994, hardcover, 330.
26. The demonizing appears in R. Wright 2000, 266, while Hofstadter is cited in *ibid.*, 392.
27. Schweikart and Allen 2004, 446. Unfortunately, another huge drawback of this book is its disgustingly mealy-mouthed ambivalence on the subject of Franklin Roosevelt's forcible internment of Japanese-Americans for racial reasons during the Second World War. Schweikart and Allen fail to admit that FDR's action was unequivocally worthy of condemnation.
28. Bannister 1988, 129. An example he gives of an intellectual attacking eugenics as "social Darwinism" is Iankny Aleksandrovich Novikov, who would later go by the name Jacques Novicow in France (1849–1912).
29. That term is used for government-imposed eugenics in Fichman 2002, 113; J. M. Hobson 2004, 236; and G. Robinson 2001, 9.
30. Barkan 1992, 71; Kevles 1985, 72, 96; G. Robinson 2001, 16-17, 30; and Spiro 2009. M. Ridley 1999, 290, writes, "The Immigration Restriction Act of 1924 was a direct result of eugenic campaigning."
31. Kevles 1985, 101; and M. Ridley 1999, 290.
32. Karl Pearson is identified as one of the leaders of the government-imposed eugenics movement in Allan Chase 1980, paperback, 14; and G. L. Mosse 1964, 98.
33. Barkan 1992, 151; Allan Chase 1980, paperback, 9; Kevles 1985, 23; and G. L. Mosse 1985 trade paperback, 79.
34. G. L. Mosse 1985, trade paperback, 79.
35. NoBeliefs.Com, "Hitler Myths: Myth #3," updated September 29, 2010, http://www.nobeliefs.com/hitler-myths.htm#myth3, accessed Monday, June 9, 2014.
36. Bannister 1988, 166.
37. G. L. Mosse 1964, 98.
38. Bannister 1988, 168; and T. C. Leonard 2003, accessed online Monday, February 13, 2012. Fukuyama 2002, 27, and M. Ridley 1999, 288, inaccurately state that Galton coined *eugenics* in 1885.
39. Francis Galton, qtd. by A. Tone 2001, 140.
40. Galton 1908, 323, qtd. by T. C. Leonard 2003, 690–91, accessed online Monday, February 13, 2012.
41. T. C. Leonard 2003, 690, accessed online Monday, February 13, 2012.
42. Francis Galton, qtd. by T. C. Leonard 2003, 690, accessed online Monday, February 13, 2012.
43. Plato, *The Republic,* Jowett Number 459 in Plato 1892b vol. 3, http://oll.libertyfund.org/titles/plato-dialogues-vol-3-republic-timaeus-critias, accessed Wednesday, August 20, 2014.
44. Brad Abelle narrating, T. Atkinson and J. Baran prods. 1998.
45. Andrew Kupfer with Karen Nickel, "When Lester Thurow Talks, Democrats Listen," *Fortune,* July 20, 1987, http://money.cnn.com/magazines/fortune/fortune_archive/1987/07/20/69279/index.htm, accessed Monday, November 7, 2011.
46. L. Thurow 1997, 249.
47. Edward Pearce, "Nietzsche Is Radically Unsound," *The Guardian,* July 8, 1992, page 20, qtd. by J. Goldberg 2007, 447 n. 22.
48. Eric Foner in his introduction to the 1992 edition of *SDAT*, in Hofstadter 1992, x-xi.
49. Hofstadter, qtd. by Eric Foner in his introduction to the 1992 edition of *SDAT*, in Hofstadter 1992, x–xi.
50. Hofstadter 1992, xxix. T. C. Leonard 2009, 38, accessed online Sunday, November 6, 2011, states that *SDAT*'s bias was, as Hofstadter "later admitted, refracted through a New Deal [il]liberal's conception of reform"
51. T. C. Leonard 2009, 38, accessed online Sunday, November 6, 2011.
52. R. Perlstein 2001, 452, quoting Richard Hofstadter, "A Long View: Goldwater in History," *New York Review of Books*, October 8, 1964.
53. D.W.B. 1946, 124.
54. Herbert Spencer, letter, to Kentaro Kaneko, August 26, 1892, in "Herbert Spencer: Three Letters to Kaneko Kentaro (1892)," ed. by Roderick T. Long, Molinari Institute website, http:// praxeology.net/HS-LKK.htm, accessed Friday, April 20, 2007.
55. Also, Spencer's laissez-faire view did not extend beyond peacetime. He believed that if England were invaded by a hostile power, the U.K. government would be justified in imposing controls on everyday life. For that, see Spencer 1978 vol. 2, pt. 4, ch. 10, sec. 293, para. 2, vol. 2, pt. 4, ch. 12, sec. 302, para. 3; vol. 2, pt. 4, ch. 15, sec. 316, para. 2; vol. 2, pt. 4, ch. 17, sec. 310, para. 2, accessed Tuesday, January 22, 2013; in respectively Spencer 1978 vol. 2, 79, 102, 131, 139.
56. Kevles 1985, 7.
57. Lombardo 2008, 84, 32, 159.
58. Kevles 1985, 72.
59. Spiro 2009, 182.
60. That Davenport wrote to Galton and even spoke to him in person on occasions is noted in Black 2003, 34–35.

61. Dewar 2004, 26. That John D. Rockefeller II financed Davenport is further confirmed in L. Thompson 1994, 60.
62. Spiro 2009.
63. J. Marks 2002, 120.
64. Spiro 2009, 353.
65. Lombardo 2008, 10, 47–48.
66. Spiro 2009, 211.
67. G. Robinson 2001, 30.
68. Zubrin 2012, 52, 60.
69. For this reference to Ford by name, see A. Hitler 2001, paperback, 639.
70. J. Greenberg 1990, 85.
71. M. Parenti 2004, 118.
72. That Carnegie paid for the founding of Charles Davenport's eugenics laboratory is in M. Ridley 1999, 289.
73. Black 2003, 56.
74. Hofstadter 1959, 45, 49.
75. Carnegie 1901, 4, accessed online Wednesday, June 6, 2007.
76. Bannister 1988, 77, 76.
77. Hofstadter 1959, 45.
78. Carnegie 1986, 327, 326.
79. Black 2003, 87, 57.
80. Spiro 2009, 127.
81. Kevles 1985, 54; and Spiro 2009, 127.
82. Kevles 1985, 54.
83. "Pittsburgh Pirate," *The Economist*, January 30, 2003, http://www.economist.com/node/1559629, accessed Saturday, November 5, 2011.
84. P. Krass 2002, 112-13, 551 n. 2.
85. Both of these books are online, in their entirety, in the *Online Library of Liberty*. You can run a keyword search for both of them.
86. Robert C. Bannister, personal communications with author over e-mail, July 30, 2013.
87. D. Wiltshire 1978, 197.
88. S. LeDrew 2013, 33. This source also cites page 81 of the 1965 Albert Jay Nock-edited edition of *The Man Versus the State* as the source of the quotation.
89. Bannister 1988, 83, 86.
90. Carnegie, qtd. by Kolko 1963, 173.
91. Carnegie 1901, 6–23, 15, 18–19, accessed online Wednesday, June 6, 2007. Interestingly enough, while Carnegie did not get this argument from Spencer, Carnegie's entire philosophic case—that a rich man does not truly own his own fortune, but should still be allowed to retain custody of it as a "public trustee" who manages it for the social collective—is identical to an argument laid out by Spencer's ideological adversary, socialist French philosopher Auguste Comte in Comte, *Système de Politique Positive* vol. 1: *A General View of Positivism*, ch. 6: "The Religion of Humanity," originally published 1851–54, in Comte 1983b, 386–88; and Comte, *Système de Politique Positive* vol. 2: *Social Statics, or the Abstract Theory of Human Order*, ch. 5, originally published 1851–54, in Comte 1983c, 427.
92. J. L. Outman and Outman 2003b, 15.
93. S. Watts 2005, 189.
94. H. Barnard 2002, 77, quoting from the author's interview with Madeleine Yaw née Couzens.
95. Hitler 2001, 434–35. A much less violent man also advocated this economic theory. That was Mohandas K. Gandhi in Gandhi, *Harijan*, June 3, 1939, p. 145, republished in Gandhi 1960, ch. 2; and Gandhi, *Harijan*, March 31, 1946, pp. 63–64, republished in Gandhi 1960, Ch. 2, accessed online Tuesday, March 11, 2008. Recall from Book One that Gandhi noticed that the government was violent, and he said he wanted to do with violence. Gandhi's call for government control over people's wealth contradicts his putative pacifism.
96. James T. Adams 1931, 383.
97. Calvin Coolidge, article for the *Sunday Advertiser and American*, September 1, 1918, in C. Coolidge 1919h, 133.
98. Carnegie 1901, 11, 27–28, accessed online Wednesday, June 6, 2007.
99. Qtd. by Chernow 1998, 314.
100. Chernow 1998, 313–14.
101. Qtd. by Chernow 1998, 313–14.
102. Spiro 2009, 388, 58, 61.
103. A. Lee 1980, 15.
104. S. Watts 2005, 379–382.
105. N. Baldwin 2001, 106. *Ibid.*, 89, notes that Jay Gould was actually a Presbyterian.
106. R. Lacey 1986, 219.

107. J. Greenberg 1990, 85.
108. R. Lacey 1986, 223.
109. Qtd. by R. Lacey 1986, 223.
110. R. J. Herrnstein and C. Murray 1996, trade paperback, advances these claims.
111. See S. Kühl 1994.
112. Spiro 2009, 347. Spiro 2009, 265, points out that Wickliffe Draper modeled his explicit political philosophy after New York Zoological Society founder and eugenicist Madison Grant's. All of Spiro 2009 is about Madison Grant being a conservationist and big-game hunter à la his colleague Theodore Roosevelt.
113. Bannister 1988, 171.
114. Lombardo 2008, 56, quoting Alexander Graham Bell, "How to Improve the Race," *Journal of Heredity* vol. 5, January 1914, 6.
115. Bannister 1988, 171, quoting Alexander Graham Bell, "Upon the Formation of a Deaf Variety of the Human Race," National Academy of Sciences, *Memoir*, year 1883, pp. 179–180.
116. Lombardo 2008, 56.
117. C. W. Ackerman 1930, 467.
118. Chernow 1998, 216, 240–42.
119. W. T. Anderson 1996, 105.
120. Spiro 2009, 128.
121. Samuel P. Hays, *Conservation and the Gospel of Efficiency*, qtd. by Ekirch 1974, 145.
122. Spencer 1981d, 254–55, letter 10, para. 2-3, accessed online Wednesday, January 11, 2012.
123. Kevles 1985, 20–21.
124. Lombardo 2008, 50.
125. The phrase "in the land of Horatio Alger" is quoting Diane Paul. I learned about this from Spiro 2009, 161.
126. Bannister 1988, 178.
127. Sumner's statement of support for open immigration can be seen in Sumner, "Liberty and Responsibility, Part 3 of 9: Liberty and Discipline," January 16, 1890, republished as Sumner 1969-I, 321.
128. P. Maier et al. 2003, 621, 659, 662.
129. T. Parsons 1932, 325, qtd. by G. Hodgson 2004.
130. Hofstadter 1959, vii, 74, 4, 176.
131. Hrdy 1981, 13.
132. D. F. Bowers 1945, 104.
133. T. C. Leonard 2009, 44, accessed online Monday, November 7, 2011.
134. D. S. Wilson 2007, 11.
135. Hrdy 1981, 13.
136. Spencer 1906 vol. 2 of 3, 610.
137. An example of someone making the equivocation that nineteenth-century free-market economists were just like totalitarian eugenicists on account of their promoting "social inequality" is that of David Kolodney, introduction to Noam Chomsky's "IQ Tests: Building Blocks to the New Class System," republished in Karier ed. 1975, 393–395.
138. The quotation of Galton is from F. Galton 1873, 209. Galton did not use the exact phrase *nature versus nurture,* though he did set up *nature* (meaning one's inborn biological traits) and *nurture* (traits conditioned into someone) as being contrasting explanations for what causes a person to behave in the manner he does. In his *Inquiries Into Human Faculty and Development*, Galton said, "Nature is far stronger than Nurture within the limited range that I have been careful to assign in the latter." For that, see Galton 1883, 235.
139. J. M. Hobson 2004, 236.
140. Tudge 1995, 42.
141. W. Petersen 1979, 224.
142. Spencer 1851, "Introduction—The Doctrine of the Moral Sense," Lemma 2, sec. 5, para. 2-3, accessed online Wednesday, November 2, 2011.
143. Spencer 1851, pt. 2, ch. 9, sec. 9, para. 2, accessed online Tuesday, July 17, 2007.
144. Spencer 1851, pt. 2, ch. 16, sec. 2, para. 5, accessed online Wednesday, November 2, 2011.
145. Spencer 1851, "Introduction—The Doctrine of the Moral Sense," Lemma 2, Sec. 5, Para. 2-3, accessed online Wednesday, November 2, 2011.
146. Spencer 1981g, 73, Para. 3, accessed online Friday, April 27, 2007.
147. R. L. Carneiro 1981, 178-79.
148. Herbert Spencer to Rev. J. Llewelyn Davies, letter, July 24, 1890, para. 3, reproduced in Spencer, *The Principles of Ethics* vol. 2, Appendix, accessed online Tuesday, January 22, 2013, in Spencer 1989 vol. 2, 446.
149. Qtd. in Fiske 1940, 93. Also qtd. by G. P. Winston 1972, 3. Re-quoted in Bannister 1988, 62.
150. B. Werth 2009, 43, xvi.
151. Carl Schurz, qtd. by L. Schwarzschild 1947, 188.

152. Bannister 1988, 77. Schurz was not a consistent supporter of laissez faire, however. He supported the notion that the federal government should seize statutory ownership over forests and prevent private entrepreneurs from owning them. For that, see Carl Schurz, "The Need for a Rational Forest Policy in the United States," 1889, quoted in C. Schurz 1964, 89-91.

153. Bannister 1988, 76.

154. R. L. Carneiro 1981, 179.

155. Sumner, "The Conquest of the United States By Spain," *Yale Law Journal* vol. 8, January 1899, 168–193, republished as Sumner 1969-C, 277–78.

156. Hofstadter 1959, 202. Hofstadter is also qtd. on this in Bannister 1988, 4.

157. Hofstadter 1959, 5–6, 46–47, 103, 138, 161-67.

158. G. Hodgson 2004.

THREE

The Equivocation Infects the Intellectuals

The Consequences of Hofstadter's Equivocation

As an outgrowth of the misconceptions about social Darwinism, mainstream intellectuals ignorantly identify the aims of Herbert Spencer and William Graham Sumner and Inventive-Period industrialists with those of government-enforced eugenicists. Neoconservative political scientist Francis Fukuyama, for instance, first applies the social Darwinism label to Herbert Spencer[1] in his 1992 bestseller *The End of History and the Last Man*, and then to Karl Pearson and the state-implemented eugenics movement ten years later with *Our Posthuman Future*.[2] In 1999 Fukuyama pejoratively referred to "social Darwinist theories of racial and ethnic hierarchy."[3] Then in 2011 Fukuyama blared, "Spencer saw human societies as engaging in a competition for survival, in which superior ones came to dominate their inferiors. Non-European societies were ones whose development was stunted or arrested. Indeed, developmental theory in the immediate post-Darwinian period succeeded in justifying the existing colonial world order, with northern Europeans occupying a place at the top of a global hierarchy that stretched through various shades of yellow and brown down to black Africans at the bottom."[4]

Adding to this, biochemist and renowned science-fiction author Isaac Asimov thunders that "Spencerian thought had its effect on history, for during the decades prior to World War I, it gave extreme nationalists and militarists a chance to speak of war being 'good,' since it insured the survival of those most fit."[5] "Spencerian thought" inspired men to "speak of war being 'good'"? Isaac Asimov should have bothered to read chapter 21 of Spencer's *Social Statics*. Had he done so, he would have detected that Spencer penned the following words: "Unquestionably war is immoral."[6] Spencer also verbalized, "To me the cry—'Our country, right or wrong!'—seems detestable."[7] In the period of 1842 to 1843 Spencer penned a series of articles clarifying that insofar as "we remember . . . the horrors of battle," we must acknowledge that such are to be avoided. He recommended that his own nation take the lead in abstaining from any military invasions. "Never need we expect to see all nations abandon war at the same time. One must lead the way. Let England be that one. Let Britain first hold up the fair flag of peace." Other nations would follow the British lead, and "appeal to arms would become less and less frequent . . ." He pretty much laid out his position on the subject in enumerated points, which I quote below:

1. That war is a great evil, and that the fact of its exclusion by a proposed definition, is a powerful argument in favour of that definition.

2. That depriving our rulers of the power to make war, would be one of the most effectual means within our reach, of hastening that period, when "nations shall not lift up sword against nation."
3. That resistance to invasion is the only war that has any claim to the title of necessary, and that we have little need to fear its requisition.[8]

Spencer then pronounced that "it is our duty . . . to adopt all feasible means of putting an end' to war, 'and that restricting governments, to the fulfillment of their primitive functions . . . would be the most effectual means of preventing it."[9] According to the words in Andrew Carnegie's autobiography, Spencer cursed his fellow "countrymen" for being "such fools that the military and naval element in Britain could stampede the masses, frighten them, and stimulate militarism."[10] Once again to quote the text of *Social Statics*, Spencer reviled English efforts to "justify our colonial aggressions by saying that the Creator intends the Anglo-Saxon race to people the world."[11] This same book devoted an entire chapter to denouncing European imperialism. Hither, the free-market evolutionist happily foretold the day on which this drive for "state colonization"—the British Empire's "craving for wider possessions"—would be "destroyed by the conviction that territorial aggression is as impolitic as it is unjust."[12] Far from being an enemy of Mohandas Gandhi's, Spencer happily anticipated the day on which the U.K. granted independence to its Asian colonies.

Arthur A. Ekirch, Jr., ascribes socially Darwinian sentiments to Spencer, predictably. But Ekirch then contradicts Asimov. Ekirch cites social Darwinist convictions as the cause behind Spencer's incessant antiwar rhetoric. "Many now believed, with Herbert Spencer, the English Social Darwinist, that war was an anachronism, incompatible with the forces of modern technological and industrial progress."[13] From 1879 to 1881, British Parliament debated over whether to invade Egypt. Spencer leaned on everyone in his circle, including New Liberal and socialist Leonard T. Hobhouse and liberal Members of Parliament, along with other intellectuals, to mount a campaign against the invasion. Spencer wrote to Darwin himself for public support. Darwin replied that he would not join such a group or support it publicly until he learned what specific activities its campaign would entail, but that he already agreed with the general thrust of Spencer's argument.[14] Suffice it to say that Spencerian thought was far from the militaristic jingoism that Asimov carps about.

And what did other "Spencerians" have to say about militarism? Carl Schurz, the aforementioned Union soldier and fan of Spencer's, vehemently opposed the forcible annexation of foreign lands to the United States. Consider, from the previous chapter, that Schurz came to this position as a result of fervently agreeing with what Spencer had said on the subject. As a U.S. Senator, Schurz voted against President Ulysses S. Grant's (1822–1885) initiatives to annex the Dominican Republic, which was then known as Santo Domingo. Barry Werth writes this of the Spencerian's clash with the president. "Grant wanted the island nation not just for its rich resources and markets but also as a redoubt against British and French adventurism in the Caribbean, as well as an inroad for further 'necessary' expansion in the region and an outlet for freedmen driven out of the South." But, Schurz "denounced the scheme as an imperialist land grab, anathema to party and national principles. A member of the Committee on Foreign Relations, he revealed in public hearings that Grant's aide and a friend stood to make a fortune through annexation and had secretly pledged to keep Santo Domingo's corrupt ruler in power with military support. . . . Along with committee chairman Charles Sumner, Schurz maneuvered to thwart the treaty, presenting Grant with the most embarrassing challenge of his presidency." Schurz orated on the floor of Congress, "Suppose we annex the Dominican Republic, will there be an end to our acquisitions? Having San Domingo, Cuba and Porto Rico, you will not rest until you possess also the other West Indies Islands; and what then? . . . The grave question arises: is the incorporation of that part of the globe and the people inhabiting it quite comparable with the integrity, safety, perpetuity, and progressive development of our institutions which we value so highly?" Werth explains that Schurz's anti-imperialism resulted from his conviction that he had "adapted Spencer's arguments from the Law of Equal Freedom—'Every man is free to that which he wills, provided he infringes not the equal freedom of any other man'—to the business of governing."

There were disciples of Spencer's other than Schurz who applied Spencer's views on foreign policy to their own public actions. William Graham Sumner, writes Barry Werth, "served as a vice president of the Anti-Imperialist League, laboring despite his weakened state" of health "to block U.S. expansion because he feared it was an evolutionary step back toward militarism."[15] Richard Hofstadter himself admits in *SDAT* that his own ideological nemesis, Sumner, "attacked the imperial impulse with practically all the weapons in the arsenal of the anti-annexationists." Hofstadter comments that Sumner's moral damnation of the 1898 Spanish-American War "had an unquestionable ring of sincerity, particularly since it once again put in jeopardy" his Yale professorship.[16]

To quote Sumner himself, ". . . expansion and imperialism are at war with the best traditions, principles, and interests of the American people, and . . . they will plunge us into a network of difficult problems and political perils . . ." After damning warfare as "criminal aggression," Sumner ranted that "we are told by all the imperialists" that the men of poorer countries "are not fit for liberty and self-government; that it is rebellion for them to resist our beneficence; that we must send fleets and armies to kill them if they do it; that we must devise a government for them and administer it ourselves . . ." What is that, Sumner asked, but the horrifyingly immoral policy that the Spanish Empire had toward its colonies during the Renaissance? He judges that we Americans "are brought by our new conquests face to face with this dilemma: we must either hold them as inferior possessions, to be ruled and exploited by us after the fashion of the old colonial system, or we must take them in on an equality with ourselves . . ." Sumner continued that imperialism raises the vagary over

> whether the Constitution of the United States extends over all men and territory owned by the United States, or whether there are to be grades and planes of rights for different parts of the dominions over which our flag waves. . . . The expansionists have recognized this fact by already casting the Constitution aside. . . .
>
> The question of imperialism, then, is the question whether we are going to give the lie to the origin of our own national existence by establishing a colonial system of the old Spanish type, even if we have to sacrifice our existing civil and political system to do it. I submit that it is a strange incongruity to utter grand platitudes about the blessings of liberty, etc., which we are going to impart to these people, and to begin by refusing to extend the Constitution over them, and still more, by throwing the Constitution into the gutter here at home. If you take away the Constitution, what is American liberty and all the rest? Nothing but a lot of phrases.

This "new policy" of imperialism, he concluded, is "destined to thrust a sword into every joint in our historical and philosophical system."[17] In this essay, Sumner does express some prejudices against foreigners. He says that the morally proper alternative to treating nonwhites as second-class citizens or noncitizens would be for the USA to recognize the rights of these nonwhites to vote and participate in the republican legislative process. He assumes that such nonwhites are not ready for liberal republicanism and therefore to recognize them as American citizens would allow these nonwhites to destroy U.S. liberalism by introducing illiberal legislation. What is pertinent, though, is that whatever racial prejudices Sumner held, he did not favor political inequality among races, nor did he condone militarism or a bellicose foreign policy.

Sumner's staunch anti-imperialism is the same "Spencerian thought" that Asimov accuses of rationalizing war and conquest. Sumner is, of course, incorrect to assume that it is inherently detestable for a freer country to introduce greater liberalization on a penurious dictatorship through military might, against the consent of the majority of persons in the overthrown dictatorship. I already argued against Sumner's position in Book Two. What is relevant here is the injustice in how the Left has depicted this peacenik as an apologist for Western colonialism. Interestingly, Geoffrey Hodgson writes, "Despite their support for capitalist competition," Spencer and Sumner "were anti-militarist and anti-imperialist,"[18] as if that were some sort of internal contradiction on the part of Spencer and Sumner—as if supporting the Rule of Peace normally spurs someone to become a warmonger.

In *Inventing America*, Daniel Kevles and his scholarly coauthors do admit that "many social Darwinists were anti-expansionists . . ." Nevertheless, these scholars still attempt to destroy the reputations of antiwar laissez-faireists by remarking that such "social Darwinists" opposed expansionism "because they feared adding 'inferior' races to American society . . ." And besides, add the PC scholars, lots of other social Darwinists did support imperialism.[19] In other words, if a conservative right-winger supports colonialism, these scholars denounce him as a racist. On the converse, if a conservative right-winger opposes colonialism, these scholars surmise that this must be because . . . this conservative right-winger is a racist.

Don't Confuse Them with Social Darwinists

Regrettably, Asimov's insults appear tame in comparison to those that astrophysicist Carl Sagan (1934–1996) casts. Sagan warns about how

> the Darwinian insight can be turned upside down and grotesquely misused: Voracious robber barons may explain their cutthroat practices by an appeal to Social Darwinism; Nazis and other racists may call on "survival of the fittest" to justify genocide. But Darwin did not make John D. Rockefeller or Adolf Hitler. Greed, the Industrial Revolution, the free enterprise system, and corruption of government by the monied are adequate to explain nineteenth-century capitalism. Ethnocentrism, xenophobia, social hierarchies, the long history of anti-Semitism in Germany, the Versailles Treaty, German child-rearing practices, inflation, and the Depression seem adequate to explain Hitler's rise to power. Very likely these or similar events would have transpired with or without Darwin. And modern Darwinism [i.e., sociobiology or evolutionary psychology] makes it abundantly clear that many less ruthless traits, some not always admired by robber barons and Fuhrers—altruism, general intelligence, compassion—may be the key to survival.[20]

I resent that distortion. I refer to how Sagan morally equates John D. Rockefeller, Sr., with Nazis. Sadly, this same equivocation manifested itself in a rather well-known controversy. That controversy came to the fore when Harvard entomologist and left-wing environmentalist Edward Osborne Wilson brought new popularity to the burgeoning scientific field of sociobiology, now more widely known as evolutionary psychology.[21] This is the same evolutionary psychology I discussed in Book Two. In the 1970s Wilson and other sociobiologists sought to apply biology to the social sciences to discover which modern social practices, institutions, and customs could be at least partially explained by what scientists currently know about evolution, genetics, psychiatry, and neurobiology.

As expected, a number of professors recognized that Herbert Spencer, William Graham Sumner, and the eugenicists had already taken upon themselves that same mission of applying evolutionary theories to the social sciences. Hence those PC professors, including the Harvard evolutionary biologist Stephen Jay Gould (1941–2002), reacted indignantly. The PC scholars presumed

1. Modern sociobiologists might come to conclusions at least remotely similar to those of Spencer or eugenicists.
2. The hard sciences could consequently lend their imprimatur to such conclusions.[22]

Hence these PC academicians took action by forming an organization called the Sociobiology Study Group of Science for the People—SSGSP or "Science for the People" for short. Science for the People fired off an angry letter published in no less than the November 17, 1975 *New York Review of Books*. It demanded that no one should so much as consider studying sociobiology, as the field could easily "provide a genetic justification . . . of existing privileges for certain groups according to class, race, or sex. . . . For example, John D. Rockefeller, Sr., [sic! –S.H.] said, 'The growth of a large business is merely a survival of the fittest. . . . It is merely the working out of a law of nature and a law of God.' These theories provided an important basis for the enactment of sterilization laws and restrictive immigration laws by the United States between 1910 and 1930 and also for the eugenics policies which led to the establishment of gas chambers in Nazi Germany."[23] Recount from Book Two that Standard Oil's founder is not the source of that quotation that the SSGSP attributed to him. That quotation is actually from John D. Rockefeller, *Junior*, who spent his adult life as a professional philanthropist and not a business executive. As mentioned in Book Two, Hofstadter misattributed that quotation to Standard's founder, and generations of academicians unquestioningly cited him in quoting it.[24] And indeed there was nothing unusual about Rockefeller speaking of the survival of the fittest in a context wherein he did not mean that he expected people to die. As mentioned in Book Two, George Washington Carver also described his pursuit of his inventive career as the survival of the fittest. Also consider an important inventor affecting most readers as of this writing. In the early days of automobiles, such as that of the Model T, a car had to be started by the motorist getting underneath the vehicle to crank it. That changed with the self-starter engine—the mechanism whereby you start the vehicle's engine by slipping the key into the ignition and turning it. That process was invented by Charles F. Kettering (1876–1958). He, too, employed the expression *survival of the fittest* in a benign reference to how separate organizations compete among one another in the economic realm.[25]

Science for the People received a response from E. O. Wilson and his colleague, University of Toronto biologist Charles J. Lumsden. Mortifyingly, the rebuttal from Wilson and Lumsden offered an equally illiterate interpretation of the past. Wilson and Lumsden wrote that they resented their sociobiological research being equated with what they considered "Social Darwinism and the attendant evils favored by capitalist-imperialist regimes." The pair of sociobiologists intimated that they were already cognizant that such social Darwinism carried a legacy of "support for racism, sexism, and the status quo . . ." They assured their readers that they did not expect their findings to do the same. Then the pair scolded the SSGSP, "*To link opponents with Rockefeller and Hitler is to call for their exile from the dovecotes of academia.* The purpose is not so much to answer the arguments of the authors under attack as to destroy their credibility. This is particularly true at Harvard University, where a professor accused of fascist sympathies is in roughly the same position as an atheist in a Benedictine monastery"[26] (emphasis added). Let us examine this. Wilson and Lumsden took umbrage at any attempt to discredit a professor by arbitrary comparisons of him with Hitler. Nevertheless, they themselves approved of the SSGSP's arbitrary equivocation of Rockefeller with the Führer, going as far as referring to "capitalist-imperialist regimes" as if capitalism and imperialism invariably go together. E. O. Wilson and his academic critics have, much as Carl Sagan has, betrayed an equally shallow understanding of their social-science heritage. They are in unanimous, tacit agreement that the financial transactions of Rockefeller were on par with the tyranny and genocide perpetrated by Nazis.

As Robert Bannister assesses the situation, "Wilson himself could not resist the temptation to label" earlier theories about the application of evolutionary biology to society "to exonerate" his own biases.[27] And Stephen Jay Gould's participation in the SSGSP's campaign against "Social Darwinism" is funny, considering that he himself would go on to say that Darwin's theory "was essentially Adam Smith's economics read into nature."[28] It does not surprise me that Isaac Asimov, Carl Sagan, E. O. Wilson, Science for the People, and other governists on the political Left would promote this "non-religious free-marketers are eugenicists" equivocation. What truly horrifies me is that ostensible free-market advocates, having been misled by Hofstadter's followers, do so as well.[29]

While admitting that such measures were "not advocated by Spencer" himself, right-wing economist Mark Skousen nonetheless writes that Spencer's "thinking led directly to the eugenics movement, which, in its most extreme manifestation in Nazi Germany, called for compulsory sterilization and selective breeding to improve the quality of the population."[30] As is the case with many Republicans sympathetic toward the evangelical Right, Dr. Skousen conflates Spencer's pro-capitalist evolutionism with villainy. Skousen's rationale is not that he sympathizes with Hofstadter's animosity toward big business. In Skousen's case, it is that he prefers that people reject secular arguments for free enterprise, and accept a justification more compatible with the Judeo-Christian religious tradition. Hence Skousen inveighs that "there's a dark side to [Ayn] Rand's teachings. Her defense of greed and selfishness, her diatribes against religion and charitable sacrificing for others who are less fortunate, and her criticism of the Judeo-Christian virtues under the guise of rational Objectivism have tarnished her advocacy of unfettered capitalism." Skousen thus upholds Adam Smith as a superior espouser of capitalism, on order that "Smith's theme" allegedly "echoes his Christian heritage, particularly the Golden rule, 'Therefore all things whatsoever ye would that men should do to you, do ye even so to them' (Matt. 7:12).'"[31]

It cannot be denied that the mainstream altruist paradigm, popularized by Christianity, pressured Smith to air specific assumptions. First would be that the entrepreneur's profit motivation could be rationalized by the benefits that the entrepreneur's activities provided to others, but that the profit motive could not be justified by the entrepreneur's peaceable self-interest alone. Nonetheless, it is misleading for Skousen to cite Smith's "Christian heritage" in the absence of acknowledging that Smith was not a Christian but a deist. À la Thomas Jefferson, Smith believed in a God while simultaneously rejecting the Abrahamic theological tradition. The God of Smith and Jefferson was "Nature's God"—a god who allegedly created Existence but did not intervene in human affairs. And an apologist attitude mirroring Skousen's is expressed by right-wing pundit Lowell Ponte in David Horowitz's *Front Page Magazine*. There, Ponte implies that once a woman accepts the veracity of natural selection, that acceptance steers her toward evil. The recognition of biological mutations, writes Ponte, "had in earlier decades begotten 'social Darwinism,' a worldview in which the 'fittest' survived and those with inferior genes were supposed to die out. . . . This neo-Darwinian thinking would reach its peak in both the death camps and 'Aryan' breeding programs of socialist Adolf Hitler."[32]

Another right-winger who takes advantage of the social Darwinism straw man to denounce his political opponents is inventor-engineer Robert Zubrin. In a book that criticizes environmentalists and abortion activists, Zubrin pronounces that nineteenth-century industrialists "readily understood that Darwinism gave them precisely the scientific and ethical justification" for their own callous ruthlessness "that they desired, and they lost no time putting it to work." Copying the left-wingers before him, Zubrin conflates evolutionary economics with state-imposed compulsory sterilization. "Initially, like Darwinism and Malthusianism before it, eugenics found its warmest reception among Britain's *laissez-faire* liberals."[33]

Polemicist George Gilder, another member of the Religious Right who espouses some free-market rhetoric, charges, "Both Nazism and communism were inspired by Darwinism."[34] Patrick J. Buchanan (b. 1938) is a religious fundamentalist and business-thrashing governist politician and essayist. Despite Buchanan doing well to make his viewpoint known, the media wantonly mischaracterize him as an opponent of the welfare state. What is relevant is that Buchanan takes his approaches of Christian zealotry and anti-capitalism to their logical conclusion. In his writings he repeats every left-wing, Hofstadter-inspired untruth about Spencer founding social Darwinism and governist eugenics. Thither Buchanan proclaims that it was Darwin's discovery of biological evolution that prodded Spencer to originate his vile doctrines. From this, Buchanan expects his readers to conclude that for you to accept the reality of biological evolution is to render you loathsome and hateful, whereas moral and humane behavior is contingent upon an arbitrary refusal to accept the veracity of evolutionary theory. Buchanan rasps that Darwin's discovery "suited the purpose of the eugenicists and Herbert Spencer, who preached a survival-of-the-fittest social Darwinism to robber baron industrialists exploiting 19th-century immigrants . . ." Then he adds, "Darwin suited Adolf Hitler's purposes, too." Upon quoting a religious fundamentalist book trying to pin warfare and genocide on Darwin's statement of facts, Buchanan concludes, "Darwinism is not science. It is faith. Always was."[35]

Indeed, in a single controversy we can observe both the Religious Right and secular Left engage in the same sort of defamatory tactics regarding social Darwinism. In *Expelled: No Intelligence Allowed*—a propagandistic documentary that argues for Biblical Creationism and against atheism—Hollywood actor and conservative activist Ben Stein (b. 1944) invokes the memory of government-imposed eugenics and tries to blame Darwinian theory for it. Then Stein confronts evolutionary biologist Richard Dawkins about the ethical ramifications of Darwinian theory. As is typical of propaganda from the Religious Right, *Expelled* insinuates a directly proportional relationship between the extent to which someone (a) recognizes evolutionary theory's validity and (b) concludes that government-imposed eugenics is ethically righteous. To wit, Ben Stein's movie argues that to enact governist eugenics is simply to take an acceptance of evolutionary theory to its logically inevitable and necessary conclusion. The movie blames Charles Darwin for Naziism and the Holocaust. The implication of the argument is that there is but one method for us to avoid becoming eugenicists: we must deny any acceptance of evolutionary theory.

Interestingly, the film does not mention Herbert Spencer. Spencer's name could not go unmentioned for long, of course, in a debate of this sort. On the website of his not-for-profit educational foundation, Richard Dawkins responded to *Expelled* in the manner that people normally respond when accused of being social Darwinists. That is, Dawkins took it at face value that social Darwinism was an actual ideological movement and that it was populated by right-wing, uncharitable bigots. Dawkins fired back, *I am not a right-wing, uncharitable bigot. Therefore, by definition I cannot be a social Darwinist.* These are Dawkins's words: "...I want to construct a society which is about as un-Darwinian as we can make it. I approve of looking after the poor (very un-Darwinian). I approve of universal medical care (very un-Darwinian). . . . If we look at . . . history, the closest representatives you'll find to Darwinian politics are uncompassionate conservatives like Margaret Thatcher, George W. Bush, or Ben Stein's own hero, Richard Nixon. Maybe all these people, along with the Social Darwinists from Herbert Spencer to John D. Rockefeller, committed the is/ought fallacy and justified their unpleasant social views by invoking garbled Darwinism."[36]

Similarly, left-wing governists of Dawkins's mind relish their abilities (1) to link social Darwinism with free-marketers like Ayn Rand, and (2) to link social Darwinism with governmental eugenics. This ability has provided governists the opportunity to smear such free-marketers as Rand as eugenicists. Pursuant to denouncing Sumner and Herbert Spencer as social Darwinists,[37] Susan Jacoby associates them with state-run eugenics and then tries to associate every one of them collectively with twentieth-century free-market advocates. Susan Jacoby says that in the late 1800s, Darwin's discovery

> was twisted by Sumner and his followers into a social philosophy . . . that enshrined competition and validated the worthiness of whoever and whatever came out on top. . . . Sumner declared emphatically that the business titans of the Gilded Age were "a product of natural selection . . . just like the great statesman, or scientific men, or military men." . . . Social Darwinism has never died: it manifested itself as a bulwark of eugenics until the Second World War; in the tedious midcentury [*sic;* it should be "mid-century" with a hyphen] "objectivist" [*sic;* "Objectivist" is capitalized] philosophy of Ayn Rand. . . . All of the theories included in the general category of social Darwinism may be summed up in the immortal line uttered by the hero of Rand's *The Fountainhead* (1943): "The only good which men can do to one another and the only statement of their proper relationship is 'Hands off!'"[38]

Jacoby resorts to a specific fatuous insinuation. It is that Sumner attributed a businessman's success to his genes rather than to the businessman having exercised conscious reason to adapt to his market through satisfying consumer demand. We have quashed that untruth already. Predictably, Jacoby quotes Rand out of context in order to besmirch her. Following the same playbook as her right-wing opponent Arthur Brooks, Jacoby tries to prejudice the reader into inferring that a Rand quotation indicates that Rand thought that every individual should consciously refrain from performing any action that might benefit someone else. That Rand quotation comes from a speech given by her character Howard Roark. Roark edifies in that speech that, rather than have the State behave as a dictator controlling everyone, it would be better for the government to have a more laissez-faire, hands-off approach. Here is a fuller quotation of that same speech that is in context. The line that Jacoby quoted is in boldface.

> Men have been taught that their first concern is to relieve the suffering of others. But suffering is a disease. Should one come upon it, one tries to give relief and assistance. To make that the highest test of virtue is to make suffering the most important part of life. . . .
> In all proper relationships there is no sacrifice of anyone to anyone. An architect needs clients, but he does not subordinate his work to their wishes. They need him, but they do not order a house just to give him a commission. Men exchange their work by free, mutual consent to mutual advantage when their personal interests agree and they both desire the exchange. If they do not desire it, they are not forced to deal with each other. . . . This is the only possible form of relationship between equals. . . .
> The "common good" of a collective—a race, a class, a state—was the claim and justification of every tyranny ever established over men. Every major horror of history was committed in the name of an altruistic motive. . . . The most dreadful butchers were the most sincere. They believed in the perfect society reached through the guillotine and the firing squad. Nobody questioned their right to murder since they were murdering for an altruistic purpose. . . .
> **The only good which men can do to one another and the only statement of their proper relationship is—Hands off!**[39]

The first paragraph clarifies that Rand believed in aiding deserving individuals in need; the paragraphs before her "hands off" statement establish that she wants governists of Jacoby's ilk to cease their manhandling of private, peaceful individuals. That matters little in Jacoby's writings, which exhibit far greater concern for demonizing free-market organizations as social Darwinists—"Because Sumner was able to invest his pseudoscientific theories with scientific authority and an aura of rationality, . . . he must be ranked not only as one of the most influential academics of his day but as the philosophical forefather of the right-wing public intellectuals who have exercised similar influence in American society since the early 1980s. Sumner's ideas would fit perfectly today in the position papers of the Heritage Foundation and the American Enterprise Institute."[40] In other words, Susan Jacoby is insinuating that those who agree with the consensualist policy proposals of the Heritage Foundation and American Enterprise Institute (AEI) must be social Darwinists who revile the hungry. I find it amusing that AEI's president as of this writing—Arthur Brooks—does not dispute Jacoby's mischaracterization of Spencer and Sumner. The USA's contemporary Left and Right are quite united in their shared misconception on this issue of historiography.

The Misconception Appears in Nonpartisan Nonfiction

It gets worse. Several intellectuals speak about a nonexistent Herbert Spencer-eugenics connection when they are not even intending to score points for some ideological argument. They exhibit no overtly left-wing or religious agenda. Nay, they just uncritically swallow the misinformation the academic mainstream has spoon-fed them. One example can be found in *The Roots of Desire*—an otherwise fun book by Marion Roach who, as stated in the preceding chapter, is most famous as a commentator on National Public Radio's *All Things Considered*. *The Roots of Desire* explores the scientific reasons why some individuals sport red hair, and also provides a historical survey of the depiction of redheads in folklore and popular culture. Unlike the works of Hofstadter and Skousen, this one does not appear to be trying to advance any political cause. Yet, due to the piles of propaganda that were dumped on the authoress as fact in college classrooms and treatises, her *Roots of Desire* nevertheless conveys to its readers most of the old myths about Spencer encouraging government-inflicted genocide.

> . . . from Spencer's work comes the term "social Darwinism" . . . For a while, social Darwinism tidily explained why people with advantages beget children with the same advantages, deeming those advantageous far fitter than others and by extension, more suited to survive. But this theory worked to explain the rungs of society only until those who had it all had began to be outnumbered by those who had nothing. . . .
>
> It was in this climate that the new science of genetics was used by some to identify which of the world's people might become a burden to society. . . . How to identify these people was key.
>
> One method was by applying social Darwinism to eugenics. . . . At its worst, [government-decreed eugenics] was Hitler's hateful science in which the fittest race was granted biological sanctity and bestowed a superior status. But Hitler did not formulate the ideas he embraced. He and his scientists adopted these ideas as part of a huge movement of social Darwinist eugenics that burned through the first thirty years of the twentieth century. The German version was adopted from the American model [started by Charles Davenport].[41]

Hofstadter's influence is additionally rampant in one part of the otherwise excellent work *War Before Civilization* by anthropologist Lawrence Keeley of the University of Illinois. It will be recollected that I cited this work in Book Two about the warlike tendencies of hunter-gatherers. Keeley's anthropological data are well-verified. But, depressingly, when Keeley condemns the ideological factors that motivated the nineteenth-century's European colonialism, Keeley releases the old Social Darwinism bugbear. "In the second half of the nineteenth century...sociologists and anthropologists united the neo-Hobbesian perspective with something quite foreign to Hobbes' careful arguments for human equality: Social Darwinism and racism. Imperialists had long been troubled by the common and often violent refusal of native peoples to acknowledge the superiority of European culture and religion or adopt them willingly. The new doctrines of the struggle for existence and survival of the fittest provided a cornucopia of explanations and justifications." Consonant with the academic cliché, Keeley considers a certain British ideologue to be culpable for this: "The spread of Western civilization and Europeans at the expense of other cultures and races became a splendid illustration of *Spencer's* survival of the fittest" (emphasis added). As the procedure normally commences, Keeley devotes not a word to acknowledging that Spencer publicly denounced imperialism wholesale. Keeley then prattles on that from Spencer's ideology, "imperialists thus discovered a moral duty and a biological right to wrest dominion of the earth from such less-favored peoples." Finally in the endnotes section we are treated to the requisite straw man against the political partisans of laissez faire. "Nonracist Social Darwinism universally remains a theme in conservative political thinking . . ."[42]

Keeley's disparagement of Spencer is ironic in consideration of the very thesis of Keeley's book *War Before Civilization*. The point of this book, as well as of Steven Pinker's *The Better Angels of Our Nature*, is that preindustrial hunter-gatherers were much more violence-prone generally than are members of twenty-first century Western Civilization. We considered the evidence for this claim in Book Two. Yet the notion so heavily argued in Pinker's *Better Angels* and Keeley's *War*—that human beings began as violence-prone hunter-gatherers and are gradually evolving toward a more consistently peaceful society—had long been argued by Spencer. Preindustrial violent communities were what Spencer called "militant societies," whereas the much-more-peaceful, liberalized societies of the West are what he dubbed "industrial societies." This is George H. Smith's paraphrasing of Spencer's viewpoint—"In a primitive society where warlike qualities (strength, cunning, etc.) are necessary for survival, feelings of pleasure accompany the exercise of warlike faculties. As society evolves, as voluntary cooperation becomes the dominant mode of interaction, new emotions suited to the new environment evolve as well. Brutality generates abhorrence, and pleasure is derived from peaceful activities."[43] Keeley disparages Spencer for crudity when Spencer actually argued a variant of Keeley's thesis before Keeley did.

I can identify this sort of inconsistency in the quality of scholarship in the work of futurist Alvin Toffler as well. His book, *The Third Wave*, was published in 1981, and it is impressive in various respects. I laud its accuracy in predicting the information technology that would proliferate a whole seventeen years subsequent to the book's inaugural publication. Despite that, Toffler's account of the history of imperialism exposes itself as misinformed in its attribution of the imperialists' ideology to be rationalized somehow by the theories of evolution and market economics. "Darwin was chiefly concerned with biological evolution, but his ideas had distinct social and political overtones that others were quick to recognize. Thus the Social Darwinists argued that the principle of natural selection worked within society as well, and that the wealthiest and most powerful were, by virtue of that fact, the fittest and the most deserving. It was only a short leap to the idea that whole societies evolve according to the same laws of selection. Following this reasoning, industrialism was a higher stage of evolution than the non-industrial cultures that surrounded it. . . . Just as Social Darwinism rationalized capitalism, this cultural arrogance rationalized imperialism."[44]

Worse, there is a still more fearsome repercussion of Hofstadter's smear. There is a small group of intellectuals who are familiar enough with Spencer's writing to admit that he was never a full-blown social Darwinist. Yet the majority of these same intellectuals have still been bamboozled into believing that governist eugenics is a logical extension of night-watchman-state economics.

Take, for instance, Robert Wright, a contributing editor to *The New Republic* magazine. To his credit, he concedes in his book, *The Moral Animal*, that although Spencer is regarded to be the "classic proponent of social Darwinism," it is "in some ways unfairly." Elsewhere in this volume Wright elaborates that Spencer does not deserve a reputation for being "heartless," as this man "put lots of emphasis on the goodness" of "sympathy, and he was a pacifist."[45] Yet on the same page of this treatise, Wright maintains that John D. Rockefeller, Sr., and most free-market intellectuals were social Darwinists who unconditionally hated the underclass. In an opus that is otherwise well-versed in history and economics, *Nonzero*, Wright affirms that "social Darwinists" tried "to justify things like racism, imperialism, and laissez-faire indifference to poverty."[46]

Wading through so many inaccurate portrayals of Spencer sours me deeply. Once in a while, thankfully, it appears that sometimes the truth does come out. I recognized this in the governist-biased *New York Times*. On May 5, 2007, this newspaper predictably reported on its front page, "Victorian-era social Darwinists like Herbert Spencer adopted evolutionary theory to justify colonialism and imperialism, opposition to labor unions and the withdrawal of aid to the sick and needy. Francis Galton based his 'science' of eugenics on it."[47] Fortunately the paper partially retracted this error. On May 12, the *Times*'s website appended the article thusly: "A front-page article last Saturday . . . erroneously included one social Darwinist among Victorian-era social Darwinists who adopted evolutionary theory to justify colonialism and imperialism. Herbert Spencer opposed both." The derogatory label of social Darwinism remains, but at least the publication acknowledges that Spencer's position is distinct from those that left-wingers routinely presume him to hold.

On the Matters of Holmes and Kidd, the Left Must Be Kidding

The silliness of the ploys to group Spencer together with his political adversaries is more transparent in other instances. Indiana University English professor Patrick Brantlinger (b. 1941) reproofs "the social Darwinism of Spencer, Benjamin Kidd, and Karl Pearson . . .", as if these three men argued the same position.[48] Almost as improper as the comparison between Spencer and socialist Pearson is the comparison of Spencer with American sociologist Benjamin Kidd (1858–1916). John M. Hobson, too, lumps Spencer together with Kidd.[49] Kidd's claim to fame was his 1894 treatise *Social Evolution*, which, as Robert Bannister notes in *Science and Myth*, "attacked Spencer on two grounds." Being devoutly religious, Kidd resented Spencer's aggressively secular worldview, and chastised Spencer's thoughts on cosmology as too difficult to comprehend.[50] But the biggest lapse in comparing Spencer to Kidd is that the latter's *Social Evolution* explicitly called for more government regulations over the economy. Before it acquired the Marxian bias for which it has become famous, *The Nation* magazine evaluated that "the general trend" in Kidd's thought, "at least, in all but the earlier chapters" of *Social Evolution*, was "unmistakably that of the so-called Christian socialist."[51] Concurrently, *Popular Science* magazine's review evaluated that Kidd "seems to anticipate great and beneficial results from a vast extension of socialistic legislation."[52]

Bannister observes that Kidd's "was the mood which in America produced progressivism . . ." This places Kidd in the same class as other such "American reformers." Kidd conceded the communists' point that capitalism exploits workers. Moreover, Kidd tried to shoot down the idea that entrepreneurship is the primary ingredient of economic growth. *SD-SMAAST* further elaborates that in Kidd's philosophy, "Regulation was the means, social efficiency the end; and humanitarianism the guiding spirit."[53] In Kidd's own words, it "may be noticed that the characteristic feature" of the new "legislation" that Kidd wants enacted "is the increasing tendency to raise the position of the lower classes *at the expense* of the wealthier classes" (emphasis Kidd's). In a just social system "the general tendency must be expected to be toward state interference and state control on a greatly extended scale . . ."[54] If an unwavering trust in laissez faire is the defining characteristic of social Darwinism, as Robert M. Rees chants, then neither Kidd nor Pearson can justly be deemed social Darwinists.

The equivocations then take a more sinister turn. Edwin Black tries to hold Spencer accountable for a pro-eugenicist ruling by U.S. Supreme Court Justice Oliver Wendell Holmes, Jr. (1841–1935). This was in the 1927 case *Buck v. Bell*. In Holmes's opinion, it was constitutional for U.S. states to enforce laws coercively sterilizing the mentally ill and their blood relatives as well. Black crafts this passage in order to imply that Holmes's eugenicism was shaped by an admiration Holmes supposedly held for Spencer: "the eighty-six-year-old Holmes was in many ways defined by the Civil War and ethically shaped by the nineteenth century. While recovering from the wounds of Chancellorsville, his reading included Spencer's *Social Statics*, the turning-point tract that advocated social Darwinism . . . Spencer argued the strong over the weak, and believed that human entitlements and charity itself were false and against nature. Indeed, Holmes' 1881 lecture series in *The Common Law* also asserted that the idea of inherent rights was 'intrinsically absurd."[55] Here, Black insinuates that Holmes learned from Spencer's *Social Statics* that because "the weak" deserve to be exploited by "the strong," "the idea of inherent rights" must be "intrinsically absurd." Recall from Book Two my quotation of Edwin Black falsely accusing Herbert Spencer of opposing private charity. Were one to accept Black's accusation, one would likely infer that Spencer's ideas had informed Justice Holmes's bizarre exclamation that philanthropy is "the worst abuse of private ownership."[56] Other writers have baselessly faulted Spencer's teachings for Holmes's eugenicism. Citing Vermont Law School historian Sheldon M. Novick (b. 1941), Georgia State University law instructor Paul A. Lombardo inaccurately asserts that Holmes was a close hewer to Spencer's doctrines.[57] Yet an actual reading of *Social Statics* reveals that Spencer pledged intransigent loyalty to Locke's theory of individual rights. In *Social Statics*, Spencer writes that adults "must have rights to liberty of action"[58] —that is, the freedom of peaceful action. Were I sold on Black's imputation that Spencer convinced Holmes that the strong could ethically abridge the rights of the weak, then it would be difficult to reconcile with verbatim quotations of Spencer. *Social Statics* declares that "the taking away of a man's property against his will, is an infringement of his rights..."[59] Spencer made it extraordinarily clear to anybody willing to listen to him that he categorically opposed spoliation. "Aggression of every kind is hateful to me," he said.[60]

Anyone who cracks open Volume II of Spencer's *Principles of Ethics* and bothers to read it carefully will discover chapters detailing "the right to physical integrity," "the rights of free belief and worship," "the rights of free speech and publication," "the rights of women," and "the rights of children."[61] More than seven years prior to the debut of *Social Statics*, Spencer promulgated that the proper "office of government" is "simply to defend the natural rights of man—to protect person and property—to *prevent the aggressions of the powerful upon the weak*—in a word, to administer justice"[62] (emphasis added). The previous quotation by Spencer merits an immediate second glance. Again, Black asserts that "Spencer argued the strong over the weak."[63] Yet Spencer's own words are that for the government to "administer justice," it must "prevent the aggressions of the powerful upon the weak . . ."

Softening his unfair denunciation, Paul A. Lombardo ends up relenting that Spencer "believed in a radical libertarianism and freedom from state control" whereas "Holmes did not . . ." Consequently, Lombardo realizes that "Spencer never called for the execution of the socially deficient, as Holmes later would. In fact, Spencer condemned those like Holmes who delighted in 'passing harsh sentence on his poor, hard-worked, heavily burdened fellow countrymen.'"[64] And Black does not deign to admit that Holmes vocally opposed Manchesterism. Justice Holmes ruled in favor of having the U.S. Justice Department forcibly break up the Standard Oil Company of John D. Rockefeller, Sr.[65] This is the exact same Rockefeller whom Hofstadter and his followers finger as a social Darwinist. Justice Holmes was not a pal to Rockefeller, Sr., or other such profiteers who wished to keep the government's hands off their private fortunes. Despite his own governist premises, Rockefeller denounced the personal income tax to a reporter in 1914, "When a man has accumulated a sum of money, accumulated it within the law, the Government has no right to share in its earnings."[66] By contrast, a news reporter famously paraphrased Justice Holmes telling him that taxes "are the price we pay for a civilized society."[67]

Another consideration demolishes Black's insinuation. Spencer's *Social Statics* could not have inspired Holmes's pro-eugenicist rulings, as Holmes actually publicly disparaged that book. This was visible in the 1905 case *Lochner v. New York*. In the case, the state of New York had attempted to micromanage the labor practices of baker Joseph Lochner, dictating over the employment hours that he and his staff members would otherwise negotiate freely on their own.[68] Lochner owned a small bakeshop, and he hired Polish immigrants who were willing to work much longer hours, and for lower remuneration, than what was often expected of native-born bakeshop employees. Labor unions and Lochner's much-larger, wealthier competitors teamed up in lobbying for regulations to limit the work hours of everyone. That would deprive Polish immigrant workers of the ability to work long hours, their willingness to brave such hours being their competitive advantage against native-born labor. Lochner and his employees would not tolerate such government force impeding their business relationships. For that reason, Lochner appealed this case to the U.S. Supreme Court. In a vote of five to four, the Court came down on Lochner's side.[69] In the majority opinion, Justice Rufus Peckham noted that in Joseph Lochner's management decisions, Lochner's business was "not dangerous in any degree to morals, or in any real and substantial degree to the health of the employee." The majority of Supreme Court Justices, against Holmes's wishes, cited the Fourteenth Amendment in their ruling that private individuals may peaceably run their businesses as they choose. The ruling majority did not want state governments impinging upon such rights. Such rights, Peckham stated, "cannot be prohibited or interfered with, without violating the Federal Constitution."[70]

This was actually a great victory for liberty. Recollect, from Book One, the U.S. Supreme Court case *Buchanan v. Warley*. It is no accident that Moorfield Storey and Clayton Blakey cited the Lochner decision in their own 1917 brief for the case, wherein they argued that individual property rights must be prioritized above any collectivists' attempt to forbid a white man from selling land to a black man.[71] It will be recalled that it was this argument of Storey's and Blakey's that successfully struck down the ordinance-enforced segregation of states across the nation. The Lochner precedent helped render this possibility.[72]

But in his dissenting opinion for *Lochner*, Holmes wrote, "The liberty of the citizen to do as he likes so long as he does not interfere with the liberty of others to do the same, which has been a shibboleth for some well-known writers, is interfered with by school laws, by the Post Office, by every state or municipal institution which takes his money for purposes thought desirable, whether he likes it or not." To Holmes, the mere fact that people adapted to the condition of such governmental spoliations taking place proved to be justification for governmental regulation as such. Then Holmes sniped, "The Fourteenth Amendment does not enact Mr. Herbert Spencer's *Social Statics*."[73] Hofstadter here admits that Holmes uttered these words in "protest" against Spencer's sociopolitical-economic philosophy.[74]

Holmes's stand is not unexpected to those who know, as Robert Bannister does, that Holmes "typically mentioned Spencer only to criticize him . . ."[75] This Supreme Court Justice sided with the aims of *Lochner v. New York*'s pro-regulation defendants, which was not unusual for a governist eugenicist like himself.[76] It should not startle anyone, as Holmes once confided to a friend, "All my life I have sneered at the natural rights of man."[77] Consequently, John Dewey expressed strong approval for Holmes's opinion on the matter, commending Holmes for being one to "remind" the other justices that the U.S. Constitution did not enshrine a laissez-faire liberal doctrine.[78] On this matter, legal scholar Thomas A. Bowden evaluates that Holmes has "uttered" a terribly "damaging slur" in "this offhand reference to Herbert Spencer's *Social Statics*."[79] Commensurately, Justice Holmes expressed his eugenicist convictions when on May 2, 1927, he explained why in *Buck v. Bell*, he supported the compulsory sterilization of those who had mentally ill relatives. "*We have seen more than once that the public welfare may call upon the best citizens for their lives.* It would be strange if it could not call upon those who already sap the strength of the State for these lesser *sacrifices*...in order to prevent our being swamped with incompetence. It is better for all the world, if instead of waiting to execute degenerate offspring for crime, or to let them starve for their imbecility, society can prevent those who are manifestly unfit from continuing their kind. The principle that sustains compulsory vaccination is broad enough to cover cutting the Fallopian tubes. Three generations of imbeciles are enough"[80] (emphases added). In opposition to Spencer's politics, Holmes's eugenicism prescribes the elevation of the social collective's welfare above the individual's own private rights. Take a second gander at my quotations of Holmes. Holmes's position was logically formulated from his premise—one that is shared, in part, by Edwin Black—that the welfare of the social collective justifies State impingements upon individuals' Lockean rights to private property. In Holmes's case, the property rights being violated pertained to Carrie Buck's ownership over her organs.

One might consider Holmes to be a capitalist Darwinist on account of his writings on the late-nineteenth-century gas-stokers case. In the year 1872, gas stokers went on strike in England. Strikes are not *per se* inimical to laissez faire. In this case the gas stokers had agreed, in their written contracts, not to conduct a strike in the manner that they did, and the employers went to court with the workers for contract breach. The court ruling mostly came down in favor of the employers.[81] In his writings, Holmes sided with this ruling, though not because he inexorably sided with corporation managers over their employees or labor unions. Nor was it on account of any loyalty to laissez faire; Holmes's coming down on the correct side was merely coincidental. Rather, on account of Holmes' rejection of free-market economics, Holmes presumed that economics was naturally a zero-sum game, wherein someone's gain perforce had to come at someone else's forcible loss. On that premise, Holmes found it inevitable that in any legal dispute between employer and employee, a court had to rule in favor of one financial class to the spoliative detriment of the other financial class. In fact, in Holmes's argument in favor of the court ruling, he mentioned Spencer by name in an effort to riposte him. In 1873 Holmes hammered in print Spencer's "tacit assumption of the solidarity of the interests of society is very common, but seems to us to be false. The struggle for life, undoubtedly, is constantly putting the interests of men at variance" with one another. "The more powerful interests must be more or less be reflected in legislation; which, like every other device of man or beast, must tend in the long run to aid the survival of the fittest."[82] Yes, Holmes invoked the phrase *survival of the fittest*, but in an attempt to denigrate Spencer's laissez-faire outlook. These facts about Holmes's governism give the lie to depictions of Holmes being too much of a laissez-faire social Darwinist. It discredits such depictions by Cornell University historian R. Laurence Moore. Both "William Graham Sumner and Oliver Wendell Holmes, Jr. . . ." Laurence Moore asserts, "vigorously opposed"[83] the need for "government intervention . . ." This suggests that Dr. Moore needs to take a closer look at Holmes's rulings in *Lochner v. New York, U.S. v. Standard Oil,* and especially *Buck v. Bell.* I would certainly identify those as forms of government intervention.

In contrast to Moore, some governists loudly praise Holmes for his *Lochner* dissent and cite it as evidence that Holmes was an opponent to social Darwinism. "How," Thomas C. Leonard rhetorically questions, "does an opponent of Spencerian Social Darwinism come to endorse coercive sterilization of the unfit?" As "a matter of history," Leonard understands, "there is no contradiction in the views that underwrite" Holmes's respective opinions in *Lochner* and *Buck.* It is not simply that the statutes being contested in both cases "proposed to subordinate individual rights to a putatively greater social good. Progressive thought, it turns out, did not have to travel far when it moved from labor statutes conceived as protecting society from Social Darwinism to eugenic legislation conceived as protecting society from persons deemed biologically unfit. . . . In fact, in the Progressive Era especially, eugenic treatment of those deemed biologically inferior was promoted *as a means* to the end of uplifting the industrial poor"[84] (emphasis Leonard's).

Some Enlightenment for Those Who Mistake Nazism for Rationality

There is another case of someone without an obvious political vendetta repeating the myth that free-market advocates were social Darwinists who inspired Naziism. This case is particularly maddening for me. This myth is stated by Abraham Foxman (b. 1940)—a Holocaust survivor[85] who serves as the Anti-Defamation League's national director. He correctly supports U.S. publication of *Mein Kampf* so that people can read it and gain some insight as to what exactly that wretched manifesto espouses. Sadly, the Anti-Defamation activist unknowingly defames many innocent people when he blames thinkers in the Enlightenment tradition—Herbert Spencer implicitly among them—for fueling Nazi ideology. In his *Mein Kampf* introduction, Foxman writes that the "glue that Hitler used" to hold Naziism together

> was an extreme form of race-oriented social Darwinism. . . . The modern "science" of race had evolved with the Enlightenment when the Aristotelian distinctions between the "cultured" and the "barbaric" races were revived. . . . By positing that certain races were inherently "primitive," white men of the Enlightenment were able to justify both their continued toleration of black slavery and their imperialist designs on places such as Africa. . . .
>
> Racial theories became increasingly radical as they incorporated aspects of Darwinism, which swept the Western world in the mid- to late 1800s. Applied to race, the ideas of evolution and "survival of the fittest" turned the history of humanity . . . into a story of racial conflict. When coupled with nationalism, racial (social) Darwinism led to the development of national archetypes; thus educated people at the end of the nineteenth century could seriously claim that the distinctive cultural characteristics of the English, French, Americans, and Germans were biological. Eugenics movements with the goal of improving national or racial "stock" through selective breeding (which later became inextricably linked with the Nazi regime in popular perception) arose in England, Scandinavia, and the United Sates.[86]

My heart sinks that the Enlightenment movement is here so crudely implicated for laying the philosophic foundation for totalitarian Naziism. Herbert Spencer, it will be recalled, was one of the foremost intellectuals at the tail end of the Enlightenment period. This was the period that Doron Swade, assistant director and head of Collections at the Science Museum in London, aptly identifies as the apex of the "age of reason," as it "witnessed a ferment of scientific invention . . ."[87]

But right-wing comedian Evan Sayet (b. 1960) agrees with Foxman. Sayet dumps on the eighteenth-century rational philosophes, as they undermined faith and tried to remove Christian domination from the public sphere. Sarcastically, as Sayet will say it, "The Enlightenment was all about 'pure science.' If something couldn't be measured, quantified, or recreated in the laboratory, then it simply wasn't true." Consequently, one of the few philosophers to stand up to this worrisome trend—Jean-Jacques Rousseau—"noticed the high price" that cold rationality exacted on mankind. "Only when the Enlightenment reached its inevitable soulless extremes—when Darwin's scientific theory of 'the survival of the fittest,' untamed by the extra-scientific qualities of morality . . . manifested itself in the Holocaust—did the Enlightenment disqualify itself as the philosophy that might finally bring peace and universal prosperity. To the contrary, the Holocaust suggested to many that reason is the most monstrous of all of man's faculties." Sayet then issues his own perspective on the matter: "...it was not science that had led to the Holocaust, but rather science untamed by the extra-scientific—science without God—that had. But as the shaken world looked to rise from the literal ashes of the Enlightenment," the secularists still refused to "resurrect God as a means of taming pure science."[88] That, according to Sayet, identifies Enlightenment-era secularism as the pathogen responsible for modern social ills.

In *Voltaire's Bastards: The Dictatorship of Reason in the West*, businessman-turned-philosophic-essayist John Ralston Saul sounds more forceful than Sayet. Employing *reason* and *rationality* as pejoratives, Saul propounds, "The Holocaust was the result of a perfectly rational argument." He invites readers to contemplate the "systematic, scientific" methodology whereby the gassings of the Jews had been "carried out. . . . The massacre was . . . 'well managed'" and an "act of pure logic carried out in a rational manner . . ." Observe that Saul equivocates anything that he considers methodical or orderly to be *rational*, for he later admits that the mass murder furthered no self-interested purpose on the Nazis' part. "There was no practical reason for it. No property was gained, as it had all been expropriated already. No territory was at stake. The killings were a money-losing proposition—Nazi Germany was destroying a slave population capable of great production at a time when Aryan males had been sent away to fight." Clearly Saul does not comprehend the inextricable connection between reason and its justification, self-interest. Remember from Book One that this is the same man who proclaimed the Spanish Inquisition to be the consequence of reason: "The Inquisitors were the first to formalize the idea that to every question there is a right answer. . . . Being already in possession" of what they held to be "the truth, they were interested in the rational demonstration of it by each victim."[89] I find it utterly shameful that Camille Paglia praised Saul's *Dictatorship of Reason* as "a rich, rewarding, highly original book that casts a fresh perspective . . ."[90]

John Gray—a philosopher of the London School of Economics, not to be confused with the American psychologist of the same name who authored *Men Are From Mars, Women Are From Venus*—does not go as far as Saul; Gray articulates a more nuanced view. Peculiarly, though, Gray still delivers some puzzling proclamations about the Enlightenment somehow influencing Naziism. The "Nazi project of racial hierarchy," he writes, "continued some influential strands of Enlightenment thinking. Nazi 'scientific racism' had precedents in the Positivist plans for a science of society grounded in physiology, and in theories of human inequality and eugenics promoted by nineteenth-century Enlightenment thinkers such as Francis Galton and (in more explicitly racial terms) Ernst Haeckel."[91] Given Galton's arbitrary presumptions about superiority, and Haeckel's explicit mysticism, Gray's assertion that they are Enlightenment thinkers rings hollow. We will delve more into why that is so, in chapter 9.

And we also recall from Book One that University of Leeds sociologist Zygmunt Bauman advanced this Enlightenment-caused-the-Holocaust interpretation. We also recall from Book One that it was the Frankfurt school of cultural Marxist thought, led by Theodor Adorno and Max Horkheimer, which most prominently popularized this interpretation. Unsympathetic to that interpretation, UCLA political scientist and historian Anthony Padgen summarizes it: "Deluded by the conviction that human beings could rise above their condition through reason, the Enlightenment had visited all kinds of horrors on the world. . . . All the intellectual revolutions of the nineteenth century, . . . the collapse of the human sciences" into rationalized "racism—the 'self-assertion of the bourgeois individual . . .' [as Adorno and Horkheimer say it]—and finally neo-Darwinian eugenics . . . were, in Horkheimer and Adorno's view, but the inevitable outcome of the eighteenth century bid to live by reason alone."[92] But contrary to Foxman and Saul and Sayet and Gray and Bauman and the Frankfurt school, the Nazis actually rejected the Enlightenment thinkers' elevation of reason above faith and sorcery. This is for a multitude of reasons.

First, the ideas of racial supremacism and chattel slavery have existed from pre-Biblical times on every inhabited continent. During the Middle Ages, the Chinese and the Japanese each proudly proclaimed their own superiority over that of every other culture.[93] The transatlantic slave trade was one consequence among many of Christopher Columbus's contact with the New World, and such European exploration grew out of the European Renaissance-era emphasis on secular enterprise and the quest for knowledge. This still predated the Enlightenment era. Prior to the Renaissance, Christian and Islamic societies brazenly condoned chattel slavery. "Local ethnic slavery," Kevin Kelly recognizes, "was practiced…in prehistoric times…, and still continues in various remote areas; its overall diminishment globally is due to the technological tools of communication, law, and education."[94]

"Africans knew slavery long before Europeans sailed to their shores," disclose historians Dorothy Schneider and Carl J. Schneider. ". . . Like all other people in the long history of slavery, Africans enslaved outsiders—members of other tribes or erring members of their own tribes." Intertribal warfare proved rampant, and slaves were normally "prisoners of war," but "sometimes a tribe raided its neighbors to take slaves, and sometimes kidnappers stole children from other tribes to enslave them." Likewise, "Slavery existed in North America long before the arrival of the European explorers and settlers. The idea of slavery did not outrage Indians any more than it outraged Europeans or Africans. In North America, as in Africa and Europe, it was a condition of life that few questioned." As was the case with African tribes, Amerinidian tribes normally "enslaved—sometimes for a limited period, sometimes for life—male prisoners of war whom they did not torture or kill outright. . . . These practices led to slave raids on neighboring tribes, particularly among the Illinois and Iroquois, as young braves vied to prove their skills as warriors by taking scalps or captives." When the Iroquois did not outright kill a male enemy, they would sometimes humiliate him by taking him as a slave, lopping off a part of his foot to prevent him from escaping. American Indians generally believed that if a man died in war, at least he died honorably. Oppositely, such a man would be spiritually dishonored if enslaved and forced to perform what was conventionally considered women's work. Thus, the Iroquois practiced slavery not so much for economic gain but to humiliate their opponents as spitefully as possible. "Before the Europeans came, most tribes refused to accept back a member who had been captured and enslaved." Sometimes "a tribe might barter slaves to placate an unfriendly neighboring tribe or in exchange for its own captured members."[95] I mention that Africans and American Indians practiced slavery prior to European contact not to downplay the atrocities that white Westerners committed in connection with slavery—those cannot be minimized—but as a reminder that white Westerners, as a culture, were far from being that institution's exclusive practitioners. Moreover, although slavery continued to be practiced during the European Renaissance and Enlightenment, those two historical periods were but a fraction of the entire span of history throughout which human beings practiced slavery, a practice that probably began prior to 5,000 B.C. Thus we should discern here, already, how fatuous it is to attribute racist, pro-slavery ideas of the Nazis to the Enlightenment.

We can concede that some Enlightenment figures exhibited hypocrisy on this issue. Thomas Jefferson did not write into his will that his slaves be emancipated upon his death. John Locke purchased £600 worth of shares in the Royal African Company, which traded in slaves,[96] and, when he assisted his patron, the first Earl of Shaftesbury, in setting up the British colony in the Carolinas of North America, Locke wrote explicitly in the colony's charter that slavery was to be legal in the colony.[97] In his *Second Treatise of Government*, Locke says explicitly that when England captures soldiers from an army that initiates war on England, the English may rightfully enslave those enemy soldiers for life.[98] Still, University of Maryland historian Holly Brewer points out that, in his old age and under the rule of King William III, Locke did indeed fight for measures to undermine the slave trade. When the British government set up its Carolinian colony, it opted to provide land grants to settlers, and the size of the land grants would be directly proportional to the number of slaves the plantation holder imported, thereby incentivizing a plantation owner to maximize the number of slaves imported. Locke argued against this measure, vocalizing that it incentivized an inhumane practice.[99]

Moreover, consider the plethora of Enlightenment-era figures who spoke out against white enslavement of blacks. That list includes A.L.C. Destutt de Tracy,[100] Jean-Baptiste Say,[101] Charles Montesquieu, Marquis de Lafayette,[102] Adam Smith,[103] Thomas Paine,[104] Jacques Turgot,[105] Manchester-based free-trade advocate John Bright, industrialist William Gregg,[106] Erasmus Darwin (Charles Darwin's paternal grandfather),[107] Josiah Wedgwood (Charles Darwin's maternal grandfather),[108] Adam Ferguson,[109] and Nicolas Condorcet.[110] These thinkers did much to start the movement to eradicate chattel slavery in Europe and North America. Joseph Priestley considered himself a religious man and he opposed slavery,[111] but his religious views were nonetheless rejected as heresy by mainstream Christians.

Some apologists, such as those in the Acton Institute—a Catholic conservative group—try to hand every last bit of credit for the abolitionist movement to Christianity, as the Quakers eventually came to take unofficial lead of the abolitionist movement. Indeed, one Quaker in particular, Anthony Benezet of Philadelphia, did much to recruit Enlightenment philosophes into the cause. Pulitzer Prize finalist and non-conservative Philip Dray admits that the movement's leaders included such evangelicals as William Wilberforce and Granville Sharpe. However, had Christianity truly been the factor most responsible for abolitionism, it would fail to explain why, from the first century A.D. to the Renaissance, the cultural leaders of Christendom did nothing to end slavery. It was not until the ascension of such secularist Enlightenment philosophers as Jean-Baptiste Say that abolitionism became popular among nominal Christians. This suggests that Enlightenment liberal philosophy, more than any other impetus, catalyzed the West's reformation in attitude about slavery. Consider an early abolitionist—Benjamin Franklin. Franklin did not start off with the cause; for much of his life he was a slaveholder. His attitude began to change as a young adult when he became a deist, meaning that he believed in a God but did not believe in a literal interpretation of the Bible; he believed in a non-intervening God of Nature. Franklin struggled with this issue throughout much of his life. In his twilight years, Franklin joined in abolitionism, finally taking his Enlightenment philosophy to its necessary, logical conclusion. As Philip Dray also observes, it was the Enlightenment's *philosophes* who "gave the antislavery cause its intellectual heft . . ."[112] Indeed, as preachers in the northern United States campaigned against slavery, their counterparts in the South were citing Bible passages to prop up the notion that Christianity endorsed the practice. As both the pro-slavery and anti-slavery priests cited arbitrary mythology in their arguments over the ethics of slavery, neither side could outdebate the other. It was up to rational argumentation to demonstrate the wrongness of human bondage. And it is no accident that the antislavery movement's biggest champions, including William Lloyd Garrison and Frederick Douglass, were big defenders of laissez-faire economics. It is therefore fallacious for Abraham Foxman to fault the Enlightenment for rationalizing one race's subjugation of another. Wiser is how John Gray admits that the Nazi outlook "repudiated Enlightenment values of human equality" of rights and "universal emancipation . . ."[113]

Secondly, in contrast to secular liberal Enlightenment ideology, Hitler and most of his lieutenants proved to be strong believers in fideism and mysticism. The Nazi philosophy's hostility to Enlightenment reason is better understood in light of explanations provided by University of Wisconsin historian George L. Mosse (1918–1999). He traces the Nazi movement's emphasis on racism and governmental control to what he dubs the more general "Volkish thought" that flowered in the nineteenth century and wilted only upon the Nazis' wartime defeat. *Volk*, remember, refers to the collective People or tribe. There are common threads tying the nineteenth-century Volkish thought with its final product, twentieth-century Naziism. Those threads are as follows.

1. Main priority is placed on the German race. The Volkish movement romanticized, with great nostalgia, its image of the German race as it was in the primitive preindustrial era: a collectivist tribe.[114]
2. Having the German race—the Volk—feel a strong, emotive, faithful bond with the native landscape.[115] This idolization of the landscape is the source of the Volkish movement cultivating ideas that would be nurtured by the West's twentieth-century environmentalist movement.[116] The commonalities between the Volkish movement and twentieth-century environmental ideology will be further discussed in chapter 9 of this book.
3. Hostility to persons who reject Volkish mysticism and collectivism—persons who instead rely on unflappable reason and pecuniary self-interest. The Jews are stereotyped as possessing these traits, and thus the Volkish movement harbors prejudice against the Jews. This scapegoating of rational, self-interested, capitalist Jews will be studied in Chapters 7 through 9.

This Nazi ideology—Volkish thought[117] or Volkism—was heavily influenced by a philosophical movement that swept England, Germany, and the rest of Europe from the eighteenth century onward. This movement was known as philosophic Romanticism. Romanticism, in this context, does not refer to the nineteenth-century's aesthetic Romanticism movement, wherein poets and composers stressed heroic and large-scale themes. Philosophical Romanticism was a philosophy that rejected technology, industrial progress, and reason in favor of mysticism and worship of a wilderness untrammeled by man.[118] Some thinkers and artists, such as Percy Bysshe Shelley and Mary Shelley, dabbled in both philosophic and aesthetic Romanticism simultaneously.

While there is much to praise in aesthetic Romanticism, a significant share of the destructiveness of Nazi philosophy can be traced back to philosophic Romanticism.[119] Joachim Fest, author of one of the most prestigious biographies of Adolf Hitler, documents this well. He observes that Germany's philosophic Romantics "treated the facts that were held up to them with haughty contempt and roundly ridiculed 'one-eyed reason.'" They were explicit in having "no regard for logic"—in their opinion, to be called *illogical* was not perforce an insult. Fest realizes that the Romantic movement's "anticivilizational sentiments" holding influence over the Nazis "was to have grave consequences." The German Romantics—and, by consequence, Hitler—"saw no good in the liberal Western society which traced its beginnings to the principles of the Enlightenment..."[120] Unsurprisingly, though he was not German, Jean-Jacques Rousseau is considered one of the luminaries of the Romantic movement.[121] Karl Popper remarks quite sympathetically, "From Rousseau onward, the Romantic school of thought realized that man is not mainly rational."[122] Popper also discloses, "While I see Kant as the defender of the Enlightenment, he is more often taken as the founder of the school which destroyed it—of the Romantic School of Fichte, Schelling, and Hegel."[123] Popper is wrong and his detractors are correct—philosophic Romanticism indeed trekked a path blazed by Kant. Germany's Romanticism displayed thematic commonalities with the Romanticism of other parts of Europe. Consider Mary Shelley's *Frankenstein*, a work of both aesthetic and philosophic Romanticism. It is from *Frankenstein* that our modern culture has gained the notion that any proud man who seeks to apply scientific knowledge to create a grand new technology will run afoul of nature, provoking nature to punish him violently for such hubris. Some modern critics deny that *Frankenstein* is primarily a fable warning against technological progress; they point out that the monster in the story was mistreated by his maker, and that the monster might have turned out genteel and humane if not for the neglect and persecution he suffered. Still, Mary Shelley stated herself that she judged her tale to be a parable protesting the selfish science she had witnessed asserting itself—"supremely frightful would be the effect of any human endeavour to mock the stupendous mechanism of the Creator of the world." [124]

Oxford University administrator Nicholas Goodrick-Clarke, Ph.D., concurs with me that Germany's nineteenth-century romantic "*völkisch* ideology" was a backlash challenging rational "modernity. . . . The survival of pre-capitalist attitudes and institutions in these countries meant that modernization" was distressing to the Germans. "Many people despised modernization," for they resented the "mushrooming industries . . . Liberalism" and rationality were distrusted "because they tended to demystify time-honored institutions and to discredit accepted beliefs and authorities."[125]

Wayne Biddle paraphrases the German Romanticist notions that informed Naziism: ". . . the romanticized *Volk* needed to be protected from the corruption of Western *Zivilisation*, especially the degenerate *Amerikanismus* whereby financial profits determined the path of progress. To accomplish this, rationality would have to mix with mythology, contrary to the Enlightenment notion that the two are mutually exclusive." This mindset, Biddle adds, was known as "thinking with the blood."[126] *Blood*, in the context of this expression, as in the Nazi slogan *Blood and Soil*, refers to blood relatives—loyalty to one's family or tribe. Such prioritization of blood over reason was visible in Hitler's rebuffs against Carl Bosch's appeals to reason. It will be recalled from Book Two that Bosch was the German corporate executive who assisted Fritz Haber in developing the process of synthesizing nitrogen fertilizer. Haber converted to Christianity, but the Nazis found themselves unable to overlook that he was born a Jew. That was enough for them to dismiss and ban him from the Max Planck Institute. In March 1933, Bosch met Hitler and informed him that if he succeeded in exiling Haber and other Jewish-born scientists from universities, the state of Germany's physics and chemistry would be set back a century. To this, Hitler merely snapped, "Then we'll work a hundred years without physics and chemistry." Upset by the industrialist's impudence, Hitler refused to speak with Bosch ever again or so much as be in the same room as him.[127]

Philosophy professor and Naziism advocate Alfred Baeumler (1887–1968), being well-read on how the Volkish movement was infused with Romanticism, said of himself and his fellow Nazi Party members, "Romanticism opened our eyes to the night, the past, our ancestors, to the mythos and the Volk."[128] Recall the quotation from Book One of the Nazi officer Robert Ley about faith—"We have faith in Adolf Hitler and in his Idea. The faith is converted into obedience. Whoever does not obey has no faith!"[129]

Now consider Alan S. Milward (b. 1935), a historian at the London School of Economics. He likewise observes, "The basis of Fascist and National Socialist political and economic thought was the rejection of the ideas of the eighteenth-century Enlightenment." [130]

As important, there is Fritz R. Stern (b. 1926)—like Richard Hofstadter, a professor of history at Columbia, but, unlike Hofstadter, also a former provost of this institution. À la Milward, Stern acknowledges that the Nazi movement was a revolt against the Enlightenment and rationality, a revolt that accepted "neither Western civilization nor Eastern barbarism..." [131]

Despite his own rejection of reason and his own sympathies for Romanticism, Karl Popper admits, "The fascist appeal . . . is to our passions, to our collectivist mystical needs . . . this appeal may be called the *cunning of the revolt against reason*. . . . it is a typical tribalist ideal" [132] (emphasis Popper's).

"By calling the fascist and Nazi experiments of the period between the two world wars antirationalist," writes Rice University historian Gale Stokes (b. 1933), "I mean to suggest that the leaders of these movements of race and rejection craved the technological power put into their hands by the industrial revolution, but at the same time they disavowed the rationalizing intellectual and social concomitants of the Enlightenment" that had rendered this very industrial revolution possible. "The Nazis believed . . . that the universe held 'a primal, nonrational force' that can be grasped only by the intuitive power" of divine men. Accordingly, "Nazism and fascism rejected reason for power, individuality" for nationalist supremacy, ". . . transparency for obscurantism," and "...objectivity for prejudice... The horrible end of the Nazism in the holocaust of World War II clearly demonstrated the bankruptcy of the antirational experiment." [133]

George L. Mosse details, "The intellectual and ideological character of Volkish thought was a direct product of the romantic movement of nineteenth-century Europe. . . . Volkish ideas showed a distinct tendency toward that irrational and emotional." Volkish romantics felt that the Enlightenment era's inductive reason "had been discredited. The patient experimentation and intellectual discipline of the Enlightenment" was abandoned, replaced by mystical animism. "The rapid process of European industrialization was indeed bewildering to them . . ." To find happiness, the German romantics looked to something bigger than themselves—a mystical cosmos of spiritual energy. In their collective unity, the German romantics would mitigate "the social chaos of industrialization and urbanization . . ." The Nazi movement, therefore, "was an ideology which stood opposed to the progress and modernization that transformed nineteenth-century Europe. It used and amplified romanticism to provide an alternative to modernity, to the developing industrial and urban civilization" that alienated the average German psychologically.[134] "Irrationalism," states Mosse, characterizes the Nazi mindset.[135] And John Gray admits that Naziism obviously held the characteristics of a "neo-primitivist cult of 'thinking with the blood,'" and admits that other intellectuals resoundingly recognize that Naziism emerges "from Counter-Enlightenment ideas..." [136] I wish Gray understood that fact as well as the other intellectuals to whom he alludes understand it.

Hans Schemm, Bavaria's first Minister of Culture under the Nazis, articulated it well: "We are not objective, we are German." [137]

For the previously mentioned reasons, Hitler himself verbalizes that he could not stand nerds who adhered to the secular reason that the Enlightenment had brought forth. Among such nerds, the Führer complained, nothing but "knowledge is prized. . . . What is needed is instinct and will." [138]

The Three Prophets of the Third Reich

Holding particularly strong sway with the Nazis were three Romanticist German intellectuals: Arthur Moeller van de Bruck (1876–1925), Paul de Lagarde (1827–1891), and Julius Langbehn (1857–1901).[139] Moeller happens to be the one who coined the expression *Third Reich*.[140] In the estimate of George L. Mosse, Lagarde publicly preached against rationality, "industrialism," liberal republicanism, and "urbanism." Lagarde "attempted to preserve the organic character of the Volk, and accordingly denounced . . . modernity . . . and socio-economic progress." To summarize Lagarde's 1878 book *German Writings*, "Neither . . . economic growth nor national prosperity reflected a unity and contentment among the general population." Capitalism and eudemonia were false idols "which had been imposed without regard for the spiritual character . . . of the German nation." Mosse also describes Lagarde's fideism and social collectivism: "History, for Lagarde, was the expression of a religious spirit. . . . Like all Volkish thinkers, Lagarde thought that the individual" must be subordinate to "the larger unit, the Volk. Personal revelation could come only within the confines of the community, a concept justified by Christ's proclamation of the Gospel to the community of the Apostles. . . . God's kingdom was the Volk. . . . When fused with the Volkish concept of nature, such a religion would guide men out of the discord of modernity. . . . For inspiration and inner contentment, Lagarde wrote, man should listen to the voice of nature unfolding within the trees of the wood. . . . The unity of the nation would, correspondingly . . . be re-established according to . . . the concrete expression of the common spiritual, emotional, and mystical qualities of the German people." Lagarde's assertions can be chalked up to his revulsion for the "industrial age and modern values. . . . Rather than the tradition-defying bourgeois-capitalist individual," it was the collective unity of the German race, and their forging their bonds with the undeveloped landscape, that would fulfill the Volk. Julius Langbehn corresponded with Lagarde and, in published works, expanded upon Lagarde's fideist thought.[141]

Besides their hatred for capitalism, observes Fritz Stern, the triumvirate of Moeller-Lagarde-Langbehn were romanticist thinkers—*feelers*, actually—in rebellion against modernity. These three bemoaned "the era's loss of religious faith . . ." They loved "unreason" and hated "secularization." What Stern finds particularly striking about these three are "their violently 'anti-Western' sentiments," and "their rejection of the ideas of 1789 and the Enlightenment."[142] The anti-capitalist, environmentalist German Youth Movement, which came to be the model for Hitler Youth, indoctrinated its adolescent members with the doctrines of Lagarde and Langbehn.

Notions of the occult and the supernatural had currency among the Volkish philosophers and a considerable number of their Nazi successors. Paul de Lagarde's publisher, Eugen Diederichs, himself became an author influential in Volkish circles. Diederichs named the ideology, which he held in common with Langbehn and the Volkists, *New Romanticism*. One of the most significant aspects of the Volkish romanticism was faith in the *Geist*, which roughly means *spirit*. Note that *geist* sounds like *ghost* and is part of the longer German term *poltergeist*. The *Geist* that Diederichs extolled was a supernatural spirit that every member of the German race held collectively, and which could not be shared by profit-accumulating Jewish industrialists.[143] Sometimes the term enunciated was *Volk-geist*. Mosse writes that in the philosophy of another Volkish intellectual, Johan G. von Herder, every ethnicity, except for the Jews, had its own *Volkgeist*, which "encompassed the whole community."[144]

Diederichs, Mosse verbalizes, detested technological commerce. "The specter confronting him was the generation of economic empire builders, in mid-passage in Europe as well as the United States. In reaction to this gilded age, Diederichs and his New

Romantics...called for a return to the higher, transcendent reality, and for intuition as a vehicle of communication with that world." His spirituality could save the race from "industrial materialism . . ." Diederichs briefly considered following the Theosophy movement of Rudolf Steiner and Jiddu Krishnamurti, which I shall elaborate further upon in chapter 9. Diederichs eventually rejected that movement, though, while remaining interested in spirituality. Throughout his quest for ethereal understanding, Diederichs told the New Romantics of the *Geist*, the collective spirit of their ethnicity, which could be comprehended by no means other than vague intuition. To Mosse, the New Romanticism of such Volkish writers as Diederichs was intertwined with their racism. "The formula called for the subordination of science, mechanization," and "modernization...to a religious racial goal."[145]

Hitler and several of his lieutenants went so far in their fideism that they came to embrace the occult, astrology, and supernaturalism. As it had with Diederichs, theosophy (more about it in chapter 9) influenced Nazi officers.[146] The swastika has been an ancient sun symbol seen throughout the Old World, and it was adopted as a symbol of Theosophy by Guido von List. The Nazis appropriated the swastika for their own emblem as an homage to their Theosophist roots.[147] More influential over the Nazis still, was an early-twentieth-century occultist movement deemed Ariosophy, popular throughout Germany and Austria. According to Nicholas Goodrick-Clarke, Ariosophy professed "Aryan-German racial excellence, anti-liberalism," and revolt against economic progress.[148] Nazis who considered themselves believers in Theosophy included Himmler, Alfred Rosenberg, Walther Darré, and Hitler.[149]

Paul Roland reports, "Hitler . . . was extremely superstitious by nature. A prophetic dream during the First World War which saved him from almost certain death was only one of several incidents which convinced the Führer that his life was protected by providence and that it was confirmed he had a divine mission to fulfill."[150]

Nazis actually made key military decisions according to what their astrologer advisors told them.[151] George L. Mosse discusses the most "notorious believer in spiritual forces" in the NSDAP. Heinrich Himmler regarded himself to be the reincarnated soul of Henry the Fowler, and professed that this was a result of Indian karma. "Indeed his whole thought was saturated with . . . mysticism."[152] To Himmler, reports Paul Roland, "astrology was the guiding principle of life. . . . Hitler's blue-eyed protégé, Richard Heydrich, once summed it up by saying: 'Goering is worried about the stars on his chest, Himmler about those in his horoscope.'"[153] Other Nazi followers of astrology included Joseph Goebbels, Rudolf Hess, and Alfred Rosenberg.[154]

The other Nazis entertained various sorts of paranormal belief. Frau Goebbels, wife to Joseph Goebbels, has publicly stated that her husband proudly interpreted various Nostradamus quatrains as proof that the Renaissance seer had foretold the Third Reich's victory in World War Two.[155] Moreover, the Nazi regime fancied, as theoretically possible, the notion that the Earth's center was not filled with molten magma, but was actually hollow and inhabited by a lost civilization. More astonishing, it was not merely in the lore of *Indiana Jones* movies, but in real life, that the Nazis ruminated on the idea that they could obtain a strategic advantage in war by claiming possession of the Holy Grail.[156] Indeed, they seriously proclaimed that their ancestors, the original Aryans, inhabited the mythical continent of Atlantis, and undertook expeditions to Tibet to unearth artifacts from their Atlantean forebears.[157]

Another Nazi exemplar of fideism is the existentialist philosopher Martin Heidegger. Strongly related to Heidegger's enthusiasm for totalitarianism is his mystical attack on individualism and free will. Throughout his book *The Essence of Truth*, Heidegger appeals to what he considers the collective "spirit" of his ethnicity—*spirit* referring, in this context, literally to the supernatural.

As University Paris Quest-Nanterre La Défense professor Emmanuel Faye paraphrases Heidegger's book *Being and Time*, "authentic existence has nothing of an individual being about it. It can be accomplished only as a common destiny (*Geschick*) in 'the historicizing of the community, of the people' (*das Geschehen der Gemeinschaft, des Volkes*). . . . the ideas that are at the very foundation of National Socialist doctrines are...those of a community of destiny [*Schicksalsgemeinschaft*] and of a community of the people [*Volksgemeinschaft*]... The destruction of the individual and the human *I* in order to make room for the communal destiny of the people is . . . a 'political project' embedded in the very foundations of National Socialism, with its *Volksgemeinschaft* doctrine." Unsurprisingly, Heidegger and his fellow Nazi eugenicists, such as Hans F. K. Günther, took Plato as their model, emulating Plato's anti-reason epistemology and, quite appropriately, his governist politics as well.[158]

Professional scientists serving under the Third Reich, also, had explicated a decidedly unscientific, romanticist view. Let us revisit the words of Nobel Prize-winning physicist and early NSDAP supporter Philip Lenard, whom we quoted in Book One. Lenard, who joined the NSDAP in 1924 and stuck with it to the end, complained that in his own era, "the successes of technology have produced a special form of arrogant delusion. . . . The actualization of practical possibilities opened up by a greater comprehension of nature gave rise to the notion of the 'mastery' of nature. 'Man has slowly become the master of nature.' Such utterances on the part of the spiritually impoverished 'grand technicians' acquired a great influence because of the impressive display their new techniques and inventions made possible. And that influence has been even strengthened by the all-corrupting foreign spirit"—he means secular Jewish scientists—"permeating physics and mathematics."[159]

Also consider the espousals of Bruno Thüring, an astronomer and professional mathematician in the Heidelberg Association of Students in Science. Thüring says, "The feeling" for the untamed wilderness "possessed by Nordic man, who strives to comprehend nature . . . with his heart . . . , are here opposed by a concept of nature which aims to set up the intellect alone as the cognitive principle in the investigation of nature and which consequently disregards the possibility of conceptions geared to our spirit . . ."[160]

Understanding the thrust of the philosophies adopted by Lenard and Thüring, scholar Rene Denfield remarks that Nazi belief was "grounded in wishful thinking, and dismissive of objective science."[161]

As University of Wisconsin historian George L. Mosse summarizes, Naziism was characterized by a "fascist appeal to irrationality . . ."[162] In conclusion, "Obviously there is a connection between Hitler's reading and the New Romantic ideology. . . . What a contrast that supposed Jewish materialism formed with such a world view!"[163]

The now-discredited accusation that Naziism naturally sprung from the Enlightenment began with Theodor Adorno and Max Horkheimer of the Frankfurt school of Marxian thought. "Of course," says Smithsonian Institution historian Arthur Herman (b. 1956), "one could point out that the Enlightenment's supposed final product, Nazi Germany, was actually the self-declared enemy of Enlightenment liberalism and all its works, just as it championed" romanticist "*Kultur*" (culture) over rational "*Zivilisation*." Yet Adorno and Horkheimer arbitrarily rejected such considerations.[164]

Naziism was a rebellion against the general category of Enlightenment thinkers, a group which implicitly includes Spencer and Sumner. That does not deter Richard Hofstadter from issuing his mischaracterization. He goes on to argue that if Spencer and Sumner personally opposed colonialism, the free-market theories they espoused will—when taken to their logical conclusion—still ultimately instill imperial ambitions into their readers. *SDAT* provides shabby evidence for such an implication. *American Thought* purports that it was his enthusiasm for Spencer's evolutionism that motivated a certain turn-of-the-twentieth-century American philosopher to become a cheerleader for militarism. Says Hofstadter, "The writings of John Fiske, one of the earliest American synthesizers of evolutionism, expansionism, and the Anglo-Saxon" supremacy "myth, show how tenuous could be the boundary between Spencer's ideal evolutionary pacifism and the militant imperialism which succeeded it." In Fiske's hands, says Hofstadter, "evolutionary dogma issued forth in a bumptious doctrine of racial destiny." The reason is that Fiske "had long believed in Aryan race superiority" and accepted the 'Teutonic' theory of democracy. This doctrine sanctified any conquest incidental to Anglo-Saxon expansion."[165] *SDAT* is once again guilty of twisting the truth.

Robert Bannister issues the correction: "Fiske disliked the term 'Anglo-Saxon . . ." For that reason, Fiske "thus substituted" that expression with the "'English race,' but insisted that the latter term was more cultural than biological since as a race the English had 'shown a rare capacity' in assimilating others"—those *others* being other ethnicities. Yes, Fiske exalted the English as supreme. But Fiske did this on the grounds that he thought the English had the best cultural institutions. Such institutions were private property and rule of law. Fiske's Anglophilia was not based upon his having the idea that the English held any biologically inborn virtues. Also, it is true that Fiske concocted utopian dreams of a one-world government that reigned according to Lockean principles. However, he wanted this one-world federation to be formed not by having America or England forcibly annex other territories, but by having every territory ratify its own willing annexation to the global federation by means of republican voting. "Ironically," Bannister informs us in *Science and Myth*, an excerpt of Fiske's that is often "quoted as evidence of his imperialistic tendencies and even his social Darwinism—a reference to the 'great and glorious future...of the Anglo-Saxon race'—was meant to be a caricature" of the typical "expansionist view" that pushed for violent military conquest. Such imperialist fervor was a view that Fiske "did not share."[166] Once the global federation was formed, Fiske envisioned that goods and services would cross national borders unfettered.

Hofstadter lets this slip when he paraphrases Fiske's dream as, "If the United States would only drop its shameful tariff and enter into free [economic] competition with the rest of the world, it would exert such pressure, peacefully of course, that the states of Europe would no longer be able to afford armaments and would finally see the advantages of peace and federation. Thus, according to Fiske, would man finally pass out of barbarism."[167] As Hofstadter does not bother to state the truth so openly, I will: for the United States to "drop its shameful tariff" is hardly the same as overthrowing the government of another country and then annexing it.

Hofstadter's influence can be noticed in how New America Foundation senior fellow and *New Republic* contributor Michael Lind (b. 1962) darkly interprets a passage of John Fiske's March 1885 *Harper's* magazine article. In this passage, Fiske states, "The work which the English race began when it colonized North America is destined to go on until every land on the earth's surface that is not already the seat of an old civilization shall become English in its language, in its political habits and tradition, and to a predominant extent in the blood of its people. The day is at hand when four fifths of the human race will trace its pedigree to English forefathers, as four fifths of the white people in the United States trace their pedigree today."[168] The mention of the spread of people of English blood across the continent does sound creepily ethnicity-centric to our modern sensibilities, though there is no obvious call to violence against any nonwhites. Nonetheless, Michael Lind breathlessly intones that this very same passage "justified the expansion of American territory and power in terms of Social Darwinism—the influential if scientifically illiterate attempt to describe world history in terms of racial competition. In Fiske's essay, American expansionism degenerated into American exterminism—other races would actually become extinct, allowing the Anglo-Saxons to repopulate the world from the British islands and North America. . . . The racism of Fiske is hardly distinguishable from the genocidal racism of Hitler a few generations later."[169] It is a huge leap to compare Fiske's passage to the Holocaust. In reality, Fiske is talking about free migration. In his vision, under a peaceful global regime, English people would migrate to other lands and intermarry peaceably with those other lands' inhabitants.

It cannot be denied that, as a consequence of the Enlightenment period's prestige, a number of racists in the nineteenth century attempted to disguise their irrational prejudices by employing the vernacular of science. That racists exploited scientific-sounding terminology to provide a pretense of credence to their emotionalism, though, does not demonstrate the Enlightenment to be culpable for the racism. For instance, were Mr. Nobody to put on a flimsy plastic Halloween mask depicting Mr. Celebrity, and were Mr. Nobody then to go on a killing spree, it would be irrational to say Mr. Celebrity undertook the murders. Yet the principle is similar when someone attributes the atrocities committed in the name of governist eugenics to the Enlightenment. As Steven Pinker puts it, "The idea that the Holocaust was a product of the Enlightenment is ludicrous, if not obscene."[170]

No, Abraham Foxman and Evan Sayet, the Enlightenment did not contribute to the Holocaust, nor did those Darwinians who applied secular Enlightenment philosophy to the social sciences—Spencer and Sumner. Nonetheless, the situation grows more atrocious in consideration of another fact. It is that the same Progressive-Era intellectuals, whom left-wing writers have valorized as opponents of social Darwinism and eugenics, were usually the people who most enthusiastically campaigned for eugenicist legislation. One such historical figure was the botanist turned sociologist Lester Frank Ward. We have examined the philosophies of various historical figures, and shall continue that procedure on him. For Ward, we go forward.

NOTES

1. Fukuyama 1992, 68; and Fukuyama 1999, 74.
2. Fukuyama 2002, 27, 85.
3. Fukuyama 1999, 74.
4. Fukuyama 2011, 50.
5. Asimov 1964 paperback, 70-71. For other examples of ignorant attempts to tie Spencer or free-market economics with government eugenics, see Corballis 1991, 15; Degler 1991, ix, 11–15, 69; B. Friedman 2005, 122–26; Ravitch 2000, 27, 76; M. Roach 2005, 108-110; and E. O. Wilson 1998, 182-84.
6. Spencer 1851, pt. 3, ch. 21, sec. 8, para. 1, accessed online Friday, April 20, 2007.
7. Qtd. by G. H. Smith 1978, 14-18; and G. H. Smith 1991-K, 247.
8. Spencer 1981f, 214–17. See Letter 5, para. 5, 6, 8, accessed online Wednesday, January 11, 2012.
9. Spencer 1981f, 261. That is Letter 12, para. 3, accessed online Wednesday, January 11, 2012.
10. Carnegie 1886, 323.
11. Spencer 1851, pt. 2, ch. 16, sec. 3, para. 1, accessed online Monday, November 7, 2011.
12. Spencer 1851, pt. 3, ch. 28, sec. 2, para. 1, accessed online Monday, November 7, 2011.
13. Ekirch 1974, 209–210.
14. B. Werth 2009, 257–58.
15. B. Werth 2009, 43–44, 306. *Ibid.*, 44–45, goes on to state that Schurz was racist in that he believed that nonwhite races were congenitally less civilized than whites, and that he therefore believed that Dominicans would never assimilate willingly into the USA's constitutional republican system of government. Such a belief on Schurz's part is indeed consistent with the genetic-determinist theories of eugenicists, but that does not undermine my purpose in bringing up Schurz. The accusation of Isaac Asimov and the academic Left is that agreement with Herbert Spencer's social theories will necessarily lead someone to support Western imperialism. That accusation is refuted by the fact that Schurz's very intent in opposing Western imperialism was that he wished to apply Spencer's social theories in a logically consistent fashion.
16. Hofstadter 1959, 195.
17. Sumner 1969-C, 295, 269, 278–282, 285, 287.
18. G. Hodgson 2004.

19. P. Maier et al. 2003, 659.
20. C. Sagan 1995, 260.
21. E. O. Wilson coined the term "sociobiology," according to M. Bressler 1981, 181.
22. C. Lumsden and E. O. Wilson 1983, 39–40.
23. Qtd. by C. Lumsden and E. O. Wilson 1983, 39. Jonathan Beckwith authored this quotation with fourteen cosigners, including Stephen Jay Gould. Again, Beckwith's letter appeared in the November 13, 1975, *New York Review of Books*.
24. The fact that the quotation comes from professional philanthropist John D. Rockefeller, Jr., rather than his tycoon father, was pointed out by University of Wisconsin-Parkside chancellor Irvin G. Wyllie, according to Bannister 1988, 274 n. 70. As noted in Fosdick 1956, 130-31, John D. Rockefeller, Jr., spoke those words at a 1902 address at Brown University. Hofstadter attributes that quotation to a "John D. Rockefeller," citing Ghent 1902, 29. Hofstadter does not specifically say "Rockefeller, Senior." However, the implication is that he is talking about Rockefeller, Sr., given that Hofstadter said that the man who uttered that quotation was among the "business entrepreneurs." Rockefeller, Sr., is known primarily for being a business entrepreneur. Conversely, the man who actually uttered that quotation—John D. Rockefeller, Jr.—is not known primarily for being an entrepreneur, but for being a professional full-time philanthropist for most of his life, even in young adulthood. Derber 1998, 26–27, only says "John D. Rockefeller," and not "Rockefeller, Senior," when attributing this quotation to him. But that he attributes this quotation to Rockefeller, Sr., rather than John D. Rockefeller, Jr., is seen in that he attributes to the same "John D. Rockefeller" the quotation "God gave me my money," which were words spoken by Rockefeller, Sr. However, Allan Chase 1980, paperback, 8, very explicitly ascribes the quotation to "John D. Rockefeller, Sr." For other examples of authors falsely ascribing the quotation to John D. Rockefeller, Sr., see Commager 1950, 209, 455; M. Fichman 2002, 186; A. Trachtenberg 1982, 84–85; and J. A. Yoder 1975, 32–33.
25. H. G. Bowen and C. F. Kettering 1958, 67.
26. C. Lumsden and E. O. Wilson 1983, 39–40.
27. Bannister 1988, xxix.
28. This quotation is from M. Ridley 1997, 252.
29. This can be seen in Kasun 1988, 158.
30. Skousen 2001, trade paperback, 212
31. Skousen 2007, accessed online Friday, October 28, 2011.
32. L. Ponte 2002, accessed online Wednesday, May 16, 2007.
33. Zubrin 2012, 35, 40.
34. Qtd. by P. Cohen 2007, accessed online Saturday, May 26, 2007.
35. P. Buchanan 2009, accessed online Wednesday, July 1, 2009.
36. R. Dawkins 2008, accessed online Thursday, October 31, 2013.
37. S. Jacoby 2008, 61–62, 69–72, smears Sumner. *Ibid*., 70–79; and S. Jacoby 2004, 140, smears Spencer.
38. S. Jacoby 2008, 62, 80. Up until the part saying "military men," I am quoting S. Jacoby 2008, 62. Staring with "Social Darwinism has never died . . .", I am quoting S. Jacoby 2008, 80.
39. Rand 1992, paperback, 680–83.
40. S. Jacoby 2008, 71.
41. M. Roach 2005, 107–109.
42. L. Keeley 1997, paperback, 166, 222 n. 8.
43. G. H. Smith 1981, 129.
44. A. Toffler 1981, paperback, 101.
45. R. Wright 1994, hardcover, pictures section in the middle of the book, 330.
46. R. Wright 2000, 266.
47. P. Cohen 2007, accessed online Saturday, May 26, 2007.
48. Brantlinger 1977, 234.
49. J. M. Hobson 2004, 236-37.
50. Bannister 1988, 151.
51. *The Nation* 68 (1894), p. 294, qtd. by Bannister 1988, 155.
52. *Popular Science* 45 (1894), p. 558, qtd. by Bannister 1988, 155. Of course, because millionaire socialist Henry Demarest Lloyd supported the welfare state even more fervently than Kidd did, he criticized Kidd for being too laissez faire (Bannister 1988, 155). That the description "socialist" fairly applies to Henry Demarest Lloyd is evinced in Ekirch 1974, 48.
53. Banister 1988, 155.
54. B. Kidd 1894, 233, 237, qtd. by Bannister 1988, 156.
55. Black 2003, 119.
56. That Holmes thought private philanthropy "the worst abuse of private ownership" is a quotation coming from Oliver Wendell Holmes to Frederick Pollock, letter, April 23, 1910, in M. D. Howe ed. vol. 1, 169–171.
57. Lombardo 2008, 163, citing S. M. Novick 1992, 709–712.
58. Spencer 1851, pt. 2, ch. 4, sec. 3, para. 1, accessed online Tuesday, July 7, 2007, 2007
59. Spencer 1851, pt. 3, ch. 19, sec. 1, para. 1, accessed online Tuesday, July 17, 2007.

60. Spencer 1910b, 116. This is also qtd. by Bannister 1988, 51.
61. Spencer 1978 vol. 2.
62. Spencer 1981d, 187, and online at http://tinyurl.com/2kybvr, para. 10, accessed Saturday, May 5, 2007.
63. Black 2003, 119.
64. Lombardo 2008, 164.
65. Justice John Marshall Harlan, "Opinion Concurring in Part, Dissenting in Part," *Standard Oil Co. of New Jersey v. United States* 173 Fed. Rep. 177, http://tinyurl.com/32njod, accessed Wednesday, April 4, 2007.
66. Qtd. by Chernow 1998, 566.
67. Mark P. Painter, "Justice Oliver Wendell Holmes, Jr. (1841–1935)," *From Revolution to Reconstruction*, May 5, 2003, http://odur.let.rug.nl/~usa/B/oliver/oliverxx.htm, accessed Wednesday, April 4, 2007.
68. D. E. Bernstein 2011.
69. D. W. Root 2007, accessed online Saturday, September 1, 2012.
70. Qtd. by D. W. Root 2007, accessed online Saturday, September 1, 2012.
71. For Blakey's and Storey's citation of *Lochner v. New York*, see C. Blakey and Mr. Storey 1918, 62.
72. D. W. Root 2007, accessed online Saturday, September 1, 2012.
73. Oliver Wendell Holmes, *Lochner v. New York*, 198 U.S. 45 (1905), "Mr. Justice Holmes Dissenting," published at http://www.tourolaw.edu/patch/Lochner/HOLMES.asp, accessed Wednesday, April 4, 2007.
74. Hofstadter 1959, 47.
75. Bannister 1988, xvi.
76. Confirmation of Holmes's explicit support for the eugenics movement is found in Allan Chase 1980, paperback edition, 275, 315–16, 353, 368, 619; Degler 1991, 47; Fukuyama 2002, 85; M. Kaku 1997, 256; Kevles 1985, 111; F. Osborn 1968, 92; and A. Tone 2001, 144.
77. Qtd. in R. A. Posner ed. 1992, xxv.
78. J. Dewey 1987b, 16.
79. T. A. Bowden 2009, 30.
80. *Buck v. Bell* 274 U.S. 200 (1927), republished in Lombardo 2008, 287.
81. J. F. Stephen 1883 vol. 3, 225.
82. O. W. Holmes 1873, 582, republished as O. W. Holmes 1988, 50.
83. R. L. Moore 1974, 41.
84. T. C. Leonard 2003, 687–88, accessed online Monday, February 13, 2012.
85. "Group Blasts PETA 'Holocaust' Project" 2003, http://www.cnn.com/2003/US/Northeast/02/28/peta.holocaust/, accessed Thursday, May 31, 2007.
86. Abraham Foxman, Introduction to *Mein Kampf*, in Hitler 2001, xx–xxi.
87. D. Swade 2001, 12.
88. E. Sayet 2012, 8-10. I learned this from Alexander Marriott in his podcast—Alexander Marriott and Daniel Roberts, "Episode 41: Back to the Past," *History Guys*, June 22, 2013, the 63 minute mark, hosted on *BlogTalkRadio* online, http://www.blogtalkradio.com/history-guys/2013/06/22/episode-41-back-to-the-past, accessed Monday, September 2, 2013.
89. J. R. Saul 1993, paperback, 74, 41.
90. Blurb on the front cover of J. R. Saul 1993, paperback.
91. J. Gray 2013, 7, accessed online Thursday, January 24, 2013.
92. A. Padgen 2013, 18–19.
93. D'Souza 2002, 49.
94. K. Kelly 2010, 39.
95. D. Schneider and C. J. Schneider 2007, 1, 216.
96. P. Gay 1969 vol. 2, 409; A. Hochschild 2005, 87; and J. Powell 2008, 26.
97. J. Powell 2008, 26; and J. K. Taylor 1974, 242–43.
98. John Locke, *Second Treatise of Government*, ch. 7, sec. 85, para. 1, accessed online Saturday, October 25, 2014, in J. Locke 1764 rpt. J. K. Taylor 1974, 242–43, brought this to my attention.
99. Ambrose Evans-Pritchard, "Towards a Radical New Theory of Anglo-American Slavery, and Vindication of Free Markets," *The Telegraph*, August 7, 2013, http://www.telegraph.co.uk/finance/comment/ambroseevans_pritchard/10229570/Towards-a-radical-new-theory-of-Anglo-American-slavery-and-vindication-of-free-markets.html, accessed Wednesday, August 20, 2014.
100. D. P. Kotzin 2007 vol. 2, 681; L. J. Kern 1994, 172; and A. L. C. Tracy 1817, ch. 10, para. 20, Para. 42–45, accessed online Monday, June 10, 2013.
101. J. B. Say 1823, 60; and E. Schoorl 2012, 19–20, 126–133, 194.
102. P. Dray 2005, 200.
103. N. Chanda 2007, 214; and J. Powell 2008, 27.
104. J. Powell 2008, 31, points out that Thomas Paine advocated abolitionism in *Postscript to the Pennsylvania Journal and the Weekly Advertiser*, published on March 8, 1775.
105. J. Powell 2008, 36, points out that Turgot denounced slavery in his letter to Richard Price on March 22, 1778, and in *Reflections on the Formation and Distribution of Riches*.

106. Yafa 2005, 167, 133, 166.
107. D. King-Hele 1999, 231–32; and Uglow 2003, 412–13.
108. J. Powell 2008, 81; D. King-Hele 1999, 231–32; and Uglow 2003, 411–13, discuss Wedgwood's abolitionism.
109. J. Powell 2008, 27.
110. J. S. Schapiro 1963, 148-150, points out Condorcet's abolitionism. *Ibid.*, 87, likewise observes that Condorcet argued for women's suffrage in *On the Admission of Women to the Right of Suffrage*. J. S. Schapiro 1963, 179, mentions Condorcet's opposition to organized religion.
111. Uglow 2003, 411, 413, points out that Joseph Priestley opposed slavery. Uglow 2003, 411, further notes that Priestley stated in this condition, slaves were "deprived of every advantage of their rational nature."
112. P. Dray 2005, 200.
113. J. Gray 2013, 7, accessed online Thursday, January 24, 2013.
114. G. L. Mosse 1964.
115. G. L. Mosse 1964, 19-20.
116. Anna Bramwell and Simon Schama.
117. G. L. Mosse 1964, 13.
118. Buruma and Margalit 2005, softcover, 28, describes the qualities of the German Romanticism movement.
119. Romanticism being the preeminent philosophy of nineteenth-century Germany is noted in I. Berlin 1980b, 19.
120. Fest 1974 U.S. version, 96–97.
121. T. Kealey 1996, 310; C. Paglia 1990, 232; K. Popper 1971b vol. 2, 74; and M. Specter 2009, 33.
122. K. Popper 1971b vol. 2, 74.
123. K. Popper 1996a, 127.
124. M. Shelley 2011a, 441.
125. Goodrick-Clarke 1992, paperback, 4.
126. W. Biddle 2009, 37.
127. P. Choate 2005, 134.
128. A. Baeumler 1966, 97, adapted from A. Baeumler 1937, 283–85, 288–294.
129. E. Michaud 2004, 179, citing Robert Ley, "Organisieren heist: Wachen lassen!"
130. A. Milward 1977, 5.
131. F. Stern 1987, 155. Galbraith 1987, 92, admits that Adam Müller (1779–1829) was a German romanticist and advocate of governism who had an influence on Nazi ideology and Adolf Hitler.
132. K. Popper 1971b vol. 2, 74.
133. G. Stokes 1993, 5–6.
134. G. L. Mosse 1964, 13–14, 16–17.
135. G. L. Mosse ed. 1966, xxx.
136. J. Gray 2013, 7, accessed online Thursday, January 24, 2013.
137. G. L. Mosse ed. 1966, xxxi, citing B. Lochmüller 1935 vol. 1, 40.
138. Adolf Hitler, speech in Munich, Germany, on April 27, 1923, qtd. in Hitler 1966, 10, adapted from W. Siebarth ed. 1936, 132.
139. Goodrick-Clarke 1992 paperback, 4; and F. Stern 2006, 226–28.
140. I first learned of Moeller van den Bruck coining *the Third Reich* from Buruma and Margalit 2005 paperback, 8. F. Stern 2006, 191, 226, confirms that Moeller van den Bruck coined *Third Reich* with the title of his 1923 book.
141. G. L. Mosse 1964, 32–36, 39, 43.
142. F. Stern 2006, 227–28, 191.
143. G. L. Mosse 1964, 173–74, 39, 54.
144. G. L. Mosse 1985, trade paperback, 36–37.
145. G. L. Mosse 1964, 54, 97.
146. Goodrick-Clarke 1992, paperback, 26.
147. P. Roland 2008, 25.
148. Goodrick-Clarke 1992, paperback, 5.
149. G. L. Mosse 1964, 307.
150. P. Roland 2008, 141.
151. J. Randi 1982a, 62–63.
152. G. L. Mosse 1964, citing Felix Kersten, *The Kersten Memoirs*, London, no date, pp. 296, 65.
153. P. Roland 2008, 141.
154. T. Atkinson and J. Baran prods. 1998.
155. Lemesurier 1997, 146–47.
156. T. Aktinson and J. Baran prods. 1998; Eric Meyers narrating I. Chapman prod. 2012; and P. Roland 2008, 153, 158.
157. T. Atkinson and J. Baran prods. 1998.

158. E. Faye 2009 U.S. version, 140, 16–17, 68.
159. P. Lenard 1966, 204–05, adapted from P. Lenard 1936, 1–2, 11–13.
160. B. Thüring 1966, 210, adapted from B. Thüring 1936, 706–711.
161. R. Denfield 1996, 251.
162. G. L. Mosse 1975, 11.
163. G. L. Mosse 1964, 306.
164. A. Herman 1997, 307–08. G. L. Mosse 1975, 11, also discusses the way in which political leftists have tried to claim that Naziism was merely an extension of capitalism. In that discussion, Mosse cites Renzo De Felice, *Le interpretazioni del fascismo*, Bari, 1971, p. 51ff.
165. Hofstadter 1959, 176.
166. Bannister 1988, 228.
167. Hofstadter 1959, 177.
168. M. Lind 2005, 309, citing H. W. Brands 1998, 16. The article is John Fiske, "Manifest Destiny," *Harper's*, March 1885.
169. M. Lind 2005, 309.
170. S. Pinker 2011, 643.

FOUR

The Camouflaging of Eugenicists as Eugenicism's Opponents

The Academic Left's Canonization of Lester Ward and Richard T. Ely

There is a gaggle of historians that includes John A. Garraty, Carl Degler, Diane Ravitch, Arthur Ekirch, George Mowry, Richard Hofstadter, and, once again, R. Laurence Moore. This gaggle presumes that two historically significant intellectuals—sociologist Lester Frank Ward and economist Richard T. Ely—could not have been eugenicists or social Darwinists. The gaggle presumes as much simply on account of Ward's antipathy toward Herbert Spencer and Ely's animus against Spencer[1] and William Graham Sumner.[2] Subsequent to haranguing Spencer and Sumner for being social Darwinists,[3] these authors proceed to cast Ward and Ely as social Darwinism's most ardent foes.[4] Hofstadter gushes that Ward was "a champion of the masses . . ."[5] Hofstadter's protégé Eric Foner parrots the sentiment. From Foner: "the penetrating criticism" supplied by "Ward . . . and others" contributed to the middle class's eventual rejection of the social Darwinism preached by Spencer and Sumner. This thankfully prodded the population, Foner continues, to "adopt a more reform-minded outlook in the Progressive era."[6] Yale's Daniel Kevles, Harvard's Alexander Keyssar, and MIT's Pauline Maier and Merritt Roe Smith share authorship in a book that propounds that "social Darwinism . . . did not go unchallenged. Among the most important dissenters was Frank Lester Ward [sic] . . . he was the most influential as a critic of social Darwinists and as a leader in the new field of sociology."[7]

Similarly, New York University's Diane Ravitch praises Ward as a "remarkable polymath" who drove back that menace, "Spencer's laissez-faire philosophy."[8]

She shows how wonderful Ward was by disclosing that historian Henry Steele Commager (1902–1998) called him "the philosopher, the protagonist, even the architect, of the modern welfare state."[9] One would hardly think of Ward as a social Darwinist upon reading Hofstadter's assertion, "Ward had felt all too keenly the sting of his lower-class origin, and the aristocratic innuendos of social Darwinism as it found expression in the 1870's and 1880's offended his democratic sensibilities." Ward resembles a social Darwinist still less when Hofstadter dubs him "the first and the most formidable of a number of thinkers who attacked the unitary assumptions of social Darwinism and laissez-faire individualism." The Columbia luminary explicates that Ward "evolved a twofold criticism of social Darwinism," and exposed social Darwinist texts as nothing more than "upper-class apologies for social oppression and misery."[10]

Arthur A. Ekirch hails Ward as a "staunch critic of competition and laissez faire . . ." Moreover, Ward was a crusader for "social reform" who "came to rival . . . Social Darwinists like Spencer and Sumner" as "a social theorist." Ward wisely repudiated Spencer's and Sumner's nonsense and "offered his concept of a dynamic, applied sociology in opposition" to their "static sociology." Indeed, Ekirch praises the entire Progressive movement, paraphrasing its agenda in this manner: "Instead of accepting natural or biological evolution in terms of hands-off or laissez faire, men might cooperate to shape the environment to meet their social needs." Elsewhere Ekirch happily explains, "Progressives, who did not accept the mechanistic determinism of Darwin's evolutionary system or Marx's economics, sought to control progress and sustain human values."[11]

Right after blasting Sumner and Justice Holmes for being too laissez faire for his tastes, R. Laurence Moore lauds "Lester Ward and Richard Ely" as "two pioneers in the revolt against Social Darwinism . . ."[12] George E. Mowry (b. 1909) of UCLA cheers Ward on as well. Mowry christens Ward "the father of American sociology . . ." What he likes most about Ward, Mowry elaborates, is that Ward led the destruction of Sumner and social Darwinism. He announces, "The first sustained theoretical attack against Sumner and his mentor Spencer came from a group of sociologists, including Lester Ward, . . . Charles H. Cooley, and Edward A. Ross." This group spearheaded the "first sustained theoretical attack" on "the social Darwinism" mentality. Indubitably, this social Darwinism mentality "assumed that the fierce competition in nature was the natural and good thing for both animal and man." Ward "challenged the philosophic heart" of the ideology spouted by Sumner and Spencer.[13]

Comparable sensibilities are expressed by John A. Garraty (b. 1920), who authored the best-selling textbook *The American Nation*, and who, like Hofstadter, taught at Columbia University. Garraty rhapsodizes that Ward was an "outstanding" and "phenomenally learned man . . ." Ward put his knowledge to good use, Garraty coos, as Ward "effectively demonstrated" the "dubious character" of "American social Darwinists, most notably William Graham Sumner and John Fiske . . ."[14]

And Sidney J. Blumenthal, an adviser to President Clinton, keeps this fable alive. In his 1986 history of the conservative movement, Blumenthal emphasizes the identity of the scoundrels whom Ward had fought. ". . . Herbert Spencer, the Victorian philosopher of Social Darwinism, propounded the iron 'law' of nature: the wealthy were 'fittest' to enjoy their wealth and the poor had earned their poverty." Blumenthal continues with what he regards to be the corrective to this phenomenon. "The roots of both modern American sociology and economics can be traced to the nineteenth-century assault on the Social Darwinist viewpoint. In 1884, Lester F. Ward, the father of American sociology, published a piercing essay entitled 'The Laissez Faire Doctrine Is Suicidal.'" Then Blumenthal exclaims that it "was in this spirit that the discipline of sociology in America was born." He then goes on to commend Richard T. Ely and John Maynard Keynes for being enemies of social Darwinism.[15] Insofar as one considers support for eugenics legislation to be support for social Darwinism, Blumenthal's assessment of Ward, Ely, and Keynes poses a dilemma, as we shall examine shortly.

Governists are not alone in believing and disseminating such inaccuracies. Brink Lindsey, once vice president of research at the libertarian Cato Institute, contends that Sumner's "social Darwinist" writings could not withstand logical criticisms from "Lester Ward" and other great "social scientists . . ." Lindsey is grateful that Ward "scoffed at the notion" that free markets "reflected in their every detail the immutable workings of natural law and were thus beyond improvement."[16]

And, as with these others, the Pulitzer Prize-winning Carl N. Degler (b. 1921) of Stanford University congratulates Ward for being Spencer's "longtime ideological enemy" and a "social reformer" who fought for his "belief in the equality of people . . ."

Degler then proceeds, "Invidious comparisons among social groups, for example—the very stock-in-trade of eugenics—could not help but challenge, if not threaten, the socially democratic and reformist values of sociologists and anthropologists. The nationally prominent Lester Frank Ward, for instance, as early as 1907 pointed out that there was no reason to consider the lower classes as any less worthy genetically than the upper classes."

Note the following points. Degler's paragraph insists that Ward repudiated what was "the very stock-in-trade of eugenics." Degler continues that Ward "pointed out that there was no reason to consider the lower classes as any less worthy genetically than the upper classes. . ." From these points, one would infer that Ward held not a whit of sympathy for eugenics. In Degler's words, Ward warned that eugenics was a "true danger to human progress because it tinkered with nature's own forces for good."[17]

Then let us move on to a book by Henry Steele Commager that was published by Yale University Press and accorded public accolades by Arthur Schlesinger, Jr, in *The Nation*. In this book, Commager sings hymns to Ward for spotting "the error of the Spencerian system."[18] In Commager's phrasing, this error is that "Spencer lost sight of the first law of evolution." As for Spencer's American protégé, "Sumner's logic was dubious and his history tendentious . . ." In the long run "logic revealed" the intellectual "bankruptcy of Sumner." In short, "The stubborn laissez faire of Sumner was discredited . . ." [19] Commager minces no words: "Lester Ward was the first major scholar to attack the inadequate science, the dubious logic, and the specious rhetoric of the Spencer-Sumner school, and he remains the ablest. To the study of sociology he brought immense resources of scientific and philosophic learning and a firm grasp of the meaning of evolution to social development."[20]

Why, ". . . Ward was the first truly evolutionary sociologist." Ward's nobility derives from his having "protested equally the inconsistency and insincerity of the doctrine of laissez faire as applied in America. . . . He knew that laissez faire was a rationalization rather than a first principle . . . Laissez faire was rather the validation than the inspiration" of the rapacious "economic conduct of the age of big business." As with the heroic Ely, Ward "saw how archaic notions of property, contract, and due process of law had taken on the character of Laws of Nature, and how the term 'liberty' had been perverted to assure immunity from government interference to the practices of corporations." Indeed, Ward consistently "saw what . . . Sumner failed to see . . ." Thanks to this botanist, the "old distrust of the State, deeply rooted in American experience, gave way before the inescapable realization that only the State was prepared to act effectively in many of the crises of national affairs . . ." Commager summarizes that Ward "was both a pathfinder and a prophet." As if there were not centuries' worth of propaganda against the night watchman state, Commager hyperbolically proclaims Ward the "first major scholar to challenge the doctrine of laissez faire on scientific grounds and to articulate social with natural evolution . . ." Ward "was the first, too, to accept the full implications of pragmatism and to give sociology a philosophic foundation. . . . He inspired a whole generation of scholars and reformers to believe that it was possible to remake society along happier lines, and a new generation that did not know him worked with his tools and fought with his weapons. He was the prophet" of Woodrow Wilson's "New Freedom and the New Deal, of all those movements looking to the reconstruction of society and economy through government intervention which is the most striking development in the political history of the last half-century of America. . . . In the rugged vigor of his mind, the richness of his learning, and the originality of his insights, the breadth of his conceptions, he takes place alongside William James, John Dewey, and Oliver Wendell Holmes as one of the creative spirits of twentieth century America." Commager appreciates that Ward wrote in a letter to Edward A. Ross, "I would probably go further toward populism than you. No one is more anxious to throttle the money power."[21] William James himself advocated governist ideas. His father, Henry James, was a great admirer of collectivist commune leader Charles Fourier, and William adopted many of his father's ideas. That explains William's derision of capitalism as "the bitch goddess success."[22] As for Ward, Commager delivers no words about how Ward also helped bring governist eugenics advocacy to academic sociology.

Upon venerating Ward as the man who saved the Earth from social Darwinism in general, *SDAT,* likewise, broadcasts the idea that Ward despised governist eugenics in particular. Hofstadter writes about a 1906 American Sociological Society conference in which Ward was horrified that the speaker preceding him "had presented a social-Darwinist thesis advocating careful elimination of the unfit and dependent, chiefly by eugenic methods. In reply Ward branded the doctrine presented as 'the most complete example of the oligopocentric world-view which is coming to prevail in the higher classes of society and would center the entire attention of the whole world upon an almost infinitesimal fraction of the human race and ignore the rest.' He would not be contented, Ward continued, to work in so small a field as the education and preservation of a select few of the higher classes. 'I want a field that shall be broad enough to embrace the whole human race, and I would take no interest in sociology if I did not regard it as constituting such a field.'" Hofstadter then quotes Ward as saying at that conference that the poor are "by nature the peers of the boasted 'aristocracy of brains' that now dominates society and looks down upon them, and the equals in all but privilege of the most enlightened teachers of eugenics."[23]

Sweeping Ward's Eugenicism under the Rug

There is something unsettling about the manner in which these governist authors have whitewashed Ward as the eugenicists' *bête noir. Science and Myth* takes notice of the irony of Ward having the gall to "blast Darwinized-social theory while simultaneously grounding" his own position "in a Darwinian vision of struggle . . ."[24] University of Illinois-Urbana historian Clarence J. Karier contributes to Bannister's critique, exposing Ward as an explicit advocate of eugenics, white supremacism, and military conscription.[25] Bear in mind for the upcoming quotations that *stiripculture* was an early synonym for government-imposed eugenics that Lester Ward had employed. On his desire to create a master race, Dr. Ward stated in his 1891 address to the Biological Society of Washington,

> It is the right and the duty of an energetic and virile race of men to seize upon every great principle that can be made subservient to its true advancement, and . . . to apply it. Natural selection is the chief agent in the transformation of species and the evolution of life. Artificial selection [that is, selective breeding conducted by human beings] has given to man the most that he possesses of value in the organic products of the earth. May not men and women be selected as well as sheep and horses? From the great stirp [i.e., a line of descendants from a common ancestor, according to *Dictionary.Com*] of humanity with all its multiplied ancestral plasm—some very poor, some mediocre, . . . only a comparative few very good . . .—from all this, why may we not learn to select on some broad and comprehensive plan with a view to a general building up and rounding out of the race of human beings? At least *we should by a rigid selection stamp out of the future all the wholly unworthy elements.* Public sentiment should be created in this direction, and when the day comes that society shall be profoundly shocked at the crime of perpetuating the least taint of hereditary disease, insanity, or other serious defect, as it now is [toward] the comparatively harmless crime of incest, the way to practical and successful stirpiculture will have already been found [emphasis added].[26]

Some of Ward's opinions about eugenics changed over time. But on into the last year of his life, he persisted in a paper in *The American Journal of Sociology* that persons beleaguered by mental illness or hereditary diseases in their families should be sterilized against their will.[27] In that very same paper, Ward very explicitly applied the word *eugenics* to describe his own position. "Are we to accept that modern scientific fatalism known as *laissez faire*?", that paper rhetorically probed. Obviously, Ward's answer was no. The paper continued that, given that "the end and aim of the eugenicists cannot be reproached, it is therefore a question of method rather than of principle."[28] One wonders if *this* is what Arthur Ekirch meant when he cheered that Ward "took an optimistic view of the lessons to be drawn from the natural world."[29] Recall, also, that Carl Degler (1) pronounces Ward to be one who rejected every form of social inequality, and (2) insists that eugenics is entirely about inequality, "the very stock-in-trade of eugenics." The inevitable deductive conclusion to draw from Degler's claims is that, following in the logic of his supposedly egalitarian premises, Ward rejected eugenics. Note that in applying the *eugenics* label to himself, Ward directly contradicts Degler's characterization of him. In the final estimation, Ward believes that the oppressive measures he suggests will carry out evolution and the survival of the fittest in a manner superior to peaceful market enterprise.

As for the subject of race, Ward is normally hailed as someone who rejected racial bigotry, who would not countenance that there was an inherent difference in the races in terms of how they behaved. George Mowry proclaims that as far as Ward "could see, ability, intelligence, and energy were all distributed equally among men, and the near monopoly of these characteristics was due simply to their social advantages."[30] Now consult Ward's actual words on the subject of inborn racial differences. Ward proclaimed that "the Chinese" are "less brilliant" than whites, while "the African is probably less energetic and less intellectual."[31] Remember that *SDAT* identifies racism, imperialism, and eugenics as forms of social Darwinism.[32] In light of this information, *SDAT* contradicts its own characterization of Ward as social Darwinism's nemesis. *American Thought* acknowledges, "Ward is sometimes classified among the social Darwinists" on account of his own belief that social progress emerges when one race conquers another.[33]

Dr. Karier renders it unambiguous that Ward "did not believe in the equality of the races. He was convinced," as were many of his left-wing "[il]liberal followers, that while the black race may be superior in feelings and sentiment, they were inferior to the white race in cognitive abilities. Ward, in *Pure Sociology*, suggested that the black man who rapes a white woman does so not only out of lust, but out of an almost unconscious desire 'to raise his race to a little higher level.' Combining a bit of male chauvinism with his racism, Ward asserted that it is more permissible for a male of a inferior race to have sexual relations with a female of a superior race, because in the first instance it would be a matter of 'leveling up' while in the latter case it would be a matter of 'leveling down.'"[34] Consider that Henry Steele Commager lionized Ward as one who "ranged himself unreservedly on the side of the plain people, fighting their battles with weapons more formidable than even those which Sumner and his disciples could muster."[35] Insofar as the ethnic minorities that Ward denigrated can be considered plain people, Commager's praise for Ward amounts to empty assumptions on Commager's part. According to Yale University historian Ralph Henry Gabriel, Ward championed the "the American concept of the planned society . . ."[36] Were that not bad enough, Ward championed this abomination "under a rubble heap of ponderous phrases and technical words."[37] Hofstadter, too, admits that Ward "was a forerunner of social planning,"[38] though the Columbia University historian hardly intends that as a criticism. Similarly, Ward receives plaudits from Ekirch for being "an early exponent of a . . . planned collectivistic society" and "positive social and economic planning,"[39] and likewise from Mowry for his "social engineering."[40]

Still more thrilled than Hofstadter and Ekirch about this is Henry Steele Commager. Commager is elated that Ward assessed that there was not too much governism in the USA, but too little. In Ward's words, "Modern society is suffering from the very opposite of paternalism—from under-government." Commager approvingly quotes Ward saying, "Individual freedom can come only through social regulation."[41] Refusing to face the Orwellian self-contradiction in that statement, Commager finds it perfectly logical, proclaiming, "Ward cherished liberty as sincerely" as anyone else.[42]

Yet Ward's explanation for his support for a socialized system of education undermines Commager's assumption that Ward "cherished liberty . . ." As we did in Book One, we quote Ward himself on his opinion about tax-funded, legislation-mandated government schooling at its best. Referring to a baby by the pronoun of *it* rather than *he* or *she*, Ward proclaims, "Every child born into the world should be looked upon by society as so much raw material to be manufactured. Its quality is to be tested. It is the business of society as an intelligent economist to make the very most of it," *it* being the child.[43] In *Dynamic Sociology* from 1907, Ward recommends that government schools shield their students from learning any ideas that Ward dislikes. Likewise, these students should be deprived of the option of airing opinions that Ward dislikes. He rationales, "The forcible suppression of the utterance or publication in any form of unwelcome opinions is equivalent to withholding from all undetermined minds the evidence upon which such views rest; . . . It is simply that true views may as easily be created by this method of exclusion as false ones, which latter is the point of view from which this fact is regarded. The more or less arbitrary exclusion of error, i.e., of false data, is to a great degree justifiable. . . . This, however, is the essence of what is here meant by education, which may be regarded as a systematic process for the manufacture of correct opinions"[44] (emphasis Ward's).

It should be mentioned that Ward's avowed progressivism did not stop him from looking down on the penurious and uneducated, whom he described as "wretched beings" and "the unfortunate victims of social imperfection."[45] Karier's revelations about Ward's views on blacks, the Chinese, and the financially downtrodden bring to light the astonishing inaccuracy of Diane Ravitch's depiction of him. Ravitch describes this racial elitist as a "passionate egalitarian"[46] who "insisted that not only all classes but all races were equally capable of learning and employing the social achievements of mankind." Nor do the facts line up with Mowry's assertion that Ward believed that "ability, intelligence, and energy were distributed equally among all men. . . ."[47] If *social Darwinism* alludes to eugenics, elitism, racism, ethnic conflict, and "classism," then it would appear that the term *social Darwinist* more accurately characterizes Ward's position than that of either Spencer or Sumner. Yet this is the same Lester Ward whom Pulitzer Prize-winning Carl Degler compliments for being the antidote to social Darwinism on account of Ward's "belief in the equality of people."[48]

Not Even Ward Likes the Social-Darwinism Label

Notably, though, unlike the Ward apologists I have quoted, Ward himself eventually came to realize that, just as I am presently arguing, the social Darwinism tag is mostly a form of name-calling. Ward knew full well that political activists describe someone as a social Darwinist in order to associate him with now-unpopular political ideologies to which he does not truly subscribe. Bannister narrates that at first in the late 1800s, "Ward persistently warned contemporaries of the dangers of conservative"—in this context, Bannister means *free-market*—"Darwinism, or as he himself finally called it by its modern name, social Darwinism." But, later, Ward found that the social Darwinism epithet that he once affixed to others was soon being exploited to debunk his own eugenicism. *Science and Myth* continues, "In 1905 and 1906, Ward learned that others had their own version of the same charge—social Darwinism—directed not only against laissez faire, but also against certain eugenic arguments and against the view that international struggle and warfare produced progress. The irony, and the cause of Ward's concern, lay in the fact that he had argued each of these latter positions himself."[49]

Starting from the late 1890s, Ward became a zealous espouser of what alternately came to be known as *race-conflict theory*[50] and *struggle sociology.*[51] This theory was initially propounded by the Austrian sociologists Gustav Ratzenhofer and Ludwig Gumplowicz, both of whom I quoted in Book One about how government is violent by its nature. According to struggle sociology, the first sedentary city-state emerged as a consequence of one racial group conquering another and then extracting tribute from the conquered tribe.[52] In the eyes of Ratzenhofer, Gumplowicz, and Ward, race-based tribal warfare thus spurred a form of beneficent social progress. This is the reason for which Hofstadter ended up partially conceding that sometimes it appeared that Ward could be counted among social Darwinists.[53] "Not surprisingly," writes Robert C. Bannister, the struggle sociology's academic enemies "quickly found in the rhetoric of Darwinism a convenient weapon for discrediting them."[54] Thither, Jacques Novicow lobbed the *social Darwinist* epithet at Ward. Novicow, explains Thomas C. Leonard, "regarded race-conflict theory a wrongful defense of war . . ." In response, "Ward denied the charge but did not retreat from race-conflict theory."[55] Ironically, one of the earliest times in history that the expression *social Darwinism* was used, the epithet was directed against Ward, not Spencer nor Sumner. *Science and Myth* observes that it was in the year 1905 that Novicow "singled out Ward, Ratzenhofer, and Gumplowicz as leading exponents of *le Darwinisme social.*" Here, *Science and Myth* continues, Ward "was in a quandary. . . . How could Ward maintain his positions . . . and at the same time escape the social Darwinist charge?" It was on account of controversies such as this one, that Lester "decided that the term social Darwinism must be eliminated."[56]

Thus Ward conceded the point that I am articulating. In 1907 in *The American Journal of Sociology*, Ward groused, "With this vague notion in their minds," political activists and academicians "have invented the phrase 'social Darwinism' and have set it up as a sort of 'man of straw' in order to show their agility in knocking it down." He went as far as acknowledging that it is "wholly inappropriate to characterize as social Darwinism the *laissez faire* doctrine of political economists." For once, Ward and I agree. Interestingly, whereas I consider the imaginary relation between Spencer's free-market economics and government-mandated eugenics to be an insult to Spencer's philosophy, Ward held the opposite conclusion. Ward considered the false equation to be an insult to his own governist eugenicism. He elaborated, "That *laissez faire* doctrine is false and not sustained by biological principle I freely admit and have abundantly shown,"[57] and that he must protest "in the strongest possible terms against the application of the term Darwinism to the race struggle" that so appealed to Ward.[58]

Bannister wisely questions Ward's condescension toward Spencer's economics. "Had Ward also created a straw man? Did not allegations of social Darwinism merely state explicitly what he and others had been saying for some time?" Ward's own derision toward Spencer and Sumner "strongly suggest that he too was fashioning from his own concerns a portrait of the opposition that was more effective than it was accurate."[59] Essentially, Ward's character received a convenient rewrite from distinguished professors from Harvard, Yale, MIT, Stanford, Columbia, NYU, and UCLA.

On some level, Hofstadter might have recognized that his admissions about Ward's eugenicism contradicted his overall portrayal of Ward as the antithesis of social Darwinism. As an attempt to dodge this reality, Hofstadter blames Ward's eugenicism on a particular free-marketer whom Hofstadter evidently loves to hate. "While Ward's dualism of the genetic and telic was in effect a departure from what William James called the 'block-universe' of Spencer," Hofstadter proclaims, "the Spencerian virus remained in his blood." *SDAT* maintains that it was because Ward "admired Spencer's version of universal evolution" in his younger days, that Ward "could lapse" into the elitist and classist mindset of such "sociological nature-worshippers" as Spencer and Sumner.[60] By implication, Hofstadter would have it that insofar as Ward espoused eugenicism—in contrast to Spencer—Ward's eugenicism must be attributed to some latent influence Spencer continued to exert over Ward. It is interesting to note that Ward actually gave a speech at the November 1882 Delmonico's banquet thrown for Spencer. Defying decorum, Ward quite consciously assured his audience of Spencer-admirers that his own model of social science proved superior to Spencer's, taking evolutionary theory to its logical conclusion by expecting an endlessly wise government to manage the economy.[61] Incidentally, recall from Book One that Ward extolled the institution of government schooling on account of the schools' ability to instill civic indoctrination into impressionable youths. Further recall Ward's proclamation, "The individual has reigned long enough." Ward's compulsory eugenicism is consistent with these other governist sentiments.

Yes, Hofstadter partially acknowledges some of the story behind Ward's own racialist theories. But, as Thomas C. Leonard notes, Hofstadter mostly "dismisses it."[62] Hofstadter assures us that struggle sociology "found but a small place in his work."[63] Upon digging up Ward's corpus, though, Leonard realizes that struggle sociology was actually instrumental to Ward's career. Leonard explicates that "in a three-part article surveying contemporary sociology" from the year 1902, Ward "wrote that 'the doctrine of the struggle of races,' was the most important subject in sociology."[64] In this article, Ward exclaims that race-conflict theory "forms so large a part of my own conception of sociology that it will be necessary to deal with it extensively elsewhere."[65] Ward focuses a considerable proportion of his 1903 opus on this theory, where he states unequivocally that "war has been the chief and leading condition of human progress." Moreover, "when races stop struggling progress ceases."[66] Hofstadter swept this fervor of Ward's under the rug.

Interestingly, Ward resented being called a socialist, instead proclaiming himself a critic of both socialism and laissez faire. When he touted his own version of governist supremacy, he did not call it socialism[67] but "sociocracy"[68] . . . as if that makes for a major difference. Ward attempted to distinguish his sociocracy from socialism by defining sociocracy as "the scientific control of the social forces by the collective mind of society."[69] That sounds very familiar. Perhaps it is time for us to move on to a governist eugenicist less embarrassed to place himself under the banner of socialism: Richard T. Ely.

Do Not Elide the Truth about Ely

When you first consider that *Mammon* is the New Testament's demonic personification for greed,[70] Richard T. Ely's characterization of free-market advocates as "mammon worshippers" gives you an idea of what he thinks of those who disagree with him.[71] In deriding capitalists as Mammon, Ely alludes to the two greatest influences on his beliefs about ethics and political economy—Christianity and socialism. Ely was a self-proclaimed adherent to a movement calling itself Christian Socialism.[72]

Think back to Book One, where we quoted Ely saying, "It is difficult to define *laissez-faire* categorically, because it is so absurd that its defenders can never be induced to say precisely what they mean." In that same 1886 essay where he pronounces such, Ely at least elucidates on what *he* truly means to be the mission of his American Economic Association—". . . we hold that there are certain functions which do not belong to the individual, certain functions which the great co-operative society, called the state—must perform. . . . We hold the doctrine of *laissez-faire* is unsafe in politics and unsound in morals. . . . In other words we believe in the existence of a system of social ethics; we do not believe that any man lives for himself alone. . . . All have duties as well as rights, and, as [Ralph Waldo] Emerson said several years ago, it is time we heard more about duties and less about rights."[73] Henry Steele Commager hails this set of statements as "a veritable declaration of independence from the tyranny of the *laissez-faire* philosophy."[74] Commager joyfully reports that "two years after the publication" of Lester Ward's "*Dynamic Sociology*, a group of younger economists, including Richard T. Ely, . . . Simon N. Patten, . . . and . . . Woodrow Wilson, met at Saratoga Springs to organize the American Economic Association and draft what proved to be a declaration of independence" from the stodgy capitalist ideology of "Manchester."[75]

Given that George Mowry denigrated William Graham Sumner and glorified Lester Frank Ward, it hardly raises an eyebrow to read Mowry's praise for Ely. Mowry gives a thumb up to Richard T. Ely and Simon Patten for stirring up the academic "revolution in economics" that displaced Sumner and the other "conservatives" pathetically hanging onto their "laissez faire."[76] Philip Taft Award-winning financial journalist and labor historian Steve Fraser predictably sneers at Spencer and Sumner for having "espoused" what he deems "Orthodox social Darwinism."[77] Not surprisingly, Fraser likewise snipes at Ayn Rand for having authored "philosophical fictions in defense of pure-blooded, unimpeded capitalism . . ." As a foil to Spencer and Sumner and Rand, Steve Fraser sets up Richard T. Ely. Fraser seeks comfort in how Ely "and other reform-minded economists . . . worried about what they spied as the inherent conflict of interest between investors and the speculating promoters who midwived" the robber-baron industrialists' mergers during the Inventive Period. He finds it laudable that Ely took it upon himself to "subject the . . . phenomenon of publicly traded corporation to close inspection."[78]

And observe how John Kenneth Galbraith sets up progressives Richard T. Ely and John R. Commons as sainted foils to capitalist William Graham Sumner, whom he introduces as "the most eminent of American Social Darwinists." Indignant that Sumner "opposed anything that subtracted from the income of the rich or preserved or uplifted the poor," Galbraith visibly approves that "Sumner was mentioned by Richard T. Ely, the leading figure in the founding of the American Economic Association, as the kind of economist he would not wish to see join" that association. Aha—Ely's AEA discriminates against Sumner's villainy. Upon establishing that, Galbraith tells us about two morally superior figures. "John R. Commons (1862–1945) . . . is the American companion figure to Bismarck and Lloyd George." Commons was a "brilliant" man who found a compatriot in "Richard T. Ely (1854–1943), himself a pioneer figure in American economic dissent . . . Ely eventually brought Commons to the University of Wisconsin, where he wrote learned books" of the utmost importance. The "greatest achievement" of Commons and Ely was "in assembling and leading a brilliant and devoted coterie of colleagues and students who," not having been deceived "by the classically orthodox views" of free-marketers under Spencer's tutelage, such as Sumner, "set out in a highly practical way to redress the evident social grievances of the time. Their primary instruments were the Wisconsin state government in Madison, which was conveniently adjacent to the university, and its governing family, Robert La Follette and his two sons." Galbraith happily reports on the results of the advocacy of Ely and Commons: "a pioneer state civil service law, the effective regulation of public utility rates, a limit on usurious interest . . ., support to the trade union movement, a state income tax and finally, in 1932, a state unemployment compensation system. This last had a penetrating effect on economic and political attitudes; nothing else contributed so directly to the federal legislation on the subject three years later. And it was the Commons and Wisconsin economists who again led in the federal design."[79]

Although Clarence Karier exposed Ward's racism, he still casts a favorable eye upon Ely. Karier expresses gratitude to "such men as Herbert Croly, Richard Ely, Edward

Ross," and "John R. Commons" for "increasing the standard of living for a larger middle class." They "all thought and worked toward a larger, more orderly corporate society utilizing knowledgeable experts to ameliorate the many varied problems of that society."[80]

American Thought is almost as complimentary toward Ely as Steve Fraser is. Upon identifying Ely as the founder of the American Economic Association (AEA) and a follower of the Bismarckian "German historical school" of political ideology, *SDAT* rah-rahs Ely for having "attacked classical" free-market "economics for its dogmatism and simplicity, its blind faith in laissez faire, and its belief in the adequacy of self-interest as an explanation of human conduct." Then, to establish that Ely and his AEA cohorts were not themselves social Darwinists, Hofstadter clarifies that "the German historical school rather than Darwinism was their model" of how to evaluate social issues.[81]

Ely's worship of governism cannot be overemphasized. He cried that "there is no limit to the right of the State . . . Duty . . . is co-extensive with power. The State is a moral person."[82] His zealotry for governism was of religious proportions. "Now it may rationally be maintained that, if there is anything divine on earth, it is the State, the product of the same God-given instincts which led to the establishments of the Church and the Family. It was once held that kings ruled by divine right, and in any widely accepted belief, though it be afterwards discredited, there is generally found a kernel of truth. In this case it was the divine right of the State."[83] Ah, Ely is more open than most about his faith in *GOD*-vernment. Ely's faith went so far that, during the First World War, Ely beat the drums for military conscription. Then he wrote to Allyn A. Young that any professor who expressed "opinions which hinder us in this awful struggle," the military effort, should be "fired" if not "shot."[84] As a collectivist, he prattled, "The only right point of view is the social . . ."[85]

Ely stated very explicitly that he supported governist eugenics—the same governist eugenics that Hofstadter loathes as Darwinism. "As a part of the preparation of our human material," wrote Ely in 1918, "we shall give increasing attention to eugenics . . . there are certain human beings who are absolutely unfit and who should be prevented from a continuation of their kind."[86] Much as Garrett Hardin would in pursuant decades, Ely pronounced that industrial society's reduction of the West's death rate had actually produced deleterious effects. Ely considered this a poor influence on the Western population's gene pool. In words that would not ordinarily be intended as a complaint, Ely fumed, "Philanthropy and science keep alive men who would otherwise perish." In the very next sentence, Ely indicates that he disapproves of that fact. "These men reproduce their kind, and the result is an enfeebled progeny. Reproduction goes on, and as heredity determines chiefly the characteristics of those who live, we have a feebler parentage leading to a feebler race of men."[87]

On this issue, Ely received words of support from another progressive, the academic economist Simon Patten.[88] Thomas C. Leonard explains, "Patten's influence in Progressive circles derives from his role as a protectionist Wharton School leader who viewed big business as an enemy of society, and, most especially, as an activist who argued that professional economists had an obligation to lead reform movements—to leave the library" and "write for newspapers. . . Students attracted to . . . Patten's reform activism include many leading Progressive voices . . ."[89] One of those students was Rexford Tugwell, who would become an architect of the New Deal as President Franklin Roosevelt's advisor. Tugwell took Patten's courses at university and considered Patten's own governist collectivism to be a major influence on his own thinking about statecraft.[90]

On the matter of eugenics, Patten worried, "Every improvement" in industrial technology "which simplifies or lessens" the exertions required by "manual labor" hereby "increases the amount of the deficiencies which the laboring classes may possess without their being thereby overcome in the struggle for subsistence that the survival of the ignorant brings upon society."[91] In other words, Patten and Ely consider it *awful* that technological progress has reduced the death rate among the handicapped.

Richard T. Ely continues that, given that industrialization's slashing of the death rate "is leading to an increasing number of an increasingly feeble population, should it not be checked?" Ely's answer is that of course it should be, by forcible means. He determines it essential that there be an "effort. . .to prevent" the "reproduction" of "the degenerate classes." He regrets not merely that competition for employment between the mentally challenged and the abler-bodied workforce allegedly drives down wages; he is offended by the very "existence of these feeble persons." On the matter, he therefore concludes, "The problem is to keep the most unfit from reproduction, and to encourage the reproduction of those who are really the superior members of society."[92] For the AEA's founder, "The great word is no longer natural selection, but social selection."[93] Again observe that the eugenicist—in this instance, Ely—characterizes eugenics not as some branch of social Darwinism, but as the repudiation of someone who would leave circumstances to be settled by the market's peaceable version of natural selection.

Simon Patten agrees: "Social progress is a higher law than equality, and a nation must choose it at any cost." Hence the State ought to embark on campaign for the "eradication of the vicious and inefficient"—Patten's words for the handicapped.[94]

A predictably racist strain ran throughout this governist eugenicism. In 1911, Ely opined, "The negro race, while endowed with a splendid physique and with great power for work, is neither progressive nor inclined to submit to the regularity of toil, such as an industrial civilization demands."[95] On racist and eugenicist grounds, Ely expressed dismay at any voluntary effort in the West to raise funds to mitigate starvation in India in 1903. He proposed that "the famine continue for the sake of race improvement . . ."[96] To the degree that Hofstadter considers governist eugenics to be social Darwinism, he should have conceded that Ely supported social Darwinism. As usual, Hofstadter offered no such concession, but provided an impression at variance with the truth.

Incidentally, I have something to add about Henry Steele Commager's proclamation that Ely's derision of free enterprise amounted to an overthrow of the "tyranny of the *laissez-faire* philosophy." That remark is Orwellian on numerous counts. This supposed "tyranny of the laissez-faire philosophy" is an oxymoron. Tyranny amounts to encroachments by government far beyond the scope of the night watchman state. The laissez-faire philosophy, insofar as that expression alludes to the night watchman state, is the opposite of tyranny. And that Commager would compare Ely's apologia for governmental intrusiveness to the "declaration of independence" is really a slap in the face of Thomas Jefferson. In Ely's call for greater government control, he was advancing the very form of intrusive government from which the Declaration's signers sought independence.

Social(ist) Darwinism: Do Left-Wing Eugenicists Count As Social Darwinists?

It is objectionable enough that Hofstadter coaxed hundreds of historians into bowdlerizing eugenics proponents Lester Ward and Richard T. Ely as eugenics' opponents. Sadly, Hofstadter did something worse. Hofstadter wrongly pegged an anti-capitalist eugenicist as an exponent of free-market competition. That anti-capitalist is not simply a eugenics proponent, but the one who coined the very term *eugenics*—Sir Francis Galton.[97] To be sure, Hofstadter later coughs up a minor, partial concession that some American public figures argued for eugenics on a basis consistent with their left-wing progressive ideology. Nonetheless, Hofstadter mostly propounds that, thanks to Sir Francis's influence, eugenics started out as—and predominantly remained—a logical corollary to Spencer's conviction that some individuals displayed more moral merit than others. "Early eugenicists," *SDAT* historicizes, "tacitly accepted the identification of 'fit' with the upper classes and the 'unfit' with the lower that had been characteristic of the older social Darwinism." These rich people happen to be Hofstadter's caricature of such laissez-faire social Darwinists as Spencer and Sumner. "Their almost exclusive focus upon the physical and medical aspects of human life helped to distract public attention from the broad problems of social welfare. . . . The social preconceptions of Sir Francis Galton were not seriously questioned by the early eugenicists; and Galton, like . . . Sumner, . . . had postulated the free competitive order in which awards are distributed according to ability. He was convinced 'that men who achieve eminence, and those who are naturally capable, are, to a large extent, identical.'"[98] *SDAT*'s insinuations aside, Sir Francis's political philosophy was the opposite of the one that Spencer and Sumner popularized. Journalist Kenneth Silber reveals that Sir Francis "thought" that it would be wonderful "if the state played a central role in eugenics . . ." He "envisioned the hereditarily well-endowed receiving official [tax-paid] subsidies to reproduce . . ."[99] Those who would mistake Galton's program for something based on Spencer's teachings forget a salient fact. I cite Spencer's moral objection to the practice of taxpayer subsidies going to any enterprise. Spencer, as I do, propounded that government operations should be funded through contractual user-fees instead of compulsory taxation. Fortunately, historian Daniel Kevles acknowledges that the government eugenics movement's leaders were—as with Justice Holmes, Pearson, and Ward—openly antagonistic toward free enterprise.

Incidentally, there is unintended humor in that, in its further effort to depict Ward as the archenemy of social Darwinism, *SDAT* focuses the reader's attention on Ward's public criticism of Francis Galton. We are to interpret this as Ward publicly criticizing governist eugenics. In actuality, Thomas Leonard informs us, when Ward criticized Galton, it was not on account of the brutality of eugenics policy nagging at Ward's conscience. Nay, Ward's quibble was over a technical issue. Sir Francis's version of governist eugenics made heavy use of Gregor Mendel's discoveries about genetic genotypes, phenotypes, dominant genes, and recessive genes. Ward rejected the much-more-scientifically-accurate Mendelian genetics in favor of the older, less-accurate interpretation of evolution advanced by Jean-Baptiste Lamarck. And Ward expected Galton to come over to his side, away from Mendel and toward Lamarck. This, Thomas C. Leonard sighs, is "a distinction that Hofstadter elides."[100] Upon deciding whose political philosophy was similar to his own, it was Ward to whom Sir Francis leaned toward.

Despite his misrepresentations of Spencer, which we read about in chapter 2, science journalist Colin Tudge comes forward as yet another intellectual willing to concede that the governist's eugenics movement was peopled by members of the political Left. Tudge, too, contradicts Hofstadter's presumption that governist eugenics was largely a right-wing phenomenon. Recall that after noting that eugenicists "differed" from free-market advocates in failing to "draw sweeping laissez-faire conclusions," *American Thought* still insists that both camps—the governist eugenicists and the Manchesterite evolutionists—were still "almost equally conservative in their general bias . . ." Recall also Hofstadter's insistence that governist eugenics was ultimately "fundamental conservatism."[101] In what seems an indirect indictment of Hofstadter, Edwin Black, and their academic followers for such misrepresentations, Tudge writes, "It has been customary, in the late decades of the twentieth century, to decry the eugenics movement" as nothing more than a "right-wing plot: an attempt by the ruling class to beat the lower orders into shape . . . But history does not bear this out." A number of celebrities thought to "belong to the left," Tudge continues, "wrote chillingly and frankly" about their governmental eugenic aims.[102] The University of Houston's James H. Jones (b. 1943), too, reports that governist eugenics "commanded wide support among [left-wing welfare-statist] liberals and conservatives alike."[103] Bannister knows that "eugenics had a right and a left wing . . ." And he adds that eugenicists "on the 'right'" likewise employed the social Darwinism label "to stigmatize laissez faire."[104] Along with these others, Matt Ridley affirms that evaluation.[105]

Besides Pearson and Ward, examples of self-avowed socialists among the State-run eugenics movement include British Labour Party chairman Harold Laski (1893–1950),[106] activist "Red" Emma Goldman (1869–1940), Nobel Prize-winning geneticist Hermann Joseph Muller (1890–1967),[107] genetics expert J. B. S. Haldane, and, as a young scientist, World Wildlife Fund founder Julian Huxley (1887–1975).[108] Laski wrote of his fretting over "the fostering of the unfit at the expense of the fit, and their consequent over-propagation." For that reason, Laski stated, Laski wanted "natural selection" supplanted with governmental "reproductive selection . . ." This governmental control would construct a master race intelligent enough to "build a strong political superstructure" of State-driven collectivism.[109]

Playwright George Bernard Shaw (1856–1950) of *Pygmalion* fame,[110] and polemicists Sidney (1859–1947) and Beatrice Webb (1858–1943)[111] were socialists who each pledged allegiance to eugenicism, compulsory sterilization, and immigration restriction. Sidney Webb wrote these complaints in *Fabian Tract* about England being infested with new ethnicities he did not feel comfortable around. "In Great Britain at this moment, . . . children are being freely born to the Irish Roman Catholics and the Polish, Russian, and German Jews, on the one hand, and the thriftless and irresponsible—largely the casual laborers and the other denizens of the one-roomed tenements of our great cities—on the other. . . . This can hardly result in anything but national deterioration; or, as an alternative, in this country gradually falling to the Irish and the Jews. Finally, there are signs that even these races are becoming influenced. The ultimate future of these islands may belong to the Chinese!"[112] Webb refuses to tolerate the possibility that persons of East Asian descent could hold money and power.

In keeping with their governist eugenicism, Shaw and the Webbs also seriously prevailed upon the British Empire to acquire more underdeveloped nations as colonial territories.[113] This conspicuously contradicts the contemporary stereotype that radical leftists categorically oppose colonialism and militarism. Shaw clarified, "The problem before us, is how the world can be ordered by Great Powers of practically international extent. . . . The notion that a nation has the right to do what it pleases with its own territory, without reference to the interests of the rest of the world, is no more tenable from the International Socialist point of view—that is, from the point of view of the twentieth century—than the notion that the landlord has a right to do what he likes with his own estate without reference to the interests of his neighbours."[114] It will be remembered from Book Two that in the early twentieth century John A. Hobson and Vladimir Lenin wrote of imperialism being evil. Yet the prevailing notion among socialists from the 1800s to World War Two had been in favor of militarism. It was not until the United States and the United Kingdom coming into conflict with the Soviet Union and other communist states that Western leftists—in sympathy for the communists—urged the West to refrain from war.

Despite his prior criticisms, Arthur Ekirch at least recognizes Sumner as an "anti-imperialist" in contradistinction to Shaw and the Webbs. As a corollary, Ekirch concedes a specific point of Sumner's. This point is that much nineteenth-century rhetoric urging European imperialism amounted to a logical extension of the socialists' and progressives' conceit that the State should control other people's lives. Shaw and the Webbs wanted the government to control people within their native Britain. By extension, Shaw and the Webbs would have it that this State control be extended beyond national boundaries. "The contemporary American rationale for imperialism, as professors like Sumner . . . complained, seemed to anticipate and parallel Progressivist ideology. Thus the . . . government sought by Progressives, it was believed, could also prove beneficial abroad. Exhortations of reform and self-sacrifice at home as the means of simple justice for the less fortunate classes of society were readily translatable into demands that the American people should shoulder the white man's burden and embark on missionary crusades overseas."[115]

"Far from being deliberate tools of capitalist interests, as critics charged," notes Arthur Herman, "American colonies and dominions in far-off tropical places" were considered by socialists and Progressives to be "a salutary counterweight to those measures at home."[116] Despite his generally blinkered outlook, Sidney Webb did understand one fundamental truth that Spencer's modern critics ordinarily skate over—"no consistent eugenicist can be a 'Laissez Faire' individualist unless he throws up the game in despair. He must interfere, interfere, interfere!"[117]

Socialist Novelists and Their Not-So-Novel Ideas

There was also science-fiction writer and self-described "Christian Socialist"[118] Edward Bellamy (1850–1898). Bellamy advanced his own brand of eugenics,[119] though it was a sort less invasive than that of Shaw and the Webbs. This assessment of Bellamy may come as a shock to anyone who read what Hofstadter wrote about him in *SDAT*. Here is yet another case where *SDAT*—which, in no uncertain terms, christens eugenics "the most enduring form of social Darwinism"[120] —dresses up a vocal eugenics supporter as a eugenics critic. And, though Hofstadter at least conceded that aspects of Lester Ward's public statements coincided with the aims and premises of eugenics, Hofstadter discloses no such qualifications about Ward's friend Bellamy.[121] Just as he did with Ward, Hofstadter paints a picture of Bellamy as an implacable enemy of social Darwinism and thus, by implication, a crusader against eugenics. Hofstadter narrates, "In an address before a Boston audience Bellamy declared that 'the final plea for any form of brutality these days is that it tends to the survival of the fittest; and very properly this plea has been advanced in favor of the system which is the sum of all brutalities.' . . . . the competitive system apparently causes the unfittest to survive, not in the sense that the rich are worse than the poor, but that the system encourages the worst in the character of all classes."[122]

American Thought quotes from a "Declaration of Principles" that Bellamy authored in support of nationalizing industry. "The principle of the Brotherhood of Humanity is one of the eternal truths that govern the world's progress on lines which distinguish human nature from brute nature. The principle of competition is simply the application of the brutal law of the survival of the strongest and most cunning. Therefore, so long as competition continues to be the ruling factor in our industrial system, the highest development of the individual cannot be reached, the loftiest aims of humanity cannot be realized."[123] Bellamy most notoriously pressed this point in his 1888 utopian propaganda novel *Looking Backward, 2000–1887*. In it the narrator from 1887, Julian West, goes into suspended animation and awakens in a perfect socialist society in the year 2000. This is not some sort of satire targeting "perfect societies" in the spirit of *Brave New World* (more on that book's author later); it is the sort of civilization to which Bellamy genuinely aspires. In the story, the reanimated Julian West is greeted by Dr. Leete, the benevolent representative of the ideal civilization. Leete informs Julian that humans achieved heaven on Earth once they abandoned the profitable commercialism that nineteenth-century economists of Spencer's ilk defended. "Selfishness," Dr. Leete ruefully meditates, was the free-market economists' "only science, and in industrial production selfishness is suicide. Competition, which is the instinct of selfishness, is another word for the dissipation of energy, while combination is the secret of efficient production."[124]

Have another glance at Bellamy's explicit philosophy—a glance closer than the one that Hofstadter gives us. In *Looking Backward* Dr. Leete first states that the men and women of his day marry for "love" rather than for the sake of financial security. Yet Dr.

Leete then adds that people in his times are healthier and more sexually attractive than those of the nineteenth century because the implementation of socialism has encouraged men and women to engage in a consensual type of eugenics. "You were speaking . . .," Dr. Leete says to Mr. West "of the physical superiority of our people to your contemporaries. Perhaps more important than any of the causes I mentioned then as tending to race purification has been the effect of untrammeled sexual selection upon the quality of two or three successive generations. I believe that when you have made a fuller study of our people you will find in them not only a physical, but a mental and moral improvement. . . . a profound moral sentiment has come to its support. . . . To-day this sense of responsibility, practically unrecognized in all previous ages, has become one of the great ethical ideas of the race, reinforcing, with an intense conviction of duty, the natural impulse to seek in marriage the best and noblest of the other sex." This can be translated into simpler words. In deciding on a husband, a woman of Dr. Leete's era considers each potential suitor's genes. The man ascertained to have the most advantageous genes is the one chosen; this is the main criterion. Yet this criterion contradicts what Dr. Leete said mere paragraphs earlier—that men and women form pair bonds based on love. Dr. Leete cheers that under his present system, "sexual selection, with its tendency to preserve and transmit the better types of the race, and let the inferior types drop out, has unhindered operation. . . . The attributes that human nature admires are preserved, those that repel it are left behind."[125]

Bellamy's eugenics is a step far above, ethically, those of most other governist eugenicists. Bellamy's is better than Pearson's, as Pearson propagandized for compulsory sterilization and immigration control. Nevertheless, Bellamy's proposal still counts as eugenics. And, tellingly, Bellamy's proposal gained support from yet another socialist eugenicist—the man who co-discovered natural selection with Charles Darwin, Alfred Russel Wallace (1823–1913).[126]

We should not jump to the conclusion that Hofstadter is right to judge Bellamy as ethically superior to Spencer, especially not on the premise that Bellamy champions "social equality" whereas Spencer does not. There is something else to ponder. Bellamy proposes the ratification of laws forbidding women from holding down specific jobs. Dr. Leete boasts, "The heavier sorts of work are everywhere reserved for men, the lighter occupations for women. Under no circumstances is a woman permitted to follow any employment not perfectly adapted, both as to kind and degree of labor, to her sex. . . . The men of this day so well appreciate that they owe to the beauty and grace of women the chief zest of their lives and their main incentive to effort, that they *permit* them to work at all *only* because it is fully understood that a certain regular requirement of labor, of a sort adapted to their powers, is well for body and mind, during the period of maximum physical vigor"[127] (emphases added). By contrast, Hofstadter gives no mention of Bellamy recommending such restrictions upon what women can do with their own bodies and minds and lives. But *SDAT* at least admits that Herbert Spencer wrote a defense of the equal "rights of women" that was decidedly "radical" for its day.[128] In the nineteenth century, a woman was considered not the exclusive property of herself, but of her husband or father. Spencer bore no patience for such chauvinism. In 1851 he wrote, "Equity knows no difference of sex. . . . The law of equal freedom manifestly applies to the whole race—female as well as male. . . . Hence the several rights deducible from that law must appertain equally to both sexes." Spencer further pressed,

> Married life under this ultimate state of things will not be characterised by perpetual squabbles, but by mutual concessions. Instead of a desire on the part of the husband to assert his claims to the uttermost, regardless of those of his wife, or on the part of the wife to do the like, there will be a watchful desire on both sides not to transgress. Neither will

> have to stand on the defensive, because each will be solicitous for the rights of the other. . . .
> There is nothing Utopian in this. . . . An attitude like that described is not uncommonly maintained in the dealings of honourable men with each other; and if so, why should it not exist between the sexes? Here and there, indeed, may be found, even now, a wedded pair who preserve such a relationship. And what is at present the exception may one day be the rule.

And then Spencer went on to preach, "The extension of the law of equal freedom to both sexes will doubtless be objected to, on the ground that the political privileges exercised by men must thereby be ceded to women also. Of course they must; and why not? Is it that women are ignorant of state affairs? Why then their opinions would be those of their husbands and brothers; and the practical effect would be merely that of giving each male elector two votes instead of one. Is it that they might by-and-by become better informed, and might then begin to act independently? Why, in such case, they would be pretty much as competent to use their power with intelligence as the members of our present constituencies. We are told, however, that a woman's mission is a domestic one—that her character and position do not admit of her taking a part in the decision of public questions—that politics are beyond her sphere. But this raises the question—Who shall say what her sphere is? . . . Who now will tell us what woman's sphere really is?" Here, Spencer anticipates the argument that women being treated equally with men would allow women to participate in the political process. Critics of women's rights scoff that this would be wholly unfeminine and unattractive. In rebuttal to such an objection, Spencer writes, "However much . . . the giving of political power to women may disagree with our notions of propriety, we must conclude that, being required by that first pre-requisite to greatest happiness—the law of equal freedom—such a concession is unquestionably right and good." The British thinker then closes the chapter by asserting that "the rights of women must stand or fall with those of men; derived as they are from the same authority; . . . demonstrated by the same argument."[129] These words arrive via Spencer's *Social Statics*. This is the same philosophic treatise that PC academicians universally condemn as a right-wing reactionary's apologia for old-fashioned traditionalism—an apologia that was close-minded toward any suggestions about how society may be improved. Yet, subsequent to conceding the radicalism of Spencer's thoughts on gender equity, Hofstadter feels obliged to pronounce *Social Statics* "ultra-conservative."[130] This piques my curiosity as to what Hofstadter means by *ultra-conservative*. Spencer's words were not within an inch of being conservative by the standards of 1851. If the Science-for-the-People organization that chastised E. O. Wilson's sociobiology considers Spencer a social Darwinist, then that does not comport with its accusation that social Darwinism per se attempts to present a "genetic justification" for "existing privileges" on the basis of "class" or "sex."[131] In any case, Spencer's support for female rights and women's suffrage poses a challenge for PC, Spencer-hating, Bellamy-approving academicians who consider themselves feminists. Would they rather live in a world engulfed in the gender equity extolled by the Spencer they despise? Or would those alleged feminists prefer a society that implemented polices prescribed by a fellow progressive—Bellamy? Recall that Bellamy's society would impose restrictions on what sorts of occupations women may choose for themselves.

Likely ignorant of these facts, evolutionary psychologist and Rice University anthropologist Sarah Blaffer Hrdy blithely remarks that the grand ayatollah of social Darwinism—Herbert Spencer—"thought females never had been inherently equal to males and could never be; subordination of women was not only natural but, in his view, desirable." Hrdy fulminates that Spencer's horrendous social Darwinism "became popularly associated with evolutionary biology. . ." She believes this "helps to explain why feminists have steadfastly resisted" the lessons that can be drawn from sociobiology. When Hrdy delivers her farcically inaccurate characterization of Spencer's views on women, she cites *SDAT* as her source.[132] This is in spite of *SDAT*'s explicit admission that *Social Statics* presents a radical case for gender equity.

Bellamy is far from being the sole socialist novelist to form eugenicist convictions. The illustrious H. G. Wells advocated a much worse sort of eugenics.[133] Wells warned, "It has become apparent that whole masses of human population are, as a whole, inferior in their claim upon the future. To give them equality is to sink to their level, to protect and cherish them is to be swamped by their fecundity."[134] The government, Wells deduced, had to do something about this scourge. Sure, he anticipated the rejoinder that the choice of other human beings to raise their own families was none of his business. He thus snapped, "The children people bring into the world can be no more their private concern entirely than the disease germs they disseminate or the noise a man makes in a thin-floored flat."[135] The author of *The War of the Worlds* proposes that the government should execute people experiencing psychiatric disorders. Were that act performed, he explains, then future generations would not inherit their disorders. Worse, he wanted to cleanse humanity of nonwhite races. The "swarms of black, and brown, and dirty white, and yellow people . . . will have to go."[136]

"'Go' meant 'die'. . .", translates Colin Tudge.[137]

Foreseeing objections, Wells qualified that such a policy would not be inhumane. "All such killing," he rationalized, "will be done with an opiate."[138]

Wells propounded that his brand of governmental eugenic enforcement was far nobler than the methods that Bellamy endorsed. On May 16, 1904, Wells stood up at a eugenics conference and announced, "It is the sterilization of failures, and not the selection of successes for breeding, that the possibility of an improvement of the human stock lies." On top of that, the prolific sci-fi figure implored his readers in one essay, "We want fewer and better children. . .and we cannot make the social life and the world-peace we are determined to make, with the ill-bred, ill-trained swarms of inferior citizens that you inflict upon us."[139] Related to this, filmmaking legend Ridley Scott (b. 1937), director of *Alien* and the Academy Award-winning *Gladiator*, made an episode of his documentary series, *Prophets of Science Fiction*, which lionized Wells for a specific reason. The show discussed Wells's novels *The Invisible Man* and *The Island of Dr. Moreau*, and pointed out that they were cautionary tales about scientists misusing technology for nefarious ends. The episode tried to convey that Wells supported ethical responsibility in science, but brooked no mention of Wells's hard-line eugenics advocacy.[140] I think the advocacy of State-enforced eugenics exemplifies Wells himself misusing the imprimatur of science to advance his own irrational, harmful agenda. In support of Well's eugenicism, Well's pal George Bernard Shaw recorded himself on film stating a similar argument. In the film, Shaw propounded that any peaceful woman should be euthanized if she does not consider herself a contributor to the community. "You must all know half a dozen people at least who are of no use in this world, who are more trouble than they are worth. Just put them there and say, 'Sir, or madam, now will you be kind enough to justify your existence? If you can't justify your existence, if you're not pulling your weight in the social boat, . . . then, clearly, we cannot use the organization of our society for the purpose of keeping you alive, because your life does not benefit us and it can't be of very much use to yourself.'"[141]

Novelist Aldous Leonard Huxley (1894–1963)—Julian Huxley's brother—did not openly advocate the same brutal governist eugenics policies as H. G. Wells or Shaw. Nevertheless, he expressed fears similar to Wells's.[142] This news may come as a shock. Aldous's seminal novel, *Brave New World*, is normally taken as the ultimate satire against government-managed eugenics. Despite that, Aldous expressed sympathies toward the eugenics movement throughout his life. In his nonfiction *Brave New World Revisited*, Aldous sounded the alarm about the opposite of eugenics—cacogenics. As *-genics* is the root word for "gene," *caco-* is "bad," in contrast to the "good" that is *eu-*. Writing in 1958—over a decade subsequent to the Nazis' racial polices having soured world opinion over the eugenics movement—Aldous brazenly wrote,

> In the *Brave New World* of my fantasy eugenics and dysgenics [a common misnomer for *cacogenics*] were practiced systematically. . . . in our random and unregulated way we are not only over-populating our planet, we are also, it would seem, making sure that these greater numbers shall be of *biologically poorer quality*. In the bad old days children with considerable, or even with slight, hereditary defects rarely survived. Today, thanks to sanitation, modern pharmacology and the social conscience, most of the children born with hereditary defects reach maturity and multiply their kind. Under the conditions now prevailing, every advance in medicine will tend to be offset by a corresponding advance in the survival rate of individuals cursed by some genetic insufficiency. In spite of new wonder drugs and better treatment (indeed, in a certain sense, precisely because of these things), the physical health of the general population will show no improvement, and may even deteriorate. And along with a decline of average healthiness there may well go a decline in average intelligence. . . . "Under conditions that are both soft and unregulated," writes Dr. W. H. Sheldon, "our best stock tends to be outbred by stock that is inferior to it in every respect . . ." [Emphases added.]

That consideration motivates Huxley to implore us to contemplate "congenitally insufficient organisms"—meaning humans possessing birth defects "whom our medicine and our social services now preserve so that they may propagate their kind. . ." Huxley frets that the "wholesale transmission to our descendants of the results of unfavorable mutations, and the progressive contamination of the genetic pool from which the members of our species will have to draw . . ."[143] Aldous implies that some kind of government-imposed solution will be optimal to save mankind from this alleged depreciation of the world's gene pool. True, Aldous is most likely aware that he would receive horrendous flack if he openly advocated a return to eugenics regulations. He knows about the bad press such regulations gained in the aftermath of the World War Two. Probably bearing that in mind, Aldous chooses to be conveniently cryptic about what he wants society, as a whole, to do about this alleged problem. That government intervention is the "middle way" of which he speaks—a middle way between (1) a laissez-faire system in which individuals are free to procreate peaceably as they choose, and (2) a governist system that dictates over who may have children and who may not. As we shall consider in the upcoming chapter, it would take a Garrett Hardin to describe the specific eugenicist legislation that Aldous wanted but would not go into detail over.

Aldous was more explicitly enthused about governist eugenics in his younger days. *Nash Magazine*'s April 1934 issue contained an article of his, entitled "What Is Happening to Our Population?" This article states that the key to saving the human race is in governmental efforts "encouraging the normal and super-normal members of the population to have larger families and in preventing the sub-normal from having any families at all."[144] As the sci-fi author's British-born biographer, Nicholas Murray, summarizes the piece, Aldous "advocated . . . use of the family allowance system to encourage some groups and not others to bear children. But he went further in proposing sterilisation of 'certified defectives,' pointing out that eugenic sterilisation was already legal in half the states of America."[145] Aldous conceded in this article that governist eugenics may be "another attempt on the part of the rich to bully the poor . . ."[146] But in Nicholas Murray's words, Aldous's article nevertheless favored governist eugenics "because democracy needed intelligent citizens. . . He also believed that, on humanitarian grounds, 'defectives' made poor parents and were responsible for many acts of cruelty according to the NSPCC," the United Kingdom's National Society for the Prevention of Cruelty to Children.[147]

There are but a few socialists whom Richard Hofstadter admits are anti-pacifist eugenicists. Among them is another novelist, Jack London (1876–1916).[148] Hofstadter quotes a telling 1904 article in which Jack London cheers that impending military conflicts will be based on genetic ancestry. "The possibility of race adventure has not passed away. We are in the midst of our own. The Slav is just girding himself up to begin. Why may not the yellow and brown start out on an adventure as tremendous as our own and more strikingly unique?"[149] Curiously—and conspicuously . . . and conveniently . . . —Hofstadter does not bother to disclose that Jack was enthused enough about socialism to found the Intercollegiate Socialist Society in 1905,[150] but a year subsequent to his call for race war.

And among the members of the political Left, the early twentieth-century's partisans of state-supported eugenics were not limited to socialists. Throughout the first half of the 1900s, state-enforced eugenics had a place in the heart of Americans who would have resented being called socialists—progressives, such as Theodore Roosevelt, Louis D. Brandeis, and the initial proponents of federal minimum wage legislation.

NOTES

1. Hofstadter 1992, 34.
2. R. H. Gabriel 1956, 248.
3. For characterizations of Herbert Spencer in such a way, see Degler 1991 ix, 11–15; Ekirch 1974, 23–24, 209–210, 219; Hofstadter 1959, 31–50, 176, 198–99, 202; Mowry 1958, 17; and Ravitch 2000, 27, 76. The castigation of Sumner as a social Darwinist is in Ekirch 1974, 22–24; Hofstadter 1959, 51–66; and in R. L. Moore 1974, 41–42. To see both Spencer and Sumner be criticized as social Darwinists at the exact same time, see Ekirch 1974, 23–24.
4. Degler 1991, 20–1, 110; Hofstadter 1959, 68, 70–76, 79–82, 84; Mowry 1958, 21; Ravitch 2000, 28–9, 33.
5. Hofstadter 1959, 83.
6. Eric Foner in the Introduction to the 1992 edition of *SDAT*, in Hofstadter 1992, xvi.
7. P. Maier et al. 2003, 621–22.
8. Ravitch 2000, 27.
9. Ravitch 2000, 27, citing Commager 1967, xxii.
10. Hofstadter 1959, 68, 71, 78–79, 81.
11. Ekirch 1974, 22–24.
12. R. L. Moore 1974, 41.
13. Mowry 1959, 20–21, 50.
14. Garraty ed. 1968, 237.
15. S. Blumenthal 1986, 89–90.
16. B. Lindsey 2007, 53–54.
17. Degler 1991, 21–22, 145.
18. Commager 1950, 206. B. Folsom 2008, 8, points out the unsurprising fact that Commager also wrote a hagiography for Franklin D. Roosevelt.
19. Commager 1950, 202, 208, 217, 225.
20. Commager ed. 1964 2d ed., 337.
21. Commager 1950, 206, 208-09, 212–13, 215–17.
22. The quotation and information come from R. H. Gabriel 1956, 337.
23. Hofstadter 1959, 82–83.
24. Bannister 1988, 126.
25. Karier ed. 1975, 140, 148, 190, 194.
26. That was originally in Lester Frank Ward, *Glimpses of the Cosmos* vol. 4, originally published in 1891, p. 295, and it was qtd. by Karier ed. 1975, 194–95. The definition of "stirp" I provided in the brackets comes from "Stirp," *Random House Unabridged Dictionary* 2006, (New York, New York: Random House, 2006), republished in Dictionary.Com, http://dictionary.reference.com/browse/stirp, accessed Wednesday, May 2, 2007.
27. He argues as much in L. F. Ward 1913, 738. Karier ed. 1975, 195, brought this to my attention.
28. L. F. Ward 1913, 746–47.
29. Ekirch 1974, 23.
30. Mowry 1958, 50.

31. L. F. Ward 1975, 148.
32. Hofstadter 1959, 170-200.
33. Hofstadter 1959, 78. For his summary of Ward's views on racial warfare, he cites Ward 1903, 215–26, 237–240.
34. That is from Karier ed. 1975, 190, and the passage Karier cites can be found in Ward 1903, 359–360.
35. Commager 1950, 215.
36. The words "American concept of the planned society," in reference to what Dr. Ward believed, were first used by Gabriel 1956, 215, and then qtd. by Karier ed. 1975, 139.
37. R. H. Gabriel 1956, 298.
38. Hofstadter 1959, 83.
39. Ekirch 1974, 23, 27.
40. Mowry 1958, 21.
41. Lester Frank Ward, qtd. by Commager 1950, 211.
42. Commager 1950, 211.
43. L. F. Ward 1975, 150–51. Clarence J. Karier gained access to it with the help of Adelia Peters.
44. L. F. Ward 1907a vol. 2, 547.
45. Karier ed. 1975, 275.
46. Ravitch 2000, 28. On page 464, she calls him a "passionate proponent of intellectual egalitarianism."
47. Mowry 1958, 50.
48. Quoting Degler 1991, 22.
49. Bannister 1988, 127–130.
50. T. C. Leonard 2009, 42, accessed online Sunday, November 6, 2011.
51. Bannister 1988, 127–28.
52. Bannister 1988, 127–30; Hofstadter 1959, 78; and T. C. Leonard 2009, 42, accessed online Sunday, November 6, 2011. According to R. L. Carneiro 1981, 182–83, Herbert Spencer partially agreed with this theory. He believed that when primitive hunter-gatherer tribes were warring against each other, the desire to win wars provided an impetus for these tribes to innovate in terms of technology and social customs. Spencer concluded, however, that by the time a society progressed to liberal industrialization, for that society to go to war would spur no social progress in it, but merely cause retrogression in terms of quality of life.
53. Hofstadter 1959, 78.
54. Bannister 1988, 129–130.
55. T. C. Leonard 2009, 42, accessed online Sunday, November 6, 2011.
56. Bannister 1988, 129–130.
57. L. F. Ward 1907b, 290–93. Bannister 1988, 130, and T. C. Leonard 2009, 42, brought this statements to my attention.
58. Ward, qtd. by Bannister 1988, 130.
59. Bannister 1988, 130–31.
60. Hofstadter 1959, 68, 84.
61. Bannister 1988, 79.
62. T. C. Leonard 2009, 42, accessed online Sunday, November 6, 2011.
63. Hofstadter 1959, 62.
64. T. C. Leonard 2009, 42, accessed online Sunday, November 6, 2011.
65. Ward 1902, 759. T. C. Leonard 2009, 42, accessed online Monday, November 7, 2011, brought this statement of Ward's to my attention.
66. Ward 1903, 238. T. C. Leonard 2009, 42, accessed online Monday, November 7, 2011, brought this passage to my attention.
67. Bannister 1988, 135.
68. Bannister 1988, 126, 238; and Commager 1950, 220.
69. Qtd. by S. Fine 1956, 263. T. C. Leonard 2003, 706, accessed online Tuesday, February 14, 2012, brought this to my attention.
70. *Random House Unabridged Dictionary* 2006, qtd. in "Mammon," *Dictionary.Com,* http://dictionary.reference.com/browse/mammon, accessed Thursday, January 10, 2008.
71. That Ely called free-market advocates "mammon worshippers" is visible in the quotation from Seldes 1960, 232.
72. J. C. Cort 1988, 251, identifies Ely as a Christian Socialist. *Ibid.*, 327, points out Ely's early advocacy of "the living wage."
73. R. T. Ely 1964, 336.
74. Commager ed. 1964 2d ed., 335.
75. Commager 1950, 234.
76. Mowry 1958, 22.
77. S. Fraser 2005, 274–75.
78. S. Fraser 2005, 590–91, 258–59, 312–13.
79. Galbraith 1987, 123, 165, 214–15.

80. Karier ed. 1975,8.
81. Hofstadter 1992, 146–47.
82. R. T. Ely 1889, 92.
83. R. T. Ely 1885, 73.
84. B. G. Rader 1966, 181–191.
85. R. T. Ely 1914, 496.
86. R. T. Ely 1918, 115.
87. R. T. Ely 1903, 165. The statements of Ely's came to my attention through Thies and Daza 2011, 152, accessed online Sunday, February 12, 2012.
88. According to Henry Steel Commager's editorial note in R. T. Ely 1964, 335, Ely's derision of laissez faire held "the approval" of Simon Patten.
89. T. C. Leonard 2003, 694 n. 10, accessed online Tuesday, February 14, 2012.
90. T. Russell 2010, 248.
91. Simon Patten, qtd. by D. Ross 1991, 197.
92. R. T. Ely 1903, 165, 173, 163, 139.
93. R. T. Ely 1901, 61.
94. Simon Patten, qtd. by D. Ross 1991, 199.
95. R. T. Ely 1914, 34-35.
96. R. T. Ely 1903, 165-66. The statements of Ely's came to my attention through Thies and Daza 2011, 152, accessed online Sunday, February 12, 2012.
97. Bannister 1988, 168; and T. C. Leonard 2003, accessed online Monday, February 13, 2012, point out that Galton coined the expression in 1883. Fukuyama 2002, 27, and M. Ridley 1999, 288, inaccurately state that Galton coined "eugenics" in 1885. While Herbert Spencer cannot be quoted on defining *social Darwinism*, Galton can be quoted as defining *eugenics*.
98. Hofstadter 1959, 163–67.
99. K. Silber 2005, accessed online Wednesday, March 7, 2007.
100. T. C. Leonard 2009, accessed online Monday, November 7, 2011.
101. Hofstadter 1959, 164, 167.
102. Tudge 2000, 285.
103. J. H. Jones 1997, 194.
104. Bannister 1988, 253 n. 5.
105. M. Ridley 1997, 253.
106. Allan Chase 1980 paperback, 14, 316, 644; J. Goldberg 2007, 249; and Tudge 2000, 285. He and Oliver Wendell Holmes, Jr., corresponded with one another about their mutual support for government-imposed eugenics. See Black 2003, 120. Laski met FDR through Felix Frankfurter (1882–1965) and became one of FDR's most vocal British supporters, according to J. Goldberg 2007, 251. J. C. Cort 1988, 11, points out the unsurprising fact that Laski reviled Adam Smith's *Wealth of Nations*.
107. Kevles 1985, 64, 190.
108. That Huxley was a eugenicist in his younger days is in Barkan 1992, 177, 184–187, 235, 255; Kevles 1985, 126, 138; A. Tone 2001, 207; and Tudge 2000, 286. That Huxley founded the World Wildlife Fund for Nature is in J. Goldberg 2007, 250. Julian and Aldous Huxley were descendents of Thomas Henry Huxley, a governist and sometimes-friend/sometimes-foe to Herbert Spencer. Thomas Henry Huxley was additionally a friend of Charles Darwin's and he helped win scientific acceptance for Darwin's theory.
109. Lombardo 2008, 165, quoting Harold Laski, "The Scope of Eugenics," *Westminster Review* vol. 174, 1910, 25–34, at pages 25–28, 32, 34.
110. R. Bailey 2000b, 60-61, accessed online Wednesday, March 7, 2007; Fukuyama 2002, 85; J. Goldberg 2007, 249; Kevles 1985, 86; M. Ridley 1999, 292; and Tudge 2000, 285. Also, a blurb of praise from Julian Huxley appears on the front cover flap of the book jacket for the pro-eugenics Osborn 1968. Julian Huxley supported eugenics as a young adult. He changed his mind about it in his old age.
111. Allan Chase 1980 paperback, 307; Fukuyama 2002a, 85; J. Goldberg 2007, 249; Kevles 1985, 74; M. Ridley 1999, 292; and Tudge 2000, 285. Also, Sidney Webb was a good friend of Henry Demarest Lloyd's, according to Nevins 1969 vol. 2, 335.
112. Qtd. by Kevles 1985, 74.
113. Ekirch 1974, 181–82.
114. Qtd. by Ekirch 1974, 182.
115. Ekirch 1974, 186–87.
116. A. Herman 1997, 177.
117. S. Webb 1910–1911, 237.
118. J. Cort 1988, 229; and R. J. Ellis 2005, 27–28; and L. Simon 2004, 266. Edward Bellamy's jingoistic cousin, Rev. Francis Bellamy (1855–1931), who is most famous for having written the Pledge of Allegiance recited in American schools, also described his own philosophy as "Christian Socialism," according to R. J. Ellis 2005, 27-28; R. H. Gabriel ed. 1956, 265; and L. Simon 2004, 258. Incidentally, Francis Bellamy, too, was a socialist eugenicist, though he went much farther than Edward, as Francis was a "race-conscious nativist" (R. J. Ellis 2005, 32) who supported restrictions on immigration on account of his disliking

immigrants for being racially different from himself (R. J. Ellis 2005, 33). Despite Edward Bellamy considering himself a "Christian Socialist," however, Mowry 1958, 25, reports that he and his immediate family did not attend church.

119. Bannister 1988, 33, 90, 171-74; and B. Friedman 2005, 126.
120. Hofstadter 1959, 161.
121. Bellamy wrote at least one essay praising Ward's ideology, according to Bannister 1988, 128, and Hofstadter 1959, 114.
122. Hofstadter 1959, 114.
123. Hofstadter 1959, 113, quoting from Bellamy, *The Nationalist* vol. 1 (1889), inside cover page.
124. Hofstadter 1959, 113, quoting Bellamy's *Looking Backward*. That can be read online in Bellamy 2000, ch. 22, para. 28, accessed online Wednesday, May 2, 2007.
125. Bellamy 2000, ch. 25, para. 30-39, http://tinyurl.com/3x56lp, accessed Wednesday, May 2, 2007.
126. You can read about Wallace's support for Bellamy's version of eugenics in Bannister 1988, 90, 173-74. Wallace's socialism is noted in Edey and Johanson 1989, 74.
127. Bellamy 2000, ch. 25, para. 7, http://tinyurl.com/3x56lp, accessed Wednesday, May 2, 2007.
128. Hofstadter 1959, 40.
129. Spencer 1851, pt. 2, ch. 16, http://tinyurl.com/yst9wn, accessed Tuesday, July 17, 2007.
130. Hofstadter 1959, 41.
131. Qtd. by Lumsden and Wilson 1983, 39. This quotation was authored by Jonathan Beckwith with fourteen cosigners, including Stephen Jay Gould, and it appeared in the November 13, 1975 *New York Review of Books*.
132. S. Hrdy 1981, 12, 12 n. 31, 13, 204 n. 31.
133. R. Bailey 2000b, 60–61, accessed online Wednesday, March 7, 2007; Kevles 1985, 92–94; and Tudge 2000, 285.
134. Qtd. by R. Bailey 2000b, 60–61, accessed online Wednesday, March 7, 2007; and M. Ridley 1999, 292.
135. Qtd. by M. Ridley 1999, 292.
136. Qtd. by M. Ridley 1999, 292; and Tudge 2000, 285.
137. Tudge 2000, 285.
138. Qtd. by R. Bailey 2000b, 60-61; M. Ridley 1999, 292; and Tudge 2000, 285.
139. Black 2003, 209, 130.
140. A. Knapp prod. 2011.
141. George Bernard Shaw, qtd. by J. Walker 2010, 113, accessed online Wednesday, May 18, 2011.
142. N. Murray 2002, 200, 254, 373–74, points out that both Aldous and Julian Huxley supported governist eugenics, though Aldous was not as enthusiastic about it as his brother.
143. A. Huxley 1958, 18–21.
144. Aldous Huxley, "What Is Happening to Our Population?", *Nash Magazine*, April 1934, qtd. by N. Murray 2002, 274-75.
145. N. Murray 2002, 275.
146. A. Huxley, "What Is Happening to Our Population?", qtd. by N. Murray 2002, 275.
147. N. Murray 2002, 275.
148. H. L. Mencken, "Jack London," *Prejudices: First Series*, 1919, 236–39, republished in Mencken 2006-I, 276–77, points out Jack London's socialism.
149. Hofstadter 1959, 189.
150. R. Perlstein 2001, 72, points out that Jack London started the Intercollegiate Socialist Society in 1905.

FIVE

Progressivism

The Genesis of Eugenics

Eugenicism among So-Called Progressives

Some left-wing historians evidently enjoy expounding upon the Progressive Era. Thinking back to Book One, this was the period in the early 1900s America when the political philosophy of Progressivism, also known as reformism, grew in prominence. This trend arrived at the tail end of the Gilded Age—which, as we mentioned in Book Two, New York-based philosopher Andrew Bernstein more accurately calls the *Inventive Period*. We additionally recollect that the Progressive Era is touted as a much-needed rebellion and corrective against the relative laissez faire of the Inventive Period. Remember, also, from Book One what Boston College historian Charles Derber proclaims. He opines that the Progressive Era—and, more significantly, the New Deal that arose from it—should be praised. This governist trend "created a new American social contract." Progressivism "was a genuine alternative to Gilded Age capitalism, and the nation's most decisive repudiation of social Darwinism."[1] DePauw University historian Stanley P. Caine likewise cheers, "Progressivism began with the breaking of chains of…thought that bound Americans in the late nineteenth century to precepts and assumptions that militated against reform," such as classical laissez-faire liberalism. "So long as man believed that the enormous gap between rich and poor was the natural result of an inexorable process of selection ordaining that some would succeed but others must fail, dogma hemmed in those who sought change." To translate, Caine hails Progressivism as a foil to his social Darwinism straw man. "Until men reexamined their belief" that the free reign of "corporations was good for the country, . . . reformers had few tools to use in forging a new society." Ah, but, thankfully, "when, at last, the rigid bonds of tradition and dogma were broken," the progressive "advocates of constructive change could act effectively."[2] Thomas K. McCraw proclaims, "In securing their limited list of reforms, the progressives worked very hard… That so many of them succeeded was progressivism's chief legacy for them, and one of its many meanings for the future."[3] R. Laurence Moore, who denounced William Graham Sumner, additionally rhapsodizes about how "Woodrow Wilson, Charles Beard,… Charles Horton Cooley, E. A. Ross,… and John R. Commons tried in their respective fields to bring American scholarship of age."[4] There is a tremendous irony in this. The Progressive movement, more than any other turn-of-the-twentieth-century social and political movement, spawned the prestige of government-imposed eugenics. As Thomas C. Leonard is aware, a veritable "*Who's Who* of Progressive Era political economy appealed to eugenics...to justify the economic reform legislation so characteristic of the time."[5] To the degree that government-mandated eugenics can be called social Darwinism, the early twentieth-century's Progressives were not opponents of social Darwinism but its foremost practitioners.

And two famous progressives who have dedicated themselves to the compulsory eugenicist cause were New-Jersey-governor-turned-U.S.-President Woodrow Wilson (1856–1924) and attorney cum U.S. Supreme Court Justice Louis D. Brandeis (1856–1941).[6] Tellingly, despite its in-depth study of government-regulatory eugenics, *SDAT* elicits no note of either man's contributions to the eugenicist cause. Instead *SDAT* celebrates both of them as staunch opponents of social Darwinism, again because of their aversion to the night watchman state. Richard Hofstadter writes of the Progressive Era's intellectuals emerging in a "renaissance" of ideology who, "in sympathy with the new spirit of the Progressive Era," had "assailed" laissez-faire social Darwinism's "old scheme of thought . . ." Among the participants in this Renaissance were "Charles A. Beard, . . . John R. Commons, John Dewey, . . . Louis D. Brandeis, and Oliver Wendell Holmes." Remember the name John R. Commons; we will return to him within this chapter. Anyhow, for the aforementioned reasons, the Columbia historian beams, "Brandeis opened up new possibilities in law by drafting for the first time a factual sociological brief in defense of a state law regulating conditions of labor in private enterprise."[7]

In that same spirit is Thomas K. McCraw—a Pulitzer Prize-winning professor at Harvard Business School. McCraw venerates Brandeis as "the progressive 'People's Lawyer' who . . . played a significant part in the birth of the Federal Trade Commission."[8] R. Laurence Moore, too, commends Brandeis for having "arose before the Supreme Court to argue the constitutionality of a ten-hour law for women workers" in 1908's *Muller v. Oregon* "with cartons of data gathered by experts. . . . He made himself master of material that the masses could never understand, and, in so doing, became their servant."[9] Echoing Moore, A. A. Ekirch states, "The Progressive attorney Louis D. Brandeis, in a brief replete with sociological evidence demonstrating the adverse effects of long hours, played a key role in convincing the court, in *Muller v. Oregon* (1908), to rule in favor of maximum-hour laws." He additionally applauds Brandeis as "the liberal attorney and foe of monopoly . . ."[10]

And George Mowry summarizes a philosophy not unlike his own when he narrates, "The United States, said the future justice of the Supreme Court, Louis D. Brandeis, was rapidly becoming a nation in which the privileged few at the top had great freedom and all the rest of the people were 'more and more a class of employees.'"[11]

A similar verdict arrives via business consultant and governist ideologue Robert A. G. Monks. Monks was prominently featured in the 2003 anti-capitalist propaganda documentary *The Corporation: The Pathological Pursuit of Profit and Power*. In one of his books Monks approvingly delivers this statement from Brandeis that solidifies the justice's credentials as a progressive: "There is no such thing to my mind . . . as an innocent stockholder. He may be innocent in fact, but socially he cannot be held innocent. He accepts the benefits of the [commercial] system. It is his business and obligation to see that those who represent him" in the private company wherein he owns shares "carry out a policy which is consistent with the public welfare."[12] This item explains why Brandeis was such an avid fan of Henry Demarest Lloyd's screed against the Standard Oil Company—*Wealth Against Commonwealth*.[13] Daron Acemoglu and James A. Robinson also celebrate the man. "Another key muckraker was lawyer and author Louis Brandeis, who would later be named Supreme Court justice by President Wilson. Brandeis outlined a series of financial scandals in his book *Other People's Money and How Bankers Use It*, and was highly influential on the Pujo Committee."[14]

But recall another governist appearing in *The Corporation*, Edwin Black. This time, Edwin Black discloses a fact about Justice Brandeis that Hofstadter and Monks failed to disclose. It is this. As a U.S. Supreme Court Justice, Brandeis—who, in Hofstadter's words, "assailed" social Darwinism—voted the same ruling as Justice Holmes in 1927's *Buck v. Bell*. Both Brandeis and Holmes affirmed the legality of U.S. state governments' forcible sterilization of U.S. citizens.[15] Had Brandeis verily been the "lawyer" for all of "the people," then in this case he let his clients down. To give you an idea of how much Justice Brandeis had betrayed freedom with this decision, I will quote Matt Ridley. Siding with Holmes and seven other justices, Brandeis "ruled that the commonwealth of Virginia could sterilise Carrie Buck, a seventeen-year-old girl committed to a colony for epileptics and the feeble minded in Lynchburg, where she lived with her mother Emma and her daughter Vivian. After a cursory examination, Vivian, who was seven months old (!), was declared an imbecile and Carrie was ordered to be sterilised. . . . Vivian died young, but Carrie survived into old age, a respectable woman of moderate intelligence who did crossword puzzles in her spare time. Her sister Doris, also sterilised, tried for many years to have babies before realising what had been done to her without her consent."[16] Brandeis did not offer his own opinion for the case. Nonetheless, he did cite Holmes's words from *Buck v. Bell* to buttress his own opinion in a separate case a year afterward.[17] That later case, *Olmstead v. United States* was, ironically enough, one in which Brandeis came down on the laissez-faireist side. In *Olmstead*, the majority ruled that federal agents were allowed to perform wiretapping on telephone conversations in the absence of a warrant. Here, Brandeis correctly remarked that this violated the Fourth Amendment. With stirring eloquence, Brandeis wrote for this case, "The makers of our Constitution undertook to secure conditions favorable to the pursuit of happiness. They recognized the significance of man[: . . .] intellect," meaning the right of man to act peaceably upon his own judgment. The Founders therefore understood the need to protect, "against the Government, the right to be let alone—the most comprehensive of rights, and the right most valued by civilized men. To protect that right, every unjustifiable intrusion by the Government upon the privacy of the individual, whatever the means employed, must be deemed a violation of the Fourth Amendment."

The truth and eloquence of those words compound Brandeis's hypocrisy. Why did he not respect Carrie Buck's "most comprehensive of rights"—the "right most valued by civilized men"? That would be Carrie's right to "privacy" as an "individual." Why did Brandeis not obey the Fourth Amendment and respect the girl's "right to be let alone"? As I said, of course, Brandeis cited *Buck v. Bell*'s majority decision in this very same dissent. He cited his own participation in that decision to concede that he and his fellow justices have a rightful license to impose "regulations which, 'a century ago, or even half a century ago, probably would have been rejected as arbitrary and oppressive.'" To afford himself additional wiggle room, Brandeis asserted that "clauses" written with "individual protection against specific abuses of power" in mind still "must have a . . . capacity of adaptation to a changing world."[18] To translate, Brandeis and the other justices may capriciously re-interpret the Bill of Rights at their own convenience. Let me rephrase what Brandeis uttered about corporations. There is no such thing to my mind as a judge who can be considered innocent after having victimized a teenage girl in the manner that Justice Brandeis did.

In defiance of the truth, Brandeis's reputation as a saint lives on. *The American Heritage Dictionary* has an entry defining him as an "American jurist who served as an associate justice of the U.S. Supreme Court (1916–1939). His opposition to monopolies and *defense of individual human rights* formed the basis of many of his high court decisions"[19] (emphasis added). Something tells me that Doris Buck might beg to differ with the claim that Brandeis was a consistent defender of individual human rights.

Particularly interesting is the sort of terminology that Brandeis relied upon in his arguments for antitrust legislation. Modern governists pay homage to Brandeis especially on account of Brandeis preparing the expansion of antitrust regulation over business. What comes to my attention is this. When Brandeis wrote on this matter in *Collier's* magazine, he wrote that the usual competition among a multitude of one-man-operation businesses was "the survival of the fittest," and that this "survival of the fittest" was actually *good*.

Yet Brandeis brought into the old fallacy that this normal sort of competition, left unregulated by the government, inevitably gave way to a situation where a single monopoly could dominate an industry. Allegedly, this monopoly would arise by at least one of three methods. First, various firms in the same industry could agree to restrict output and collude on price, acting as a cartel. Secondly, these firms could officially agree to merge into a single firm. Thirdly, a single, highly capitalized firm could under-price its competitors into extinction, or purchase the smaller competitors. This assumption about laissez faire creating monopolies dates back at least as far as Karl Marx, and, as we have witnessed in Book Two, was repeated by George Orwell. And as we also remember from Book Two, there is scant historical evidence of this actually occurring.

Anyhow, Brandeis swallowed this fallacy at face value, and thus concluded that antitrust was required to maintain a competitive atmosphere. As Brandeis wrote in *Collier's*, "The purpose" of competitors combining into a single firm "has often been to curb efficiency, thus frustrating the natural law of the survival of the fittest."[20] As we recollect from *Life in the Market Ecosystem*, Hofstadter is quick to judge railroad owner James Jerome Hill as a social Darwinist primarily on the basis of Hill happily describing his industry's competition as *survival of the fittest*. When Brandeis employs that same phrase—with similarly positive connotations, and in a similar context—he escapes being labeled a social Darwinist, on account of his progressive politics.

One might misconstrue from the previous statements that Brandeis wants competition among businesses in the same industry to be maximized. Yet Brandeis presumes that there can also be too much competition. On May 27, 1913, he wrote to Robert M. La Follette, "There must be reasonable restrictions upon competition else we shall see competition destroyed."[21] Given his presumption that allegedly too much competition will result in monopoly, it follows that Brandeis approved neither of very little competition nor too much; he simply presumed that the proper amount of it could plausibly be set by the State. Actually, Brandeis delivered rather incoherent statements on several other topics as well. He opined, "Unlicensed liberty leads necessarily to despotism and oligarchy."[22] And chauvinism was implicit in the arguments that Brandeis gave for limiting the working hours of women in *Muller v. Oregon*. As journalist Damon W. Root phrases, Brandeis argued "women required special protection from the state. In fact, Brandeis argued, since women were responsible for bearing future generations, their bodies were in some sense collective property."[23] To quote Brandeis himself, "When the health of women has been injured by long hours, not only is the working efficiency of the community impaired, but the deterioration is handed down to succeeding generations. . . . The overwork of future mothers thus directly attacks the welfare of the nation."[24]

The U.S. Supreme Court agreed with Brandeis's sexism. As Justice David Josiah Brewer (1837–1910) stated in the ruling opinion, ". . . history discloses the fact that woman has always been dependent upon man. . . . As minors, thought not to the same extent," every woman

> has been looked upon in the courts as needing especial care that her rights may be preserved. . . . she is not an equal competitor with her brother. . . . she is not upon an equality. . . . It is impossible to close one's eyes to the fact that she still looks to her brother

> and depends upon him. Even though all restrictions on political, personal, and contractual rights were taken away, and she stood, so far as statutes are concerned, upon an absolutely equal plane with him, it would still be true that she is so constituted that she will rest upon and look to him for protection; that her physical structure and a proper discharge of her maternal functions—having in view not merely her own health, but the well-being of the race—justify legislation to protect her from the greed as well as the passion of man. The limitations which this statute places upon her contractual powers, upon her right to agree with her employer as to the time she shall labor, are not imposed solely for her benefit, but also largely for the benefit of all.

And Brewer remarked that this was consistent with the increasingly respected opinion that the "liberty" of free contracting "is not absolute, and extending to all contracts, and that a State may, without conflicting with the provisions of the Fourteenth Amendment, restrict in many respects the individual's power of contract."[25] The Center for American Progress's (CAP's) pamphlet *The Progressive Tradition in American Politics* lionizes Brandeis as saintly progressive and mentions none of his misogyny.[26]

The Eugenicist Argument for Regulation

Also interesting, another of the justices ruling for Carrie Buck's sterilization was former President William Howard Taft.[27] Some left-wing professors decry Justice Taft as a conservative.[28] William Howard Taft was the father to future U.S. Sen. Robert A. Taft (1889–1953) who tended to favor deregulation over governism.[29] Additionally, one might find it incriminating to Sumner that the two Tafts—both father and son—were students of Sumner's at Yale.[30] Despite his son's reputation, though, William Howard Taft was yet another Progressive. Massachusetts-based scholars Daron Acemoglu and James A. Robinson admit that Taft "prosecuted trusts even more assiduously" than did TR.[31] Arthur Ekirch admits this as well.[32] President Taft initiated twice as many antitrust suits in one presidential term as Theodore Roosevelt did in two.[33] Taft, observes Ekirch, "was by no means the service creature of big business depicted" by progressives who were farther Left of him politically.[34] Additionally, in 1909 Taft implemented the USA's first federal corporate income tax.[35]

Just as it did with Justice Brandeis, *SDAT* stresses Woodrow Wilson's bona fides as a slayer of social Darwinism. *SDAT* happily quotes this passage from President Wilson's *New Freedom*: "The man with only a little capital is finding it harder to get into the field, more and more impossible to compete with the big fellow. Why? Because the laws of this country do not prevent the strong from crushing the weak." Wilson, of course, promised that while in office he would rectify such injustice.[36]

Wilson also stated, "There has been something crude and heartless and unfeeling in our haste" to "succeed" financially "and be great. Our thought has been, 'Let every man look out for himself, let every generation look out for itself,' while we reared giant" corporate "machinery which made it impossible that any but those who stood at the levers of control should have a chance to look out for themselves." How fortunate, Wilson decides, that we "have come now to the sober second thought."[37]

For the latter set of words, Wilson wins deep praise from governist historian James Truslow Adams. "Here once more," says Truslow, "was the authentic voice of the great American democracy; here once more was the prophet speaking of the American dream, of that hope of a better and richer life for all the masses of humble and ordinary folk who made the American nation. It was the voice once more of the democratic frontier . . ." Wilson is great, pronounces Truslow, for having provided "the first graduated income tax," the Federal Reserve Act, and an antitrust provision known as the Clayton Act.[38]

Albert Einstein holds this president in high regard—"Among the most important statesmen it is probably Wilson who most represents the intellectual type."[39]

We must fill in an omission that Hofstadter and Truslow Adams leave about this Commander-in-Chief. Back when he was governor of New Jersey, Wilson implemented a state policy of compulsory sterilization of epileptics and the mentally ill beginning on April 21, 1911.[40] Governor Wilson received help from a state government agency he founded, the Board of Examiners of the Feebleminded, Epileptics, and Other Defectives.[41] Meanwhile, Arthur Ekirch concedes that Wilson was the first U.S. president to institute racial segregation protocols formally for every federal U.S. office.[42] Incidentally, Woodrow Wilson stated that the economic doctrines of his New Freedom were largely inspired by the teachings of Louis D. Brandeis.[43] More tellingly, the man whom Wilson, as New Jersey's governor, had asked to draft his sterilization law was Edwin Katzen-Ellenbogen. Years later, Katzen-Ellenbogen would move to Europe and actually serve as a doctor to the SS in Nazi-occupied France and as an adviser at the Buchenwald concentration camp.[44] When Arthur Ekirch praises Wilson as a man who "conceded that government was needed to ensure fair play for the individual,"[45] Ekirch must not have been thinking of the nonwhite and epileptic individuals whom Wilson's government had denied fair play. The Center for American Progress (CAP) admits to regretting that Wilson opposed women's suffrage and segregated government offices, but otherwise hails him as a magnificent Progressive and overlooks the compulsory sterilization initiatives. Insofar as the CAP criticizes Wilson, it criticizes him for being too right-wing and for his distrust toward the Communist Party of the USA.[46]

Nor were Wilson and Brandeis benign examples for nations across the Atlantic. I do confess that Germany's Social Democrat party deserves some commendation for being the sole political party to vote against the 1933 Enabling Act that gave Adolf Hitler supreme legislative power. Nonetheless, this same German Social Democrat party promoted governist eugenics. In 1927, the Social Democrats of Prussia unsuccessfully filed a petition for the Prussian government to procure data and statistics on the results of the U.S. state governments' compulsory sterilization policies. This was on account of the Social Democrats' desire to emulate these enactments. "The Social Democrats' initiative," sociologist Stefan Kühl informs us, "signaled the importance" of Wilson's compulsory sterilization initiative "as a role model for Germany, which also indicated that interest in such legislation" existed on "the left of the political spectrum."[47] It is indeed no accident that as the nation of Sweden began to elect self-described socialists to its Parliament, it likewise increasingly looked to the State to advance eugenicist goals. Writes Cincinnati pediatrics professor Steven Potter, "the world's second-largest sterilization program was conducted in Sweden . . ." This law went into effect in 1934,[48] the same year that the Nazis gained power,[49] and was not repealed in Sweden until 1976.[50] In the interim in which this regulation was in effect, over 60,000 Swedes were forcibly sterilized.[51] The victims of this surgery included the usual suspects—epileptics, the mentally challenged, and those working in the black market.[52] But in Sweden, not everyone in these categories was legally required to be forcibly sterilized; one could, in lieu of receiving such surgery, agree to be segregated from the rest of society until his or her childbearing years had ended.[53]

Yet another progressive eugenicist was John Maynard Keynes.[54] Remember, from Book Two, that in his much-revered treatise *The End of Laissez Faire*, Lord Keynes recommended controlling the population's size and its genetic makeup. "The time has already come," he announced therein, "when each country needs a considered national policy about what size of Population…is most expedient. And having settled this policy, we must take further steps to carry it into operation." Likewise, the "community as a whole must pay attention to the innate quality"—the biological characteristics—". . . of its future members."[55] Acting upon his advice, the lord served as the vice president of the Eugenics Society of London, and sat on its board of directors from 1926 to 1944.[56]

Keynes's Eugenics Society of London had several equivalents in the USA. One of those was the American Eugenics Society, of which a fellow Malthus admirer, Dr. Garrett Hardin, served as vice president.[57] Hardin also sat on the advisory board of the Repository for Germinal Choice. This repository was a special type of sperm bank that initially accepted genetic deposits from no one except Nobel laureates and Olympian. Moreover, it initially provided access to this sperm to no one except women of high IQs. That was due to the founder's assumption that genetics determined IQ, and that IQ likewise determined moral character.[58] I do not have much trepidation about this sperm bank's existence, as every interaction with it was strictly consensual. Sadly, Dr. Hardin eventually reached a point at which he implored that persons with low IQ scores ought to be sterilized. That way, low IQ scores would not bedevil tomorrow's generations. "There seems to be," Hardin stated, "little danger of society's being deprived of something valuable by the sterilization of all feeble-minded individuals. . . . Sooner or later . . . human population will reach a limit. . . . Sooner or later, not all the children that humans are willing to procreate can survive. Either there must be a relatively painless weeding out before birth or a more painful weeding out after birth. . . . If we neglect to choose a program of eugenics, will the production of children be nonselective? . . . People with low I.Q. are reproducing at a faster rate than those with high I.Q."[59]

Hardin tried to convince other men of high IQ that the majority of human beings are worthless. As he assumed that virtue was genetically inborn, he implored, "Look around you. How many heroes do you number among your neighbors? Or your colleagues? . . . Where are the heroes of yesteryear? Where is Sparta now?"[60] Thus we confront the perverse injustice in how Herbert Spencer and Garrett Hardin have been treated in nonfiction literature. Spencer is smeared as a eugenicist who promulgated that the destitute deserved euthanasia and nothing else. In reality he explicitly recommended private charity and, by logical implication, opposed the governist eugenicists' effort to pass laws to control population.

Conversely, Hardin actually did propagate the governist eugenicist argument that the human race would profit if low-income individuals died before breeding. He said so quite loudly. Likewise, Hardin would have it that the U.S. government proscribe people from voluntarily helping the poor. He publicly supported other eugenicist causes as well. The Malthusianism of both Keynes and Hardin were in lock-step. The latter man receives accolades, sometimes from some free-market libertarians, just because he coined that phrase they love so much: "the tragedy of the commons." Spencer did not say that poor people bear a duty to die before reproducing, yet he gets flack for saying it. By contrast, left-wing environmentalist Hardin, who did say that the poor have a duty to die, does not get flack for saying it.

I commend Princeton University historian Eric Goldman (1915–1989) on one count. In 1956, he was among the first governist intellectuals to call attention to the gulf between the free-market evolutionists (Spencer and Sumner) and the state eugenicists. Goldman noticed that, of course, both groups invoked evolutionary theory to advance their own public-policy prescriptions. More importantly, though, Goldman noticed that both groups advocated completely opposite political philosophies. Sadly, Goldman has not been very helpful when it comes to choosing labels for the two groups. He simply drubbed the ideology of Spencer and Sumner as *conservative Darwinism*. He distinguished the state eugenicists from them by affixing the new label of *reform Darwinism* to the state eugenicists. To Goldman, relative free enterprise was America's default position. Hence, Goldman inaccurately deemed Spencer and Sumner to support a free-market status quo, ignoring that pure free enterprise had never existed in the USA's history. Continuing along that line of thought, Goldman considered the Progressive eugenicists to be pushing for helpful *reforms* of the capitalist system. That is the supposed *reform* in *reform Darwinism*.[61] In hindsight, it would have been more helpful had Goldman dispensed with the social Darwinism label altogether.

Racism from the So-Called Progressives

Other calls for eugenicist legislation arrived from American journalist Ray Stannard Baker (1870–1946). Only partially forthcoming about Baker's eugenicism, George E. Mowry reveres Baker for being among the muckrakers who "attacked the evils of . . . big business, Wall Street, life insurance, . . . the food industry, child labor, women's inequality, prostitution, and the drug trade. Heavily factual in content, critical in tone, and full of righteous but optimistic indignation, the average muckrake article . . . sought to give the average citizen a scientific description of what was wrong with the varied sectors of American life. Taken as a whole, the impact of the muckrake literature was enormous. Before this journalistic crusade had run its course, few literate Americans could have any feelings of complacency about their civilization." Mowry gives Baker especial credit for churning out "articles attacking the railroads and the recalcitrant conservatives in the Senate who supported them."[62]

Recall that DePauw University historian Stanley P. Caine praises progressivism as a repudiation of social Darwinism. Caine also pegs Ray Stannard Baker as one of the "leading progressives." His heart is warmed by how Baker's journalism created publicity for a civic activism campaign by progressive businessman Jacob Sechler Coxey, Sr. (1854–1951). Coxey wished the government would hire the poor to participate in more public-works projects.[63] Basically this campaign would consist of governmental extortion of tax money to pay workers for providing services. This taxpayer subsidy would be awarded irrespective of whether such services were in demand. It would further be irrespective of whether this enterprise turned a profit or ran at a loss.

And Arthur Ekirch misleadingly asserts that a series of articles by Baker on Jim Crow regulations, *Following the Color Lines*, "reported objectively the continuing extent of race prejudice and discrimination in both the North and South in the 1900's."[64] One would gather from Ekirch's remark that Baker was an early civil-rights crusader pushing for racial equality and racial integration. In reality Baker argued that according to his (mis)understanding of natural selection, it was morally imperative to retain racial segregation laws in the U.S. South. Whites and blacks, he concluded, were, by default, inherently violent toward one another. Racial conflict, Baker surmised, was jungle law and survival of the fittest. Baker proposed forced segregation as a remedy to this Darwinian struggle.

In *Following the Color Line*, Baker characterized the travails of African-Americans under Jim Crow regulations as "the struggle of a backward race for survival within the swift-moving civilization of an advanced race." In his mind, of course, whites were that more "advanced race." If a racial war comes, Baker continued, "The fittest here . . . will survive (there is no escaping the great law!)..."[65] Since he wanted to delay that race war's arrival, Baker advised that blacks and whites be kept as far apart from one another as possible. This suggested that Jim Crow laws "are at present necessary to avoid the danger of clashes between the ignorant of both races. They are the inevitable scaffolding of progress."[66] Yes, the same journalist whom Dr. Caine praised for being among the "leading progressives" considered Jim Crow laws to be a step of "progress." As Robert Bannister summarizes it, it was within "this framework" that Baker "justified disfranchisement and Jim Crow despite the fact he had evidence that the new restrictions exacerbated struggle rather than mitigating it."[67]

It cannot be denied that William Graham Sumner and Herbert Spencer held some condescending opinions of nonwhite races. This, however, was not something they attributed primarily to genetics or inborn biology, but to nonwhite races coming from cultural backgrounds quite dissimilar from the Western Enlightenment tradition. To be sure, Spencer did believe that two parents' freely chosen customs could influence the overall makeup of their children's inborn predispositions. In the end, nonetheless, when Spencer and Sumner concluded that valid generalizations could be induced about an ethnic group's behaviors or ethical standing, those generalized behaviors resulted mostly from cultural conditioning, not mostly genetics.[68] Hence, Bannister concludes, "however objectionable to modern standards" were the condescendingly "paternalistic . . . philanthropy and education" that "Herbert Spencer and his American disciples" suggested for nonwhites, that attitude was "moderate, humane, and constructive by comparison with the theories and practices of the Jim Crow era" that Ray Stannard Baker *did* support.[69]

Also contrast Baker's attitude against that of another group of people notoriously slandered as social Darwinists—big businessmen. According to popular myth, wealthy corporate executives are prone to bigoted discrimination. Malcolm X blared, "You can't have racism without capitalism. If you find antiracists, they're usually socialists or their political philosophy is that of socialism." In fact, Southern Illinois University historian Jonathan J. Bean has amassed a collection of historical data evincing that numerous major corporations lobbied against Jim Crow and segregation. Upon failing to amend these racist policies, some of these businesses outright violated these laws. They had an economic incentive to do so. In their quest for additional dollars, they observed enormous untapped markets in the form of racial minorities whose consumer demand was not being serviced adequately by the competition. As Coca-Cola ignored the enormous potential for African-American customers, Pepsi-Cola eagerly marketed to them in the South, despite such marketing being illegal at the time. In 1898 and 1900, the Augusta Streetcar Company desired to serve black passengers and hire the ablest employees, no matter what race they may be. Consequently, the company's president, D. B. Dyer, and its counsel, Boykin Wright, wrote letters to city officials requesting the repeal of segregation.[70]

Consider an anecdote from the life of Chris Gardner (b. 1954). As he details in his memoir *The Pursuit of Happyness*, he rose from a life of homelessness to one as a millionaire. Of course, that ascension was not easy. When Gardner worked at the brokerage Bear Stearns, he handled a high-rolling client, whom in *Pursuit* he identifies as "J.R.," with whom he initially dealt only over the telephone. Gardner executed various trades with J.R. and made millions of dollars for him. But, to Gardner's consternation, it never occurred to J.R. that Gardner was black. On the assumption that Gardner was white, J.R. continually cracked bigoted jokes to Gardner about blacks and Jews. Eventually J.R. requested that they meet in person. Gardner feared that once J.R. saw him, J.R. would refuse to work with him ever again.[71] Instead, Gardner narrates to TV reporter Bob Brown, J.R. "closed every other account that he had. We did all of his business until the day he died. . . . That's when I learned in this business, it's not a black thing; it's not a white thing. It's a green thing." The capitalist attitude, Gardner paraphrases, is: "If you can make me money, I don't care what color you are."[72] That principle continues to apply throughout the market economy.

Similarly, it was business acumen, more than it was legislation, which led to social progress for women. The first time that a woman was ever hired as a detective in the United States, the decision was not made by a progressive government agency but a profit-hungry business. The business was Allan Pinkerton's (1819–1884) National Detective Agency, and the woman was Kate Warne (1833–1868).[73] When no government agency would do so, Pinkerton sought assistance from Warne on account of her being highly qualified for the position, regardless of her gender.

Although Franklin D. Roosevelt was not as explicit as his equally presidential cousin in supporting eugenics, racial discrimination figured into his administration. His abstention from support of federal anti-lynching bills arose from an unwillingness to alienate himself from southern Democratic Congressmen. And for that same reason, FDR declined any opposition to the Jim Crow-era poll tax whose sole purpose was to prevent southern blacks from voting.[74]

Further consider the ideological bait-and-switch that *SDAT* pulls on the issue of early twentieth-century efforts to restrict immigration into the USA. Hofstadter writes disapprovingly of these efforts at immigration control, and I cannot blame him for that disapproval. I support open immigration to a radical degree. What bothers me, though, is that *SDAT* obscures the identities and political ideologies of the civic leaders most vocal and effective in championing immigration control. "The influx of a large immigrant population from peasant countries of central and southern Europe, hard to assimilate because of rustic habits and language barriers, gave color to the notion that immigration was lowering the standard of American intelligence, so at least it seemed to nativists who assumed that a glib command of English is a natural criterion of intellectual capacity."[75] For these reasons, the government instituted a literacy test that would determine which immigrants could stay in the USA and which could not. Concurrently, those from the English-speaking countries, who were racially WASP-ish, would ace this test. By contrast, southern European immigrants, who were more racially dissimilar from the native-born population, had a worse command of the English language. Hence, according to statistical probability, the literacy test would keep out more southern European immigrants than otherwise would be the case. This literacy test was—in practice but not in name—an ostensive effort to apply government force discriminatively to exclude specific races from mixing with the native-born population.

Judging by Hofstadter's tone in characterizing the situation, it would be unsurprising if the reader imagined that the literacy test's creators were stereotypical conservatives, à la Rush Limbaugh and Glenn Beck. When my tenth-grade U.S. history teacher railed about these immigration restrictions, she tried to conjure up the idea that the immigration test was implemented by Spencer-influenced, racist, conservative opponents of welfare. That is definitely the image in the mind of the Marion Roach we quoted in chapter 3 about social Darwinism and immigration. Moreover, adds law professor Paul Lombardo, "Even before his rise to power in Germany, Hitler praised U.S. immigration restrictions and condemned the automatic grant of citizenship extended to every child born in Germany as 'thoughtless [and] hare-brained.'"[76]

Here is a fact that Hofstadter omitted. The federal U.S. immigration literacy test was initially proposed by a progressive economist, Edward W. Bemis, in a March 1888 issue of *Andover Review*.[77] This is the same Bemis who, along with John R. Commons, receives praise from Hofstadter for being among the "progressive economists" who were wont "to criticize individualism . . ."[78] On the matter of immigration, Ray Stannard Baker and another Progressive, Sen. Albert Beveridge, display more eugenicist sensibilities. For anyone who doubts Albert Beveridge's bona fides as a progressive, I will lay them out. First off, he supported President Theodore Roosevelt's Meat Inspection Act, which provided the groundwork for federal regulation over the safety of domestically marketed food. Moreover, he led the movement to ban child labor and limit the workday to eight hours. He pushed to have the government control specific tracts of land to preserve them as wilderness, and he menaced the railroad trusts.[79] George Mowry confesses that Baker and Sen. Beveridge "with more or less emphasis" concurred with fellow progressive William Allen White (1868–1944) that immigration should be curtailed. Further, Mowry admits that Baker, Beveridge, and White took this stand with the explicit intention of keeping especial ethnicities from entering the USA.[80] William Allen White wrote in 1910 that

Americans are "separated by two oceans from the inferior races and by an instinctive race revulsion to cross breeding that marks the American wherever he is found."[81] Baker could not dispute that. Accordingly, Beveridge and White[82] additionally cheered for Western colonialism. Beveridge clearly thought that one ethnic group was entitled to dominate every other. "We [whites]," he stated in Boston in an April 1898 speech, "are a conquering race. In the Almighty's infinite plan . . . debased civilizations and debased races" must submit "before the higher civilization of the nobler and more virile types of man."[83] Forasmuch as I am from Hawaii, it especially catches my attention that Beveridge expressed revulsion toward Native Hawaiians, Filipinos, and Mexicans: "Why is it more difficult to administer Hawaii than New Mexico or California? Both had a savage and alien population: both were more remote from the seat of government when they came under our dominion than the Philippines are today."[84]

Given the preponderance of apologias that my tenth-grade U.S. history teacher issued on behalf of labor unions, she probably would not want me to bring up another inconvenient consideration. It is what Cornell University historian Vernon M. Briggs, Jr., discloses—"At every juncture, and with no exception prior to the 1980s, the union movement either directly instigated or strongly supported every legislative initiative enacted by Congress to restrict immigration and enforce its policy provisions."[85]

To throw support to its false accusation that Inventive-Period industrialists were social Darwinists, *SDAT* cites the works of progressive sociologist Charles R. Henderson. You may recall my quoting this passage from *American Thought* in Book Two, wherein I discuss Hofstadter's tarring of John D. Rockefeller, Sr., as a social Darwinist—"Successful business entrepreneurs apparently accepted almost by instinct the Darwinian terminology which seemed to portray the condition of their existence."[86] That very sentence is followed by an endnote that leads to the source that *SDAT* cites for that claim. The source cited is Charles R. Henderson's 1896 paper "Business Men and Social Theorists."[87] *SDAT* skips over the tidbit that Henderson was a progressive who himself embraced eugenicist governist policy. In the publication of the progressive American Economic Association, under the aegis of Richard T. Ely, Henderson argued that southern Europeans are "unfit."[88] Then and there, Henderson prescribed that southern European immigrants, in addition to the psychologically handicapped, be detained and thrown into internment camps. As Henderson put it, the "feebleminded and degenerate . . . can all be easily segregated in self-supporting rural colonies."[89] Into these colonies, Henderson proposed, we are to toss the "insane, feebleminded, and epileptics . . ." As for Henderson's proposal that particular races be obstructed from entry into the USA, he indicated, "This is not an argument against immigration, but only against the immigration of persons who can never be induced to demand a civilized scale of life."[90] Once the State has removed racial minorities and the psychologically handicapped from mainstream society, "the real workers" remaining in the workforce—the WASPs—"will more rise in earning power."[91] As for whether any such government policies are ethical, the progressive maintained, "It is clearly and distinctly the right of a commonwealth to deprive" these outcasts "of liberty and so prevent their propagation of defects and thus the perpetuation of their misery in their offspring. Therefore the policy of painless asexualization is offered . . ."[92]

Richard T. Ely himself offered eugenicist arguments for barring specific races from U.S. residency. The AEA founder implored, "we must notice the deterioration in the character of our immigration and ponder well the effects which a large admixture of baser foreign elements is likely to have upon American nationality."[93] Concordantly, "the fullest unfolding of our national faculties requires the exclusion of discordant elements—like, for example, the Chinese."[94]

Jane Addams, the progressive social worker, worked closely with one of the pioneers of the U.S. eugenics movement, Harry Olson. Olson was among the first judges for America's juvenile courts, as well as the founder of the American Eugenics Society and previously the president of the Eugenics Research Association. He, too, itched to segregate "the cheaper races" from his own ethnicity.[95] Likewise, an unsigned editorial in *The New Republic*, most likely written by Herbert Croly, proclaimed, "Laissez-faire as a policy of population leads straight to perdition. . . . Imbecility breeds imbecility as certainly as white hens breed white chickens; and under laissez-faire imbecility is given full chance to breed, and does so in fact at a rate far superior to that of able stocks. . . . We suggest that a socialized policy of population cannot be built upon a laissez-faire economic policy. So long as when the state neglects its good blood, it will let its bad blood alone. . . . When the state assumes the duty of giving a fair opportunity for development to every child, it will find unanimous support for a policy of extinction of stocks incapable of profiting from their privileges."[96] The Center for American Progress's series of pamphlets extolling Progressivism, *The Progressive Tradition*, hails Herbert Croly as a wonderful Progressive[97] but acknowledges nothing of Croly's eugenics. In fact, CAP tries to attribute "new restrictions on immigration" in the 1920s to the decline of the Progressives' influence,[98] glossing over the fact that progressives initiated those immigration restrictions.

The Rough Rider's Rough Rhetoric

Of special note is President Theodore Roosevelt—sometimes called *TR* for short—who had appointed Oliver Wendell Holmes, Jr., to the Supreme Court.[99] TR, the twenty-sixth president in the history of the United States, receives overall approval from the political Left on account of his attacks on free enterprise. "Whatever may have been Roosevelt's faults," judges James Truslow Adams, "and they were many and open to all men's view, I think it cannot be denied that he left the heart of the nation sounder and more wholesome than he had found it; and that is something of which few statesmen can boast."[100] With TR as President, cheers the CAP, salutary "reform" at last "had the national-level advocate it needed."[101]

Ah, but TR had the same views on immigration and foreign policy as did Ray Stannard Baker, Albert Beveridge, William Allen White, and Charles Henderson. If Robert M. Rees is right that confidence in free enterprise is the defining characteristic of social Darwinism, then one would be unable to place President TR in the rogue's gallery of social Darwinists. Arthur Ekirch recounts that "several successful antitrust cases early in his Administration—notably the breaking up of the Northern Securities Company, a railroad combine put together by E. H. Harriman, James J. Hill, and J. P. Morgan—won Roosevelt a reputation as a trustbuster and staunch Progressive. . . . Roosevelt's leadership contributed to such important legislation in his second term as the Hepburn railroad rate bill, the Meat Inspection Act, and the Pure Food and Drug Law."[102]

But this same trustbuster presents yet another stark example of a politically mainstream progressive roughly riding on the eugenics bandwagon.[103] In Teddy Roosevelt's own words, it is "obvious that if in the future racial qualities are to be improved, the improving must be wrought mainly by favoring the fecundity of the worthy types and frowning on the fecundity of the unworthy types. At present, we do just the reverse. There is no check on the fecundity of those who are subnormal, both intellectually and morally, while the provident and thrifty tend to develop a cold selfishness, which makes them refuse to breed at all."[104] TR further worried that the rate at which nonwhites bred had accelerated, whereas contraception had slowed down the white population's. Such behavior, Roosevelt warned in a 1911 article published in *The Outlook*, "means racial death."[105] *Race suicide* is something that TR shrieked was the "greatest problem of civilization."[106]

In 1906, the twenty-sixth President speechified to a uniformly white audience at Oxford University's annual Romanes lecture, "Let us hope that our own blood shall continue in the land, that our children and our children's children to endless generations shall arise to take our places and play a mighty and dominant part in the world."[107] He said elsewhere that "it is not only undesirable but impossible that there should be racial intermingling and the result is sure to bring disaster."[108] On January 3, 1913, he wrote to Charles Davenport, "I agree with you…that society has no business to permit degenerates to reproduce their kind. . . . Some day, we will realize that the prime duty, the inescapable duty, of the good citizen of the right type, is to leave his or her blood behind him in the world; and that we have no business to permit the perpetuation of citizens of the wrong type."[109] Finally, on the same issue, TR publicly stated in 1914, "I wish very much that the wrong people could be prevented entirely from breeding."[110] But anteceding TR's expression of that wish, Woodrow Wilson worked to bring that wish into reality. Due to their anti-capitalist rhetoric and governist actions, H. L. Mencken characterizes Theodore Roosevelt, William Jennings Bryan, and Woodrow Wilson as "demagogues,"[111] and admits that the emergency powers granted to Wilson during World War One made him a "dictator" in that period.[112] "Of all the men who lived in my time," Mencken wrote in his memoirs, "the most repellant to me was the late Woodrow Wilson; he was the archetype of every human trait that I detest and despise. . . . Next to him I rank Theodore Roosevelt . . ."[113]

In terms of ideology, TR kept some other interesting company. It will be recalled from chapter 2 that for his advocacy of governist eugenics, American zoologist Madison Grant received fan letters from Adolf Hitler. It should here be mentioned that Grant received fan letters from TR as well, and for the same reason.[114] In his own fan letter to Grant, TR waxed about the zoologist's manifesto, *The Passing of the Great Race*. The book advocated that white Europeans militarize against presumably and reflexively untrustworthy Asians. *Passing*, remarked the Rough Rider, is "a capital book; . . . in grasp of the facts our people [whites] most need to realize. . . . It is the work of an American scholar and gentleman; and all Americans should be sincerely grateful to you for writing it."[115] Castleton College historian Jonathan Peter Spiro names another trait that TR shared with Grant—the reason for which Grant "embraced eugenics was that the movement appealed to his progressive instincts." Interestingly, exactly as Grant beseeched the State to oppress specific groups of human beings, Grant also pioneered in calling for the State to preserve endangered species at the forcible expense of human landowners. Grant was good friends with Sierra Club founder John Muir, and the two collaborated in their activism. As we shall examine in chapter 9, that Grant wanted both to oppress specific races and to elevate endangered species above human property owners, is not as ironic as we may initially assume.

American Thought classifies Theodore Roosevelt as a social Darwinist. This contradicts Robert M. Rees's definition of a social Darwinist as being someone who evangelizes on laissez faire's behalf. *SDAT* apparently takes issue with the faith TR placed in militarism. "In his historical work, *The Winning of the West*," Hofstadter assesses, "Roosevelt drew from the story of the frontiersman's struggle with the Indians the conclusion that the coming of the whites was not to be stayed and a racial war to the finish was inevitable. . . . American development represents the culminating achievement of this mighty history of racial growth." *SDAT* asserts the "idea of inevitable Anglo-Saxon destiny figured" in TR's philosophy quite strongly. Subsequently, Hofstadter disapprovingly quotes from TR's 1899 speech, "The Strenuous Life," which pronounces, "We cannot avoid the responsibilities that confront us in . . . Cuba, Porto [sic] Rico, and the Philippines. All we can decide is whether we shall meet them in a way that will redound to the national credit. . . . The . . . over-civilized man, who has lost the great fighting, masterful virtues, . . . whose soul is incapable of feeling the mighty life that thrills, 'stern men with empires in their brains'—all these, of course, shrink from seeing the nation undertake its new duties. . . . I preach to you, then, my countrymen, that our country calls not for the life of ease but the life of strenuous endeavor. The twentieth century looms before us big with the fate of many nations. If we stand idly by, if we seek merely swollen, slothful ease and ignoble peace, if we shrink from the hard contests where men must win at hazard of their lives and at the risk of all they hold dear, then the bolder and stronger peoples will pass us by, and will win for themselves the domination of the world." In a shocking move for him, Hofstadter hints that, in the area of foreign policy, TR was a far bigger social Darwinist than William Graham Sumner. *American Thought*, for once, commends Sumner's pleas for world peace in the face of the famous trustbuster's saber-rattling.[116] Yet, having categorized Sumner as, in the final estimate, a conservative stooge, Hofstadter proves unable to provide a plausible explanation for the reasoning behind Sumner's numerous clashes against conservatives and then-mainstream Republicans. Hofstadter merely offers this quip—"we may wonder whether in the entire history of thought, there ever was a conservative so utterly progressive as this."[117]

What is not so shocking is that, as it excoriates TR for being a social Darwinist, *SDAT* leaves the progressivism of TR's domestic policies almost completely unexplored. Nor, conveniently, does *SDAT* observe that TR adduced decidedly progressive talking points when calling for more imperialism. As Texas A&M University historian H. W. Brands illuminates in his Pulitzer Prize-winning biography of the man, TR's 1906 Oxford speech "put a Rooseveltian spin on Social Darwinism, drawing analogies between the biological and historical spheres. The great nations . . . would remain great only as long as they served the interests not merely of themselves but of less advanced nations and peoples. Roosevelt called for imperialism with a conscience."[118] In Teddy's own words, "In the long run there can be no justification for one race managing or controlling another unless the management and control are exercised in the interest and for the benefit of that other race."[119] In its hagiography of TR, the CAP proves guilty of the same omission—TR is vaunted as terrific and no word is uttered about his explicit eugenics.[120]

To my vexation, Université de Québec à Montréal historian Greg Robinson (b. 1966) names particular free-market ideologues as those he deems worthy of blame in transmitting eugenicist ideas to Teddy Roosevelt. "Like most educated Americans of the period," Robinson proclaims, "TR's views on society were heavily marked by the work of Herbert Spencer, William Graham Sumner, and other Social Darwinists, who adapted Charles Darwin's biological theories of evolution and natural selection to the study of human society." Robinson then implicates these very same evolutionists of having "used the Darwinian principle of 'survival of the fittest' to justify notions of white supremacy . . ." The insinuation is that it was Spencerian white supremacy that persuaded TR that anyone of Japanese ancestry was inherently violent.[121] Ekirch similarly ventures, "An offspring of the Darwinian generation, Roosevelt accepted both Social Darwinism's concept of competition and struggle and Reform Darwinism's belief in man's ability to control his environment and effect beneficial changes."[122] And Teddy's younger cousin, Franklin D. Roosevelt—whom the CAP classifies as a "full-throated progressive"[123] —himself assimilated eugenicist premises.[124] FDR abided by a eugenicist rationale for opposing intermarriage between whites and those of Asian ancestry. FDR based his internment of Japanese-Americans on the presumption that every last one of them was likely treasonous. Greg Robinson considers FDR's internment order to be contradictory to the man's otherwise philanthropic New Deal. To Robinson, it is "perplexing" that an action as inhumane as internment could have been imposed by "a President justly celebrated for . . . his dedication to creating government programs to serve the needs of ordinary Americans."[125] Actually, there is philosophical consistency in both the internment and the New Deal's economic controls. Both the internment and the New Deal's economic controls involve governist subordination of individual rights to the ostensive security of the greater community. Fortunately, Ayn Rand understood the reality that Greg Robinson elides. Archivist Jeff Britting informs us, "Rand regarded the internment" of Japanese-American "citizens during World War II as an injustice arising from collectivist New Deal policies."[126]

It may seem tempting to write off regulatory eugenics as a field reserved exclusively to white supremacists. Yet Ray Stannard Baker possessed a nonwhite counterpart who, near the twilight of his life, likewise called for racial separatism and segregation. Ironically, this was also an African-American who had previously pioneered in the civil rights movement and had helped Oswald Garrison Villard (1872–1949),[127] owner of the by-then illiberal *Nation* magazine at the time, co-found the NAACP (National Association for the Advancement of Colored People).[128] This was W. E. B. Du Bois.

The Left's Ploys Concerning Du Bois

Should we trust what he has written on the matter, W. E. B. Du Bois would not approve of the manner whereby PC historians have lionized him. "One is astonished," he wrote, "in the study of history at the recurrence of the idea that evil must be forgotten, distorted, skimmed over." To the degree that the moral failings of revered historical figures have been minimized or whitewashed, "history loses its value as an incentive and example; . . . it does not tell the truth."[129] For argument's sake, let us assume that Du Bois means his own words and would have us apply his explicit principles as we examine Du Bois himself. Du Bois's remark about the bowdlerizing of revered historical figures is strangely prophetic, for they anticipate how Du Bois has come to be treated among practitioners of the social sciences. He was not the sainted crusader for political equality and morally just race relations that he has been vaunted to be.

Pulitzer Prize-winning Rutgers University historian David Levering Lewis informs his readers that Du Bois found himself on "the brink of demonizing white Americans" in his old age.[130] Du Bois's 1920 book *Darkwater: Voices from Within the Veil* contains a chapter titled "The Soul of White Folk." In this chapter, the author articulates his general hostility toward Americans of European ancestry. As is common with a multitude of the present's PC academicians, Du Bois gets away with generalizing that Caucasians, as a whole, are congenitally racist. "On the pale, white faces. . . . I see . . . hatred, a deep and passionate hatred, vast by the very vagueness" of the "expressions" on those faces. His words dripping in sarcasm, he says of Euro-Americans, "These super-men and world-mastering demi-gods listened...to no low tongue of ours ['we' being nonwhites], even when we pointed silently to their feet of clay. . . . We have curled our lips in something like contempt as we have witnessed glib apology and weary explanation" from whites in general. Accordingly, Du Bois anticipates the rejoinder that surely a U.S. citizen cannot be prejudged as evil simply because he is Caucasian. Du Bois preempts the rejoinder, "The number of white individuals who are practising with even reasonable approximation the democracy and unselfishness of Jesus Christ is so small and unimportant as to be fit subject for jest in Sunday supplements and in *Punch* [a humor magazine], *Life, Le Rive,* and *Fliegende Blätter.*"[131] Yes, Du Bois decided, by 1920, that most white Americans were innately bigoted. Accordingly, Du Bois gave up on his earlier calls for racial integration. By 1935 he reversed direction, proposing that blacks *should* be segregated from whites. This would shove blacks into an enclave and penalize interracial interaction. Du Bois huffed, "There exists...a chance for the Negroes to organize a cooperative State within their own group. . . . No sooner is this proposed than a great fear sweeps over the older Negroes. They cry 'No segregation'—no further yielding to prejudice and race separation. Yet any planning for the benefit of American Negroes on the part of a Negro intelligentsia is going to involve organized and deliberate self-segregation." Should Du Bois's will being done impel "an increase in segregation and prejudice, then that must come."[132]

A year earlier Du Bois concluded, "The thinking colored people of the United States must stop being stampeded by the word segregation . . ." Indeed, "there should never be any opposition to segregation pure and simple . . ."[133] Du Bois's writings overlook that *self* refers to an individual human being, not the collective that an entire race embodies. To describe the "self-confidence" of an entire race of persons—each person having his or her own unique level of confidence—is a contradiction in terms. If Du Bois's jaundiced opinion about Euro-Americans had not been clear by 1935, he spelled it out in plain English by 1962. "I am quite frank: I do not pretend to 'love' white people. I think that as a race they are the most selfish of any on earth."[134] He lamented that whites had "overrun the earth and brought not simply modern civilization and technique, but with it exploitation, slavery, and degradation to the majority of men."[135] This man eventually exhorted U.S.-born blacks to go "Back to Africa."[136] Here, Du Bois overlooks an essential fact. If a black person is born in the USA and lived here his whole life, and then makes a first visit to Africa, such a trip would not constitute a *return* to that continent.

Worse, Du Bois came to judge that the United States, in its venal doings, has "out-Hitlered Hitler . . ."[137] Du Bois's reasoning was that America was too laissez faire for his liking. "Capitalism," he growled, "cannot reform itself; it is doomed to self-destruction. No universal selfishness can bring social good to all."[138] And thus he rhapsodized that Karl Marx was "a colossal genius of infinite sacrifice and monumental industry . . ."[139] Du Bois also took pride in having "spent four hours" in the company of "Mao Tse-tung."[140] And his affinity for dictators extended to those of previous centuries as well; he considered Otto von Bismarck a personal hero.[141]

When Josef Stalin finally expired, it was quite possibly the most merciful act the Soviet strongman ever committed. Nevertheless, Du Bois eulogized Stalin as follows, evincing none of the sarcasm we detected in "The Soul of White Folk": "Joseph Stalin was a great man; few other men of the 20th century approach his stature. He was simple, calm and courageous. . . . But also—and this was the highest proof of his greatness—he knew the common man, felt his problems, followed his fate. . . . His reward comes as the common man stands in solemn acclaim."[142] Du Bois, the so-called champion of racial equality, conceded no mention of Stalin's persecution of Jews.[143] It appears that while Du Bois disliked white people in general, exemptions from such contempt were given to Stalin, Marx, and Bismarck.

Du Bois cannot be accurately pegged as a consistent advocate of racial equality, except in one respect. Du Bois applied eugenicist ideas both to whites and to blacks. Jonah Goldberg discloses that Du Bois "shared many of the eugenic views held by white progressives. His 'Talented Tenth' was itself a eugenically weighted term."[144] According to Du Bois, those who were the Talented Tenth were "the best of the race," who must "train and breed for brains, for efficiency, for beauty."[145] Talented or not, my tenth-grade U.S. History teacher demanded that I show gratitude to Du Bois. In her inaccurate account of history, Du Bois heroically spent his final days fighting for the equal rights of every ethnicity to coexist. My teacher conveniently neglected to acknowledge any of the aforementioned facts about Du Bois. Regrettably, she is not alone in culpability for such conspicuous evasions, though. Recall that *The Age of American Unreason* is the book in which Susan Jacoby unleashes jeremiads against "social Darwinism" and "eugenics." In this very same book, Jacoby sings praises both to FDR and to Du Bois, overlooking any recognition of their eugenicism.[146] And *SDAT* glides over any mention of the racial separatism pushed by these same two historical figures.

An interesting twist took place after Du Bois's pal Oswald Villard, had taken over the once-quasi-consensualist *Nation* magazine and changed its editorial tone to a thereafter governist one. Once Villard was in charge, the magazine predictably vaunted Madison Grant for his eugenicist views. The pages of *The Nation* sang that the *Passing of the Great Race* exhibited "distinct qualities of originality, conviction, and courage."[147] It is a bit peculiar that the brains behind the NAACP's founding had knowingly published a piece that glorified a white-supremacist tract.

Edward Ross, Social Control, and the Control of Ross's Reputation

In fact there is but one progressive eugenicist[148] whom *SDAT* openly identifies as both a "Progressive thinker" and someone who found the eugenicists' "biological data" to be "authoritative." This lone figure is sociologist Edward Alsworth Ross. Even then, to whitewash his own progressive ideological sympathies, Richard Hofstadter separates himself from Ross by qualifying that, beneath the progressive veneer, Ross must still have been "conservative" in his "general bias."[149] Subsequent to conferring so much praise upon Lester Ward for his anti-capitalism, Hofstadter, Ravitch, and Ekirch each confirm Ross to be an admirer of Ward's arguments against free enterprise.[150] Ross eventually married Ward's niece. Describing him as both a "supporter of Populism in his early days" and "later a friend of many muckrakers," Hofstadter writes that Ross "expressed in his formal writings the aggressive spirit of protest and reform. . . . In his early work Ross tore down the analogy between natural selection and the economic process, and condemned it as 'a caricature of Darwinism, invented to justify the ruthless practices of business men.' "[151]

George E. Mowry counts Ross among the great sociologists who assisted Ward in his "sustained theoretical attack against Sumner and his mentor Spencer" for their "social Darwinism."[152] Ekirch reports that Edward A. Ross's 1901 monograph *Social Control*, which was "dedicated to Ward and admired by Theodore Roosevelt, called for the modification of the Darwinian struggle for survival through orderly reform and organization. Laissez faire, he maintained in his later work *Sin and Society* (1907), was obsolete in the modern interdependent world." Ekirch expresses gratitude that "sociologists like Ward and Ross gave important intellectual encouragement to Progressivist ideals." Then Ekirch further adds that Ross, like Ward, was among the Progressives who "sought to control progress and sustain human values."[153] *Human values* is an interesting phrase to attach to what Ross and Ward advocated. Recall from Book One that Diane Ravitch quotes from Ross's explanation that his desire was for government schools "to collect little plastic lumps of human dough from private households and shape them on the social kneading board. . . . And so it happens that the role of the schoolmaster in the social economy is just beginning."[154]

Moreover, Ross argued on eugenic grounds for closing off the borders that foreigners crossed to reach the United States. "The progressive sociologist Edward A. Ross . . ." Ravitch scribes, "warned about the danger to American society of continued immigration from southern and eastern Europe and the Orient."[155] Ross specified in his own words the particular types he found objectionable: "Latins, Slavs, Asiatics, and Hebrews."[156] Ross warned that such groups "lack the ancestral foundations of American character, and even if they catch step with us they and their children will nevertheless impede our progress.'"[157] Newspapers quoted Ross zealously advocating, "Should the worst come to worst, it would be better for us if we were to turn our guns upon every vessel bringing [Asiatics] to our shores rather than permit them to land."[158] In fact, observes Stefan Kühl, it was Ross who coined the expression *race suicide* in 1901, and it was Theodore Roosevelt who soon adopted it.[159] When Ross was publicly criticized in his own lifetime for supporting governist eugenics, yet another famous progressive—economist Richard T. Ely—came roaring to his defense.[160]

I am therefore befuddled when *SDAT* describes both Ross and progressive sociologist Charles Horton Cooley as men who would not "worship at the shrine of the fittest" and who "refused to look upon the poor as unfit . . ."[161] This claim by *SDAT*, notes Thomas C. Leonard, "was nonsense. Ross, especially, was quite happy to worship at the shrine of the fittest, provided the fittest were selected by state experts."[162] And Ross most certainly dismissed, as "unfit," the hundreds of penniless families coming in from the Old World trying to attain a better life.

Comment must also be cast concerning Charles Cooley, whom Arthur Ekirch also happily identifies as a progressive. Ekirch writes that Lester Frank Ward's "social theories were given greater precision in works like Charles H. Cooley's *Human Nature and the Social Order* (1902) . . ."[163] R. Laurence Moore joins Ekirch and *American Thought* in the hagiographic treatment of Cooley. Subsequent to denouncing Sumner as a social Darwinist and setting Sumner up as progressivism's foil, Moore moves on to Cooley the progressive. Moore pronounces that Cooley, unlike free-market evolutionists, "knew the difference between higher and lower human instincts as they emerged in a social context, and his goal was a society that allowed men to live a rational and moral existence."[164] But Leonard observes that Cooley promoted governist eugenics.[165] Interestingly, despite their disagreements on politics, Cooley expressed admiration for Herbert Spencer. Cooley commented that Spencer's *Descriptive Sociology* volumes "are much less known than they deserve to be."[166] And, incidentally, a mere four pages subsequent to proclaiming that Ross did not worship at the shrine of the fittest, *SDAT* states that the "biological data" of pro-eugenics research "were convincing to men like E.A. Ross, who had thoroughly repudiated Spencerian individualism."[167]

In his *In Search of Human Nature*, Pulitzer Prize-winning Carl Degler similarly contradicts himself about Ross. First Degler lauds him as a "social reformer" who "set forth" a thoroughly "forthright repudiation of social Darwinism." Later in that same book, though, Degler rues that Ross "supported eugenics around the time of the First World War." Degler then censures Ross for "his racism." Some time afterward Degler continues, "As late as the 1920s, Ross was still accounting for invidious differences in behavior between blacks and whites on grounds of biology or heredity." Ross further "lent his name and reputation in support of certain eugenic pieces of legislation."[168] Yet nowhere in his own book does Degler confess that Ross learned about the importance of eugenics from the exact same Lester Ward whom Degler so highly reveres. This is the same Ward whom Degler's descriptions disguise as an anti-eugenicist. It is quite outlandish that *SDAT* and Degler initially portray Ross as social Darwinism's enemy, only to deride him as a eugenicist in subsequent pages. This consideration flies in the face of Hofstadter's statement that "eugenics . . . has proved to be the most enduring aspect of social Darwinism."[169] By that standard is Ross a social Darwinist or not?[170] Interestingly, Ross was one of the few academicians to use the phrase *social Darwinists* prior to 1947. In 1903, he criticized the convictions of free-market advocates, and thus spoke of "the master error of the social Darwinists . . ." That is, this arch-eugenicist referred to free-market advocates as social Darwinists in order to criticize them.[171]

By now it should be evident that it is inaccurate to characterize a large number of the Progressive Era's most vaunted reformers as the friends or champions of poor people in general. Most often, they beat the drums of reform for the ostensive gain of poor white Protestants, simultaneously expressing disdain and distrust for penurious non-Aryans.[172]

None of this is to imply that those who presently call themselves progressives, as of this writing, are necessarily racist on account of the progressives of the prior century being so. Nevertheless, it is a perverse distortion of the historical record that Spencer and Sumner are reviled as racists, whereas the original progressives are lionized as the foremost opponents of social Darwinism and, by implication, lionized as opponents of racism.

The Philosophic Roots of Government-Sponsored Eugenics

The hidden history of governist eugenics also uncovers the dark underbelly of minimum wage legislation. In Book Two, we went over how minimum-wage legislation contributes to unemployment. It particularly contributes to the unemployment of young, unskilled members of racial minorities. Some conservatives aware of this effect of minimum-wage legislation presume that the contribution to the unemployment rate counts as an "unintended consequence" of such regulation. In reality, the disemployment of racial minorities was an intended consequence of the Progressives who passed the earliest minimum wage laws at the turn of the twentieth century. The original crafters of U.S. minimum wage laws had openly stated their knowledge that their legislation would throw unskilled, low-income, racial minority members out of work. They justified these consequences on explicitly eugenicist and racialist grounds.

A case in point resides in John R. Commons, a self-described "socialist" economist of the University of Wisconsin[173] and a protégé of Richard T. Ely's.[174] Recall that *American Thought* hailed Commons as a participant in what began as a backlash against Sumner and eventually became a "renaissance in American social thought . . ."[175] There are various other sources listing Commons' credentials as a luminary of progressivism. Commons has been inducted into the Labor Hall of Fame. The Bureau of Labor Statistics' website contains an essay by Wisconsinite industrial relations professor Jack Barbash that hails Commons as a "progressive" and a "pioneer of labor economics."[176] One historian proclaims, "More than any other economist," Commons "was responsible for the conversion into public policy of reform proposals designed to alleviate defects in the industrial system."[177] R. Laurence Moore praises "John R. Commons and Richard T. Ely" for their quest to "seek to understand the mechanism of social behavior and organization and derive social thought from a description of human nature as it actually existed," in contrast to free-marketers whom, Moore implies, cared nothing for facts.[178]

What goes unmentioned is the dark side of Commons's agenda. In 1890 Commons estimated that 5.5 percent of whites in the USA were genetically "defective" and that 2 percent of whites ought to be forcibly segregated from the rest by law.[179] The laborite wrote in 1907 that when multiple ethnicities compete for employment, it is "competition of standards of living. The reason the Chinaman or the Italian can save three days' wages is because wages have been previously fixed by the greater necessities of the more advanced races. But competition has no respect for superior races. The race with lowest necessities displaces others." Commons thereupon concluded that "the Jewish sweatshop is the tragic penalty paid by that ambitious race." As for blacks, Commons judged that they were inherently incompetent and that nothing could prepare them for hard work except for their having been enslaved throughout prior centuries. In "the entire circuit of the globe," Commons stated, "those races which have developed under a tropical sun are found too indolent and fickle. . . The negro could not possibly have found a place in American industry" if he arrived in this nation "as a free man. . . . if such races are to adopt that industrious life which is a second nature to races of the temperate zones, it is only through some form of compulsion."[180] With these pretexts, Commons supported the minimum wage, as that would preclude unskilled black workers and Eastern European immigrants from underbidding native-born whites seeking employment. This sentiment also informed Commons's support for the Davis-Bacon Act, as less-skilled minority workers would be priced out of the market if employers were required to pay out higher prevailing wages.[181] This same Commons served as an adviser to Robert M. Lafollette, Woodrow Wilson, and Theodore Roosevelt.[182]

Similar arguments about labor law come from Royal Meeker, who served as U.S. Commissioner of Labor under President Woodrow Wilson. "It is much better to enact a minimum wage law," proclaimed Meeker, "even if it deprives" Eastern European immigrants "of work."[183] This deprivation of employment is exactly what socialist Sidney Webb aimed for in agitating for minimum-wage regulations in his native England. Speaking of racial minorities and the disabled, Webb stated, "Of all the ways of dealing with these unfortunate parasites, the most ruinous to the community is to allow them unrestrainedly to compete as wage earners."[184] Besides racial minorities, the list of persons whom Sidney and Beatrice Webb wanted forcibly removed from the workforce was quite long: "the aged, and the child-bearing women . . . the crippled, . . . the epileptic, the blind and the deaf and dumb, . . . and all those who are actually 'morally deficient.'" And the list includes persons "incapable of steady or continuous application, or who are so deficient in strength, speed or skill that they are incapable . . . of producing their maintenance at any occupation whatsoever."[185] Henry Seager, a progressive at Columbia University, stated, "The operation of a minimum wage requirement would merely extend the definition of defectives to embrace all individuals, who even after receiving special training, remain incapable of work." Logically, this raises the question of what should be done with immigrants who would remain unemployed as a consequence of the minimum wage being too high. Seager proposed the run-of-the-mill governist eugenicist deus ex machina. "If we are to maintain a race that is to be made up of capable, efficient and independent individuals and family groups we must courageously cut off lines of heredity that have been proved to be undesirable by isolation or sterilization."[186] To the extent that governist eugenics is unethical, the first arguments for the minimum wage law's establishment cannot be attributed to good intentions.

Moreover, writing in 2009, Thomas C. Leonard notices something interesting about this—"from *SDAT* one would know almost nothing of the leading role played by progressive economists (and other reformers) in advocating eugenically motivated immigration and labor legislation."[187] Tell me about it. Finally, let us not overlook Frances Perkins, the Secretary of Labor under FDR who was architect of the Fair Labor Standards Act (FLSA). It was the FLSA that forbid every U.S. state from having a minimum wage lower than that which was set at the federal level. It is by more than a mere coincidence that Frances Perkins was ideologically influenced by the same Mary Rumsey *née* Harriman who tirelessly promoted progressive, governist eugenics.

Indeed, there is a trait common to every left-wing eugenicist, from Lester Ward to H. G. Wells to Justice Brandeis to FDR and beyond. Each of them is fixated on providing social welfare to society. This is both in terms of collectivist economics and in improving the human species' collective genetic makeup. This consideration gives the lie to *SDAT*'s assertion that the governist eugenics movement was predominantly a right-wing ploy "to distract public attention from social welfare."[188] A slew of supporters of government-mandated eugenics in both the U.S. and Western Europe assumed that it was morally peremptory for governments to draft such legislation. They knew full well that they were spoliating individuals in their compulsory sterilization and immigration quotas. They accepted such spoliation as morally defensible because they, as with Hofstadter himself, took for granted that it was ethically imperative to sacrifice individual rights for the commonweal's social welfare.

Meanwhile, as these left-wing luminaries thrust eugenics upon the masses, we should take note of the biologists who fought governist eugenicism. Two of them were Ronald Fisher (1890–1962) and Raymond Pearl (1879–1940). They initially supported governist eugenics, but later turned against it. In their anti-eugenics phases, they were described as anti-racist "conservatives" who railed against the New Deal.[189] Ironically, one of the few socialists to rebuke State-sponsored eugenics in public was a scientist named Herbert Spencer Jennings (1868–1947).[190] He received this name as a result of his father being enthralled with evolutionary theory. For this same reason Herbert Spencer Jennings's brother was complementarily named Charles Darwin Jennings.[191]

The diametric clash between Spencerism and Progressive eugenics came to a head in a legislative battle that one British politician held against Karl Pearson and other socialist eugenicists. Government-enforced eugenics was first promulgated by Charles Darwin's cousin Francis Galton, and endorsed by Charles's own son Leonard Darwin (1850–1943).[192] Yet I would not be hasty in faulting the Darwin family, as a whole, for this movement. It was another cousin of Charles Darwin's who combated compulsory sterilization. That man was a Member of Parliament—Josiah Wedgwood IV (1872–1943), the First Baron Wedgwood.[193] He first served in the free-market Liberal Party and then the socialist Labour Party,[194] and was described by Matt Ridley as an overall "libertarian."[195] In 1912, Prime Minister Winston Churchill (1874–1965) and Dr. Pearson introduced to Parliament a bill—the first of its sort in England—for compulsory eugenic sterilization. Ridley notes that Wedgwood fortunately thwarted them because he objected to the bill "on the grounds of individual liberty." He was "the individual against the state." He emerged victorious over the bill then and again when it was revived the subsequent year.

The exchange between Wedgwood and Dr. Pearson over this issue encapsulates the fundamental nature of the conflict. Wedgwood adduced the same Lockean individual rights that Spencer did. Pearson thus retorted to him as would many other philosophically consistent socialists and progressives—"What is *social* is right, and there is no definition of right beyond that" (emphasis mine). Ridley rightly concludes that the state-imposed eugenics movement is not so much "an example of letting science, genetics especially, out of control . . ." Nay, it is really "an example of the danger of letting government out of control."[196]

As with Ridley, Swarthmore College's Robert Bannister remains unmoved by the gambit of those who would have it seem that governist eugenics is simply pro-capitalist ideology taken to its logical end. Bannister remarks that the governist eugenics movement is not to be confused with Manchesterism. Bannister recognizes that state eugenics either resulted from—or was at least empowered by—the same progressive governism that sought to shackle peaceful enterprise. Moreover, it was the same progressive governism with whom Hofstadter usually sided. Such twentieth-century historians as Hofstadter, observes the Swarthmore scholar, "have tended to picture" regulatory eugenics "as a perverse afterglow of earlier conservative"— i.e., free-market "uses of Darwinism. Thus, the eugenics movement becomes the final proof of the reactionary effects of the Darwinian revolution. But, as with other varieties of reform Darwinism"—i.e., progressive ideology—"this approach distorts . . . the place of the social Darwinist slogan in eugenic debates." Throughout the 1900s, the contention of left-wing historians "that eugenics was simply an extension of…the industrial laissez faire" that "eugenicists despised," has "blurred" the fact that governist eugenicists only invoked Darwin's name to promote State regulation of society's genetic stock. Progressives did not view the "social justice" of the welfare state as "necessarily incompatible" with governist "social control." Indeed it was "the progressive era" that "bred a variety of proposed controls, from immigration restriction to new sanctions against nonwhite Americans," and other forms of "eugenic legislation." For this reason, governist "eugenicists from the start vied with each other in condemning laissez faire, a term they applied to anyone who opposed *them*"[197] (emphasis Bannister's).

For instance, to quote British Fabian socialist Annie Besant, "*laisser aller* in marriage is no wiser than in other parts of life"[198] (emphasis hers). And a more conservative eugenicist stated that "the prevention of the multiplication of the defective classes . . . is so obvious a duty and so feasible a project that the continuation of our present *laissez faire* policy is nothing short of a crime to society." [199]

Paul A. Lombardo likewise identifies progressivism as a governist movement intending to "delegate the control of social welfare programs to a professionally trained class of experts." Additionally, Lombardo grants that this progressivism "was a key underpinning of [governist] eugenic activities . . ."[200] And for Jonathan Peter Spiro, it stands unequivocal that the pseudo-scientific ideology that influenced Hitler violently clashed with the politics of Spencer and Sumner.

> Like all progressive reform movements, eugenics called for widespread . . . state action (and this is where the eugenicists parted company with the [laissez-faire] social Darwinists[)] . . . The eugenicists needed an activist state—so providentially strengthened under TR [Theodore Roosevelt] and the progressives—to intervene in society if the programs were to succeed. . . . negative eugenics definitely required the coercive power of the state to prevent dysgenic marriages, segregate the unfit, restrict immigration, and implement involuntary sterilization.
>
> Thus, eugenics meshed well with the [phony] scientific and reformist ethos of American progressivism. It was not an accident that soon after [Madison] Grant and his colleagues in the interlocking directorate [of Progressive organizations] formulated the original principles of [government-directed] wildlife management, Grant and another group of colleagues created the organized eugenics movement in the United States. For if wildlife management was the penultimate progressive idea, eugenics was the *ultimate* progressive idea [emphasis Spiro's].[201]

As Spiro mentions, the American eugenicist movement held an affinity for the imposition of environmentalist regulations by the State. At the 1932 Third International Congress of Eugenics, the Eugenics Congress's vice president, Henry Fairfield Osborn, Sr. (1857–1935)—director of the American Museum of Natural History and the man who bears the distinction of giving Tyrannosaurus Rex its name—provided a keynote speech laden with what would become talking points for the late-twentieth-century Green movement. In words that would be approvingly reprinted in the journal *Science*, Osborn proclaimed the need for eugenics legislation on account of overpopulation, which led to what he deed the six *over*'s. They included:

> Over-destruction of natural resources, now actually world-wide;
>
> Over-mechanization, in the substation of the machine for animal and human labor, rapidly becoming world-wide;
>
> Over-construction of warehouses, ships, railroads, wharves and other means of transport, replacing primitive transportation;
>
> Over-production of both the food and of the mechanical wants of mankind, chiefly during the post-war speculative period [the 1920s].
>
> Over-confidence in future demand and supply, resulting in the too rapid extension of natural resources both in food and in mechanical equipment;
>
> Over-population beyond the land areas, or the capacity of the natural and scientific resources of the world, with consequent permanent unemployment of the least-fitted . . .
>
> I have reached the opinion that over-population and underemployment may be regarded as twin sisters. From this point of view I even find that the United States is over-populated at the present time [with its 125 million people] . . .[202]

Adding to everything, Thomas C. Leonard states what should by now be unequivocal—Hofstadter, "who had portrayed reform as the polar opposite of social Darwinism . . . does not discuss or even identify the many influential reformers who were drawn to eugenic explanations of social and economic life."[203] As a consequence thereof, Leonard apprehends that insofar as one wants to label free-market Spencerism as social Darwinism, "Progressives certainly opposed Social Darwinism . . . But Progressives did not reject 'survival of the fittest' ideology. What distinguishes the reform Darwinism of Progressives" from Spencerism "is the Progressive belief that the state . . . could do better than 'nature' in the essential Darwinian task of weeding out the unfit."[204]

In short, though these left-wing luminaries called themselves socially *progressive*, every one of their favorite regulations did nothing but impede social progress. Throughout part I, I have argued that the early twentieth century's governist eugenics movements in the USA and U.K. came not from Spencer and Sumner, but from governist activists. Still, a doubter may proclaim that this does not detract from his notion that Adolf Hitler's eugenicism still can be traced back to Sumner or Spencer. In part II, I will examine whether Hitler's entire governmental system was a tribute to Spencerian capitalism or a willful rejection of it. We might consider that Hitler did not fashion himself after any popular images of what a right-wing social Darwinist was supposed to be. Rather, it may be that before the German public the Führer scapegoated his enemies, the Jews, as the actual capitalistic social Darwinists.

NOTES

1. Derber 1998, 96–97.
2. S. P. Caine 1974, 11.
3. T. K. McCraw 1974, 200.
4. R. L. Moore 1974, 43.
5. T. C. Leonard 2009, 37, accessed online Sunday, November 6, 2011.
6. Degler 1991, 42, 47.
7. Hofstadter 1959, 168–69.
8. T. K. McCraw 1974, 196–97.
9. R. L. Moore 1974, 48.
10. Ekirch 1974, 86, 170.
11. Mowry 1958, 54.
12. Qtd. by R. Monks and N. Minow 1991, 4.
13. Chernow 1998, 341.
14. Acemoglu and J. A. Robinson 2012, 323.
15. Black 2003, 120; and Degler 1991, 47–48.
16. M. Ridley 1999, 290.
17. Degler 1991, 47–48.
18. Louis D. Brandeis in dissent, *Olmstead v. United States*, 277 U.S. 438 (1928), in T. R. Bruce ed. 2009, accessed online Saturday, July 25, 2009.
19. "Brandeis," *The American Heritage Dictionary of the English Language* 4th ed. 2006, republished as "Brandeis," https://web.archive.org/web/20070307162723/http://dictionary.reference.com/browse/Brandeis, accessed Thursday, October 1, 2014.
20. Brandeis qtd. in "The Third Party and the Trusts," *The Nation* 95, September 19, 1912, pp. 253–53. Bannister 1988, 88, brought this to my attention.
21. Qtd. by T. K. McCraw 1984, 102.
22. Qtd. in G. Seldes 1960, 112.
23. D. W. Root 2010, accessed online Tuesday, September 3, 2013.
24. Brandeis 2013, 47, accessed online Tuesday, September 3, 2013.
25. Justice David Josiah Brewer, opinion of the Court, *Muller v. Oregon*, 208 U.S. 14 (1908), in http://www.law.cornell.edu/supct/html/historics/USSC_CR_0208_0412_ZO.html, in T. R. Bruce ed. 2013.
26. Teixeira and Halpin 2010, 10, accessed online Tuesday, September 3, 2013.
27. Lombardo 2008, 158.
28. For an example of a governist author deriding Taft for being too much of a laissez-faire conservative, see James T. Adams 1931, 332–33. Here the author derides supporters of free enterprise as "reactionary."
29. J. T. Patterson 1972, 5, mentions that Robert Taft was William Howard Taft's son, and J. T. Patterson 1972, 309, 384, mentions Robert Taft's reputation for being relatively laissez faire.
30. That William Howard Taft was Sumner's student is mentioned in Lombardo 2008, 158. J. T. Patterson 1972, 33, mentions Robert Taft being Sumner's student.
31. Acemoglu and J. A. Robinson 2012, 320–23.
32. Ekirch 1974, 143.
33. Cannadine 2006, 220. Chernow 2001, trade paperback, 130, also mentions Taft initiating more antitrust suits than Theodore Roosevelt.

34. Ekirch 1974, 168.
35. C. Webber and A. Wildavsky 1986, 420.
36. Hofstadter 1959, 120, taking the quotation from W. Wilson 1913, 15, also in W. Wilson 2005, ch. 1, http://www.gutenberg.org/files/14811/14811.txt, accessed May 2, 2007. Here is an extension of that quotation.
37. Qtd. by James T. Adams 1931, 335.
38. James T. Adams 1931, 335–37.
39. Einstein 2005d, 102.
40. Black 2003, 68, 322.
41. J. Goldberg 2007, 255.
42. Ekirch 1974, 232.
43. Bannister 1988, 88.
44. J. Goldberg 2007, 447 n. 17.
45. Ekirch 1974, 170.
46. Teixeira and Halpin 2010, 9-11, accessed online Tuesday, September 3, 2013.
47. S. Kühl 1994, 24. On this matter, Kühl cites M. Schwartz 1992 and P. Weingart et al. 1988, 105–114.
48. S. Potter 2010, 138.
49. J. Goldberg 2007, 448 n. 34, points this out.
50. S. Potter 2010, 138.
51. J. Goldberg 2007, 449 n. 34, and S. Potter 2010, 138.
52. S. Potter 2010, 138.
53. J. Goldberg 2007, 449 n. 34.
54. R. Bailey 2000b, 60–61, accessed online Wednesday, March 7, 2007; M. Ridley 1999, 292; J. L. Simon 1998, trade paperback, 553; and Tudge 2000, 285. Interestingly, on this same page, Tudge writes that Keynes "bore the flags of socialism . . ."
55. Keynes 2009a, 42.
56. J. L. Simon 1998 trade paperback, 553.
57. Both Hardin's position as vice president of the American Eugenics Society, and his admiration for Malthus, are mentioned in Allan Chase 1980 paperback, 78.
58. Plotz 2005, 32. This Repository's founder, multimillionaire inventor Robert Klark Graham (1906–1997) was unlike Dr. Hardin in that he sought to achieve his eugenic goals through getting consenting adults of high IQs to "selectively breed." Under Graham's management, the Repository did not openly advocate the same pro-spoliation government regulations that Hardin advocated.
59. Hardin 1951, 613–18, qtd. by Allan Chase 1980, paperback, 374–75.
60. A. Schroeder 2008, 642-43, quoting Hardin 1963.
61. E. F. Goldman 2001, 73, 346. Ekirch 1974, 22, also makes this distinction between "social Darwinists" (free-marketers) and "reform Darwinists" (progressives).
62. Mowry 1958, 65, 202–203.
63. S. P. Caine 1974, 11, 21–23.
64. Ekirch 1974, 62.
65. Qtd. by Bannister 1988, 198.
66. Bannister 1988, 180, 198–99. That Ray Stannard Baker and many other Progressives supported government-imposed segregation is noted in McGerr 2003, 191.
67. Bannister 1988, 198.
68. Bannister 1988, 180–81; R. L. Carneiro 1981, 178–79.
69. Bannister 1988, 181.
70. J. Bean ed. 2009, 6, 113-17; and J. Roback 1986.
71. C. Gardner et al. 2006, paperback, 281–82.
72. Chris Gardner, interviewed by Bob Brown, in L. Redmond, prod. 2003.
73. J. D. Horan 1967, 29, 52.
74. B. W. Folsom 2008, 206, 209.
75. Hofstadter 1959, 139. T. C. Leonard 2009, 46, accessed online Monday, November 7, 2011, brought this passage to my attention.
76. Lombardo 2008, 200.
77. E. W. Bemis 1888. T. C. Leonard 2009, 46, accessed online Monday, November 7, 2011, brought this to my attention.
78. Hofstadter 1992, 107.
79. J. Goldberg 2007, 91.
80. Mowry 1958, 93.
81. William Allen White, *The Old Order Changeth* (New York, NY, 1910), qtd. by Mowry 1958, 93.
82. Ekirch 1974, 202, 188.
83. I found the quotation in A. Herman 1997, 177, citing B. Tuchman 1970, 177.
84. Albert Beveridge, "The March of the Flag" speech, September 16, 1898, http://www.fordham.edu/halsall/mod/1898beveridge.html, accessed Thursday, April 18, 2013.

85. V. M. Briggs 2001, 1–11.
86. Hofstadter 1959, 44.
87. The citation is found in Hofstadter 1959, 44, 222 n. 45. The paper that Hofstadter cites is C. R. Henderson 1896, 385–86. T. C. Leonard 2009, 46, accessed online Monday, November 7, 2011, brought to my attention the fact that Hofstadter cited Henderson to denounce nineteenth-century industrialists as social Darwinists, in the absence of mentioning Henderson's own involvement in the governist eugenics movement.
88. C. R. Henderson 1909, 228–29. T. C. Leonard 2009, accessed online Monday, November 7, 2011, brought this to my attention.
89. C. R. Henderson 1900, 253.
90. C. R. Henderson 1909, 228–29, 232.
91. C. R. Henderson 1900, 253. T. C. Leonard 2003, 698, accessed online Tuesday, February 12, 2012, brought this and the other aforementioned statements of Henderson's to my attention.
92. C. R. Henderson 1909, 228–29. T. C. Leonard 2009, 46, accessed online Monday, November 7, 2011, brought this to my attention.
93. R. T. Ely 1894b.
94. R. T. Ely 1894a. These statements of Ely's came to my attention through Thies and Daza 2011, 150, accessed Sunday, February 12, 2012.
95. J. Goldberg 2007, 448 n. 24.
96. J. Goldberg 2007, 448 n. 24, quoting *The New Republic*, March 18, 1916.
97. J. Halpin and C. P. Williams 2010, 2, http://www.americanprogress.org/wp-content/uploads/issues/2010/04/pdf/progressiveintellectualism_execsumm.pdf; and Teixeira and Halpin 2010, 1, accessed online Tuesday, September 3, 2013.
98. Teixeira and Halpin 2010, 24, 12, accessed online Tuesday, September 3, 2013.
99. Black 2003, 118.
100. James T. Adams 1931, 336.
101. Teixeira and Halpin 2010, 7, accessed online Tuesday, September 3, 2013.
102. Ekirch 1974, 132–33.
103. Degler 1991, 42, and Kevles 1985, 94.
104. Theodore Roosevelt, letter to the Committee to Study and to Report on the Best Practical Means of Cutting off the Defective Germ-Plasm in the American Population, a.k.a. "The Van Wagenen Committee," January 14, 1913, qtd. by Allan Chase 1980 paperback, 127–28.
105. A. Tone 2001, 141. And Black 2003, 209, also ascribes the phrase "race suicide" to TR on this subject.
106. T. Roosevelt 1907, 550, qtd. by T. C. Leonard 2003, 696, accessed online Tuesday, February 14, 2012.
107. H. W. Brands 1997, 663.
108. Theodore Roosevelt, letter to Arthur Balfour, December 18, 1906, qtd. in G. Robinson 2001, 16. G. Robinson cited G. Sinklair 1971, 320–323.
109. Black 2003, 99, quoting Theodore Roosevelt to Charles Davenport, letter, January 3, 1913, Charles Davenport Papers, American Philosophical Society, Philadelphia, Pennsylvania. The final sentence quoted, but not the first, also appear in Lombardo 2008, 32; and M. Ridley 1999, 289. A fraction of the final sentence appears in Kevles 1985, 85.
110. Lombardo 2008, 32, quoting Theodore Roosevelt, "Twisted Eugenics," *The Outlook* vol. 106, January 3, 1914, 32.
111. Mencken 1926, 23. Mencken 1990-J, 72, brought this statement of Mencken's to my attention.
112. H. L. Mencken, "Down With Boils," *Baltimore Evening Sun*, April 26, 1937, republished as Mencken 1990-A, 367. Mencken also accuses Woodrow Wilson of having a wartime "dictatorship" in H. L. Mencken, "A Time to Be Wary," *The Baltimore Evening Sun*, March 13, 1933, qtd. in Mencken 1990-N, 408–09.
113. H. L. Mencken, unpublished autobiography 1925, 200, quoted in Mencken 1990-OO, 408.
114. J. Marks 2002, 120.
115. Spiro 2009, 158, quoting Theodore Roosevelt to Madison Grant, letter, October 30, 1916, reel 386, series 1, Theodore Roosevelt Papers, the Library of Congress.
116. Hofstadter 1959, 163-64, 170, 175, 179-180, 195.
117. Hofstadter 1959, 8. T. C. Leonard 2009, 41, accessed online Sunday, November 6, 2011, brought this quotation to my attention.
118. H. W. Brands 1997, 663.
119. Theodore Roosevelt, qtd. by H. W. Brands 1997, 663.
120. Teixeira and Halpin 2010, 8–9, accessed online Tuesday, September 3, 2013.
121. G. Robinson 2002, 15–16.
122. Ekirch 1974, 133.
123. Teixeira and Halpin 2010, 14, accessed online Tuesday, September 3, 2013.
124. Allan Chase 1980 paperback, 353, and G. Robinson 2001, 34, 121.
125. G. Robinson 2001, 42, 6.
126. Britting 2004, 80.

127. The life of Oswald Garrison Villard is very interesting in light of what it reveals about *The Nation* magazine, which, today, is famous for its advocacy of radical anti-capitalism. Surprisingly, it often advocated pro-capitalist views in its early days, having been founded in 1865 by Edwin Lawrence Godkin (1831–1902). Later, its ownership fell into the hands of "robber baron" Henry Villard (1835–1900), who ran the Northern Pacific Railroad and merged Thomas Edison's (1847–1931) electric company with Thomson-Houston to create General Electric, of which he served as president until 1893. *The Nation* was passed on to Henry's governist son Oswald Garrison Villard, who gave it the radical leftwing bias with which it is associated today. See D. L. Lewis 1994, paperback, 197, 295.

128. That Du Bois helped O. G. Villard start the NAACP is confirmed in D. L. Lewis 1994, paperback, 387. That O. G. Villard was the leader of the NAACP is further confirmed in N. I. Painter 1987, 279. According to P. Sullivan 2009, 6-7, the NAACP started with Oswald Garrison Villard publishing an article on February 12, 1909, titled "The Call," inviting fellow civil rights leaders to hold a conference to be held from May 31 to June 1, 1909, in New York. P. Sullivan 2009, 15, states that it was a later such conference, from May 12-14, 1910, where the attendees formed the NAACP.

129. Du Bois qtd. by J. R. Stephens 1997, 259.

130. David Levering Lewis's editorial notes in Du Bois 1995, 439.

131. Du Bois, "The Soul of White Folk" in Du Bois 1995, 456–57, excerpted from Du Bois 1920.

132. Du Bois 1935, republished in Du Bois 1995, 569.

133. Du Bois "Segregation," *The Crisis*, January 1934, republished in Du Bois 1995, 557.

134. Du Bois, "Whites in Africa After Negro Autonomy," 1962, republished in Du Bois 1995, 686-87. This essay originally appeared in Roback ed. 1962, 243–255.

135. Du Bois qtd. by A. Herman 1997, 206.

136. Du Bois 1923, 539–548, republished in Du Bois 1995, 333–339.

137. Du Bois, "A Program of Reason, Right and Justice for Today," *National Guardian*, May 23, 1960, republished in Du Bois 1995, 426.

138. Du Bois, "Application for Membership in the Communist Party of the United States of America," *Worker*, November 26, 1961, republished in Du Bois 1995, 632.

139. Du Bois, "Marxism and the Negro Problem," *The Crisis*, May 1933, republished in Du Bois 1995, 538.

140. Du Bois, "The Vast Miracle of China Today," *National Guardian*, June 8, 1959, republished in Du Bois 1995, 627.

141. A. Herman 1997, 194.

142. Du Bois "On Stalin," *National Guardian*, March 16, 1953, republished in Du Bois 1995, 796-97.

143. Stalin's persecution of the Jews is mentioned in J. Goldberg 2007, 75.

144. J. Goldberg 2007, 258.

145. J. Goldberg 2007, citing quotations of Du Bois from D. English 2000, 297, 293.

146. The favorable words about FDR are in S. Jacoby 2008, xiii, 3–5, 88, 93, 106, 181, 255, 281, 310; and the praise for Du Bois is in *ibid.*, 147-48.

147. Qtd. by Spiro 2009, 159.

148. Black 2003, 23, 209, also confirms E. A. Ross to be a supporter of government-imposed eugenics.

149. Hofstadter 1959, 160, 163–64.

150. Ekirch 1974, 23-24; Hofstadter 1959, 70, 82, 156, 160, 163-64; and Ravitch 2000, 80.

151. Hofstadter 1969, 156, 160.

152. Mowry 1958, 20-21.

153. Ekirch 1974, 23-24.

154. Qtd. by Ravitch 2000, 80.

155. Ravitch 2000, 140.

156. E. A. Ross 1906. T. C. Leonard 2009, 47, accessed online Monday, November 7, 2011, brought this to my attention.

157. Black 2003, 23. Ross's hatred for immigrants is also mentioned in Degler 1991, 48.

158. Qtd. in M. Furner 1975, 235–36. T. C. Leonard 2003, 697, accessed online Tuesday, February 14, 2012, brought this to my attention.

159. Kühl 1994, 16. According to Kühl 1994, 114–15 n. 9, the first time that Edward A. Ross said *race suicide* was in E. A. Ross, "The Causes of Race Superiority," *Annals of the American Academy of Political and Social Science* 18 (year 1901 A.D.): 85–88.

160. J. Goldberg 2007, 261.

161. Hofstadter 1969, 160.

162. T. C. Leonard 2009, 47, accessed online Monday, November 7, 2011.

163. Ekirch 1974, 23.

164. R. L. Moore 1974, 44.

165. C. H. Cooley 1909, 296. T. C. Leonard 2009, 47, accessed online Monday, November 7, 2011, brought these statements of Cooley's to my attention.

166. Cooley 1920, 144. R. L. Carneiro 1981, 164, brought this statement of Cooley's to my attention.

167. Hofstadter 1959, 164. T. C. Leonard 2009, 47, accessed online Monday, November 7, 2011, brought this to my attention.
168. Degler 1991, 13–14, 19, 44, 145.
169. This Hofstadter quotation comes from Hofstadter 1959, 161.
170. J. Goldberg 2007, 261.
171. G. Hodgson 2004.
172. T. C. Leonard 2003, 704, accessed online Tuesday, February 14, 2012.
173. J. Goldberg 2007, 264.
174. Barbash 1989, 44, accessed online Tuesday, February 7, 2012.
175. Hofstadter 1959, 168.
176. Barbash 1989, accessed online Tuesday, February 7, 2012.
177. J. Dorfman vols. 4–5, 377. J. Goldberg 2007, 449 n. 32, brought this statement of Dorfman's to my attention.
178. R. L. Moore 1974, 44–45.
179. J. Goldberg 2007, 265; and T. C. Leonard 2003, 704, citing J. R. Commons 1897.
180. J. R. Commons 1907, 151, 148, 136.
181. J. Goldberg 2007, 265.
182. D. W. Root 2010, accessed online Tuesday, September 3, 2013.
183. Qtd. by T. C. Leonard 2009, 47, accessed online Friday, November 4, 2011.
184. S. Webb 1912, 992. J. Goldberg 2007, 264, and T. C. Leonard 2003, 702–03, accessed online Tuesday, February 7, 2012, quoting from S. Webb and B. Webb 1920, 785.
185. T. C. Leonard 2003, 703, accessed online Tuesday, February 7, 2012
186. Qtd. by T. C. Leonard 2009, 47, accessed online Friday, November 4, 2011.
187. T. C. Leonard 2009, 46, accessed online Friday, November 4, 2011.
188. Hofstadter 1959, 163.
189. Ronald Fisher is described as an "antiracist conservative" and enemy of the eugenics movement in Kevles 1985, 170. Raymond Pearl is confirmed to be an anti-racist opponent of the New Deal and a man reputed to be the bane of the eugenics movement in Barkan 1992, 219. Tudge 2000, 286, states that Ronald A. Fisher started out as a "keen advocate" of eugenics but that he "did turn against the idea, however . . ."
190. For information on Herbert Spencer Jennings's opposition to the governist eugenics movement, see Kevles 1985, 69, 132. Spiro 2009, 334–335, also mentions this, but only refers to Jennings as "Herbert S. Jennings," failing to disclose that his full name was Herbert Spencer Jennings. For information on Herbert Spencer Jennings being left-of-center, see Kevles 1985, 127.
191. Barkan 1992, 191; and Kevles 1985, 124.
192. M. Ridley 1999, 294.
193. M. Ridley 1999, 293–97; and Tudge 2000, 284.
194. M. Ridley 1999, 294.
195. Ridley calls him a "libertarian" in M. Ridley 2000, accessed online Wednesday, May 9, 2007.
196. M. Ridley 1999, 294–95, 297, 300.
197. Bannister 1988, 166, 178, 278 n. 4, 165, 174.
198. Qtd. by Bannister 1988, 174.
199. Qtd. by Bannister 1988, 174.
200. Lombardo 2008, 17.
201. Spiro 2009, 137–38.
202. H. F. Osborn 1932; and H. F. Osborn 1934, 30. Zubrin 2012, 63–64, brought this statement to my attention.
203. T. C. Leonard 2009, 47, accessed online Monday, November 7, 2011.
204. T. C. Leonard 2003, 707, accessed online Saturday, February 18, 2012.

II

The Governism of the Third Reich

SIX

Is Naziism the Final Stage of Capitalism?

The Free-Market Evolutionists' Nonexistent Influence in Germany

We have learned that calls for government-imposed eugenics in the USA and U.K. did not originate from any supposed laissez-faire social Darwinism movement. These calls were made by governists of the left-wing progressive variety. Still, a politically-correct governist might retort that it is more than possible that Naziism came about from a laissez-faire social Darwinist ideology that manifested among German intellectuals. This chapter shall examine the level of plausibility in such an allegation. This Part of this book—part II—examines the degree to which the Nazi regime might justly be considered governist philosophy taken to its final implementation.

There were, to be sure, eugenicist theorists in Germany in the early 1900s whose writings provided a veneer of scientific respectability for the racist and genocidal policies that the Third Reich would enact. A reliable authority on this piece of history is Alfred Kelly (b. 1947), Hamilton College's Edgar B. Graves Professor of Modern European History. Kelly has documented these intellectuals' applications of biological theories to social science in his 1981 monograph *The Descent of Darwin: The Popularization of Darwinism in Germany, 1860–1914*. This survey provides no mention of either Spencer or Sumner. That is not an oversight on Kelly's part; Spencer and Sumner exercised no significant influence on the policy prescriptions the German eugenicists promulgated. The vast majority of the early-twentieth-century German intellectuals who applied evolutionary theories and genetics to the social sciences had argued for more governism, not laissez faire.

Alexander Tille (1866–1912) was a German philosopher who, in the late nineteenth century, opined that the poor were obligated to die without procreating. In Alfred Kelly's words, Tille thought that "the horrors of the slums were actually good because they purged the nation of useless citizens." Tille served as deputy business director of the Organization of German Industrialists and subsequently became a lobbyist for Saarbrücken's industry trade groups. Accordingly, some might hastily infer that Tille was a laissez-faire Social Darwinist who aspired to "give a scientific justification to what may be denounced as the evils of capitalist society." But such a conclusion falls flat. Tille abhorred the night watchman state. According to Kelly, Tille wanted the government to "intervene actively to 'help' nature by killing the cripples and lunatics" and to expend tax money on "giving more food to the gifted members of society."[1]

Kelly offers other reasons to judge it superficial to tie Manchesterism with Naziism. This is evinced in how German steel industrialist Friedrich Alfred Krupp (1854–1902) tried to promote the application of genetics to social science. In 1900 Krupp sponsored a contest wherein he would award 10,000 marks for an essay that could answer the question[2] "What lessons can be drawn from the principles of the theory of heredity in its relation to the evolution of international politics and state legislation?"[3] Wilhelm Schallmeyer, a contributor to governist Alfred Ploetz's eugenics journal[4] and a physician from Munich, submitted the winning entry.[5] His piece was *Verebung und Auslese im Lebenslauf der Völker* ("Heredity and Selection in the Life Courses of Peoples").[6]

The mere fact that Krupp was a wealthy businessman is often inaccurately cited as proof that Nazi eugenics is compatible with capitalism. Despite his frequent admissions that the Volkish writers who inspired Nazi eugenics were explicit socialists, Marxists, or progressives, George L. Mosse inaccurately pronounces that, in contrast to favoring governism, "the usual attitude" among Germany's nineteenth-century eugenicists "was to praise the capitalist system as beneficial for the process of natural selection. . . . Capitalists themselves responded favorably to Darwinism." Mosse proceeds to cite Friedrich Krupp's essay contest as a case in point. In purporting that the contest's winner won on account of his pro-capitalist bias, Mosse proclaims that Dr. Schallmayer, the contest's winner, "rejected socialism . . ."[7]

Professor Alfred Kelly, thankfully, demonstrates that Schallmeyer's winning essay was actually not inclined to promote free enterprise over governism. In Kelly's words, Schallmeyer's essay argued that "the state had to take an active role in preserving the best racial elements. For example, marriage laws ought to encourage the racially superior upper classes to marry earlier and have more children."[8] I fault French historian Léon Poliakov for contributing to the false interpretation that Herbert Spencer inspired governist eugenicism. It particularly ticks me off that Poliakov cites Richard Hofstadter's *Social Darwinism in American Thought* in doing so. On the topic of the Krupp essay contest, though, Poliakov redeems himself somewhat by being more careful with the truth on this other topic. Poliakov observes that Schallmayer's essay was one of the least consistently racist of the entries. In advancing race-based policies, writes Poliakov, Schallmayer evinced some relative "moderation . . ." As an example, he "was severely critical of the extravagance of pan-Germanism" which, he "feared, might lead to wars with harmful eugenic results."[9]

Sixty essays were entered in the contest,[10] and, as Alfred Kelly writes, "advocates of unrestrained capitalism would have been disappointed" to read them.[11] Poliakov elaborates, "all but one or two of the competitors placed their hopes in State intervention . . ."[12]

Kelly continues, "Most of the essayists advocated . . . state socialism, calling for a regulated economy with an expansion of worker protection laws, sickness and old-age insurance, and other forms of social welfare." An overview of the contents of the essays contradicts "the conventional view of Social Darwinism as a bourgeois defense of the free-enterprise system. . . . This is an important point to keep in mind when drawing parallels between Social Darwinism and Nazism." In Kelly's estimate, any inference that Naziism sprang from free-market bionomicist thinking is necessarily "problematic."[13]

Consider the competition's runner-up, Ludwig Woltmann. To his credit, Mosse does mention that Woltmann, who would later become a major player in the Third Reich's eugenics propaganda, was a Marxist.[14] Besides that, Woltmann was a "neo-Kantian."[15] Mosse discloses that this particular thinker "tried to effect a marriage of Marx . . . and Kant" into a coherent ideology that implored the Germanic people to unite, as a mass, against a common racial enemy.[16] Perhaps this might explain Woltmann's explicitly extreme anti-capitalism. In Poliakov's words, Woltmann believed that saving the Aryan race required "a vast programme" of government-imposed "social reforms." Such reforms included the forcible redistribution of land parcels "among peasants of good stock" and "the protection of the workers against 'the anti-selectionist tendencies of capitalism.' He also believed in promoting the working class politically."[17]

Also consider Albert Schäffle, a political science professor from the University of Vienna. Schäffle, too, has been falsely characterized as a social Darwinist apologist for capitalism who would somehow spark the Nazis' interest in eugenics. As Kelly paraphrases Schäffle, "Unbridled capitalism . . . was base and materialistic and did not bring out man's nobler characteristics." The German went as far as writing an essay clarifying that his ideology was based on "Darwinism and Social Welfare." Kelly evaluates Schäffle's writing to be "a good example of the limitations of the common view that Social Darwinism is merely an apology for the worst aspects of capitalism."[18]

Upon the contest's completion, the judges summarized the consensus formed among the entrants: ". . . our times demand a policy of social progress, only possible if there exists a strong central power which is bold enough to limit individual liberty in the name of the common good and which is able to intervene in the mechanics of economic life.'"[19] To wit, Krupp, Schallmayer, and everyone else connected to this competition vocally opposed laissez faire.

Compounding the discrediting of left-wing mythology, the militarist intellectuals of Germany seldom derived inspiration either in Spencerian free-market economics or in discoveries about genetics and natural selection. Kelly tells us that militarists in Deutschland from 1860 to the Nazi era "have been tagged with the label 'Social Darwinist' solely by virtue of the fact that their bellicose utterances succeeded the publication" of *The Origin of the Species.* Helmut von Moltke and Heinrich von Treitschke are one example. The militarists' "glorification of war and the power state are only very superficially Darwinian. When Treitschke remarks in his *Politics* that 'Brave peoples alone have an existence, an evolution or a future; the weak and the cowardly perish, and perish justly,' we are minded of Darwin by the choice of words. But Treitschke does not depend on or elaborate on Darwinism; rather, he merely restates the ancient 'might makes right' argument in contemporary language."[20] In fact Treitschke held the same revulsion toward Semitic entrepreneurs as did the rest of the Germans who would eventually come to support the Nazis. George Mosse reports, "Treitschke attacked the east European Jewish immigration into Germany as the shock troops of a foreign invasion destined to dominate the stock exchange and the newspapers."[21] This extreme anti-capitalist attitude, so pervasive among the Nazi Party's supporters, will be explored in the next chapter.

The "Scientific Racists" Who Most Influenced Hitler

One may wonder which particular thinkers had originated Hitler's governist eugenicism if such thinkers did not include Herbert Spencer, William Graham Sumner, or any other consensualist. Our answer arrives from Jonathan Peter Spiro, Nicholas Goodrick-Clarke, and biographer William L. Shirer. Spiro combed through the chronology of eugenicist ideas in Europe. Researching independently, Spiro and Goodrick-Clarke and Shirer each deciphered that Hitler's ideas about genetics and racism originated from sources entirely dissimilar from Spencer and Sumner. Among the thinkers are Joseph Arthur de Gobineau, Georges Lapouge, and Houston Stewart Chamberlain. George L. Mosse repeatedly refers to their scientific-sounding rationalizations for racism as "social Darwinism,"[22] but not once does he mention Spencer or Sumner. As it was in the case of Alfred Kelly, Mosse's reason is that Spencer and Sumner commanded no more than a minimal influence over Germany's scientific racialists. One philosopher who did hold a major influence over them, though, was Immanuel Kant. "Philosophically," Mosse announces, "Volkish authors and racial theorists...were inspired in part by certain tenets in Kant's speculations on race which seemed to give them license to elaborate further."[23]

We start with Joseph Arthur de Gobineau (1816–1882).[24] A French diplomat, Gobineau aired some disturbing assertions. First he asserted that he had uncovered solid evidence that racial categories entailed particular congenital character traits. Following this, Gobineau asserted that some races were congenitally more peaceful and technologically inclined than others. Finally, Gobineau asserted that blond-haired, blue-eyed Nordics were the greatest ethnic band.[25] His influences include Kant, Hegel, Fichte, utilitarian Jeremy Bentham, and the socialists William Godwin (father to *Frankenstein* author and romanticist Mary Shelley) and Charles Fourier.[26] No shill for capitalism, Gobineau explicated his hostility to commerce. In a letter to his sister, he lamented, ". . . money is the criterion for judging the esteem due to men." Far worse, France had become a "kingdom of bankers. . . . How I despair of a society which . . . has no heart left." To sum up his opinion, Gobineau shrieks, "MONEY HAS KILLED EVERYTHING"[27]—the capitalization is his. Besides Jews, Gobineau despised Asians, whom he thought of as too greedily commercialist as well. As George L. Mosse writes, Gobineau believed the "yellow race" has been "taken up with 'a steady but uncreative drive towards material prosperity.'"[28] Mosse paraphrases Gobineau's idea: "The yellow race was destined to fulfill itself through commerce and trade. This race had all those characteristics which Gobineau fastened upon the bourgeoisie, whom he blamed for having destroyed the true France. . . . Clearly the yellow race possessed none of the virtues that characterized true nobility, and so was analogous to the French bourgeoisie."[29]

Gobineau's mantle was then taken up by another man. One was born in England and university-educated in Switzerland before marrying a German and settling down in her native country. This Gobineau disciple was Houston Stewart Chamberlain (1855–1927).[30] To Gobineau's school of thought, Chamberlain added the pretense of biological evidence and a fixation on accusing Jews of being genetically predisposed toward ruthlessness.[31] As a foil to the Jews, Chamberlain characterized those who were blond-haired and blue-eyed—whom he called *Teutons*—as the ideal.

Theodore Roosevelt harbored some mixed feelings about Chamberlain's racialist theories. Though he strongly criticized some points of Chamberlain's seminal work of racial ideology, *The Foundations of the Nineteenth Century*, Roosevelt nonetheless held the man himself in high regard. "Fundamentally," TR considers, "very many of Mr. Chamberlain's ideas are true and noble. . . . a man who can write such a really beautiful and solemn appreciation of true Christianity, of true acceptance of Christ's teachings and personality, as Mr. Chamberlain has done, . . . a man who can warn us as clearly as he has warned about some of the pressing dangers which threaten our social fabric"—nonwhites— ". . . represents an influence to be reckoned with and seriously to be taken into account."[32]

Among his philosophic influences, Chamberlain counted the German romanticist and racialist Richard Wagner,[33] a friend to Gobineau.[34] The love for Wagner is something that Chamberlain and Nietzsche had in common. In fact, Chamberlain's German wife was actually one of Wagner's daughters.[35] In contrast to the Kant-hating Nietzsche, though, Chamberlain branded himself a staunch ally of Kant's. Brian and Rebecca Copenhaver, a married couple of philosophy professors, disclose that Chamberlain was "a serious student" of Kant's.[36]

George L. Mosse divulges that Chamberlain's *Foundations* "drew a semblance of respectability" from the "Kantian philosophy—particularly for arguments in favor of the existence of intrinsic values. Chamberlain held that Kant . . . postulated an essence of things, the *Ding an sich*, which was outside empirical comprehension. Chamberlain used Kant to the following effect: On the one hand, there was a German science which determined, with the utmost accuracy, that which existed empirically; on the other hand, there was a Germanic religion, which bestowed infinite vistas upon the German soul. In terms of importance, the religion took priority, since it alone could fathom the true essence of things. As it also functioned in the realm of ideas, it served to keep science within its proper limits . . ."[37] The endgame for Germans on the path to enlightenment would be for them to reach "Kant's 'pure ideal.'"[38]

It is telling that Chamberlain cited Kant in such an admittedly irrational claim. Think back to chapter 3 wherein Abraham Foxman of the Anti-Defamation League incorrectly claimed that Enlightenment-era philosophy could be properly blamed for the Nazis' seemingly scientific rationalizations for their racism. Chamberlain did indeed pay lip service to science in delivering his rationalizations. Granting that, not only was there nothing scientifically valid about such rationalizations, but Chamberlain derided human reason as congenitally inadequate. Following Kant and the romantics, Chamberlain insisted that human beings acquired objective conceptual knowledge by means other than observational rationality. That he is wrongly remembered as an advocate of reason, when he explicitly adhered to fideism, is a commonality that Chamberlain shares with Kant. Mosse writes that Chamberlain openly invoked "mysticism." Chamberlain had been inculcated with the "New Romanticism," and, in "the end, it was the mysticism that won out, eventually doing violence to the scientific method."[39] Given that information, it is understandable that Chamberlain thought of himself as a follower of Georg Hegel as well.[40]

As did Gobineau, Chamberlain reviled commerce. Although Chamberlain conceded that entrepreneurs exercised great creativity in their quest for profit, Chamberlain, in his own words, deemed it essential to carp that "capitalism . . . causes hardships . . ."[41]

I quote from Chamberlain's *Foundations of the Nineteenth Century*. Volume 1 of this work pronounces the usury charged by Jewish moneylenders to be "a parasitic condition. . . . I hope we will hear no more of the nursery tale that the Jews...only became usurers in spite of themselves during the Middle Ages, because they were cut off from every other occupation; if we read the prophets" of the Bible "carefully we shall see how often they complain of usury, which serves the rich as a means of ruining the peasants . . ." Yes, he states, "Already in ancient Babylon," the Hebrews imposed "the frightful extraction of usury . . ." And on into the high Middle Ages and Renaissance, royalty found themselves "under the thumb of Jewish usury," an inarguably "shameful usury." He sputters maniacally over the Jews' "greed" and "business cunning."[42]

In his *Political Ideals*, Chamberlain approvingly cites Kant as the source for his attack on rational self-interest: "Kant . . .," writes Chamberlain, "forbade not only that a man use another as a 'means,' but that man may not even use himself so, may not live in such a way that his existence remains without a higher content . . ." This German-by-choice writer proclaims in his prose that a person is duty-bound to "reach out beyond his limited ego and its interests into the surrounding welfare of the community. . . . To use a personality—no matter if it be one's own or another's—merely as a means to the acquisition of a fleeting, egoistic goal . . . is the sin, that alone."[43] Chamberlain repeats the anti-capitalist cliché that every person's existence can be none other than a zero-sum game, where one person's gain inexorably entails someone else's loss. And then Chamberlain quotes Kant to support that case: "Man must either work himself or others for him; and this work will rob others of as much of their happiness as he wishes to raise his own above the average." And he again quotes Kant: "Man is indeed sufficiently unholy."[44]

It is interesting to note that Kant did not merely precede Chamberlain in his opposition to self-interest, but also in his anti-Semitism. You will recall from *Life in the Market Ecosystem* that Kant seemed partially pro-capitalist in that he conceded that the more goods and services peaceably flowed between two liberalized republics, the more discouraged those republics felt from warring upon one another. Kant understood that entrepreneurs generally find it against their interest to have their profits be taxed to finance the invasion of other countries and the subjugation of their peoples. Nonetheless, Kant anticipated Chamberlain in his conventional denunciation of Jews. Kant, too, believed the stereotype that Jews are too usurious: "The Palestinians who live among us have the well-merited reputation of being sharpers, owing to the spirit of usury which holds sway amongst most of them. It is true that it seems odd to imagine a nation of swindlers but it is equally difficult to imagine a nation of merchants," for such merchants "do not aspire to civic virtue but wish to compensate their shortcoming in this respect by the benefits they derive from deceiving the people who grant them protection . . ." But that is exactly what Kant believes Jews to be—"a nation composed merely of merchants, that is to say of unproductive members of society . . ."[45]

Foundations availed Chamberlain much acclaim in Germany. In part due to the work's anti-Semitism, it won the man an audience with Kaiser Friedrich Wilhelm. Then, in Chamberlain's twilight years, he and Hitler developed a mutual admiration for one another and corresponded. Hitler considered one of the greatest moments of his life to be his coming face-to-face with a wheelchair-bound Chamberlain.[46] On that critical moment in history, Chamberlain expressed to Hitler his gratitude for seeing through Chamberlain's work to its natural end.[47] In kind, the future Führer kissed the academician's hands.[48] *Mein Kampf* itself cites Chamberlain by name. The Nazi manifesto wails that Germany lost the first World War for the reason that those "who had the Government of the country in their hands were . . . indifferent to the principles of civil wisdom laid down by thinkers like Houston Stewart Chamberlain . . ."[49]

The ideology of both Gobineau and Chamberlain found a path into the head of French anthropologist Georges Vacher de Lapouge (1854–1936).[50] Lapouge, who also influenced Hitler, hated especial ethnic groups on account of these ethnicities being stereotyped as merchant peoples. Mosse writes that in the view of Lapouge, "the inferior races such as the yellow race and the Jews were without scruples and had no sense of values, being wholly commercial. The bourgeoisie in de Lapouge's racial analogy seems the enemy once more . . . the Jew loves speculation for its own sake. . . . the Aryan middle class can only survive by destroying the Jewish bourgeoisie." Lapouge judged that the social problems that came with "urbanism and plutocracy" could be ascribed a "'Judaized' economic system."[51] Noticing that anti-capitalism and racism were traits Lapouge shared with various members of the political Left, Mosse admits, "De Lapouge and other theoreticians of race" were "not so far removed from the contemporary Fabian Socialists, particularly Sidney and Beatrice Webb. . . . For this meant that the one race which was best fitted to build socialism would be overrun by mongrels. . . . Socialism and eugenics were not inherently conflicting concepts, and neither occasionally were socialism and race."[52] Poliakov points out that Ludwig Woltmann followed Lapouge's ideas, and that Woltmann himself served as a role model to the Third-Reich-appointed eugenics policy officers Eugen Fischer, Fritz Lenz, and Otmar von Verschuer.[53] Poliakov writes that Fisher, "who was the first to apply Mendel's laws to 'racial hygiene,' declared in 1934 that he had been the first to propagate Woltmann's ideas in a university faculty and to have 'inflamed young hearts with enthusiasm for racial science.'"[54]

Meanwhile, Francis Galton's teachings on eugenics gained an audience among both American Charles Davenport and German biologist Ernst Haeckel (1834–1919). Haeckel, though still unaware of Chamberlain's work, had already concluded that biology proved the inherent superiority of native-born Germans over Jews.[55] Citing Gobineau, Haeckel proclaims that the "lower races (such as the Veddahs or Australian negroes) are psychologically nearer to the mammals (apes or dogs) than to civilized Europeans; we must, therefore, assign a totally different value to their lives."[56]

In 1900, Haeckel sat on the six-member panel of judges that evaluated the sixty essays submitted in Alfred Krupp's eugenicist essay contest. When in 1905 Alfred Ploetz founded the Society for Racial Hygiene,[57] he named Haeckel its honorary chair,[58] with

Francis Galton named another honorary member.[59] Haeckel, too, was governist and collectivist. Writes Arthur Herman, Haeckel aimed to have his Romanticist philosophy "replace selfish individualism with a new ethical monism" that prioritized "the interests of the community . . ."[60] The writings of Gobineau, Chamberlain, and Lapouge shaped the fanatical racialist attitude of New York zoologist Madison Grant. As we discerned in the previous chapter, Grant helped Davenport popularize the advocacy of eugenicist governism in the United States.

Galton, Haeckel, Davenport, and psychologist Henry H. Goddard each additionally contributed to Grant's eugenicist theories. In every fiber of their thinking, writes Jonathan Spiro, these men "were progressives, in the sense of wanting to use the state to enact scientific reforms that would utilize the nation's germ plasm more efficiently." Hitler directly read, enjoyed, digested, and cited the works of Gobineau, H. S. Chamberlain,[61] and Madison Grant.[62] Additionally, from the 1920s to his death in 1937, Madison Grant acted as an unofficial adviser and mentor to two of the Third Reich's most prominent racial theorists, Eugen Fischer and Hans F. K. Günther.[63] Fischer served as the director of the Institute of Berlin, where "angel of death" Josef Mengele received his medical training.[64] That is, in both direct and indirect forms, the Nazis learned their eugenicist doctrines from Progressives in the USA and socialists throughout Europe.

It is true that William Graham Sumner's brother-in-law Walter Camp, a Yale University football coach, had been a member of Madison Grant's ECUSA—Eugenics Council of the USA.[65] Nevertheless, Sumner himself avoided association with that organization. In the end, Hitler's ideology was informed not by such laissez-faireists as Sumner and Spencer, but by govenists who repudiated what Sumner and Spencer stood for. Besides the hitherto-mentioned figures, such as Grant and Gobineau, such men included Ludwig Ferdinand Clauß,[66] Ernst Krause, Otto Ammon, Ludwig Wilser, Wilhelm Bölsche, Willibald Hentschel, and that notorious socialist we discussed earlier, Ludwig Woltmann.[67] At an April 23, 1934, dinner thrown in his honor in Germany, an elderly Karl Pearson—whom, as we recall from chapter 2, consistently touted socialism in Britain, and who held strong influence over Eugen Fischer and fellow German eugenicist Alfred Ploetz—stood up to announce to the room his support for Hitler's eugenic policies.[68]

The reader might be astonished to learn the identity of another substantial contributor to Nazi eugenicist thought. It was none other than existentialist Martin Heidegger, who gave lectures in praise of Hitler and eugenics during the summer of 1933 at the Institute for Pathological Anatomy, and collaborated extensively with Eugen Fischer in his eugenics studies. Subsequent to 1945, Heidegger continued to exchange letters with Fischer regularly and to visit him in person occasionally.

Of special significance is Heidegger's choice in terminology. Most scholars who care to acknowledge Heidegger's support for state-imposed eugenics would probably refer to this support as social Darwinism. However, Emmanuel Faye's examination of Heidegger's papers reveals that Heidegger rejected the term *Darwinism* and criticized it as a doctrine exactly on account of his judging it to be a philosophy consistent with laissez-faire liberal individualism. In effect, Heidegger actually anticipated Richard Hofstadter's tactic of ideological argumentation. Heidegger and Hofstadter denounced laissez-faire liberalism by accusing it of trying to apply biological theories to the social sciences. Emmanuel Faye elucidates that when Heidegger taught classes on eugenics in the winter of 1933 to 1934, Heidegger's "critique of liberalism" was attached to his "critique of biology." There are those, such as Harmut Tietjen, who maintain that Heidegger could not have been a consistent Nazi, as Heidegger explicitly stated that the general principles of "biology" cannot be applied properly to social science. The assumptions there are (1) that anyone who refrains from applying biological theory to social studies cannot be a social Darwinist or eugenicist, and (2) someone cannot be a Nazi if he does not believe in eugenics or social Darwinism. Yet, as Emmanuel Faye points out, Tietjen's defense of Heidegger fails. Heidegger indeed proclaims that there are distinct human races, and that each race holds inborn character traits—that some races are congenitally sneakier or greedier than others. Heidegger, observes Faye, employed "the terms *Rasse* [race] and *Geschlecht* [sex] very explicitly" in the very same course where he professed not to believe in "biology." Not rejecting the terms *Rasse* and *Geschlecht*, Heidegger proclaimed that race should be evaluated "no longer on the basis of what he disdainfully refers to as 'liberal biology' (*liberalistiche Biologie*)." Given that Heidegger proclaimed that his own racial theory should be accepted in lieu of "liberal biology," we should understand what he meant by *liberal biology*. By *liberal biology*, he referred to what Heidegger deemed the laissez-faire political "liberal understanding of man and human society" that had been popular in Great Britain. That is, *liberal biology* alludes to the school of philosophy that Herbert Spencer led. Heidegger advanced his own eugenicism at the expense of the Spencerian interpretation of social science that he hated. Then, throughout the *Contributions to Philosophy* lectures he delivered between 1936 and 1937, Heidegger again railed against *biologischer Liberalismus*—biological liberalism. Yes, Faye writes that what Heidegger "continually critiques is Darwinism, because it is Anglo-Saxon and therefore 'liberal.' But he is careful to point out later in his course on Nietzsche," one of his own existentialist philosophic influences, "that 'Nietzsche's biologism is not Darwinism.'"[69]

In the face of such evidence, some may reply that Naziism still must have arisen from private-enterprise economics, as both free enterprise and Naziism are right-wing instead of left-wing. When a politically correct left-winger opines that laissez-faire ideologues and totalitarian Nazis are equally "right-wing," the term denotes nothing more than "everything of which politically correct left-wingers disapprove." Should the phrase *right-wing* be articulated in such a manner, then the right wing cannot be defined by any commitment to laissez faire, given that laissez faire is precisely what was missing from the Third Reich's regime. As this part shall demonstrate, the Nazis made government control the main aspect of life for every one of their citizens. The very word *Nazi* is short for *National Socialist*. More importantly—the defensive denials of left-wing commentators notwithstanding—the Third Reich showed its determination to implement the "socialism" half of the label. Thus the Nazis lived up to their party's full name—the National Socialist German Workers Party (NSDAP, for short). Although it was mostly framed in a racial or nationalist context, the term that the *Nazis* applied to the Germans—*volk*, as in "folk"—carried socioeconomic connotations as well. To some degree, the Nazis used *volk* to refer to non-capitalist proletarians, just as the communists applied *proletariat* to that same class of people. *Volk*, writes the historian Richard Grunberger, "denoted both 'the people' in the radical [social] democratic sense and 'the folk' in the racial sense."[70]

Despite the tremendous philosophic differences between Anglophone free-marketers and German volkists, writes Geoffrey Hodgson, "political sentiments, from . . . liberal free traders such as Spencer and Sumner, to more militant nationalists, and racists such as Haeckel, were all conflated under the single, misleading label of Social Darwinism."[71] Fortunately for U.S. progressives, though, such Anglophone socialists as George Bernard Shaw and such Anglophone progressives as Louis D. Brandeis, who had supported eugenicist and racist legislation, were conveniently forgotten to be eugenicist, thereby exempting them from the social Darwinism tag and, as a consequence, leaving the legacy of progressivism itself untainted by the negative reputation of eugenics and social Darwinism.

Hitler a Puppet of Germany's Big Businesses?

The historical facts go unlearned by the cultural mainstream for various reasons. Among them is that left-wing governist intellectuals have successfully implanted this irrational syllogism into the minds of collegians:

> Unregulated capitalism gave rise to big business.
> Big business gave rise to Hitler.
> Therefore, unregulated capitalism gave rise to Hitler.

Historian Joachim Fest finds himself weary of this anti-capitalist myth. He laments the trite ploy to portray Der Führer "as the 'hireling' or 'sword-arm' of capitalism . . ." As the story goes, Hitler's militaristic conquests somehow "carried out the expansionist aims of his employers."[72] Among the legions who advance this false view, Pulitzer Prize-winning Johns Hopkins University instructor Wayne Biddle relates, "The National Socialists had made their breakthrough . . . to mainstream as a petty bourgeois" social movement "aided by ultranationalist and conservative establishment circles and antiunion business leaders."[73]

Despite its having some virtues in exposing the socialist elements of the Third Reich, William L. Shirer's *Rise and Fall of the Third Reich* has partially contributed to this false interpretation. The book makes a big deal out of particular industrialists supporting Adolf Hitler, and then conveys the inaccurate impression that support for Hitler's policies is somehow inherent to the nature of big business. Shirer observes that among the industrialists to endorse Hitler publicly in the early 1930s were steel magnate Fritz Thyssen, coal tycoon Emil Kirdorf, and the United Steel Works' Albert Vögler. In fact, Shirer gives quite a long list of big-business Nazis.[74] Yet these men did not represent the majority of business interests in Germany in the late 1920s or early 1930s. Additionally, big business's financial contributions to the Nazi Party in the first years of Hitler's rule should be put into perspective. Helpful in the matter is the University of Sheffield historian Ian Kershaw (b. 1943). In Kershaw's words, the anti-market apparatchiks caricature Hitler as "an agent of capitalism, a cypher for the interests of big business and its leaders who controlled him and pulled the strings of their marionette."[75] W. E. B. Du Bois levels that sort of asinine accusation about capitalism causing Naziism.[76] And the interviewees of that governist propaganda documentary I mentioned in Book Two—*The Corporation: The Pathological Pursuit of Profit and Power*—also do their best to disseminate these misconceptions. In this 2003 film Boston University's Howard Zinn, author of *A People's History of the United States*, comments, "There is an interesting connection between the rise of fascism in Europe and . . . corporate power. . . . there was a recognition that fascism rose in Europe with the help of enormous corporations."[77] Yet another one of the movie's interviewees, Noam Chomsky, seconds that when Adolf acquired the chancellorship, "investment shot up in Germany."[78]

The canard about Naziism being the spawn of Lockeanism is fallacious on multiple counts. First, to the extent that a private company lobbies for—and profits from—any spoliation a State commits, such a company's support for governmental spoliation amounts to an indiscretion wholly irreconcilable with laissez faire. Secondly, historians who have studied the funding sources for Hitler's political campaigns discern that Germany's big businesses provided relatively little capital to Hitler and the Nazi Party prior to Hitler's attainment of his dictatorial status. Joachim Fest faults the anti-capitalist "school of thought . . ." When the anti-capitalist writer Eehard Czichon says that big businesses "manipulated" the Nazis, Fest objects. Those who have opted to categorize the Führer "as conservative, reactionary, capitalistic, or petty bourgeois," says Fest, did not and do not "comprehend him."[79] Dr. Kershaw clarifies that "the view of the Left" that "the Nazi Movement was the creature of big business and sustained by its funding" is not "solidly founded."[80]

Joachim Fest has uncovered some of the tactics that academic leftists have employed to sire the misinterpretation that German big business empowered the Nazis. According to Fest, "Czichon tends to prefer general references and unpublished documents, so that his sources in many cases can scarcely be checked. Frequently, too, he indulges in apparently deliberate deceptions, inaccuracies, and faulty references. Ernst Nolte has shown that Czichon reports a payment from IG Farben to the NSDAP in such a way that the reader would think the payment had been made before the seizure of power, whereas the document itself shows that the money was paid in 1944." The Nazis' unpopularity among industrialists in general can be gauged by an examination of the actions of the few industrialists who did support him. In 1932, steel industrialist Albert Vögler teamed up with the big bankers Hjalmar Schacht and Kurt von Schröder to win more support for the Nazis among the business community. They particularly wanted to curry the favor of the chemical industry and the department store owners, industries that remained doubtful toward the Nazis' goals. Vögler, Schacht, and Schröder thus had Hitler give a fundraising speech before the Dusseldorf Industry Club on January 26. When the trio of businessmen counted the amount in donations collected from this talk, they experienced disappointment. The three then tried to circulate a petition asking German leader Paul von Hindenburg (1847–1934) to appoint Hitler as chancellor. Notes Joachim Fest, "the majority of businessmen who were approached refused to give their signatures." Indeed, he continues, "The theory of a close, pragmatic alliance between Hitler and the major capitalists also fails to explain the time lag between the explosive growth of the party and the injection of funds from industry."[81]

Kershaw concurs,[82] as does Yale University historian Henry Ashby Turner, Jr., who explicates that "the financial subsidies from industry were overwhelmingly directed *against* the Nazis"[83] (emphasis his). Ian Kershaw, Joachim Fest, Henry Ashby Turner, John Toland, and Florida State University historian Robert Gellately disclose that, prior to Hitler gaining the chancellorship in 1933, big business financial support to the NSDAP was scant, and that funding of the NSDAP campaigns came directly from enthused individuals mostly of the middle class.[84] Whereas the German big businesses paid thousands of marks to professional artists to produce their campaign posters, NSDAP volunteers spent considerable hours crafting their own signs, and numerous painters in the NSDAP contributed their considerable artistic talents to providing the NSDAP with evocative images for its posters.[85] The surprising extent of support that artists showed the NSDAP in the late 1920s shall be further examined in chapter 8. Moreover, Hitler gained such a cult following among the middle class that he was able to command huge fees from these people simply for the privilege of being granted access to one of his public speeches.[86] Some people paid as much as 7,000 marks to attend a rally to hear the man speak.[87] The middle-class NSDAP members were so devoted to their party that they made what Turner and Gellately call financial "sacrifices," such as turning over their entire life savings to the cause. These sacrifices demonstrably had adverse impacts on the donors' financial well-being.[88] Fest observes, "Some of the early party members, like Oskar Korner, owner of a small toy store . . . all but ruined themselves in the interest of the party."[89] Jack L. Snyder, an international relations professor at Columbia University, observes that far from being a tool of Germany's big business, the Nazis were voted into office as a form of protest against big business. The middle-class supporters of the NSDAP genuinely believed the party would stand up for them. According to Snyder, about a third of the German voters felt abandoned by the business-supported conservative political parties, that such pro-business parties were "indifferent to their plight. At that point, the dense network of grassroots voluntary organizations in small cities and towns turned massively to support a populist party that seemed to be one of their own, the Nazis (National Socialists)."[90]

In short, the Nazis were no less of an enthusiastic grassroots movement than the contemporary anti-globalization activists who provide massive demonstrations at the meetings of the World Trade Organization and who occupied Wall Street in the autumn of 2011. Also not unlike that superstar of the anti-globalization movement, Noam Chomsky, Hitler raised enormous sums of money from charging fees for his speaking engagements.[91] "Even in hard times when money was short," writes Gellately, "people were charged a fee to listen to what Nazis had to say. . . . The price of admission (discounted for some) . . . helped finance" the Nazis' campaign in the 1930 election. This provided a windfall for the movement—"In just three speeches Hitler gave at Berlin's Sportpalast" that year, "the Party netted at least thirty thousand marks. . . . He had become a fundraising machine. Most parties used amounts like that to finance an entire local campaign."[92] It was already paradoxical back then that speeches from famous anti-capitalists did not come cheap, and it remains paradoxical as of this writing—a single address from Chomsky will cost an organization 12,000 dollars.[93] Despite its failure to win the support of the Industry Club, by January 26, 1933, the NSDAP boasted 800,000 members and could rely on a number of voters ranging somewhere from 6.5 million to 13 million voters. "The party's strength," surmises Fest, "depended on these legions of little people, and Hitler had to keep in mind their 'enormous anticapitalist nostalgia.' . . . he was more attuned to them than to the proud, pigheaded businessmen."[94]

Getting Down to Business

Alan S. Milward confirms other significant facts. Yes, some German firms, such as I. G. Farben, eventually tried to ingratiate themselves to the Nazi regime in order to get the regime to pass laws to restrict their competitors. However, ultimately, "the support for the National Socialist party came in large measure from a section of the population whose political sympathies were in many ways antipathetic to the world of big business. . . . This support was maintained by a persistent anti-capitalist rhetoric but also by a certain amount of legislation which cannot by any shift of argument be explained by a theory which assumes National Socialism to be a stage of state capitalism." Milward identifies the legislation as "a movement of protest against modern economic development . . ."[95] German big businesses began talking with Hitler after the NSDAP grew rapidly in popularity and influence in the 1930s. And at that point they remained skittish about supporting Hitler. Kershaw historicizes that the business tycoons' conversations with the Nazis were "insufficient to remove the 'socialist' image of the party in the eyes of many businessmen." The businessmen distrusted the NSDAP on account of how it took the side of Berlin's metalworkers during their strike in the fall of 1930. The businesses were askance of the Nationalsozialistiche Betriebszellenorganisation—NSBO, or National Socialist Factory Cell Organization. A group that fancied itself a sort of trade union—and which would later become the one labor union authorized to operate under the Third Reich—the NSBO voiced support for four other industrial strikes in 1931. The majority of the industrialists therefore "retained their healthy scepticism about the Hitler Movement during 1931." The majority of industrialists abhorred Hitler. Supporters in the vein of steel magnate Fritz Thyssen were in the minority.[96]

"All across the country," states Turner, "attacks on capitalists, capitalism, and high finance" pervaded the Nazi Party's speeches and campaign literature throughout the early 1930s." Sure, these same Nazi campaigners continued "their verbal assaults on 'Marxists' and 'Marxism,'" but that gesture "provided scant comfort for big business . . ." The reason was that "the party press proposed an array of new welfare measures," including a measure to mandate the payment of compensation to industrial workers dislocated from their employment. As these initiatives "would unavoidably have necessitated tax increases, their endorsement by propagandists of the NSDAP could only reinforce the overwhelmingly negative assessment of Nazism in the business community."[97]

German big businesses did not starting pouring out large donations to the NSDAP until after the party had already attained popular influence. This was not primarily because Germany's corporate managers professed faith in the Nazi ideology. Contrariwise, it was because they wanted to hedge their bets with any political party that won office. As late as 1932, Hitler still received more fanfare from farmers and military officials than he did from wealthy industrialists. "In the final drama," writes Kershaw, "the agrarians and the army were more influential than big business in engineering Hitler's takeover."[98]

But by November of that year, adds Turner, businesses finally gave up their previous hopes of being able to defeat Hitler politically. The Nazi Party's coffers truly began to swell with corporate funding when a majority vote by the Reichstag—that is, Germany's legislature—had democratically ratified the Enabling Act, which granted Hitler unilateral lawmaking authority. At that point, major German corporations made financial contributions to Hitler out of fear of political recriminations.[99]

The major corporate funding started to roll in on February 20, 1933, when Hitler summoned the country's top industrialists to a presentation on his new economic plan. Once the presentation drew to a close, the NSDAP solicited the industrialist attendees for funds. The industrialists promised three million marks to the Führer and provided it within weeks. This offering, reports Kershaw, "was less one of enthusiastic backing than of political extortion."[100] Realizing these facts, six-time U.S. Socialist Party presidential candidate Norman Thomas (1884–1968) admits, "In no way was Hitler the tool of big business. He was its lenient master. So was Mussolini except that he was weaker."[101] Inasmuch as there was any such leniency, it was short-lived. That I will explicate in chapter 10.

Remarkably, notes Henry Ashby Turner, there were advocates for everyone's liberty in Germany, including the Jews' right to do business, "until the end." Turner cites the example of the Hansa-Bund, "a pro-free-trade, anti-cartel organization with a sizable following in banking, commercial, and manufacturing circles . . ." The Hansa-Bund stated that it considered the Nazis the "vigorous enemy of the individualist and capitalist order for which we stand."[102] Of special note is BASF executive Carl Bosch. The reader may remember from *Life in the Market Ecosystem* that Bosch assisted Fritz Haber in developing the Haber-Bosch process for synthesizing ammonium nitrate to fertilize crops, which greatly improved farm productivity in the twentieth century and ended China's famines in the 1980s. During the entire duration of Hitler's political ascension, Bosch openly spoke out against the Nazis and anti-Semitism.[103] Years into Hitler's reign of dictatorship, Bosch eventually decided to refrain from publicly airing his views and to have BASF comply with the Nazi state's orders.[104] But this newly found silence and compliance came from the desire to protect himself and his family from censorious reprisals, not from any delusion that the Nazis' interests aligned with his own interests as a businessman.

Analyzing the available evidence, Henry Ashby Turner concludes that it is by no means apart from "gross distortion" that Germany's big businesses can "be accorded a crucial, or even major, role in the downfall" of the Weimar Republic. And he goes on that the repetition of the academician's claim that Naziism was caused by industrial commerce amounts to propaganda.[105]

Stanley G. Payne is a professor of history at the University of Wisconsin at Madison. Payne likewise writes that left-wing academicians "have failed to prove that big business 'bought' Hitler. . . . That the Marxist interpretation is incorrect is however greatly to be regretted, for if German big business had managed to buy Hitler there would probably have been no general agreement for war and the world might be a better place today."[106]

Despite his efforts to deny that the Nazis were consistently anti-capitalist, Richard Grunberger admits that the Nazi Party imposed regulations that forbade specific joint-stock businesses from issuing dividends that exceeded the company's periodic expenses by 6 percent. Likewise, he admits that the Nazi Party's supporters showed loyalty to the party on account of its promise to enact "state curbs on big firms" and department stores. In spite of department stores not being completely eliminated under the Third Reich in the end, they "suffered…curtailment" to such an extent that, between 1933 and 1936, the department stores did appear "to be heading for extinction. . . . Public agencies were forbidden to let department stores . . . tender for contracts." The department stores "were subjected to a two-pronged method of attack: governmental prohibitions on their economic activities...and boycotts promoted by the Party and such affiliated formations as the League of Civil Servants. In 1935 a local official of the Labour Front could confidently threaten, 'If Frau W. is seen shopping at Karstadt's store once again I shall have her husband deprived of his military pension.'"[107]

Was Hitler a Stooge of Corporate America?

Having failed to prove that German big business put the Nazis in charge, several governists turn toward identifying large U.S. corporations as the biggest factor in placing Hitler on the path toward global hegemony. As told by another one of *The Corporation*'s interviewees, filmmaker Michael Moore, "I think one of the greatest untold stories of the twentieth century is the collusion between corporations, especially in America, and Nazi Germany," such as "how corporations from America helped" essentially to build Germany "and support the early Nazi regime, and then, as the war broke out, figured a way to keep everything going."[108] Another left-wing Michael—Parenti, professor of political science—seconds that assessment. Parenti reviles the "hypocritical quality" of U.S.-based corporations that wrap themselves in the U.S. flag. Their hypocrisy is "blatantly displayed" in their "dealings with the Nazi regime before and during World War II."[109]

One of the companies that Michael Moore targets for abuse is Coca-Cola. He propounds that during the Second World War, Coca-Cola still desired to sell soft drinks in Germany, albeit without Americans knowing that it was accruing cash from the same country they were fighting. To get around this, insists Moore, the Coca-Cola company "invented Fanta Orange for the Germans. . . . That's how Coke was able to keep the profits coming in" from Europe. "So when you drink Fanta Orange, that's the Nazi drink that was created so that Coke could continue making money when millions of people died."[110]

That the producers of *The Corporation* kept this scurrilous untruth in their film spotlights their ineptitude in fact-checking. *Snopes.Com*, the website renowned for debunking urban myths, gives the scoop on what actually happened. Prior to the war, Coca-Cola set up a German division headed by American-born Ray Powers. Abroad, Powers hired German-born Max Keith (his last name is pronounced "kite")[111] as his assistant. When Powers died from automobile injuries, Keith replaced him as the general manager of Coke's German division.

When war broke out, communication between Keith and Coca-Cola's home office in Atlanta, Georgia, had been cut off. Keith assured the Coca-Cola bottlers in neutral Switzerland that he had no intention of liquidating Coke's German subsidiary. As the war had rendered it impossible to secure ingredients for Coke production, though, Keith was able to produce and sell fewer bottles' worth than before. To keep the cash flow going, he decided that he had to market some new soft drink assembled from ingredients that were more widely available. He asked his plant's employees for suggestions on what to name the new beverage. In his suggestion, he told the works to let their *fantasie*—German for "fantasy"—be uninhibited. In response, salesman Joe Knipp immediately shouted out *Fanta*. Although Keith was a private businessman, he often found that the Nazi state gave him production orders and threatened him with punishment if he did not comply. When the war ended, Keith reestablished contact with Coke's Atlanta office and helped the company resume Coke production in postwar Germany.

Upon the war's end, Coca-Cola commissioned an investigator to look into Keith's activities in the unsupervised period. The investigation found that neither Keith nor his lawyer, Walter Oppenhoff, ever joined the Nazi Party, despite the reprisals that were promised against them on account of this refusal. The Third Reich also demanded that Keith appoint Nazi leaders to the German Coca-Cola division's board of directors, but Keith did not obey. Due to Keith's efforts, many American Coke employees were able to remain alive during the war. Keith turned over both the wartime revenue from Coca-Cola sales, and ownership of Fanta itself, to the American parent company.[112] The man practiced a remarkable code of honor throughout these hard times. In lieu of the respect that Keith deserves, we find his creation being ignorantly impugned by the likes of Michael Moore and the producers of *The Corporation*. Michael Parenti ignores the reality of this case study when he asserts, "Throughout the war, US corporate chiefs were able to maintain direct ownership and control over their German subsidiaries with minimal interference from the Nazis, who were primarily interested in keeping war production going."[113]

Another U.S.-based company that *The Corporation* accuses of aiding Naziism is International Business Machine—better known as IBM. In the propaganda documentary, Edwin Black explains, "When Hitler came to power in 1933, his goal was to dismantle and destroy the Jewish community. This was an enterprise so vast that it required the resources of a computer. But in 1933 there was no computer. What there was, was the IBM punch card system, which controlled and stored information based upon the holes that were punched in various rows and columns. Naturally there was no off-the-shelf software that there is today. Each application was custom-designed and "an engineer had to . . . configure it" directly, in a hands-on manner. "Millions of people of all religions and nationalities and characteristics went through the concentration camp system. That's an extraordinary traffic management program that required an IBM system in every railroad direction and an IBM system in every concentration camp."[114] Black adds that, rather than simply be sold to the Third Reich, such tabulating machines were leased, and they required on-site servicing once a month by an IBM employee. As IBM's German division had to supply regular maintenance for these machines, Black would have us believe, this proves that IBM's American home office was perforce aware of, and condoned, what was happening in the concentration camps. As we shall learn soon, it is misleading to assume that the German division reliably communicated with or relayed information to the U.S. home office. Still, Black claims to have in his possession a letter from October 9, 1941, addressed directly to IBM's CEO at the time, Thomas J. Watson, Sr. (1874–1956), detailing the activities of IBM's German subsidiary. Black compiled these claims in his book *IBM and the Holocaust*. Michael Parenti repeats them—International Business Machine "prospered in Germany and the occupied territories by supplying the technology needed to identify, enslave, and exterminate millions of European Jews and other victims."[115]

Reviewing Black's book in the *New York Times*, Richard Bernstein observes that it sabotages itself in that it ignores "the moral distinction between the sellers of rope and those who use rope to hang people. . . . Mr. Black's case is long and heavily documented, and yet he does not demonstrate that I.B.M. bears some unique or decisive responsibility for the evil that was done."[116] *Commentary* magazine senior editor Gabriel Schoenfeld similarly deliberates—also in the *New York Times*—that the main question is not whether

IBM leased its equipment to Germany "but whether, as alleged, it made the Final Solution part of its 'mission' and whether its relationship with Germany in any way 'energized' or significantly 'enhanced' Hitler's efforts to destroy world Jewry. On the first point, Black never even attempts to substantiate his accusation—a scandalous omission considering the gravity of the charge."[117] A double standard becomes obvious in this case. Black accuses a company of being complicit in murder if murderers take advantage of its product or service. Simultaneously, this same Black does not give commensurate praise for companies that provide goods or services that save or extend lives, such as the manufacturers of MRI machines that assist in the early detection of cancers, or the pharmaceutical companies that produce genetically engineered insulin for diabetics.

More relevantly, Black's portrayal of Watson's relationship to IBM's Germany subsidiary and the Third Reich distorts actual events. Award-winning *USA Today* technology columnist Kevin Maney sets the record straight in his historical tome, *The Maverick and His Machine: Thomas Watson, Sr. and the Making of IBM*. It will be granted that Watson did not take the strong moral stand against Naziism that he should have taken as soon as he learned of the mistreatment of Jews. Even so, it must also be admitted that Watson had limited control over what IBM's Germany subsidiary did, and limited knowledge of that, besides. The story began in 1910 when Willy Heidinger started a company in his native Germany called Deutsche Hollerith Machinen Gesellschaft, commonly abbreviated as *Dehomag*. The "Hollerith" in the German company's name refers to Herman Hollerith (1860–1929), the American inventor of the punch-card tabulator that was then IBM's flagship product. Heidinger owned this entire outfit of Dehomag outright, and he won a license to sell and lease IBM's patented tabulating machines within his own country. To wit, Dehomag was a not a subsidiary of IBM. Rather, IBM was the franchisor, whereas Heidinger and Dehomag were the franchisee. Heidinger paid IBM's home office 10 percent of the cost of the tabulating machines that Dehomag imported and a royalty of 25 percent of the gross on the rentals. When the German government's governist inflation policies ruined the national economy, Dehomag found itself unable to cover its debts. Watson then informed Heidinger that IBM itself would save his company if he allowed IBM to take a controlling interest in it. Over the course of several years IBM eventually came to own 90 percent of it whereas Heidinger held onto the remaining 10 percent, managing it as a semi-independent subsidiary. Kevin Maney notes that Heidinger still directed Dehomag "with some autonomy." And by "the late 1930s Dehomag had become a self-contained business. It operated factories, designed its own products, licensed patents from IBM New York to build other products, and ran its own sales organization. Dehomag specialized in one product that is chilling in retrospect: It made the world's best punch card machines for taking a census. Dehomag's census machines were so good, IBM in the United States stopped making census machines, preferring to import Dehomag's."

The Third Reich used machines not from IBM New York, but from Dehomag. In 1933 the Nazis had conducted a census to establish who was Jewish and who was not. Maney goes on that the IBM chief "didn't know in 1937 how the Nazis were using Dehomag or IBM machines. Though the Nazis had oppressed Jews and other non-Germans" through laws enforcing segregation and discrimination, "few outside of Germany knew about the Nazi plans for genocide, and even fewer knew about the inner workings put in place to carry out those plans."

Edwin Black mentions that on June 28, 1935, the Nazis handed Watson an honorary medal for his promotion of American-German trade as head of the International Chamber of Commerce. The anti-capitalist journalist conveniently omits mention of the wider context of the situation. The omission prods the reader to infer that the award was for Watson's allegedly intentional contribution to the Holocaust. In reality, Watson did not expect to receive that award. He met with Hitler that day to secure a promise from the

German leader that he would not plunge into any war against other European powers, but would instead promote peace through trading with them and other nations. Later that night, Watson attended a banquet thrown by Joseph Goebbels, and the award ceremony was a surprise that the Nazis sprung on Watson, intending for him to feel flattered. Watson was happy with the medal, too, until the November of 1938, amidst reports of Nazi gangs assaulting Jews and vandalizing their stores. At that point, Watson naïvely wrote letters to the Nazi regime, begging them to put a stop to the mayhem. In 1939, Watson wrote a letter addressed directly to Hitler, admitting "a loss of good will to your country" and pleading that he respect rights "in dealing with these minorities." Rather than provide that letter to the Führer, the Nazi regime returned that letter to IBM's U.S. office unopened and unread. On March 7, 1939, IBM sent off a second letter to Hitler, this one signed by Byne Waters, who worked as Watson's secretary. She enclosed the original letter with this second one. That letter was not returned to IBM in the USA.

On June 6, 1940, Watson mailed the Nazi medal back to Hitler, enclosing it with a letter wherein Watson rebuked him for having started the war in defiance of his very promise to avoid it. On the matter of the medal, the letter stated, "In view of the present policies of your Government, which are contrary to the causes for which I have been working and for which I received the decoration, I am returning it.'" Following that gesture, Watson denounced Hitler publicly. Thereon, correspondence between IBM New York and Dehomag grew increasingly rancorous until the USA's entry into the war completely severed communication between the two firms. One of Dehomag's internal memos stated, "This stupid step of Mr. Watson's opens up a number of possibilities. It is not improbable that such a step may harm the company and all of us very seriously sooner or later since it must be considered as an insult to the Fuhrer and thereby of the German people." That same memo surmised, "It appears Mr. Watson is surrounding himself with a group of Jews who fled Europe" and that such Jews must be "beginning to affect his mind and impede his judgment."[118] Edwin Black's smears on IBM fail, as do the radical Left's repeated efforts to affix capitalism with some culpability for the Third Reich's ascension. All in all, concludes Joachim Fest, "To call Hitler a tool of capitalism, as Marxist theory does, is merely to fall back on belief in demons. Marxist orthodoxy is prone to such simplifications. Such demonology is, as it were, 'the anti-Semitism of the Left.'"[119] The Nazis' anti-Semitism, indeed, can largely be attributed to the genuine hatred for capitalism among the citizens of the Axis Powers.

NOTES

1. A. Kelly 1981, 107.
2. A. Kelly 1981, 107; G. L. Mosse 1985, trade paperback, 79; and Poliakov 1974 U.S. version, 294.
3. Poliakov 1974 U.S. version, 294. A. Kelly 1981, 107, translates the topic of the contest as, "What can we learn from the theory of descent with regard to domestic political development and state legislation?"
4. G. L. Mosse 1985, trade paperback, 79, points out that Schallmeyer contributed to Alfred Ploetz's eugenics-themed periodical.
5. That Schallmayer won is pointed out in A. Kelly 1981, 107; and G. L. Mosse 1985, trade paperback, 79.
6. A. Kelley 1981, 108; and G. L. Mosse 1985 trade paperback, 79.
7. G. L. Mosse 1985, trade paperback, 79.
8. A. Kelly 1981, 108.
9. Poliakov 1974 U.S. version, 295.
10. A. Kelly 1981, 108; and Poliakov 1974 U.S. version, 294.
11. A. Kelly 1981, 108.
12. Poliakov 1974 U.S. version, 295, citing and quoting from *Natur und Staat, Beiträge zur Naturwissenschaftlicher Gesellschaftslehre, Eine Sammlung von Preisschriften*, Jena 1903, pp. 1-24.
13. A. Kelly 1981, 108.

14. G. L. Mosse 1964, 98. B. P. Copenhaver, and Copenhaver trans. and ed. 2012, 638 n. 183, also mention Woltmann's Marxism.
15. B. P. Copenhaver and Copenhaver trans. and ed. 2012, 638 n. 183.
16. G. L. Mosse 1964, 100.
17. Poliakov 1974 U.S. version, 296, citing Ludwig Woltmann, *Politische Anthropogie, Eine Untersuchung über den Einfluss der Descendenztheorie auf die Lehre von der politischen Entwicklung der Völker*, Leipzig, 1903.
18. A. Kelly 1981, 105.
19. *Natur und Staat, Beiträge zur Naturwissenschaftlicher Gesellschaftslehre, Eine Sammlung von Preisschriften*, Jena 1903, pp. 1–24, qtd. by Poliakov 1974 U.S. version, 295.
20. A. Kelly 1981, 102.
21. G. L. Mosse 1985, trade paperback, 148.
22. G. L. Mosse 1964, 88, 91; and G. L. Mosse, 1985 trade paperback, 60, 234.
23. G. L. Mosse 1964, 88-89.
24. For Nicholas Goodrick-Clarke's explanation of how it was Joseph Arthur de Gobineau who pioneered in the racialist view of the Nazis, see Goodrick-Clarke 1992, paperback, 123–24. See also I. Berlin 1970b, 47; G. L. Mosse 1964, 90; and Shirer 1990, 103–04.
25. Spiro 2009, 103–107.
26. A. Herman 1997, 51.
27. I learned of these quotations from A. Herman 1997, 50, which cites M. Biddiss 1970, 17.
28. G. L. Mosse 1985, trade paperback, 53, citing R. E. Dreher 1970, 84.
29. G. L. Mosse 1985, trade paperback, 53.
30. I. Berlin 1970b, 47; Goodrick-Clarke 1992, paperback, 124-25; and Shirer 1990, 103-09.
31. Spiro 2009, 108–113.
32. T. Roosevelt 1913b, 240–43.
33. B. P. Copenhaver and Copenhaver trans. and ed. 2012, 638 n. 183; and G. L. Mosse 1964, 93.
34. G. L. Mosse 1985, trade paperback, 104.
35. G. L. Mosse 1985, trade paperback, 105; and Poliakov 1974 U.S. version, 313.
36. B. P. Copenhaver and Copenhaver, trans. and ed. 2012, 638 n. 183.
37. G. L. Mosse 1964, 94, citing H. S. Chamberlain 1912 vol. 1, 25ff., 558; and H. S. Chamberlain 1912 vol. 2, 967.
38. G. L. Mosse 1964, 97, citing H. S. Chamberlain 1912 vol. 2, 863.
39. G. L. Mosse 1964, 94, citing H. S. Chamberlain 1912 vol. 1, 25ff., 558; and H. S. Chamberlain 1912 vol. 2, 967.
40. B. P. Copenhaver and Copenhaver trans. and ed. 2012, 638 n. 183.
41. H. S. Chamberlain 1912 vol. 2, 353.
42. H. S. Chamberlain 1911 vol. 1, 459 n. 152–53, 348, 332, 410 n.
43. H. S. Chamberlain 2005, 119.
44. Immanuel Kant qtd. by H. S. Chamberlain 2005, 118–19.
45. Immanuel Kant, *Vermischte Schriften*, qtd. by Poliakov 1974 U.S. version, 172.
46. A. Herman 1997, 74; G. L. Mosse 1964, 93; and Shirer 1990.
47. A. Herman 1997, 74.
48. G. L. Mosse 1964, 93.
49. Hitler 2001, 232.
50. Goodrick-Clarke 1992, paperback, 124; and Spiro 2009, 113–16.
51. G. L. Mosse 1985, trade paperback, 60–61.
52. G. L. Mosse 1985, trade paperback, 61, citing J. M. Winter 1974, 190–91.
53. Poliakov 1974 U.S. version, 295–96.
54. Poliakov 1974 U.S. version, 296, citing Karl Saller, *Die Rassenlehre des Nationalsozialismus in Wissenschaft und Propaganda*, Darmstadt 1961, pp. 17–18.
55. Goodrick-Clarke 1992, paperback, 13, 243 n. 43; and Spiro 2009, 122–26.
56. E. Haeckel 1904, 390–91.
57. Poliakov 1974 U.S. version, 294, 297.
58. A. Herman 1997, 136; and Poliakov 1974 U.S. version, 297.
59. Poliakov 1974 U.S. version, 297.
60. A. Herman 1997, 416.
61. Spiro 2009, 103, 116, 126, 133, 105,113.
62. J. Marks 2002, 120; and Spiro 2009, 375.
63. Spiro 2009, 357–361. E. Faye 2009 U.S. version, 18, identifies Günther as being "among the first and foremost theoreticians of the Nazi racial doctrine."
64. E. Faye 2009 U.S. version, 70.
65. Spiro 2009, 182.
66. E. Faye 2009 U.S. version, 18.
67. Goodrick-Clarke 1992, paperback, 13, 243 n. 43.

68. Karl Pearson, *Speeches Delivered at Dinner Held in University College, London, in Honour of Professor Karl Pearson*, April 23, 1934, Cambridge, England, year 1934, p. 23, cited by G. L. Mosse 1985 trade paperback, 76.
69. E. Faye 2009 U.S. version, 68, 70, 97, 261.
70. R. Grunberger 1971, 18.
71. G. Hodgson 2004.
72. J. C. Fest 1974 U.S. version, 5.
73. W. Biddle 2009, 5.
74. Shirer 1990, 144.
75. Kershaw 2000, paperback, xxii.
76. Du Bois, "Germany and Hitler," *Pittsburgh Courier*, December 5, 1936, republished in Du Bois 1995, 737.
77. Zinn interviewed in Achbar et al., prod. 2003, transcript online in the PDF at http://hellocoolworld.com/files/TheCorporation/ Transcript_finalpt2 copy.pdf. accessed Thursday, May 17, 2007.
78. Chomsky interviewed in Achbar et al., prods. 2003.
79. Fest 1974 U.S. version, 306, 756.
80. Kershaw 2000, paperback, 358.
81. Fest 1974 U.S. version, 789, 306.
82. Kershaw 2000, paperback, 357.
83. H. A. Turner, qtd. by J. Toland 1976, 277n.
84. Fest 1974 U.S. version, 166; Gellately 2007, 185; Kershaw 2000 paperback 359; J. Toland 1976, 277; and H.A. Turner 1985, 115–16.
85. H. A. Turner 1985, 115-16.
86. Fest 1974 U.S. version, 166; and H.A. Turner 1985, 115–16.
87. Fest 1974 U.S. version, 166.
88. Gellately 2007, 185; and H.A. Turner 1985, 115.
89. Fest 1974 U.S. version, 166.
90. J. Snyder 2000, 118, citing R. Bessel 1992, 404, 412–13; Fritzsche 1992, 301–04; and W. Mommsen 1993, 359, 372.
91. That Hitler raised funds from speaking engagements is in Gellately 2007, 185, and H. A. Turner 1985, 118–19.
92. Gellately 2007, 185, 202.
93. P. Schweizer 2006, 27.
94. Fest 1974 U.S. version, 306–07.
95. Milward 1977, 12–13.
96. Kershaw 2000, paperback, 358.
97. H. A. Turner 1985, 114–15.
98. Kershaw 2000, paperback, 425.
99. H. A. Turner 1985, 302, 253–54.
100. Kershaw 2000, paperback, 447–48.
101. N. Thomas 1971, 53.
102. H. A. Turner 1985, 124.
103. D. Charles 2005, 234; M. Goran 1967, 156; and T. Hager 2008, xv.
104. D. Charles 2005, 234, 244; and T. Hager 2008, 80–85.
105. H. A. Turner 1985, 340, 350–53.
106. S. G. Payne 1980, 58.
107. R. Grunberger 1971, 167–68, 18–19, 44–45, 173.
108. Moore interviewed in Achbar et al., prods. 2003, transcript online in the PDF at http://hellocoolworld.com/files/TheCorporation/Transcript_finalpt2 copy.pdf, accessed Thursday, May 17, 2007.
109. M. Parenti 2004, 117.
110. Moore interviewed in Achbar et al., prods. 2003.
111. M. Pendergrast 1993, 218, points out that Max Keith's name was pronounced *kite*.
112. F. Allen 1994, 256–264; and M. Pendergrast 1993, 220–231.
113. M. Parenti 2004, 119.
114. Black interviewed in Achbar et al., prods. 2003.
115. M. Parenti 2004, 119.
116. R. Bernstein 2001, http://tinyurl.com/368j5j, accessed Thursday, May 17, 2007.
117. Schoenfeld 2001, http://tinyurl.com/9o7ys, accessed Thursday, May 17, 2007.
118. K. Maney 2003, 204–05, 214–15, 218–222.
119. Fest 1974 U.S. version, 307.

SEVEN

Socialism and Fascism

Close Relatives

Capitalism + Murder = Fascism?

In his otherwise excellent and well-informed book *Inside Terrorism,* foreign-policy scholar Bruce Hoffman parrots a false accusation advanced by other left-wingers—that Naziism and fascism are ideologies that have "assailed" governist, left-wing "liberal social welfare policies . . ."[1] Upton Sinclair levels a still more negative appraisal toward private entrepreneurs: "Fascism is capitalism plus murder."[2] March of the year 2000 saw, in a highly acclaimed work published by Harvard University Press, sociologist Antonio Negri (b. 1933) and Duke University literature professor Michael Hardt (b. 1960) proclaiming that "...Nazi Germany is the ideal type of the transformation of modern sovereignty into national sovereignty, and of its *articulation in capitalist form . . .*"[3] (emphasis added).

Fortunately for humanity, U.C. Berkeley political scientist A. James Gregor disagrees with this conventional, PC interpretation. He writes to his fellow professors, "We have lived so long with the comfortable conviction that Fascism was of the right [wing] and as a consequence shared nothing with the left, that we are loath to consider any alternatives." Gregor, as I do, implores readers to consider such alternatives. That sort of consideration is not easy to garner among the professorate. Gregor mentions that when Italian historian Renzo De Felice (1929–1996) presented the case "that Mussolini's Fascism shared considerable affinities with the traditional and revolutionary left," the "suggestion was enough to outrage Italian intellectuals and much of the international academic community."[4] The rhetorical obfuscations of the governist Left notwithstanding, the fascists and the Nazis recognized from the beginning that the principles of laissez faire were inimical to their own ambitions.

Italian dictator Benito Mussolini (1883–1945) coined the word *fascism* in order to emphasize the collectivist basis of his political system. As Dictionary.Com explains, *fascism* comes from the root word *fascio,* which means "bundle" or "group" or "league"[5] — appropriate for a regime that emphasized the importance of the social group above the individual. In fact, *fascio* in Italy was a synonym for *labor union.*[6] As the Nazis have frequently been called fascists, it may astonish the reader that neither Mussolini nor the Nazis themselves thought it was appropriate to describe the Third Reich as fascist. Of National Socialism and his own government, Mussolini explicated that they share many important "similarities. Both are authoritarian systems, both are collectivist, socialistic. Both systems oppose [laissez-faire] liberalism."[7] But fascism, unlike Marxist-Leninism, "recognizes religion and family."[8] And, particularly relevant to our discussion, the Italian dictator summed up that in contrast to Lockeanism, National Socialism "allows no rights to the individual..." Despite the similarities between his fascist regime and the Nazis, Mussolini said he wished that people would stop applying the label of his philosophy to that of the Nazis'.[9] When he contributed an explanation of fascism to an anthology published in 1942, Il Duce espoused a doctrine that cannot be squared with free-market individualism. Fascism, he writes, goes beyond the superficiality of sensory experience, "in which man is an individual separated from all others and standing by himself, and in which he is governed by a natural law that makes him instinctively live a life of selfish . . . pleasure." Fascism, by contrast, is about "binding together individuals" to "restore within duty a higher life . . . in which the individual, through the denial of himself, through the sacrifice of his own private interests . . . realizes that completely spiritual existence in which his value as a man lies." The mindset must be considered a "reaction of modern times against the flabby materialistic positivism of the nineteenth century. . . . Against individualism, the Fascist conception is for the State. . . . It is opposed to classical [laissez-faire] Liberalism . . ." Fascism is better than laissez-faire liberalism, for the nineteenth century's relative laissez-faire liberalism "denied the State in the interests of the particular individual . . ."[10]

The definitions of capitalism and free market do not suggest that they constitute synonyms for Nazism. According to Dictionary.Com, "Capitalism encourages private investment and business, compared to a government-controlled economy."[11] Meanwhile a free market is an "economy where buyers and sellers are allowed to transact freely (i.e., buy/sell/trade) based on a mutual agreement on price without state intervention in the form of taxes, subsidies or regulation."[12] Concomitantly, Dictionary.Com calls free enterprise the "freedom of private businesses to operate competitively for profit with minimal government regulation,"[13] and laissez faire the "theory or system of government that upholds the autonomous character of the economic order, believing that government should intervene as little as possible in the direction of economic affairs."[14] By contrast Dictionary.Com notes that fascism is characterized by "stringent socioeconomic controls,"[15] and that the Nazis established "a dictatorship over all cultural, economic, and political activities of the people . . ."[16]

It further states that *Naziism* involves "state control of the economy,"[17] as it is a "form of socialism featuring racism and expansionism and obedience to a strong leader."[18] Moreover, under Mussolini the Italian government did not discriminate against Jews from 1919 to the 1930s, when the percentage of Jewish employees in the Italian government outweighed the percentage of Jews comprising Italy's general population. It was not until 1938, pursuant to Hitler's invasion of Italy, that Italy's fascists began to persecute Jews.[19]

Laudably, A. James Gregor explicates the fascist ideology's origins for us. Mussolini started out in politics as a committed socialist and played a managerial role in an Italian socialist political party. His particular brand of socialism was known as *revolutionary syndicalism*. Moreover, Mussolini's origination of the fascist ideology had nothing to do with Mussolini "selling out" his devotion to governism and economic collectivism. He maintained his opposition to capitalism throughout his life; his departure from traditional socialism, and adoption of the fascist ideology, came about from him and some of his compatriots dissenting from their fellow Italian socialist party members over political tactics.

Fascism Budding Off from Socialism

The disagreement started during the First World War. Most members of the Italian socialist party eagerly anticipated a worldwide uprising of socialists against the leaders of every European nation. For this reason, much of the leadership of Italy's socialist movement in 1914 strongly urged that the Italian socialist parties take a neutral position concerning World War One. As far as their interpretation went, every European country was much too capitalist, and it was the capitalist nature of both sides of the conflict that propelled them to clash from the outset. By contrast, Mussolini and some other Italian socialists felt that England and France held the moral high ground over Germany and Austria, and that Italy should therefore fight on the side of the former two nations against the latter. Mussolini and several of his party comrades argued that Germany and Austria were far more socially primitive than England and France, and that a German-Austrian victory would retard the historical progression toward global socialism, hopelessly setting the movement back.

For this reason, four of Mussolini's sympathizers from the Italian socialist party—Angelo O. Olivetti, Alceste De Ambris, Filippo Corridoni, and Massimo Rocca—issued a *Manifesto* on October 5, 1914, to express their opinion forthright. They maintained full agreement with the economic policies of the mainline socialists, but departed from them by supporting England and France. Thereupon they split off from the socialist hard-liners to form their own movement, which they dubbed *Fascio Rivoluzionario D'Azione Internazionalista*, roughly "revolutionary international fascist activism." To explicate this conclusion further, Olivetti launched a new ideological journal, *Fascio*. A. James Gregor writes that in the October 10, 1914 issue of this periodical, Olivetti exhorted Italy's "socialists to face the urgent issue of making the social revolution *national* in character, for events had shown that the sentiment of nationality 'superseded and influenced every other.'" In that same issue, Filippo Corridoni "argued that traditional socialism had lightly dismissed the national sentiments of the working classes while events had proved them mistaken and had dissolved the old antinationalism." That other pioneering fascist, Alceste De Ambris, "insisted that the war had become a national and revolutionary duty." This position was not completely anathema to socialism. Antonia Gramsci (1891–1937), a major figure in Italian Marxism, publicly defended the fascist splinter group's stand.

As for Mussolini himself, he "came to recognize that the ultimate interests of the proletariat were intimately and fatally linked with those of the fatherland." Accordingly, Mussolini thought "the socialism of the future could only be a *national* socialism . . ."[20] (emphasis Gregor's).

From Karl Marx's day to the early 1900s, main-line socialist ideology preened itself on being international; the socialists fancied themselves as broadminded thinkers who transcended nationalism, as socialists in every country took up common cause. By distinction, Mussolini and the first fascists still prioritized socialism, but wanted each nation to have its own socialist government, one that did not necessarily adhere to the foreign-policy stances of other socialist nations. Hence the first fascists pioneered in advocating something that was, if not a national socialism, a variant of socialist nationalism. Affirmative, Mussolini drifted from mainstream Italian socialism and gravitated into the orbit of the fascists. In that transition, he did not waver from governism or economic collectivism. Nor did he adopt any new appreciation for individualism or peaceful entrepreneurship.

University of California, Irvine, historian John P. Diggins observes that soon after Mussolini's coup in Italy, "the *Weekly People*, official organ of the Socialist Labor Party (SLP), reprinted from the *New Republic* Giuseppe Prezzolini's 'The Fascisti and the Class Struggle.' Prezzolini, who later became an avid champion of the Fascist movement, teaching at Columbia [University], gave a sanguine account of the movement, insisting that the *fascisti* had wisely stolen the syndicalist program and adapted it to Italian institutions and national character. 'Socialism will march forward under the Blackshirts, great whether it be manifested under the Fascist tri-color or under the flag of Socialism and Communism.'"[21]

Sometimes governists point out that Mussolini stated that a better name for his ideology than *fascism* was *corporatism*. When governists assert as much, they insinuate that Mussolini's corporatism was an economic system wherein big, competing, multinational corporations—very similar to the ones operating in the USA at the day of this writing—could do whatever they wanted. Robert F. Kennedy, Jr., of the Natural Resources Defense Council sounds off, "Mussolini complained that 'fascism should really be called corporatism.'" Kennedy has this as a lead-in to his screeching that, in modern times, American wealth has become "a grab bag for the robber barons."[22]

Left-wing cartoonist Ted Rall, too, cites this alleged Mussolini quotation as he affixes the fascism label to modern competition among multinational corporations. "Mussolini...famously observed that 'fascism should more appropriately be called corporatism, because it is the merger of state and corporate power' . . . "[23] The quotation did not actually originate with Mussolini, but with a philosopher who influenced him, Giovanni Gentile (1875–1944), in the Italian *Enciclopedia*.[24] Noam Chomsky and Michael Parenti then quote an 1825 letter written by Thomas Jefferson to William Giles that warns that the USA is becoming a "government of an aristocracy, founded on banking institutions, and moneyed incorporations . . ."[25] The implication of both Chomsky and Parenti is that Jefferson was warning against the dangers of laissez faire. Chomsky and Parenti then recommend that the U.S. federal government expand its own regulatory power over private business—something they imply that this particular Thomas Jefferson quotation endorses. Mussolini did speak the word *corporatism* but, contrary to the leftwingers' vociferations, it was not in a context that lends easy comparison to a system of for-profit, transnational entities.

It is silly, for several reasons, to proclaim that Mussolini calling fascism a corporate system is proof that corporations are evil. One reason is that in 1937, Mussolini also proclaimed, "Fascism is totalitarian democracy . . ."[26] To modern Western readers, such a statement must sound completely oxymoronic. Yet Mussolini did not think it so. Mussolini thought of *democracy* as referring to the will of the people as a collective. And, as fascism implemented a collectivist policy, Mussolini believed that fascism took the will of the collective of society to its final conclusion. On account of how they have tried to co-opt *democracy* as the word for their anti-capitalism, I doubt that American leftists will concede that *democracy* must be bad on account of Mussolini trying to use that word for his own ideology.

Moreover, when Mussolini alluded to *corporatism* and *corporativism*, he thought of "corporations" as the word was employed during the Middle Ages and Renaissance period, when *corporation* meant a collectivist guild and co-op. Under this economic system of heavy-handed government regulation—known as *mercantilism*—the guild received governmental protection in the form of laws that restricted any entrepreneurial startup's attempt to compete against the guilds members' industry.[27] When Thomas Jefferson, Thomas Paine, and Andrew Jackson denounced "corporations," they specifically referred to such institutions that, by legal definition, received a monopoly status from government regulations. By the nineteenth century, though, this trend was already on the wane in the United Kingdom. The British East India Company was the quintessential government-protected monopoly throughout the eighteenth century. But in 1834 the U.K. government refused to renew the corporation's monopolistic charter, thereby opening up much greater competition in the economy.[28] It is therefore misleading for Noam Chomsky and Michael Parenti to quote criticisms against "corporations" by Thomas Jefferson and other figures of the eighteenth century and early nineteenth century as if such criticisms applied to every multinational corporation existing from the 1900s onward. Worse, Chomsky and Parenti quote Jefferson's 1825 letter to Giles as if Jefferson agrees with them about the need for the federal government to expand its regulatory powers over private commerce. In fact, in this very letter Jefferson clarified that he was arguing the opposite—that he opposed an expansion of federal regulation over business. In this letter, Jefferson disputes the common rationalization that the U.S. Constitution's Commerce Clause allows for the federal government to perform any action upon private traders that goes beyond the parameters of the night watchman state. "Under the power to regulate commerce," Jefferson fumes, the governists of his day—the Chomskys and Parentis of his own time—presume that the government must have indefinite control "over agriculture and manufactures, and call it regulation to take the earnings of one of these branches of industry . . . and put them into the pockets of the other . . ."[29] Jefferson's letter mentioning the wrongness of government-regulated corporations was arguing in favor of the night watchman state, but Chomsky and Parenti misleadingly cite it as if Jefferson were opposing the night watchman state.

Mussolini's own writings on the subject clarify that his own corporatism—in common with the sort practiced in Great Britain centuries earlier—stands in direct contrast to laissez-faire capitalism. This is visible in Mussolini's 1934 review of *New Frontiers*, a book that FDR's onetime vice president, former agribusiness executive Henry A. Wallace (1888–1965), wrote to advocate New Deal policies. Mussolini praised Wallace's message as "corporativistic." As Il Duce cheered, the New Dealer tract "is both a declaration of faith and an indictment of economic [laissez-faire] liberalism. . . . Wallace's answer to what America wants is as follows: anything but a return to the free market, i.e., anarchistic economy. Where is America headed? The book leaves no doubt that it is on the road to corporatism, the economic system of the current century."[30] Note Mussolini's distinction between "corporatism" versus laissez faire. Also note that *corporatism* was the term that Mussolini applied to the New Deal's left-wing, progressive, mixed-economy, regulatory-entitlement state. *That* is actually the sort of social system that members of the Kennedy clan, including Robert Kennedy, Jr., have advocated throughout the twentieth century. Whatever grievances Robert Kennedy, Jr., may hold toward corporations, he himself is thus more of a "corporatist," by Mussolini's definition, than the consensualism proponents he disputes.

Judging first by how contemporary leftists à la Kennedy convey such negative appraisals of Mussolini, and judging secondly by how these same leftists lionize their American Progressive forebears from the early 1900s, one might assume that these American Progressive role models similarly denounced Mussolini. Such was not the case. An impressive roster of prominent American Progressives, who remain sainted at the time of this writing, publicly expressed approval for Mussolini and his fascist policies. Consider Lincoln Steffens (1866–1936) who, according to MIT economist Daron Acemoglu and Harvard political scientist James A. Robinson, "had made his name as...a muckraker journalist who had persistently denounced the evils of capitalism in the United States."[31] As will be recalled from Book Two, Steffens wrote in *The Shame of the Cities*, "The typical businessman is a bad citizen. . . . He is a self-righteous fraud."[32] While muddying the reputations of peaceful entrepreneurs, this same Lincoln Steffens built up some of history's most destructive tyrants. Although ambivalent about Il Duce's crackdown on civil liberties, Steffens nevertheless marveled at Mussolini's ability to impose whatever rules he wanted, whatever rules he shouted were necessary during the West's then-present financial doldrums. Steffens thus praised him and, in his public autobiography, remarked that God "formed Mussolini out of the rib of Italy."[33] Not surprisingly, Steffens held Vladimir Lenin's agenda in high esteem as well. Regarding his first tour of the Soviet Union in

1921, Steffens triumphantly told U.S. Ambassador William Christian Bullit, Jr. (1891–1967), "I have seen the future and it works!"[34] It was not until late in his life that Steffens let himself face the truth about the Soviets' horrifying brutality. In hindsight, I think we have learned who the self-righteous frauds really turned out to be.

Mussolini's Progressive American Contemporaries: His Apologists

Also consider Ida Tarbell, the persecutor of John D. Rockefeller, Sr., in the pages of *McClure's* magazine. Whereas she likened Rockefeller—a peaceful American citizen—to some oligarch, she was much less hard on an actual dictator such as Mussolini. At least she did not go as far as sharing Steffens's warm feelings toward the Soviets. Nonetheless, Tarbell was like Steffens in having some ambivalence about Mussolini. She held reservations about the undemocratic aspects of Mussolini's rule but was likewise charmed by his take-charge attitude. John P. Diggins writes that Tarbell, too, "became fascinated by the experimental features of the Corporate State and wrote lavishly about the moral uplift" in it. For that reason Tarbell traveled to Italy herself to interview Il Duce face-to-face. In Diggins's words, once Tarbell returned to the USA, she informed the U.S. Department of State that "the new labor laws in Italy constituted an admirable social experiment."[35]

In her 1939 memoir, Tarbell remarks on how Il Duce took power in a relatively benign fashion. "I had been amazed, and had never ceased to be amazed," that Mussolini's "dramatic march on Rome had...been carried out without bloodshed. An astonished world had seen tens of thousands of unorganized and in part unarmed men march from every point in Italy to Rome, call for Mussolini, get him by order of the King and march home again. . . . It was the most amazing transfer of government I had known of." She adds that, overall, the most ardent defenders of parliamentary republicanism felt they had no option but to accept Mussolini. "'He has saved the country,' men told me." Her visit was not what she expected. She describes how horrible economic conditions were in Italy prior to Mussolini's coup, when the streets were marked by "a clash of order and revolt. Men and women not only refused to work themselves, but to let other people work. Grains died in the fields, threshing machines were destroyed, . . . railway trains ran as suited the crew. Sunday was a day, not of rest...but of war; fêtes were dangerous, liable to be broken up by raids." On the converse, she states, under Mussolini public order was restored and economic production resumed. She rhapsodizes about what she saw in the reformed country. It was pleasing to witness a revival of "a world of work. . . . How could it be, one asked, that in so brief a time" from 1920 to her visit in 1926 that "a people should drop its clubs and pick up its tools?" Tarbell then tells the reader: "There was only one answer: Mussolini." Ah, what a savior! She clearly credits him for rescuing the nation from self-destruction. She then fawningly reminisces about her first encounter with the man:

> As I crossed the long room toward the desk Mussolini came around to meet me, asked me to take one of the two big chairs which stood in front of his desk—and, as he was seating me, was apologizing, actually apologizing, in excellent English for keeping me waiting. As he did it I saw that he had a most extraordinary smile, and that when he smiled he had a dimple.
>
> Nothing could have been more natural, simple, and courteous than the way he put me at my ease. . . . I found myself, not at all afraid to talk, but eager to do so. If he had not been as eager, I think I should have done all the talking, for luckily at once we hit a common interest—better [public] housing. . . .
>
> "Men and women must have better places to live. You cannot expect them to be good citizens in the hovels they are living in, in parts of Italy." [Is that what you would expect a Social Darwinist to say? –S.H.]

> He went on to talk with appreciation and understanding of the various [tax-funded] building undertakings already well advanced . . .
> Altogether it was an illuminating half-hour, and when Mussolini accompanied me to the door and kissed my hand in gallant Italian fashion. I understood for the first time an unexpected phase of the man which makes him such a power in Italy. He might be—was, I believed—a fearful despot, but he had a dimple.[36]

Contrary to Tarbell's assumptions, none of what she describes is charming. Rockefeller, a peaceable entrepreneur, was subjected to Tarbell's vilification whereas a genuinely violent dictator received a more congenial depiction from her. Interestingly, Tarbell also opposed women's suffrage to the very end of her days.[37] And Samuel S. McClure, the proprietor of the very *McClure*'s magazine that originally published Tarbell's denunciations of Standard Oil, agreed with his reporter's assessment. Subsequent to his own sojourn to Italy he characterized Mussolini's system of rule as "a great forward step and the first new ideal in government since the founding of the American republic."[38]

Incidentally, yet another acclaimed muckraker, the pro-segregation Ray Stannard Baker, gushed over the sacredness of the work of Tarbell and Steffens, whom he worked alongside in the newsroom. He found them a "constant inspiration" and stated, "Never shall I forget the memorable editorial discussions and conferences we had."[39]

As William Zebulon Foster (1881–1961) led the Communist Party of the USA (CPUSA), he definitely favored his own communism over Mussolini's fascism. Nonetheless, he interpreted Il Duce's transformation of Italy's government as progress hastening the arrival of a perfectly Marxian society. "By unmasking capitalist democracy," Foster stated in 1922, "and by showing it up in its true colors as a heartless dictatorship, the new rulers of Italy will do a great service by disillusioning the confused masses of the Italian proletariat. . . . The fascisti coup is a long step toward the ultimate revolution."[40] By that he meant that fascism's better qualities would reveal government control's superiority over laissez faire, whereas fascism's more unattractive aspects—such as its failure to achieve absolute economic equality—would motivate the proletariat to achieve perfection by means of a communist revolt.

Herbert Croly and the magazine he co-founded, *The New Republic*, consistently tooted Mussolini's horn. "Whatever the danger of fascism," Croly announced in 1927, "it has at any rate substituted movement for stagnation, purposive behavior for drifting, and visions of great future for collective pettiness and discouragement."[41] For that reason, Americans should be more open-minded about fascism and permit it a chance to prove itself. "Alien critics should beware," Croly stated in another *New Republic* piece, "of outlawing a political experiment which aroused in a whole nation an increased moral energy and dignified its activities by subordinating them to a deeply felt common purpose."[42]

Charles Beard (1874–1948), the collectivist historian whose works inspired Matthew Josephson's *The Robber Barons*, proclaimed that Italian fascism's central economic planning deserved imitation in the USA. "Beyond question," Beard wrote in *The New Republic*, "an amazing experiment is being made here, an experiment in reconciling individualism and socialism. . . . It would be a mistake to allow feelings aroused by contemplating the harsh deeds . . . accomplished [in] the Fascist process . . . to obscure the potentialities and the lessons of the adventure—no, not adventure, but destiny . . ."[43]

One unsigned *New Republic* editorial cheered that fascism promoted the "national cohesion and national welfare. . . ." The "conscience of the Italian people will insist on appraising the result. Thus, willy-nilly, Fascism is an experiment. . . . If the Italian people are capable of political self-education, they will preserve that part of the program which is useful to them and discard that which is not."[44]

John P. Diggins writes, "When the political scholar Robert MacIver asked the *New Republic* how it could profess to be an 'exponent of liberal principles' and at the same time support a dictatorship, and why suppression was so necessary in Italy if Fascism meant, as the *New Republic* claimed, 'mastery and self-control,' the journal replied that the traditional 'formulas' of liberalism were inadequate to appraise development either in Italy or in Russia."

In Diggins's words, another *New Republic* article—this one from 1926 by social philosophy professor and William James pupil Horace Kallen—stated that fascism's absence of liberalized, republican procedures "should not shroud the substantial accomplishments in economic, educational, and administrative reform. Living in Italy, he advised, made one realize there could be 'intolerance of liberalism also.' . . . Liberals should therefore suspend judgment until the full-grown tree of the new theory bore the fruit of social justice or the seeds of oppressive reaction."[45]

In the words of Horace Kallen's own *New Republic* piece, ". . . the Fascist revolution is not unlike the Communist revolution. . . . Each should have the freest opportunity once it has made a start, of demonstrating" its own benefits.[46]

Diggins elaborates that "Kallen's plea for patience won the backing of the *New Republic*. In an editorial the liberal journal provided a supporting preface to Kallen's arguments by elaborating further on the need to give Fascism a sympathetic hearing. In view of the bleak record of Italy's parliamentary government from 1871 to 1921, advised the editorial, it was a 'great mistake' to judge harshly and narrowly the recent regime. . . . Fascism had given the Italians a sense of unity and direction, a national self-consciousness that awakened the country's potential." To use the *New Republic*'s words, Americans must be open-minded about "the promising Fascist venture..."[47] George Bernard Shaw was another left-wing apologist for Mussolini.[48]

The shared silence among twenty-first-century left-wing activists about the early Progressives' support for Mussolini's fascism is part of a larger trend. The hard Left has buried the fact that socialism is the originator of fascism and Naziism. According to the *Online Etymology Dictionary*, "In the USSR, the terms *national socialist* and *Nazi* were said to have been forbidden after 1932, presumably to avoid any taint to the good word *socialist* . Soviet literature refers to *fascists*." [49]

It was in 1928 when the Comintern announced that anyone whom the Soviet government disliked would be labeled a *fascist* . This policy, writes John P. Diggins, declared " war not only on Fascism or capitalism but on Socialist and liberal parties everywhere. The term was taken up with gusto" by the Communist Party of the USA "and heaped upon all 'enemies' on the Right . . ."[50]

As the *Online Etymology Dictionary* continues, it got to the point where the Soviets referred to just about any enemy as "fascist"; Josef Stalin accused Soviet state cofounder Leon Trotsky of being an agent of the Nazis before having him purged and secretly assassinated. [51]

Following the Soviets' proclamations that *fascism* was the label for everything that communism opposed, others in the Marxian tradition picked up on this. When neo-Marxians in more-liberal republican areas echoed the accusation that fascism was the name for everything the far Left hated, they got that from the Frankfurt school of neo-Marxist thought [52] —the same school of thought that, as we learned previously, blamed the Enlightenment and rationality for Naziism. [53] Basically, the Frankfurt school argued that the Holocaust amounted both to Enlightenment rationality and to its result, commercial capitalism, being taken to their logical extreme. One of the most prominent Frankfurt-tradition proposers of this notion was Columbia University sociologist C. Wright Mills. As we recollect from Book Two's final chapter, Mills denigrated the notion that someone in America's relatively free enterprise could rise from poverty or middle class to wealth. As John Diggins paraphrases Mills's assessment on right-wing totalitarianism, "Advanced capitalism is not, as Marx believed, an orgy of ruinous production; instead it is planned, rationalized, and cartelized; a process, Mills pointed out, which enabled capitalists to continue to wield power in the Fascist state along with the party, the bureaucracy, and the military."[54]

It would seem that American academicians who wished for socialism to triumph over the night watchman state have likewise wanted to avoid having "their" socialism be compared with National Socialism. That is why left-wing academicians proclaim Nazis to be fake socialists. The real socialists, they insist, are the much more benign members of Western European Labour parties. Once the United States entered the second World War, and the Left realized that the Soviets they favored could not avoid war against Italy and the Third Reich, everyone from the Left to the center of the political spectrum took to referring to the Nazis as *fascists* rather than socialists. That tendency came to be normalized in the general culture, and there came a point where moderates and conservatives also came to refer to Nazis as fascists and to assume that neither the fascists nor the national socialists were real socialists. The revisionist approach to the terms *fascist* and *national socialist* does not fool author James Cross Giblin, though. In *The Life and Death of Adolf Hitler*, Giblin describes Naziism as a "combination of socialism, nationalism, and anti-Semitism."[55] Despite his own refusal to tolerate a free market in biotechnology, a similar analysis arrives via British philosopher John Gray, who acknowledges "Communism's clear affinities with Fascism," and that "in important respects Communism and Fascism were at one." An obvious affinity was that "the two were alike in viewing mass killing as a legitimate instrument of social engineering." Just as I do, he finds it disturbing that, throughout the 1900s, acknowledging the kinship between communism and fascism has been "distinctly unfashionable—in academic contexts, a . . . damningly final dismissal." Perhaps more pertinent to our discussion, Gray thankfully also identifies "the hideous Nazi vision" as "a mix of the völkisch chimera of a seamless 'organic' culture, fraudulent 'racial science' and revolutionary anti-capitalism . . ."[56] And despite his own aversion to individualist enterprise, George Orwell, as well, recognized that Naziism stood opposed to peaceable egoism and market economics, exploiting the common cultural belief of the glories of self-renunciation. Naziism, wrote the British author, espoused "self-sacrifice . . . Whereas . . . capitalism" has marketed itself as an ideology to Germans by promising them, ". . . 'I offer you a good time,' Hitler has said to them 'I offer you struggle, danger and death,' and as a result a whole nation flings itself at his feet."[57]

Nazi Anti-Capitalism

Hitler himself would scarcely recognize the laissez-faire lackey that modern anti-business scholars have depicted him to be. His combination of socialism with nationalism is exhibited in his statement from July 28, 1922: "Whoever is prepared to make the national cause his own to such an extent that he knows no higher ideal than the welfare of his nation . . .—that man is a Socialist."[58] Skeptics against volitionism might scoff that this demonstrates that Hitler's definition of socialism was too vague. The doubters can retort that when Hitler called himself a socialist, his "pseudo"-socialism jettisoned traditional socialism's (1) hostility toward capitalism and (2) codified nationalization of the utilities. Although the Führer lost interest in the second precept during his reign, he never forgot the first. On May 1, 1927, Hitler explained the Nazi Party's motives in greater detail—"We are socialists, we are enemies of today's capitalistic economic system for the exploitation of the economically weak, with its unfair salaries, with its unseemly evaluation of a human being according to wealth and property…"[59] The Führer was not alone among party members in holding this attitude. William L. Shirer identifies the vast swaths of Hitler's storm troopers, the *SA*, as "anticapitalist."[60] Niall Ferguson is therefore correct to judge that communism and National Socialism were more alike than they were different from each other.[61]

These anti-capitalist sentiments moreover find concordance in the philosophy of Hitler's propaganda minister, Joseph Goebbels. German historian Helmut Heiber writes that Goebbels maintained a lifelong "opposition to the bourgeoisie and a belief in class conflict."[62] Goebbels's newsletter, *Nationalsozialistiche Briefe* (*National Socialist Briefings*, or *National Socialist Letters*), teems with socialist buzz terms. In phrases similar to Hitler's, it announces, "We are socialists; we are enemies, mortal enemies, of the present-day capitalist economic system with its exploitation of the economically weak, with its injustice in wages."[63]

When Goebbels initially joined the Nazi movement, he expressed admiration for the Soviets on account of their socialism. As Goebbels mailed out *National Socialist Letters* from October 1, 1925, to early 1926, the periodical's readers experienced difficulty deciphering whether or not Goebbels himself should be considered a communist; his rhetoric was not discrete from theirs.[64] Goebbels stressed, "We are living in the era of the masses" and also, "I believe in the determination of the Proletariat to make sacrifices . . ."[65]

William L. Shirer concurrently finds that in this same period, Goebbels's private diary was "full of expressions of sympathy for Communism . . ." and that Goebbels wanted the Nazi Party to revolve around the proletariat.[66] Joachim Fest doubly affirms that Goebbels went "looking for formulas which could unite the nationalistic socialists and Communists," and that he "found a whole catalogue of identical attitudes and convictions."[67]

In Goebbels's words, "We will never get anywhere if we lean on the interests of the cultured and propertied classes. Everything will come to us if we appeal to the hunger and despair of the masses."

In *National Socialist Letters*, Goebbels emphasized socialism over nationalism to a considerable degree. This degree was so considerable, that he tried to reconcile Soviet communism with German national socialism.[68] Thence on January 31, 1926, Goebbels asked himself in his diary, "Where" can we Nazis "get together with the leading Communists?"[69] Goebbels worked on a speech entitled "Lenin or Hitler?" In this speech Goebbels ultimately concluded that Hitler was the better leader, but that this was something of a close call; he did not conceal his admiration for Lenin.

Goebbels declared in ink, "When Russia awakes, the world will witness a national miracle." Elsewhere he pronounced that Germans should never take up arms "in the pay of Capitalism—perhaps, in all likelihood, in the 'Holy War against Moscow.'" Rhetorically, he queried, "Could there be a greater political infamy?" He already knew the answer he wanted—no.[70]

William L. Shirer reports that, in this period, Goebbels "published an open letter to a Communist leader assuring him that Nazism and Communism were really the same thing. 'You and I,' he declared, 'are fighting one another, but we are not really enemies.'"[71] The warm feeling between Goebbels and the Soviets was mutual. Soviet writers reciprocated Goebbels's sympathy.[72]

Despite his rooting for the Nazi side in any military conflict between Nazis and the Soviets, Goebbels judged in a journal entry dated October 23, 1925, "In the final analysis, it would be better for us to end our existence under Bolshevism than to endure slavery under capitalism." Goebbels increasingly came to side with the Nazis over the Bolsheviks, proclaiming the former's superiority over the latter. Still, he regretted that the two sides kept squabbling over power. He felt they should focus on uniting against their common enemy, Jewish capitalism. He sighed, "I think it is terrible that we" Nazis "and the Communists are bashing in each others' heads."[73] When the Nazi propaganda minister finally and fully lost his last bit of faith in the Soviets, it was because he decided that they were not loyal enough to the socialist cause. "Communism," he eventually surmised, "is nothing but a grotesque distortion of true Socialist thought. We and we alone could become the genuine Socialists in Germany, or, for that matter, in Europe."[74]

In the early days of publishing *National Socialist Letters*, Goebbels had strong allies in the brothers Otto and Gregor Strasser, who shared in the fervent anti-commercialism. The Strassers' "program," George L. Mosse remarks, was "'anti-capitalist'—in that it maintained that capitalist exploitation (that is, Jewish usury) had reduced Germany to economic and financial destitution . . ." Additionally, "the Strasser brothers concretized the anti-bourgeois element in the revolutionary drive . . ." They hungered for the state to be in charge of the "division of property and profits." Mosse writes that Otto Strasser especially called for "nationalization of all land and the abolition of all unearned income," that is, usury.[75] Otto "denied the sanctity of private property and even advocated the complete nationalization of land . . . Strasser's proposals were as follows: capital property was to be apportioned in such a way that the workers would hold the rights to 10 percent, the state 41 percent, and the private owners the remaining 49 percent; profits, meanwhile, were to go 49 percent to the workers and 51 percent to the owners."[76]

Joachim Fest observes that when NSDAP members "joined in the Berlin metalworkers strike" in the early 1930s, "Hitler explained the situation tersely by telling the employers that striking Nazis were still better than striking Marxists."[77] Hitler, Goebbels, and the rest of the Nazi Party interpreted their own economic program to be a progressive movement that would correct the harms that capitalism allegedly imposed, without resorting to full-blown communism.[78]

University of Nebraska historian Alan E. Steinweis adduces that Nazi Party leaders successfully marketed their ideology by portraying themselves as the better socialists, for many Germans thought that Nazi economics "represented a 'Third Way,' an alternative" both to the laissez-faire "liberal social and economic order" and to "the Marxist model."[79] Henry Ashby Turner, Richard Steigmann-Gall, Joachim Fest, and Columbia U's Fritz Stern agree with Steinweis's assessment—that the Nazis believed themselves to be, and successfully presented themselves as, leading a "third way" between the extremes of laissez faire and communism.[80]

Among governists, talk of such a Third Way has not lost popularity. Michael Moore urges the continuance of this search for a Third Way between laissez faire and governism: "I wish somebody would invent a system that takes the best things of capitalism and socialism and puts them together. The things from capitalism that encourage individuality and creativity and ingenuity, and those things from socialism that say no one shall be left behind. Why can't we have that?"[81]

Otto Wagener (1888–1971), who served as an economic advisor to Hitler and as a major general in his army, wrote about Hitler coming to a similar conclusion in a conversation from 1930. Wagener told the Führer that "economic freedom" has "created new autocracies, in the factories and the large" business "concerns, where the workers . . . were turned even more directly into slaves than they ever had been under the scepter of a feudal lord. . . . Man's aspirations are evil—we should say, selfish—from childhood; the Bible tells us something of the sort. Perhaps Christ was one of the first to contrast man's liberalistic attitude with the socialist stance."[82] To Wagener, Hitler rejoined, "In our program, we have even given expression to this hurdle by coining the maxim, 'public need before private greed.'" The Führer recognizes that there is an individual entrepreneurial striving in almost everyone, but that this striving must be faulted for being "selfish." Hence the Nazi regime intends to take that striving and transform it, placing " it in the service of all, in the service of the whole nation—yes, perhaps in time in the service of all mankind . . . Communism results in a welfare state where the standards are averaged downward." In contrast, we national socialists are the ones who shall truly provide "for the needs of the people—that is, in the service of the community . . ." This ideal shall arise when Germans are "connected purely and simply to the promotion of the commonweal—that is, gain for all: for the Volk community for humanity."[83]

In fact, despite Hitler's eventual hatred for the Soviets, Joachim Fest discloses that the Führer "learned his most lasting lessons from Marxism. The energy he devoted to the development of a National Socialist ideology . . . testifies to the effects of the Marxist model upon him."[84] Accounting for these realities, six-time U.S. Socialist Party presidential candidate Norman M. Thomas admits that "both the communist and fascist revolutions definitely abolished laissez-faire capitalism . . ."[85]

The rhetoric about a Third Way resonated with the Nazis from the beginning, as it pervaded the romantic Pan-German Volkish movement from which Naziism emerged. George L. Mosse states, "The search for a 'third way,' as an alternative to capitalism or Marxism, . . . was an integral part of the Volkish concern. . . . Everywhere in Europe, fascism was based upon the urge toward a 'third way,' and Volkish thought here intersected with the mainstream of an international movement." Furthermore, "Volkish economic theory, in its aversion to anything materialistic and concrete, had never advanced beyond opposition to the international (bourgeois or Jewish) capitalist conspiracy, on the one hand, and the international Bolshevik conspiracy, on the other. . . . Many revolutionaries of 'the third way' fused primitive economics with the omnipresent Germanic faith in order to direct the attack . . . [on] the bourgeoisie, whose capitalist inclinations it condemned," as well as "the Jews."[86]

Recall from chapter 3 that this idea was propagated by German cultural historian Arthur Moeller van den Bruck, the same man from whom the Nazis appropriated the expression *Third Reich*.[87] George Mosse identifies Moeller van den Bruck as "the prophet of the 'third way,'" and points out that Moeller's book *The Third Reich*, published in 1923, just two years prior to his death,[88] was originally intended to be titled *The Third Way*.[89] In *The Third Reich*, Moeller rejected both free enterprise and communism as too extreme, advocating the mixed-economy welfare state as the middle path and Germany as the nation fittest to model it for everyone else. Mosse scribes that Moeller held that "German socialism was a social order forged from the union of a medieval corporatism and the cultural peculiarities of the German Volk. As he emphatically asserted, in condemning Marxist internationalism, 'every people has its own socialism.'"[90]

Keep in mind that Fritz Stern observes something held in common by Moeller van den Bruck, Paul de Lagarde, and Julius Langbehn. What is noteworthy besides their animosity toward reason and the Enlightenment is that these three affixed blame for nineteenth-century Germany's turmoil on laissez-faire "liberalism. They attacked it because it seemed to them the premise of modern society from which everything they dreaded sprang: the bourgeois life, Manchesterism," and "materialism . . ."[91]

Anti-Capitalism and Anti-Semitism

For too long, sundry left-wing writers tended to shy from mentioning the methods whereby Adolf Hitler exploited the masses' deep-seated prejudice against the Jews. Nor have they paid much mind to the reason why he so begrudged Jews from the outset. "Contrary to what is sometimes claimed," Ian Kershaw perceives, "Hitler's antisemitism was not promoted by his anti-Bolshevism"—in other words, his rivalry against Soviet communism for domination over Europe. His anti-Semitism "long predated it."[92] Instead Hitler asserted that what made Jews so spiteful was that they were greedy merchants and financiers who accrued a windfall off of Germany's defeat in the First World War,[93] while communists were really the pawns of the international Jewish banking cabal.

As Albert Einstein discerned, one of the biggest "crimes with which the Jews have been charged in the course of history—crimes which were to justify the atrocities perpetrated against them"—was that they succeeded in "the economic domination and exploitation of all mankind."[94]

In the words of Coloma College history department chair Peter Lane (b. 1925), Hitler and British fascists "blamed the world's economic problems on Jewish bankers and industrialists."[95]

In a nutshell, George L. Mosse puts it, "National Socialism did not accept the existing capitalist order . . . The enmity of National Socialism was directed toward . . . the banks and the stock exchange."[96]

Despite his attempts to downplay the Nazis' anti-capitalism,[97] Richard Grunberger admits that, under the NSDAP regime, "Jews were attacked as the embodiment of capitalism, and department stores," being predominantly owned by Jews, "were boycotted."[98]

Joachim Fest observes the 1920s stereotype of German Jews that Hitler had exploited: "The type of personality the Jews had developed corresponded . . . to the urban," reason-oriented "style of the age. That, as much as the fact that they . . . controlled virtually all the major banks in Vienna and a considerable portion of local industry—produced in the Germans a sense of danger and of being overwhelmed. . . . Generalized anxiety condensed into the charge that . . . nothing was sacred" to the Jews, "that their 'cold' intellectuality was opposed to . . . German sentiment." Fest goes on, "In fact, the equating of Judaism and modernity, like the thesis that Jews had a special talent for the capitalistic free-enterprise economy, was not unfounded."[99]

Social scientists Ian Buruma and Avishai Margalit agree with Fest—"Jews have been associated for so long, in Christian as well as Islamic societies, with trade and finance that they are almost invariably included in hostile views on capitalism." Moreover, due to their cosmopolitanism and ability to assimilate into various communities throughout the West, Jews "are also associated with Western claims to universal ideas, such as . . . secular law," which also casts them in a negative light toward many who uphold a Romanticist philosophy.[100] Throughout the nineteenth and earlier twentieth centuries, writes George L. Mosse, "The association of Jews with finance capital was Europe-wide . . ."[101]

There is a tragic irony, of course, in how Jews are associated with financial services. The assumption is that if a Jewish man could choose whatever occupation he could possibly enter, finance would definitely be his primary pick. In fact Jews were confined to the financial industry throughout the Middle Ages on account of most other occupations being closed to them. In Christian-dominated Europe, Jews were proscribed from owning land and therefore could not be farmers. Likewise, their having been barred from the artisanal guilds precluded them from becoming craftsmen. Their few opportunities arose from the fact that Christian Europe, considering money-lending to be a sin, prohibited by law any Christian from being usurious with another Christian. Jews were permitted to become usurers on the condescending pretext that they were already sinful second-class citizens. As the Jews' souls were not considered worthy of saving anyway, the Christian government deemed it not necessary to spare them from engaging in so sinful an act as usury. As Jews were allowed to be usurers throughout the Middle Ages, they honed their skills in this unique enterprise.[102]

Recall from *Life in the Market Ecosystem* that the welfare state goes back to the chancellorship of Otto von Bismarck. It is not surprising that his reign fostered anti-Semitism as well. University of Pennsylvania historian Jonathan Steinberg writes that by this time, "capitalism brought its [laissez-faire] liberal ideology and the demand for free trade, free movement of people and goods, free access to crafts and professions, banks, stock exchanges, insurance companies, and traders. Into the thriving new capitalist state, Jews emerged as its most adept practitioners and its most ambivalent symbols. . . . From 1811 . . . to July 1918," Bismarck's partisans portrayed the Jews "as draft-dodging, black marketeers . . ." As far as many Germans were concerned, Jews "represented the corrupt and dangerous fluidity of money, capitalism, and markets. They controlled a significant share of newspapers and pioneered the department stores." In response Carl Constantin Freiherr von Fechenbach started the influential Social Conservative Association which, in the words of Steinberg, was "dedicated to anti-capitalism, anti-Semitism, and state socialism which would include nationalization of basic industries." Bismarck himself targeted Jews for ridicule. He associated them with Germany's National Liberal Party, which he despised. In Steinberg's words, Bismarck detested the Liberal Party for the reason that it "believed in . . . free markets . . ."[103]

One of the Germans who played a role in the anti-Semitic, openly mystical Pan-German Volkish movement that laid the foundation for the Nazis' popularity was Georg von Schönerer (1842–1921). In 1876 he joined a pan-German mystical club in Vienna. Schönerer's activism, writes Nicholas Goodrick-Clarke, significantly empowered "a revolutionary movement that embraced populist anti-capitalism, anti-liberalism, anti-Semitism, and prussophile German nationalism."[104] This movement also branched out in the form of Otto Boeckel's League of Peasants, existing from 1885 to 1894. Besides purveying bigotry toward Jews this League also, states George L. Mosse, "put forth a social program directed against finance capitalism and all middlemen. The abolition of interest charges was important to Boeckel in order to circulate more money among the people. Indeed, he wanted to end all speculation with the 'fruits of the soil,' to nationalize foreign trade, and restrict useless industry."[105]

A comparable influence was Theodor Frisch (b. 1852). Again writes Goodrick-Clarke, Frisch detested large "firms, factories, and mass production. He sought to mitigate these threats through the new guild." Frisch's anti-corporation vendetta "was complemented by anti-Semitic attitudes. Frisch attributed the new economic order to the growing influence of Jewish business and finance in Germany." In 1902 Frisch founded an anti-Semitic magazine, *The Hammer*, and within two years various local clubs sprung up to spread its ideology. "In 1908 these groups used the name *Deutsche Erneuerungs-Gemeinde* (German Renewal Groups): their membership was interested in anti-capitalist forms of land reform designed to invigorate the peasantry, the garden city movement, and *Lebensreform*."[106] At this point, please recall the important role that Alfred Ploetz played in Nazi Germany's eugenics policies, having founded the International Society for Racial Hygiene in 1905 with advisory counsel from Ernst Haeckel and Francis Galton. Ploetz was not finished. "In 1904," writes Léon Poliakov, "Ploetz announced the foundation of an associated society inspired by the anti-semitic agitator Theodor Fritsch; this Community for German Renewal (*Deutsche Erneuerungsgemeinde*) envisaged a return" to the undeveloped wilderness land "as the principal means for regeneration—for Aryans only, of course."[107]

We learned from chapter 3 that the Nazis were an outgrowth of the nineteenth-century Volkish movement, itself inspired by the philosophic Romanticist movement, which denounced phronesis and its consequence, modern technology. As the Volkish movement balked against industrialization, it extolled, as morally superior to the urban industrialist, the country peasant. And, as George L. Mosse discovered, Volkish writers characterized the city-dwelling Jewish businessman as the German peasant's arch-nemesis. "The Jew functioned as middleman in many of Germany's agricultural regions. . . . As a moneylender he was hated most when the peasants were in greatest financial difficulties and relied on him to tide them over, as after a bad harvest. . . . The peasants were his debtors, and in bad times, no doubt, he collected his debts legally by foreclosure. There were many generous and charitable Jews engaged in such commerce; nevertheless, to the debt-ridden peasant, the Jew represented the most easily identifiable and immediately present element of the greedy power of modern capitalist civilization." Hence, writes Mosse, "The Jews were condemned for their materialism and stigmatized as having sold themselves to the 'demon capitalism.'"[108] Hermann Lietz, a pioneer in the boarding school movement, argued such a position in the early 1900s. "In terms of the economy, he recommended nationalizing the land and distributing it to the peasants. Those industries which affected the national and common good were likewise to be nationalized. He still adhered to his 'German idealism,' which he defined in terms of nature, simplicity, and Germanic Christianity."[109]

Remember from *The Freedom of Peaceful Action* our disquisition on Werner Sombart, the German Marxist who ultimately converted to Naziism, and who was among the writers to popularize the expression *capitalism* (complete with the *m* at the end). Although Werner was not as virulently anti-Semitic as most Nazis, the Volkish movement cited his writings about the history of commerce in their vilification of Jews as financiers. His 1910 treatise, *Die Juden und das Wirtschaftsleben* (*Jews and Capitalism*), became heavily cited among the pan-German Volkish movement.[110]

In 1870 Édouard Drumont, a French writer who would, decades later, be celebrated by the NSDAP, wrote of the "expropriation of society" by the holders of "finance capital... If nothing is done to arrest this process . . . , all European society" shall be subjugated by "a few hundred bankers."[111] This contributed to the Nazis' belief in the urgency of murdering the ethno-religious group they presumed held such financial dominion.

Having been influenced by these Volkish anti-Semites, Adolf Hitler wrote in *Mein Kampf* that the Jews were in thrall to Mammon.[112] Who is this Mammon referenced by the Führer? It is the same Mammon whom Richard T. Ely referenced. To recap, *Mammon* is the New Testament's term for riches, which the Good Book employed as a synonym for greed.[113]

Among the national socialists, the Reich dictator was not alone in hating Mammon. Here is Joseph Goebbels's rant: "Money is the curse of mankind. . . . Every penny is sticky with sweat and blood. . . . To me the worst day of the week is payday. They throw the money at us like bones to a dog. . . . Money is the yardstick for the values of [laissez-faire] liberalism. . . . Money and Jew—they belong together. Money is without roots. It stands above the races. . . . In its deepest sense," laissez-faire "liberalism is the philosophy of money. Liberalism means: I believe in Mammon. Socialism means: I believe in work."[114]

Nazi existentialist Martin Heidegger, too, hated Jews for being too capitalist. Parisian professor Emanuel Faye writes that Heidegger's phrase for Weimar Germany's problem—*growing Jewification*—"designates everything Heidegger fought against to the very end," particularly laissez-faire "liberalism," republicanism, and "the 'time of the *I*'. . ." Faye states that in Heidegger's rhetoric, "we see that the word 'liberalism' and the adjective 'liberal' refer, exactly as in the Nazi attacks of the era, both to Anglo-Saxon culture and to Jewish thought—whether it be politically," in screeds against laissez-faire constitutional republicanism, "or 'philosophically,' in Heidegger, to the thought of the individual and of the *I*, in what he calls the *Ich-Zeit*"[115] (emphases Faye's).

According to Ian Kershaw, the "antisemitic tirades" that propelled Hitler to national stardom "were invariably linked to anti-capitalism and attacks on 'Jewish' war profiteers" whom he faulted "for exploiting the German people and causing the loss of the war and the German war dead."[116]

Joseph Goebbels agreed that Jewish entrepreneurs had orchestrated the war. The Propaganda Minister rambled, "While the soldiers in the Great War offered their bodies for the protection of their homes and two million of them bled to death, the speculators coined gold out of their red and noble blood. And they later used this gold to cheat the returning soldiers out of house and home. The war was won by money and lost by labor. The peoples are not the victors or the vanquished. They performed only day labor in the service of money, or defended work against this slave labor. Germany battled for work. France fought for money. Labor lost the battle. Money won it."[117]

Robert Gellately reminds readers that in the speeches that Hitler gave on the Nazi Party's behalf in 1930, the bigoted politician would bring up the Jews "in the context of being behind the machinations of capitalism."

One label that Hitler applied to what he wanted eradicated was "the Jewish-materialist spirit."[118] In Hitler's words, the populace must be cautioned that "international Jewish financiers in and outside Europe" could soon "plunge the nations once more into a world war..."[119]

Kershaw discloses that there "was no mention of Bolshevism" in Hitler's famously anti-Semitic September 1919 Gemlich letter. In it, "the 'Jewish Question' is related" specifically to "the rapacious nature of finance capital."[120] This demonstrates Hitler's philosophy to be "anti-capitalist and 'national socialist,' not least antisemitic." Through and through, many members of the NSDAP—being more zealous true believers than Hitler himself—found themselves "wedded to a type of 'social revolutionary' emotive anti-capitalism."[121]

George L. Mosse paraphrases the credo of the NSDAP during the 1920s—"The abolition of the 'slavery of interest charges' would produce both social justice and national unity."[122]

Hitler originated a phrase that later became a slogan in the German army—"The Jew is not a German but merely a trader . . ."[123] Because of the commonality of this stereotype, the Nazis incrementally stepped up their persecution of Jews throughout the 1930s. This started on April 1, 1933, when Hitler began the government-enforced boycott of Jews.[124] That year Nazi official Julius Streicher (1885–1946) ordered that signs be posted throughout Germany proclaiming, "Whoever buys from a Jew is a traitor to the people."[125] Nazi groups organized major boycotts against Jewish businesses; they went as far as crafting rules on exactly what sort of signs should be hung above the entrances of Jewish stores in order to discourage gentiles from patronizing them. The SA compiled lists of Jewish-owned businesses that they would prohibit gentiles from frequenting.[126] On July 14 of that same year, the NSDAP's policy of compulsory sterilization of the diseased and mentally ill began. The situation for Jewish-owned enterprises worsened in the winter of 1937. Mosse states that it was then that the domestic embargo against Jewish businesses "hit every aspect, from banking to retail stores . . ." The following year "Jewish bank accounts and investments were confiscated."[127]

Despite his own bias toward the left-wing interpretation of Germany's history, Wayne Biddle still writes of how the "first boycott of Jewish stores occurred on March 28," 1933, "affecting thousands of shops nationwide and bringing whole sections of many towns to a standstill."[128] Likewise, when the Nazis invaded Czechoslovakia, they made certain to confiscate factories in that nation owned by Jews, such as those owned by the prominent Low-Beer family.[129] The historian Götz Aly (b. 1947) argues that the Nazis' confiscation of Jewish assets accounts for much of the financing of the Nazis' welfare state.[130] Prior to Aly advancing this argument, a number of writers had mischaracterized the Nazis' controlled economy as a highly efficient and self-sustaining one. Aly brings to light the manner in which the Nazi government's method of funding ran on borrowed time. Throughout the war, the Nazis did not produce new wealth but fed parasitically on the wealth that Jews had already produced. And that wealth was running out.

These Jew-as-covetous-merchant stereotypes were definitely not unique—Hitler simply played up prejudicial notions about Jews that have been common throughout history. Anarchist socialist Pierre-Joseph Proudhon hisses that any Jew is "by temperament an anti-producer. He is an intermediary, always fraudulent and parasitic, who operates in trade as in philosophy, by means of falsification, counterfeiting, and horse-trading."[131] Proudhon went as far as proclaiming that if the Jews could not be banished from France, the next best solution would be to slaughter each and every one of them. In Proudhon's words, "One must send this race back to Asia or exterminate it."[132]

Another French socialist espousing this view was Alphonse Toussenel (1803–1885), a disciple of pioneering socialist Charles Fourier.[133] Toussenel, elucidates George Mosse, "was influential in popularizing National Socialism. He was to write one of the most important attacks upon the inborn and irredeemable faults of the Jewish rule. *The Jews, Kings of the Age* (*Les Juifs, Rois de l'Époque*, 1845), subtitled *History of the Feudal Aristocracy of Financiers*, linked the medieval image of the Jew as usurer to the populism of a society suddenly plunged into the maelstrom of early capitalism. . . .The Jews, according to Toussenel, dominated the world through their control of finance capital. . . . Toussenel came from a rural background, and for him the Jews were also the despoilers of the countryside—a view shared by many German writers for whom the Jew was the enemy of the peasant."[134] Mosse summarizes the collective impression of Proudhon, Fourier, and Toussenel—" The Jewish race...was predatory, competitive, and without morality, and was therefore to be excluded from participation in a genuinely national and socialist community." Mosse then observes, "The social concerns of men like Toussenel and Proudhon were based upon a rejection of modernity, a hostility to civilization as an urban accomplishment. Here they agreed with Richard Wagner, their younger German contemporary, that equality among people and a nation committed to social justice meant the destruction of the 'power of gold.' Jews symbolized this power, hence the exploitation of the people among whom they lived." This socialist "was driven by his anti-finance capitalism to a racist stance."[135]

Given our recollection that Madison Grant received mutual praise from Hitler and Theodore Roosevelt for his governist eugenicism,[136] it is not startling that Grant sounded the alarm that the Polish Jews' "ruthless concentration on self-interest" is being "engrafted upon the stock of the nation."[137] Another American associate of Theodore Roosevelt's to express anti-Semitism on anti-capitalist grounds was the progressive, muckraking photojournalist Jacob Riis (1849–1914). Riis is routinely lionized as a crusader for the underprivileged who exposed the need for urban renewal with his most famous book. Stanley P. Caine slobbers that public concern "about tenement housing mushroomed after the publication in 1890 of Jacob Riis's *How the Other Half Lives* . . ." Caine is ecstatic that in the "atmosphere of discontent" that Riis originated, "important gains were made in urban reforms."[138]

Less often discussed than Riis's alleged sainthood, and the moral greatness of *The Other Half*, is that very book's insistence that Jews are inherently avaricious. In chapters called "Jewtown" and "The Sweaters of Jewtown," Riis wrings his hands over sweatshops in which both the owners and employees are Jews. He deplores the business owner as an exploiter and the low-paid employees as harmful to the livelihoods of gentiles in the workforce, advancing the false accusation that the willingness of Jews and other immigrant minorities to work for low wages inexorably bids down the average wage for everyone else. "The Jew has monopolized the business" of garmenting, Riis complains, but "not satisfied with having won the field, he strives as hotly with his own"—other Jews—"for the profit of half a cent as he fought with his Christian competitor for the dollar." Riis shakes his head at what he considers the mindset of Jews. "Money is their God," he insists.[139]

John Maynard Keynes whined, "It is not agreeable to see civilization so under the ugly thumbs of its impure Jews who have all the money and the power and brains."[140]

In 1919 the "fear of a Jewish conspiracy—capitalist or communist, depending on the situation—was . . . used extensively" by the Pan-German Volkish movement, states George L. Mosse.[141] There is no wonder then that the famous forgery, *The Protocols of the Elders of Zion*, propagated accusations about Jewish bankers controlling everything.[142]

Nor could the Jews-as-capitalist depiction escape from the playbook of Karl Marx, himself a rabbi's grandson. Marx, according to professors Ian Buruma and Avishai Margalit, "likened Jewish capitalists to lice, feeding off the poor like filthy parasites."[143] In 1843 Marx spat, "What is the secular basis of Judaism? . . . *Huckstering*. What is his"—the Jew's—"worldly God? *Money*. . . . The god of *practical need and self-interest* is money." Reflect on how alike Marx and U.S. progressive Jacob Riis sound on that point. And Marx continues, "Money is the jealous god of Israel, in face of which no other god may exist. . . . The *chimerical* nationality of the Jew is the nationality of the merchant, of the man of money in general"[144] (emphases Marx's). Anti-Semitic socialists gleefully reprinted this tract of Marx's. They said that it bore a special authenticity, for it was written by a Jew who admitted the shameful truth about his own people.[145] And do not forget that, despite his own sympathies for international trade, Immanuel Kant likewise denounced Jews as capitalists.

Earlier in Germany's past, the original Protestant, Martin Luther, mixed anti-financier sentiments with anti-Semitism, characterizing Jews as "a pernicious race, oppressing all men by their usury and rapine. We must ever keep our guard against them." Consequently, if "I were master of the country, I would not allow them to practice usury."[146]

This is how Hitler put it to Otto Wagener in a conversation from the 1930s. "The Jew is not a socialist! *Once before* he nailed to the cross the great Creator of the concept of socialist redemption! He will do so again whenever he can! For he is an individualist, an economic liberalist, an egotist . . ."[147] (emphasis Wagener's). In the words of *Mein Kampf*, "the financial aristocracy" of "unscrupulous Jewish business men" consists of "spiritual robber-barons..."[148] The archetypal Jew "came to look upon the commercial domain and all money transactions as a privilege belonging exclusively to himself and he exploited it ruthlessly." Later, "finance and trade had become his complete monopoly."[149] The cost of this, the Führer's manifesto maintains, is "the social stratification of the people." And it presses on, "By way of stock shares" the Semitic tycoon "pushes his way into the circuit of national production which he turns into a purchasable or rather tradable object. . . . Finally, the Jewish influence on economic affairs grows with terrifying speed through the stock exchange. He becomes the owner, or at least the controller, of the national labor force."[150] But the archetypal Jew's "extortionate tyranny became so unbearable that people rebelled against his control and used physical violence against him." Consequently, angry vigilante mobs "have seized Jewish property and ruined the Jew in their urge to protect themselves . . ." Hitler, expectedly, approves of this violence, which he claims to be justified retribution. Then, Hitler regrets, the archetypal Jew "paid court to governments, with servile flattery, used his money to ingratiate himself further and thus regularly secured for himself once again the privilege of exploiting his victim."[151] Note the similarity between the conclusion of Hitler's spiel, with the accusation of various corporation-haters that corporations have purchased too much influence in Western governments and transformed them into a *corporatocracy*.

In his best-selling book *Dude, Where's My Country?*, Michael Moore similarly rants that the stock exchange is a tool that capitalists utilize to manipulate the masses. By the end of the 1980s, the book says, ". . . the rich were pretty much tapped out with their excess profits . . ." At that juncture, "the game became, 'Hey, let's convince the middle class to give us *their* money and *we* can get even richer!'" The middle class did indeed get into this racket, and "it was a sham. It was all a ruse concocted by the corporate powers-that-be who never had any intention of letting you into their club. They just needed your money to take them to that next level. . . . They knew the Big Boom of the 1990s couldn't last, so they needed your money to . . . inflate the value of their companies" and have the shares of equities rise in price, "so when it was time to cash out, they would be set for life, no matter how bad the economy got." Then the stock market tanked, and the middle-class investors lost their savings. "Another trillion dollars in pension funds and university endowments is no longer there. But here's what's still here: rich people. They are still with us and they are doing better than ever. They laughed all the way to the Swiss bank over the scam of the millennium. They pulled it off . . ."[152] (emphases Moore's).

It goes without saying, of course, that Hitler took this hatred much farther than Moore does. In his October 1923 interview with George Sylvester Viereck, the Nazi summarized, "In my scheme of the German state, there will be . . . no use . . . *for the usurer or speculator* . . ."[153] (emphasis added).

Have Capitalists Co-opted Their Reputed Enemies?

If the Aryan proletariat does not instigate an insurgency against the Jewish multimillionaires, *Mein Kampf* argues, it is because Jewish capitalists have co-opted the organized labor movement and turned it into a tool of capitalism rather than proletarian social justice. While, on one hand, the plutocratic Jewish schemer "organizes capitalistic methods of human exploitation to their ultimate consequence, he approaches the very victims of his spirit and his activity and in a short time becomes the leader of their struggle against himself."[154] Furthermore, the German people have grown so afraid of being attacked by Soviet terrorists that they have come to forget the identity of their true enemies—the Jewish capitalists. Hitler began to intimate by July 21, 1920,[155] that that is actually part of the Jewish capitalists' plot. It is the Jewish capitalists who have invented the Soviet communist movement to get Europeans so afraid of communists that they will not only avoid any effort to direct their energy where it belongs—toward overthrowing Jewish capitalists—but they will run to the Hebrew capitalist establishment and plead for protection. "Marxism," *Mein Kampf* shrieks, is really "the economic weapon" used for "the enslavement of free peoples in the service of supra-state world finance Jewry." Moreover, the purpose of "Marxist sheets" is "to break the people's national and patriotic backbone and make them ripe for the slave's yoke of international capital and its masters, the Jews." The Soviets' ultimate goal "is only to erect a despotism for international world Jewish finance." Hence the USSR is composed of "the Marxist shock troops of international Jewish stock exchange capital."[156] For these reasons the Führer, in the words of Niall Ferguson, equated "Jews simultaneously with communism and capitalism."[157] As Robert Gellately phrases it, Hitler advanced "a curious argument in which capitalists supposedly prepared the ground for Communism, with only the NSDAP blocking the way."[158] Hitler stated to Otto Wagener that whenever Aryans assist self-described Jewish Marxists in their social activism, the Aryans end up being cheated. Once the feudal or capitalist power has "been removed by a revolution . . ." Hitler told Wagener, "the Jews do not actually introduce genuine socialism as they have promised—because it would wrest power from their hands again."[159] Then in spring 1932, the Reich dictator provided Wagener a rather elaborate (and confusing) explanation of how Marxism came to be a conspiracy by Jewish financiers to co-opt socialism. The Führer speechified that between 1830 and 1848,

> the first stirrings of industrialization had moved the unnatural disproportion between the ruling classes and the mass of the working people into a new, still more glaring light, a new movement [socialism] was born. . . . it was eager to take up the true doctrines of Christ. . .
>
> Bismarck was the first who used social legislation to tread the path of conciliation—that is, actually *against* the Jews. But, by the same token, he ran afoul of the conservative party of which he himself had been a founding member. . . . reaction promptly stirred . . .
>
> Foolish blindness and a shortsightedness not schooled by instinct, such as are always the characteristics of reactionaries, allowed the Great War to ignite. Its end placed the threads

> of fate in the hands of the socialists. However, they were not ready and did not know what to do with them. Quickly the Jews were on the scene. But the crucial moment had been missed, and only bourgeois revolution was possible. It found its expression in the Weimar constitution. . . .
> That is the moment in which we live! The National Socialist movement is the harbinger of the return to the will and mission of nature and the upholder of the socialist idea, as it corresponds to the Christian doctrine and the true religion. . . . The reactionaries and the Jews are the enemies of the movement, and thus of the Volk. [emphasis Wagener's.][160]

Because he figured that Jewish financiers had corrupted Marxism and Soviet communism, Hitler presented National Socialism as socialism in its truest, purest form. "This," observes Kent State University historian Richard Steigmann-Gall, is "the standard rendering of Nazi socialism as the *gemeinschaftlich* antithesis of *gesellschaftlich* capitalism and communism . . ."[161] Thus we realize how misleading it is for Michael Parenti to prattle that capitalists everywhere "welcomed" the "anticommunism" of the Führer, praising him "as a savior who would vanquish the Soviet Union and rescue Europe from Red revolution."[162]

One might argue that Hitler was not truly sincere in expressing hatred for financiers. One might proclaim that Hitler himself did not despise bankers but that he knew that much of the voting public did, and that his public denunciations were no more than a cynical ploy to curry favor with a rabble whose hatred for financiers exceeded his own. According to that interpretation, once Hitler gained supreme authority he would be able to drop that act completely. It is therefore significant that on January 30, 1939—after having maintained his dictatorial status for six years—Hitler provided a speech that in no manner relented on pummeling Jewish financiers. "If international finance Jewry within Europe and abroad," the mustached dictator threatened, "should succeed once more in plunging the peoples into a world war, then the consequence will not be the Bolshevization of the world and therewith a victory of Jewry, but on the contrary, the destruction of the Jewish race in Europe."[163] Secure in his dictatorial position, Hitler did not need to pander to any anti-capitalist sentiment in the electorate by this point. The disgust with the financial sector was something that came from within the Führer himself.

There is a frightening parallel to Hitler's assertion that Soviet communists are the marionettes of the Jewish capitalist overlords. We observe the parallel in Michael Moore's allegation that militant Islamic terrorists, too, are puppets and hobgoblins of the modern corporate elite. According to Moore, the corporations utilize the Islamic terrorists to hide from American proletarians the fact that our true enemies are the billionaire businessmen against whom we should turn. The greatest "success" in the Republican federal politicians' war on Islamic militants "has been its ability to distract the nation from the Corporate War on Us. . . . The business bandits (and their government accomplices) who have wrecked our economy have tried to blame it on the terrorists. . . . You are repeatedly told that bad, scary people are going to kill you, so place your trust in *us*, your corporate leaders, and we will protect you." In return, say the business bandits, "don't question us if we want . . . *our* tax cut. . . . This shit is so scary," concludes Moore, "of course we do what we are told, . . . and fly our little American flags to show that yes, boss, we *believe* in your War on Terror"[164] (emphases Moore's).

Foreshadowing what Noam Chomsky would insinuate about businessmen of every race, Hitler asserts that Jewish multimillionaires own the corporate media that Manufactures the Consent of the public, rendering the masses subservient by numbing their cognitive abilities. This transforms the middle class into self-indulgent consumer zombies. *Mein Kampf* puts forth, "With all his perseverance and dexterity" the Semitic businessman "seizes possession" of "the press." And, "With it, he slowly begins . . . to guide and to push all public life, since he is in a position to create and direct that power . . . under the name of 'public opinion' . . ."[165] Aside from the targeting of a specific ethnic group, that same theory is presented intact in the voice-over narration that Mikela J. Mikael provides in *The Corporation: The Pathological Pursuit of Profit and Power*—"In a world economy where information is filtered by global media corporations keenly attuned to their powerful advertisers, who will defend the public's right to know? And what price must be paid to preserve our ability to make informed choices?" Mikael's narration intones ominously, "Some of the best creative minds are employed to assure our faith in the corporate world view. They seduce us with beguiling illusions designed to divert our minds and manufacture our consent."[166] In the same film Chomsky elaborates, "The goal for the corporations is to maximize profit and market share. And they also have a goal for their target, the population. They have to be turned into completely mindless consumers of goods they do not want. You have to develop what are called 'created wants.' . . . You have to focus them on the insignificant things in life, like fashionable consumption. . . . We have huge industries—public relations is a monstrous industry—. . . which are designed" from inception to "try to mold people into this desired pattern."[167] To a live audience in Cambridge, England, Michael Moore put it more simply—"we live in a system of enforced ignorance. The way the media works . . . it's all about keeping us stupid."[168] What trivialities of life do the corporate media dupe consumers into obsessing over? Esteemed corporation-hating crusader Ralph Nader thundered in a Congressional testimony, "Every day, hundreds of companies work with one thought in mind: how to manipulate children and teenagers to purchase video games and music, to watch movies and television. In their quest for larger audiences and greater profits, the commercial media predictably races to the lowest and basest standards, with ever more blatant displays of violence, sex, crassness, and nihilism in television, cable, movies, radio, video games and music."[169]

Mein Kampf does not quarrel with Chomsky, Nader, and *The Corporation* on this issue. The national socialist manifesto complains, "Our whole public life today is like a hothouse for sexual ideas and stimulations. Just look at the bill of fare served up in our movies, vaudeville and theaters, and you will hardly be able to deny that this is not the right kind of food, particularly for the youth. In shop windows and billboards the vilest means are used to attract the attention of the crowd. Anyone who has not lost the ability to think himself into their soul must realize that this must cause great damage in the youth. . . . Sometimes the public learns of court proceedings which permit shattering insights into the emotional life of our fourteen- and fifteen-year-olds. Who will be surprised that even in these age-groups syphilis begins to seek its victims?" The ones blamable for this corruption of morals, according to Hitler, are the Semitic media moguls who offer debasing images to young consumers in effort to fleece them. Amusements from the Jewish tycoons, the Führer alleges, pre-empt any socialist's exhortations to rise up violently against the capitalist class. The gold-hoarding Jew "has his press give a picture of him which is as little in keeping with reality as conversely it serves his desired purpose."[170] Osama bin Laden voices agreement. In his Open Letter to America he rails,

1. You are a nation that exploits women like consumer products or advertising tools calling upon customers to purchase them. You use women to serve passengers, visitors, and strangers to increase your profit margins. You then rant that you support the liberation of women.
2. You are a nation that practices the trade of sex in all its forms, directly and indirectly. Giant corporations and establishments are established on this, under the name of art, entertainment, tourism and freedom, and other deceptive names you attribute to it.
3. And because of all this, you have been described in history as a nation that spreads diseases that were unknown to man in the past. Go ahead and boast to the nations of man, that you brought them AIDS as a Satanic American Invention.[171]

Here we find a huge irony concerning W. E. B. Du Bois. Du Bois recycles Hitler's argument that the corporate media indoctrinate everyone. Yet, simultaneously this same Du Bois asserts that such corporate mind-manipulation proves that American media conglomerates are similar to the Nazis. For decades in America, Du Bois writes, "Big Business monopolized news-gathering, newspapers, periodicals, and publishing."[172] He believes, "Today it is clear to all who know the facts that American industry has launched in this country the greatest propaganda effort the world has ever witnessed. *In comparison, Hitler and Mussolini fade to insignificance.* Our daily press with few exceptions is controlled in presentation of fact and expression of opinion by the organized industrial interests of the United States" [173] (emphasis added). Hitler and Mussolini fade to insignificance in comparison to U.S. corporations? In that his accusation is paranoid and inflammatory, and in that it is directed against for-profit media, Du Bois is the one sounding like *Mein Kampf*.

Echoing these sentiments is humorist George Carlin. In a monologue intended to be contemplated seriously, Carlin blathers that you are brainwashed on account of how the capitalist conspirators "own all the big media companies so they control just about all the news and information you get to hear." They are "beating you over the head in their media telling you what to believe, what to think, and what to buy."[174] In 2006 these words of Carlin's were broadcast on the cable TV channel HBO by the corporate conglomerate Time Warner. As of this writing, this monologue is found on an audio recording that is sold and marketed by major corporations. Should we entertain the notion that corporate media have a vested interest in subjecting their consumers to propaganda, one would have to take Carlin as a shill for big business.

Should left-wing intellectuals concede that Hitler ascended to domination through anti-capitalist rhetoric, though, they will probably counter that the national socialists did not implement socialist policies when they controlled the German government. That claim—that the Nazis preached socialism in theory but refrained from it in practice—must be examined. Should we discover that the Third Reich provided a massive regulatory-entitlement state, the discovery should disabuse us of the belief that the NSDAP had anything to do with what has been believed to be a laissez-faire social Darwinism extolled by Herbert Spencer or William Graham Sumner.

NOTES

1. B. Hoffman 2006 paperback, 236. Further, in *ibid.*, 318-19 n. 85, this author decries the Libertarian Party for being "far right" and tries to associate it with neo-Nazi white supremacist groups.
2. Qtd. by G. Seldes 1960, 639.
3. M. Hardt and A. Negri 2000, 110.
4. A. J. Gregor 1979, xi.
5. "Fascism," *Dictionary.Com*, http://dictionary.reference.com/browse/fascism, accessed Wednesday, October 1, 2014. N. Ferguson 2006, 228, also refers to *fascio* meaning "bundle."
6. J. Goldberg 2007, 36. According to *ibid.*, 80, Mussolini also coined "totalitarian."
7. Qtd. by J. Toland 1976, 355.
8. J. Toland 1976, 355.
9. Qtd. by J. Toland 1976, 355.
10. Benito Mussolini, "The Basic Philosophy of Fascism," ch. 35 of Oakeshott ed. 1942. This excerpt can be seen online at http://tinyurl.com/2zlftt, accessed Sunday, June 17, 2007.
11. "Capitalism," *Investopedia*, 2005, qtd. by "Capitalism," *Dictionary.Com*, https://web.archive.org/web/20070314011455/http://dictionary.reference.com/browse/capitalism, accessed Wednesday, October 1, 2014.
12. "Free Market," *Investopedia*, 2005, qtd. by "Free Market," *Dictionary.Com*, https://web.archive.org/web/20071110204654/http://dictionary.reference.com/browse/free%20market, accessed Wednesday, October 1, 2014.
13. "Free Enterprise," *American Heritage Dictionary* 4th ed. 2006, qtd. by "Free Enterprise," *Dictionary.Com*, https://web.archive.org/web/20070314011455/http://dictionary.reference.com/browse/capitalism, accessed Wednesday, October 1, 2014.

14. "Laissez Faire," *Random House Unabridged Dictionary* 2014, qtd. by "Laissez Faire," *Dictionary.Com*, http://dictionary.reference.com/browse/laissez%20faire, accessed Wednesday, October 1, 2014.

15. "Fascism," *American Heritage Dictionary* 4th ed. 2006, qtd. by "Fascism," *Dictionary.Com*, https://web.archive.org/web/20060826225024/http://dictionary.reference.com/browse/fascism, accessed Wednesday, October 1, 2014.

16. "Nazi," *Random House Unabridged Dictionary* 2014, qtd. by "Nazi," *Dictionary.Com*, http://dictionary.reference.com/search?q=nazi, accessed Wednesday, October 1, 2014.

17. "Nazism," *American Heritage Dictionary* 4th ed. 2006, qtd. by "Naziism," *Dictionary.Com*, https://web.archive.org/web/20070715010152/http://dictionary.reference.com/browse/naziism, accessed Wednesday, October 1, 2014.

18. "Naziism," *WordNet 3.0* 2006, qtd. by "Naziism," *Dictionary.Com*, https://web.archive.org/web/20070715010152/http://dictionary.reference.com/browse/naziism, accessed Wednesday, October 1, 2014.

19. J. Goldberg 2007, 26, 54–55.

20. A. J. Gregor 1979, 175–77, 209–210.

21. J. P. Diggins 1972, 213.

22. R. F. Kennedy 2005, paperback, 194.

23. T. Rall 2004, 138.

24. E. Jareski 2010, 59. J. Gray 2013, 7, http://www.the-tls.co.uk/tls/public/article1186584.ece, accessed Wednesday, January 23, 2013, also acknowledges Giovanni Gentile's influence over Mussolini and the fascist ideology.

25. The letter is Thomas Jefferson, letter to William B. Giles, December 26, 1825, in T. Jefferson 1829a, 422. For Noam Chomsky and Michael Parenti citing it, respectively, see N. Chomsky 1994, 243–47; and M. Parenti 1995 6th ed., 63. Incidentally, M. Parenti 1995 6th ed., 63, quotes Jefferson inaccurately, quoting him speaking of an "aristocracy of our monied corporations, which . . . [ellipse is Parenti's –S.H.] bid defiance to the laws of our country." For the contrast between the fake quotation that Parenti uses, versus the real quotation, see Monticello, "The End of Democracy . . . (Quotation)," *Thomas Jefferson Encyclopedia*, http://www.monticello.org/site/research-and-collections/end-democracyquotation, accessed Friday, November 8, 2013.

26. Mussolini, *Hierarchy: Monthly review of the Fascist Revolution* [*Gerarchia: Rassegna Mensile della Rivoluzione Fascista*] vol. 17, year 1937 page 131. See Mussolini 1937, 131. The original statement is, "Il Fascismo è democrazia totalitarian . . ." On January 30, 2014, I used Google Translate to translate Mussolini's words into English.

27. Acemoglu and J. A. Robinson 2012, 187–88; J. Goldberg 2007, 296-97; and C. Webber and A. Wildavsky 1986, 243.

28. W. J. Bernstein 2008, 294.

29. Thomas Jefferson, letter, to William B. Giles, December 26, 1825, in T. Jefferson 1829a, 421.

30. Schivelbusch 2006, 24, brought this statement to my attention.

31. Acemoglu and J. A. Robinson 2012, 125.

32. Qtd. by Cannadine 2006, 112.

33. *The Autobiography of Lincoln Steffens*, qtd. by J. P. Diggins 1972, 28.

34. Qtd. by Acemoglu and J. A. Robinson 2012, 125; and P. Hollander 1981, xlv-xlvi.

35. J. P. Diggins 1972, 28, citing I. M. Tarbell 1939, 380-84.

36. I. M. Tarbell 1939, 378, 380-84.

37. K. Brady 1989, 205–07.

38. Qtd. by J. P. Diggins 1972, 28.

39. Qtd. by P. Hartshorn 2011, 139.

40. Qtd. by J. P. Diggins 1972, 204.

41. Qtd. by J. P. Diggins 1972, 204.

42. Herbert Croly, "Realistic Liberalism," *New Republic*, volume LIII, November 23, 1927, pp. 5–7, qtd. by J. P. Diggins 1972, 231.

43. Charles Beard, "Making the Fascist State," review, *New Republic*, volume LVII, January 23, 1929, pp. 277–78, qtd. by J. P. Diggins 1972, 226.

44. "Liberalism Vs. Fascism," *New Republic*, volume L, March 2, 1927, p. 35, qtd. by J. P. Diggins 1972, 229.

45. J. P. Diggins 1972, 228–29.

46. Horace Kallen, "Fascism: For the Italians," *The New Republic*, pp. 212-13, qtd. by J. P. Diggins 1972, 228.

47. J. P. Diggins 1972, 228, citing Horace Kallen, "An Apology for Fascism," *New Republic*, volume XLIX, January 12, 1927, pp. 207–09. Among the Progressives to oppose Mussolini from the start were John Dewey, Oliver Wendell Holmes, Jr., and Thorstein Veblen. That is according to J. P. Diggins 1972, 235. Mussolini also received opposition from Oswald Garrison Villard in the pages of *The Nation*, according to J. P. Diggins 1972, 68, 233.

48. J. P. Diggins 1972, 58.

49. Douglas Harper, "Nazi," *Online Etymology Dictionary*, 2001, qtd. by "Nazi," *Dictionary.Com*, http://dictionary.reference.com/search?q=nazi, accessed Thursday, May 17, 2007.
50. J. P. Diggins 1972, 215.
51. J. Goldberg 2007, 77.
52. J. P. Diggins 1972, 452.
53. A. Herman 1997, 307-08.
54. J. P. Diggins 1972, 452, citing C. W. Mills 1942, 432–37.
55. Giblin 2002, 26.
56. J. Gray 2013, 7, accessed online Wednesday, January 23, 2013.
57. George Orwell, review of *Mein Kampf*, *New English Weekly*, March 21 1940, republished in Orwell 1968c, 14.
58. A. Bullock 1991, 68; and Shirer 1990, 85.
59. Qtd. by J. Toland 1976, 224–25.
60. Shirer 1990, 206.
61. N. Ferguson 2006, 423.
62. Heiber 1972, 20.
63. Qtd. by Fest 1974 U.S. version, 234.
64. Riess 1948, 24.
65. Qtd. by Riess 1948, 25.
66. Shirer 1990, 126.
67. Fest 1974 U.S. version, 234.
68. Riess 1948, 24.
69. Shirer 1990, 126.
70. Riess 1948, 25.
71. Shirer 1990, 126.
72. Riess 1948, 25. This same observation is made in G. L. Mosse 1985, trade paperback, 186.
73. Qtd. by Shirer 1990, 126.
74. Riess 1948, 25.
75. G. L. Mosse 1964, 287–88.
76. G. L. Mosse 1964, 289, citing Schüddekopf 1960, 319–320.
77. Fest 1974 U.S. version, 307.
78. The parallels between Naziism and communism are noted by historian Robert Hessen in his "Review of Ludwig von Mises's *Omnipotent Government*," *The Objectivist* vol. 9 (no. 8, August), republished in Rand 1982, 892–95.
79. Steinweis 1993, 17, 26.
80. Fest 1974 U.S. version, 270; Steigmann-Gall 2003, 41–42; F. Stern 1987, 155; and H. A. Turner 1985, 80–81.
81. Michael Moore, interviewed by Ian Hodder, "Michael Moore's Ongoing Crusade Against Corporate Greed Won Him an Audience With Big, Bad Nike," *Industry Central*, 1997, http://tinyurl.com/25mkhg, accessed Friday, June 15, 2007.
82. Otto Wagener in a 1930 conversation with Hitler, recounted in Wagener 1985, 115.
83. Hitler, qtd. by Wagener 1985, 115–16.
84. Fest 1974 U.S. version, 125.
85. N. Thomas 1971, 55.
86. G. L. Mosse 1964, 280, 286.
87. I first learned of Moeller van den Bruck coining *the Third Reich* from Buruma and Margalit 2005, paperback, 8. That Moeller van den Bruck was an advocate of the Third Way is mentioned in G. L. Mosse 1964, 280.
88. G. L. Mosse 1964, 280; and F. Stern 2006, 191, 226; confirm that Moeller van den Bruck coined *Third Reich* with the title of his 1923 book.
89. G. L. Mosse 1964, 280.
90. G. L. Mosse 1964, 280, citing F. Stern 1961, 258.
91. F. Stern 2006, 226–27.
92. Kershaw 2000, paperback, 152.
93. Hitler claimed that the Jews owned "war corporations" that oppressed the Germans, according to Giblin 2002, 21.
94. Einstein 1993d, 247.
95. P. Lane 1973, 81.
96. G. L. Mosse 1985, trade paperback, 151.
97. R. Grunberger 1971, 167.
98. R. Grunberger 1971, 44–45.
99. Fest 1974 U.S. edition, 27, 97.
100. Buruma and Margalit 2005, softcover, 33.
101. G. L. Mosse 1985, trade paperback, 152.

102. Ammann 2009, 38–39; and Y. Brook 2007.
103. J. Steinberg 2011, 396–98, 475–76. J. Steinberg 2011, 476, considers it ironic that, given Bismarck's own anti-Semitism, Kaiser Friedrich Wilhelm still ultimately dismissed Bismarck from government office out of the belief that Bismarck consorted too much with Jews.
104. Goodrick-Clarke 1992, paperback, 12.
105. G. L. Mosse 1975, 121.
106. Goodrick-Clarke 1992, paperback, 123–25.
107. Poliakov 1974 U.S. version, 297.
108. G. L. Mosse 1964, 27, 166.
109. G. L. Mosse 1964, 166–67, citing H. Lietz 1934, 25, 89–91, 196.
110. G. L. Mosse 1964, 141.
111. G. L. Mosse 1985, trade paperback, 151, quoting É. Drumont 1968, 80.
112. Hitler 2001, 234.
113. *Random House Unabridged Dictionary* 2014, qtd. in "Mammon," *Dictionary.Com*, http://dictionary.reference.com/browse/mammon, accessed Wednesday, October 22, 2014.
114. Goebbels 1966, 109-110, adapted from Goebbels 1929, 112–115, 118–120, 137–142. You can also find these words in Goebbels 1987.
115. E. Faye 2009 U.S. version, 34, 99.
116. Kershaw 2000, paperback, 21.
117. Goebbels 1966, 109–110, adapted from Goebbels 1929, 112–115, 118–120, 137–142. You can also find these words in Goebbels 1987.
118. Gellately 2007, 203, 96.
119. Hitler, qtd. by Toland 1976, 511.
120. Kershaw 2000, paperback, 21. N. Ferguson 2006, 249, also points out that Jews were hated in Germany on account of their being regarded as a financial elite.
121. Kershaw 2000, paperback, 298, 302.
122. G. L. Mosse 1985, trade paperback, 151.
123. J. Toland 1976, 406–407.
124. G. L. Mosse 1985, trade paperback, 207.
125. Toland 1976, 501. According to this same page, Hitler's then-Minister of Economics, former banker Hjalmar Schacht (1877–1970), was offended by this sign and objected to it, although he had no problem with "Jews not wanted."
126. Gellately 2001, 26–27.
127. G. L. Mosse 1985, trade paperback, 210-15.
128. W. Biddle 2009, 7.
129. Michael Gross 2003, 92.
130. G. Aly 2008.
131. Pierre-Joseph Proudhon, qtd. by Buruma and Margalit 2005, softcover, 28; and G. L. Mosse 1985 trade paperback, 183.
132. Pierre-Joseph Proudhon, qtd. by G. L. Mosse 1985, trade paperback, 153.
133. G. L. Mosse 1985, trade paperback, 152.
134. G. L. Mosse 1985, trade paperback, 152, citing A. Toussenel 1947 vol. 1, 320.
135. G. L. Mosse 1985, trade paperback, 153-54. N. Ferguson 2006, 32, also made note of how Proudhon hated both capitalists and Jews and believed that the two categories naturally went together.
136. J. Marks 2002, 120.
137. M. Grant 1918 revised ed., 16.
138. S. P. Caine 1974, 25.
139. Riis 1997, 92, 83. T. C. Lenard 2003, 702, accessed online Tuesday, February 14, 2012, brought these statements of Riis's to my attention.
140. John Maynard Keynes, qtd. by Paulovicova 2007, 44, http://ejournals.library.ualberta.ca/index.php/pi/article/viewFile/1591/1117, accessed Friday, January 30, 2011, citing C. H. Hession 1984, 225–26.
141. G. L. Mosse 1964, 242.
142. G. L. Mosse 1985, trade paperback, 118.
143. Buruma and Margalit 2005, softcover, 28.
144. K. Marx 1844, http://www.marxists.org/archive/marx/works/1844/jewish-question/, accessed Wednesday, July 29, 2009.
145. G. L. Mosse 1985, trade paperback, 154.
146. M. Luther 1848, 352–53. The statements is from the book's sections 864 and 867.
147. Hitler, qtd. by Wagener 1985, 167.
148. Hitler 2001, 235, 86.
149. Hitler 2012, 175.
150. Hitler 2001, 316, 314.
151. Hitler 2012, 175.
152. M. Moore 2003, 140–44.

153. Qtd. by Gellately 2007, 110.
154. Hitler 2001, 321, 318.
155. Kershaw 2000, paperback, 152.
156. Hitler 2001, 600, 243, 453, 622.
157. N. Ferguson 2006, 272.
158. Gellately 2007, 108.
159. Hitler, qtd. by Wagener 1985, 167.
160. Hitler, from a conversation with Wagener in 1932, qtd. by Wagener 1985, 318–19.
161. Steigmann-Gall 2003, 46.
162. M. Parenti 2004, 118.
163. Adolf Hitler, speech on January 30, 1939, qtd. in Ammann 2009, 26, and Davidowicz 1986, 33.
164. M. Moore 2003, 137–39.
165. Hitler 2001, 315.
166. Mikela J. Mikael narrating in Achbar et al., prods. 2003, transcript online in the PDF at http://hellocoolworld.com/files/TheCorporation/Transcript_finalpt2copy.pdf, accessed Thursday, May 17, 2007.
167. Chomsky interviewed in Achbar et al., prod. 2003, transcript online in the PDF at http://hellocoolworld.com/files/TheCorporation/Transcript_finalpt2 copy.pdf, accessed Thursday, May 17, 2007.
168. Qtd. by MacFarquhar 2004, http://tinyurl.com/2nz66j., accessed Tuesday, June 19, 2007.
169. "Nader Urges Hastert, Lott to Investigate the Commercial Rewards That Feed Media Violence," *Commercial Alert*, April 30, 1999, http://tinyurl.com/2hu3nh, accessed Friday, May 18, 2007.
170. Hitler 2001, 254, 316.
171. Bin Laden 2002, http://tinyurl.com/kva1, accessed Tuesday, June 19, 2007.
172. Du Bois, "A Program of Reason, Right and Justice for Today," *National Guardian*, May 23, 1960, republished in Du Bois 1995, 426.
173. Du Bois, " 'There Must Come a Vast Social Change in the United States,'" *National Guardian*, July 11, 1951, republished in Du Bois 1995, 619.
174. G. Carlin et al., prods. 2006.

EIGHT

The Führer versus Free Enterprise

Hitler's Hatred for the Industrial Revolution

Mein Kampf further establishes that Adolf Hitler not only scorns Jewish businessmen in particular, but entrepreneurs of every ethnicity and creed. It goes as far as drubbing "the bourgeois world" for having "paved the way" for the Manchesterite "doctrine which denies the state as such." Hitler's first book laments that, as the Inventive Period went on, the German nation adopted "an industrialization as boundless as it was harmful. . . . the abrupt alternation between rich and poor became really apparent. . . . Despite all the economic prosperity, dissatisfaction became greater and deeper . . ." As the rich became richer, "the gods of heaven were put into the corner as obsolete and outmoded, and in their stead incense was burned to the idol Mammon." There is that Mammon again. Continuing on this topic *Mein Kampf* rattles on, "Unfortunately, the domination of money was sanctioned even by that authority which should have most opposed it: His Majesty the Kaiser acted most unfortunately by drawing the aristocracy into the orbit of the new finance capital." The tome additionally elicits the utmost contempt for the employer who "conducts his business in an inhuman, exploiting way, misuses the national labor force and makes millions out of its sweat."

Hence the Nazi manifesto curses entrepreneurs for their "narrow-minded rejection of all attempts to better working conditions, to introduce safety devices on machines, to prohibit child labor and protect the woman . . ." Thereafter the book whines that "on innumerable occasions the bourgeoisie has...opposed demands which were justified from the universal human point of view. . . . Never can our political bourgeoisie make good its sins in this direction . . ." In the passages where it vilifies the Industrial Revolution as an age when capitalists exploited men, women, and children in factories, *Mein Kampf*'s text is almost indistinguishable from that of any history volume penned by a progressive. The masses "moved from peasant villages to the larger cities to earn their bread as factory workers in the newly established industries. The working and living conditions of the new class were more than dismal. . . . The formal transference of the old working hours to the industrial large-scale enterprise was positively catastrophic . . . the worker's health was undermined. . . . To this finally was added the miserable wages on one hand and the employer's correspondingly and obviously so vastly superior position on the other."[1] How much does that resemble the anti-Industrial-Revolution remarks from Bertrand Russell and Paulina Borsook that I quoted in Book Two? In person, Hitler sounded off to Otto Wagener—"Industrialization has deprived the individual of all liberty, placed him in thrall to capital and the machine. . . . This is most crudely evident in the Bolshevik state, with its state capitalism [?!]." The solution is for the proletarians to "seize possession of capital." This will be "*our* socialism," and a "radical removal of all the false results of industrialization and unrestrained economic liberalism, and the redirection of this line of development to the service of humanity . . ."[2] (emphasis added).

Compare Hitler's remarks with those of another mass murderer, native-born American Unabomber Theodore Kaczynski: "The Industrial Revolution and its consequences have been a disaster for the human race. . . . they have destabilized society, have made life unfulfilling, have subjected human beings to indignities, have led to widespread psychological suffering (in the Third World to physical suffering as well) and have inflicted severe damage on the natural world."[3]

Kaczynski and Hitler are unusual among twentieth-century anti-capitalists in that, when it came to ushering in violence against capitalists, these two men did the dirty work themselves. Most anti-capitalist scholars do advocate implicitly the spoliation of peaceful people, but vie to do so through the intermediary of government in the form of regulations on other people's property. As the victims of such legislation are expected to comply, never fighting back, few skirmishes ever erupt over modern U.S. regulations. Hence, most everyone upholds the pretense that spoliative legislation is "peaceful." This is "compliance" won from the point of the Invisible Gun. What is notable in Kaczynski's case, though, is that various anti-capitalist academicians voiced sympathy for the bomber's critique on the Industrial Revolution. Such scholars would never mail bombs to their enemies, to be sure, but they voiced open sympathy for the Unabomber's complaints about capitalism and industrialism. On its final day in 1997, the 49th Annual Conference on World Affairs at the University of Colorado held a four-member panel discussion titled "The Unabomber Had a Point."[4]

Passages similar to the Unabomber's are found in Hitler's second book, which he wrote in 1928 and which Holocaust historian Gerald L. Weinberg (b. 1928) uncovered and arranged for publication in the early twenty-first century. In this second book Hitler further raved that the "nineteenth century" had given the Jewish capitalist

> a dominant position within the people's economy, due to the expansion of capital loans, founded on the concept of interest. Via the detour of stock, he finally obtains possession of a large portion of the production facilities, and with the help of the stock exchange he gradually becomes ruler not only of public economic life but ultimately also political life. He supports this domination with the intellectual degradation of the peoples, assisted by . . . the work of the press which has become dependent on him. He discovers in the newly rising fourth estate of the working class the potential force to destroy the bourgeois intellectual regime, just as the bourgeoisie was once the instrument to shatter feudal rule. . . .
>
> Around the turn of the century, the Jew's economic conquest of Europe is fairly complete; he now begins with securing it politically. . . .
>
> He uses the European people's tension—most of which is attributable to their general need for space [that is, real estate to live on, of which the nation is short due to overpopulation] and the consequences that arise from it—to his advantage by systematically agitating for world war.[5]

Hitler is nowhere close to being alone in accusing Jews of being war profiteers. To an audience in Liverpool, England, Michael Moore stated that the beneficiaries of war in the Middle East are naturally the "oil companies, Israel, Halliburton" Corporation,[6] as if these separate entities are practically one and the same.

Jewish Industrialists Portrayed as "The Man"

The parallels in political rhetoric do not stop there. We have examined the Reich chancellor's rants about the machinations of Jewish capitalists. Now match them against Michael Moore's diatribes about a cabal of multinational corporate billionaires who manipulate the public. See how similar or different they sound. According to Moore,

> There is a master plan, . . . and the sooner you can get over not wanting to believe it, . . . the sooner we have a chance of stopping them. Their singular goal is to take enough control over our lives so that, in the end, we'll pledge allegiance, not to a flag..., but to the dictates of Citigroup [a financial services company], Exxon, Nike, GE [General Electric], GM [General Motors], P&G [Procter & Gamble], and Philip Morris [the cigarette manufacturer]. . . . and all that remains is the day when it will be codified onto a piece of paper, the Declaration of the Corporate States of America. . . . The takeover has happened right under our noses. . . .
>
> And it is only going to get worse. Whatever benefits you may have now are going to be whittled down to nothing. . . .
>
> So you slaved your life away, you worked long hours, you gave everything you had to help your company earn record profits. When you went into the voting booth you voted for their Republican (and Democratic) candidates just like they asked you to—and after you retired, this is the thanks you got. A senior discount...on your life [emphasis Moore's].[7]

That sounds familiar. Henry Ashby Turner elaborates that Hitler's antipathy toward the Industrial Revolution and Inventive Period "echoed the views of those critics of German industrialism who, in the great controversy at the end of the nineteenth century, opposed increased reliance on commerce and manufacturing through freer trade. . . . He deplored the setbacks suffered by that school of thought. He traced to the inroads made by the advocates of industrialism and free trade much that had gone wrong since then. . . . Throughout the 1920s he repeated in his writings and speeches the viewpoints of the turn-of-the-century critics of industrialism. . . . Like so many other critics of industrialism, he also harbored a profound distaste for the huge urban concentrations of population to which that process had given rise. . . . Hitler's negative outlook toward industry and cities carried over into his attitude toward the practitioners of big business."[8]

And if *Mein Kampf* could identify one especial geographic locus of Jewish capitalist greed, it would be the same one that the radical Left continues to kvetch about as I type these words. To quote *Mein Kampf*, "It is Jews who govern the stock exchange forces of the American Union."[9] In a conversation with Wagener in the 1930s, Hitler announced that "a new great adversary has arisen across the Atlantic—America. As a result of excessive industrialization, . . . America has no choice but to wage an imperial policy all over the world."[10] In other conversations, the mustached demagogue could not help but jeer at the United States for being what he called "a decayed country, with problems of race and social inequality . . ." He saw it as a place "where everything is built on the dollar . . ."[11] "What is America" he raged, "but millionaires, beauty queens, stupid records and Hollywood?" Nothing, he believed, since the country has long been afflicted by "the most grasping materialism."[12] The Nazis' collaborators in France concurred, mouthing that the USA was a nation of monopolies and "gangsterism," and that every American is a "vile profiteer who only respects money."[13]

In yet another discussion with Wagener, the Führer went on that the U.S. is a hotbed of licentiousness. Such depravity, Hitler maintained, resulted from Jewish industrialists having ascended to such a high position in our nation, and the influence the USA provides to the rest of the Earth. "Hamburg appears to me a model of how the highest cultural life can be stifled and destroyed by industrialism . . ." The erosion of public life will persist "unless we oppose excessive industrialization and Americanization by offering something new." One of the results of World War One is "that, wherever capitalism reigns, America has supremacy. And since America suffers from industrial overproduction, it will exploit this supremacy to dispose of its overproduction." Note that this is part of the anti-globalization theory that Lenin bolsters in *Imperialism, The Highest Stage of Capitalism,* as we discovered in Book Two. And Hitler continues that the peril of industrialization "concerns everyone. . . . Only when capitalism has been broken, abolished, replaced by something new does America's power stop. . . . From there an anti-industrial world can be erected."[14]

At this point Wagener asked him what such a de-industrialized paradise would look like.[15] Quoting Hitler's answer, it would be a society that recognized that "the economy...must be placed at the service of the Volk. That we will achieve through a structure of...supervisory state planning." Upon completing that step, the Nazis must "form economic communities of interest with the countries around us—with Poland, Czechoslovakia, Austria, Hungary, and the Balkans; further, with Denmark and Holland, and perhaps with Belgium and Luxembourg; and with Switzerland as well. Thus, a Central European economic bloc would be created. . . . Hand in hand with it goes a monetary union, followed by a customs union. . . . In this way, Europe would at last be given a strong heart, which would assume in whole or in part the principles of our social economics—and this would happen all the sooner the more successfully this form develops among us and the more severe the Communist pressure from the East, on the one hand, and the economic-imperialist pressure from America, on the other"[16] (emphasis Wagener's).

When it comes to reviling both the Soviet Union *and* American commerce as comparably evil empires, the Reich chancellor has some company. Michael Moore shouts that as far as the U.S. economy is laissez faire, it is "unfair and unjust and it's not democratic and it has to change." Referencing his final line from his movie *The Big One*—"One evil empire [the Soviet Union] down, one to go"—Moore clarifies that he means that "our system," commercial enterprise, "is the one that's got to go."[17] Moore, a native-born American, also shares in this antipathy against the USA and its international appeal. Moreover, Moore's grievances against the USA match Hitler's—the country is too capitalistic and therefore immoral. The filmmaker resents that too few Americans are as eager as he is to start an insurrection that would separate the capitalist class from its property. For that reason, we Americans "are possibly the dumbest people on the planet . . . in thrall to thieving, conniving, smug pricks."[18] Consequently Moore asks, "Should such an ignorant people lead the world?"[19] Thus he told an audience of Germans, "Don't be like us,"[20] and addressed yet another crowd in Cambridge, England—"You're stuck with being connected to this country of mine, which is known for bringing sadness and misery to places around the globe."[21]

Mein Kampf concludes that its litany of charges against businessmen in general and Jewish businessmen in particular uncovers the reasons why the "fight against international finance and loan capital became the most important point in the program of the German nation's struggle . . ."[22]

More than a handful of Hitler's ideas are recognizable in Osama bin Laden's Open Letter to America. The al-Qaeda mastermind sputters that the American stock market is nothing more than "gambling," which results in "investments becoming active and the criminals becoming rich." This is bad on the grounds that the USA is a "nation that permits Usury, which has been forbidden by all the religions. Yet you build your economy and investments on Usury. As a result of this, in all its different forms and guises, the Jews have taken control of your economy, through which they have then taken control of your media, and now control all aspects of your life making you their servants and achieving their aims at your expense . . ."[23] Bin Laden finds this so abominable that he cites it as one reason for his jihad against the United States.

Michael Moore complementarily issues his own battle cry against the forces of finance and loan capital.

> . . . these bastards who run our country . . . need to be brought down and removed and replaced with a whole new system that we control. That is what democracy is supposed to be about—we, the people, *in fucking charge*. . . .
> The only true value your life has to the wealthy is that they need your vote every election day in order to get the politicians they've funded into office. They can't do that by themselves. This damnable system of ours that allows for the country to be run by the will of the people is a rotten deal for them as they represent only 1 percent of "the people." . . . This is why they truly hate democracy: because it puts them on a distinct advantage of being in the smallest of the smallest minorities. . . . The rich have thus had a hardcore army of conservatives, right-wingers, and Christian Coalition-types to act as their foot soldiers. . . . [T]hey are more than happy to see millions of poor whites and even millions more middle-class people cheerfully pulling the lever in the voting booth for the candidates who will only screw these poor-white and middle-class people once they're in office [emphasis Moore's].[24]

The parallels are as follows. (1.) Both *Mein Kampf* and Moore posit that there exists a shadowy rich cabal that conspires to exploit the masses. (2.) Both use the tone of *us* versus *them—we* being the exploited working class, and *they* being the rich who simply want to use us. (3.) Both *Mein Kampf* and Moore aim vituperation upon persons they suspect of being members of this cabal, and issue arbitrary insinuations about the accused, impugning their motives without any substantiation. (4.) Both cry that the cabal controls the everyday workplace, the commercial media, and politicians. (5.) Both see a final solution in a Revolution that will sock it to the cabal and end its plutocratic reign.

The flaw in the left-wing intellectuals' portrayal of Hitler should be evident. We know that governist intellectuals implicate Herbert Spencer and social Darwinism for Hitler's ascension. In promoting this charge, governist intellectuals would have it seem as though Hitler were a self-consciously classist bully who stomped on the Jews because they were the underdogs. In reality Hitler's grievances against the Jews were comparable to the regular socialist's grievances against businessmen as a whole. The Führer slurred that the Jews were classists and elitists who fancied themselves as the elite Chosen Race. The Jewish capitalists' classism and elitism, Hitler's accusation continues, resulted in the Jews' oppression of the low-income Aryans. Insofar as classism and elitism can be interpreted as emblematic of social Darwinism, Hitler (a) expresses nothing but disapproval for social Darwinism, and (b) implies that it is the Jews who practice social Darwinism. As a solution, Hitler proposes a violent uprising among the low-income Aryans against the wealthy Jewish capitalists.

Just as Karl Marx prophesized that this revolution would herald a new era where the communist class would be the sole class, Hitler foresaw a revolution by low-income Aryans that would culminate in a New Age of utopian Aryan communalism. To employ the sort of slang that anti-capitalist hippies threw around in the 1960s, *Mein Kampf* identifies Jewish industrialists as the *establishment*, the *system*, the *military-industrial complex*, and, most significantly, *The Man*. They are The Man who holds us down. Once Hitler's cronies had acquired total domination, assesses Prof. Alan Steinweis, the Third Reich's *modus operandi* amounted to "squeezing wealth from German Jews . . ."[25] Hitler contended that such expropriation was justified punishment for Jewish financiers for having stomped on low-income Aryans throughout the Inventive Period. The Nazis' expropriation of the Jews' wealth in the early 1930s was not unlike the manner in which the bourgeoisie were treated when the Bolsheviks seized power in Russia in 1917.

Hitler and the Nazis Were Anti-Capitalist from Beginning to End

We thus return to the North American progressive viewpoint, as seen in *The Corporation: The Pathological Pursuit of Profit and Power*. Unexpectedly, this Canadian documentary takes private entrepreneurs to task for not being patriotic enough. Its narrator, Mikela Mikael, complains, "It should not surprise us that corporate allegiance to profits will trump their allegiance to any flag."[26]

Hitler concurs with this criticism. In his second book he expounds that commerce undermines one's loyalty to one's own country. "People who love to speak of socialism do not understand that the most socialistic organization of all was the German people's army. Thus also the fierce hatred of the typically capitalist-minded Jewry against an organization in which money does not equate with status . . . or honor . . ."[27] Note that Hitler celebrates militarism. However, Hitler's militarism does not conform to the stereotypes that the academic Left holds toward right-wing military-industrial-complex jingoists. Contrary to left-wing presumptions, the Nazi leader does not judge the military to be a tool whereby he can aid big businesses in establishing new international markets per Lenin's imperialism theory. Nay, Hitler celebrates militarism for another reason. It is that he interprets it as an institution wherein the individual's personal will must be subordinated to the group. The Führer wishes that society, as a whole, followed the same model of organization. Hence Alan S. Milward notes that Hitler genuinely thought of "war as . . . a device for eliminating the corrupting egotistical self-seeking which he saw as the concomitant of false ideas of human liberty . . ."[28]

And predicting that "the miserable army of our present-day shopkeepers" will not share in its jingoistic sentiments, *Mein Kampf* screams at such merchants, "You have but *one* concern: your personal life, and *one* God: your money! But we are not addressing ourselves to you, we are appealing to . . . those who are so poor that their personal life cannot mean the highest happiness in the world; to those who do not see the ruling principle of their existence in gold, but in other gods"[29] (emphases Hitler's). Have we not heard this before? Michael Moore snarls that "the rich, by and large, are neither conservative nor liberal, nor are they devout Christians or Jews. Their real political party is called Greed, and their religion is Capitalism."[30] By *Greed* and *Capitalism*, Moore means *Mammon*.

In March 1930 Hitler restated his opinion that businesspeople are seditious, this time in person to Otto Wagener. Hitler wants Wagener to notice "the difference between the former age of individualism"—the late 1800s in the Inventive Period—"and the social that is on the horizon," meaning for Europe for the rest of the twentieth century. "In the past—that is, for most people it is still the present—the individual is everything, everything is directed at maintaining his life and improving his existence. Everything focuses on him. . . . Everyone is a central figure, as is officially acknowledged in his vested human rights. In the socialism of the future, on the other hand, what counts is the whole, the community of the Volk. The individual and his life play only a subsidiary role. He can be sacrificed—he is prepared to sacrifice himself should the whole demand it, should the commonweal call for it." The Führer considers his current military regime to be the setter of standards for the rest of the Reich. That is why he has instilled regulations "to punish anyone who dodges military service by self-mutilation or desertion, even prescribing death for flight in the face of the enemy. Here, therefore, the basic socialist principle prevails. But in the rest of life, individualism, [laissez-faire] liberalism, egotism continue to triumph." That will have to change. He hates laissez-faire liberals, for, in their own self-interest, they wish to avoid joining in combat. "Aren't these [laissez-faire] liberals, these reprobate defenders of individualism, ashamed to see the tears of the mothers and wives, or don't these cold-blooded accountants even notice? . . . It's understandable why bolshevism simply removed such creatures. They were worthless to humanity, nothing but an encumbrance to their Volk. Even the bees get rid of the drones when they can no longer be of service to the hive. The Bolshevik procedures are thus quite natural. . . . Such a powerful upheaval requires sacrifices, it demands sacrifices, and sacrifices must be brought."[31]

Similar to books by socialists and Progressives is the manner in which *Mein Kampf* insistently remonstrates "the egoism of shopkeepers" and "the 'big capitalist'" while upholding the "the sacrifice of personal existence." I will now reproduce a quotation of Hitler's that I had already quoted in *The Freedom of Peaceful Action*. As with socialist and Progressive tracts, *Mein Kampf* upholds that the collectivist "state of mind, subordinates the interests of the ego to the conservation of the community. . . . Every worker, every peasant, . . . official, etc., who works without ever being able to achieve any happiness or prosperity for himself, is a representative of this lofty ideal. . . . In giving one's own life for the existence of the community lies the crown of all sense of sacrifice. . . . Our own German language possesses a word which magnificently designates this kind of activity: *Pflichterfüllung* (fulfillment of duty); it means not to be self-sufficient but to serve the community. The basic attitude from which such activity arises, we call—to distinguish it from egoism and selfishness—idealism. By this we understand only the individual's capacity to make sacrifices for the community, for his fellow man." And *Mein Kampf* additionally reckons, "To win the masses for a national resurrection, no social sacrifice is too great." To sum up, morality can be attained "only as long as an idealistic national community offers the necessary preconditions. And these do not lie in material egoism, but in a spirit of self-sacrifice and . . . renunciation."[32]

This was not just some performance that Hitler put on to humor the public. As Henry Ashby Turner explains, the Volk leader "displayed much the same scorn and mistrust for the practitioners of big business even in private conversations where tactical considerations of that sort played no role." In October 1941, after having ruled as a dictator for eight years, Hitler "referred to big businessmen (*die Witschaft*) as 'rogues' and 'cold-blooded money-grubbers' . . ."[33]

The other Nazis shared in this anti-capitalism. Joachim Fest writes that in its political campaigns, the NSDAP preyed upon the lower middle class's "fears of being overwhelmed economically by large concerns and department stores, and the little man's resentment of easily acquired wealth, of profiteers and the owners of capital." He emphasizes the ceaseless "Nazi anticapitalist ranting against usurers, speculators, and department stores . . ."[34] One of the Nazi Party's orators in the 1920s screamed to a crowd, "Storm the commercial banks! Take all the money you find there and throw it into the streets and set fire to the huge heaps of it! Then use the crowbars of the streetcar lines to string up the blacks and the white Jews!"[35] Fest further writes that once Hitler was dictator, the Nazi storm troopers—the *SA*—"pasted slogans on the walls" of other people's "houses and factories . . ." At other times, they "organized commando strikes against black marketers or capitalist profiteers."[36]

Sometimes I hear the extraordinary claim that though the Nazis called themselves socialists, they did not truly believe in the socialist ideology or preach it. Put that platitude to rest. The Nazi Party advocated socialism from its very inception. And, as I shall show in chapter 10, the Nazi political economy under Hitler remained anti-capitalist to the end. It is no accident that in *Mein Kampf*, Hitler explicitly names, as his political role model, Dr. Karl Lueger (1844–1910), the anti-Semite demagogue who was elected to become Vienna, Austria's, mayor in the 1890s.[37] Lueger was no sort of Herbert Spencer-loving capitalist. On the contrary, Hitler correctly identified Lueger as a self-proclaimed "Christian Socialist" and member of the Christian Socialist Party.[38] Fareed Zakaria cites the Lueger regime as a case study in what he dubs "illiberal democracy." The Christian

Social Party, George L. Mosse explains, "opposed [laissez-faire] liberalism . . . and finance capitalism."[39]

It also behooves me to tell the tale of how the Nazi Party began, and of how it recruited Hitler. The Nazi Party was started in the 1920s by a locksmith named Anton Drexler. Its first chairman was newspaper reporter Karl Harrer, and one of its earliest disciples was the construction engineer Gottfried Feder, who founded the German Fighting League for the Breaking of Interest Slavery.[40] Originally, the party was merely called the DAP—the Deutsche Arbeiter Partei (German Workers Party). The *NS* for *national socialist* would be added to the party's name in 1920. At its first-ever meeting, Drexler explicated the organization's goals. The party directives, writes Joachim Fest, expressed "resentment toward the rich" and "the price gougers. . . . The program called for annual profits being limited to 10,000 marks . . .'" Moreover, from its beginning, the party "defined itself as a classless 'socialist organization . . .' Drexler's 'inspired idea' was to reconcile nationalism and socialism."

Hitler first became acquainted with the Nazi Party on September 12, 1919, when he stumbled upon a public lecture Feder delivered on the villainy of Jewish bankers charging interest. Its title was, "How and By What Means Can Capitalism Be Eliminated?"[41] Usury was something that Feder wanted abolished, particularly when practiced by Jewish investment houses. According to William Shirer's *The Rise and Fall of the Third Reich*, Hitler "was much impressed by Feder's lecture. He saw in Feder's appeal for the 'breaking of interest slavery' one of the 'essential premises for the foundation of a new party.'" At the same talk, Drexler introduced himself to Hitler and gave him a gift—a pamphlet of his own authorship, entitled "My Political Awakening." Shirer documents that when Hitler read the pamphlet the next day, he discovered that "it reflected a good many ideas which he himself had acquired over the years. Drexler's principal aim was to build a political party which would be based on the masses of the working class but which, unlike the Social Democrats, would be strongly nationalist." Drexler was once associated with a much-less-leftist nationalist advocacy group "but had soon become disillusioned with its middle-class spirit which seemed to have no contact at all with the masses." Shirer reminds the reader that Hitler "had learned to scorn the bourgeoisie for the same reason—its utter lack of concern with the working-class families and their social problems. Drexler's ideas, then, definitely interested him."

In the early 1920s, Hitler became the officer in charge of the Nazi Party's propaganda. The party's first chairman, Karl Harrer, agreed with Hitler's denunciations of capitalism, but objected to how Hitler placed most of the blame on Germany's troubles on a particular ethnic group of businessmen—the Jews. As Harrer found that Drexler, Feder, and most other DAP members maintained the notion that Jews deserved to be hated for being too capitalist, Harrer realized that he was the odd man out in the dispute. Harrer thus resigned as chairman, his position soon filled by Drexler. As I stated in Book One, Hitler thence co-authored the Nazi Party's anti-capitalist platform with Drexler and Feder.

An apologist for socialism might rationalize that once Hitler established himself as the party's leader, he hijacked it and dispensed with any desire to implement its anti-capitalist program. One might arbitrarily accuse Hitler of betraying the man who had seduced him into the Nazi Party from the outset, Gottfried Feder. But Hitler remained loyal to Feder to the end. Soon upon reaching dictator status in 1934, Hitler appointed Feder an undersecretary in the Ministry of Economics—the area of governance where Feder most coveted control. Feder held onto that position as late as 1940, when Der Führer was at the peak of his power.[42]

In light of the Nazis' consistency in their anti-capitalist rhetoric, the rhetoric of the twenty-first century's anti-capitalist movement remains troubling. In January of 2014, billionaire and Silicon Valley venture capitalist Thomas Perkins (b. 1932) pointed out in a

Wall Street Journal letter to the editor how similar the Occupy Wall Street movement was in philosophy and activist tactics to the Volkish movement of the 1920s. He implores readers to direct their "attention to the parallels" between "fascist Nazi Germany to its war on its 'one percent,' namely its Jews," and the corresponding "progressive war on the American one percent, namely the 'rich.'"[43] The media responded by ridiculing Perkins as a paranoid, right-wing bigot. *Cracked.Com*, which touts itself as an online humor periodical, derided Perkins as an "asshole" and "the supreme douche-moneybag."[44] But the historical record bears out the verity of Perkins's assessment.

Hitler Puts Socialist Theory into Socialist Practice

It is no accident that Adolf Hitler modeled the National Socialist German Workers Party after other socialist movements. In *Mein Kampf*, the future führer resolved to enact welfare laws to mitigate the growing disparity between the incomes of the wealthiest and poorest: ". . . we must guard ourselves against an excessive differentiation of wage rates. . . . It may be that . . . gold has become the exclusive" concern of the domineering capitalists, "but the time will come when man will again bow down before a higher god." The national socialist movement "must herald a day which will give to the individual what he needs for living, but uphold the principle that man does not live exclusively for the sake of material pleasures. This must some day find its expression in a wisely limited gradation of earnings which in any event will give every decent working man an honest, regular existence as a national comrade and a man." *Mein Kampf* additionally promised State support for laborers in imposing demands upon employers.[45] By 1931 the Nazis announced that anyone caught participating in any form of strikebreaking would be kicked out of their party.[46]

Yet, in trying to portray Adolf Hitler as an enemy of authentic socialism, socialism's apologists cite a specific historical fact. On May 2, 1933, Hitler had the Nazi government raid the offices of labor unions, seize their assets, and arrest their leaders.[47] This fact is cited out of context in attempt to demonstrate that the NSDAP's leaders were stooges of big business and indifferent to labor legislation.[48] Contrary to left-wing propaganda, still, this is not evidence of the national socialists favoring entrepreneurs over their laborers. Rather, the NSDAP rationalized that because the government dictated over everything business owners would do, proletarians no longer needed private unions to assist them. Moreover, the Nazis did not want private unions to compete against the State for the laborers' loyalty. Just as the Soviets saw fit to nationalize agriculture, the Führer found it sensible to nationalize the trade unions. The State-run union was the German Labor Front (Deutsche Arbeits-Front), headed by Robert Ley, whom we quoted in chapter 3 concerning his explanation that loyalty to Hitler is a form of faith. By the summer of 1934, the State required every corporate employee to join the government-approved labor union that Ley controlled.[49] On the day that the Nazis first attacked the private labor unions, Ley had the Labor Front issue this statement: "Workers! Your institutions are sacred to us National Socialists. . . . I know the exploitation of anonymous capitalism. . . . we will build up the protection and the rights of the workers still further."

Hitler himself enjoyed reciting, "Honor and respect the worker!"[50] In every way, the Reich despot cherished the socialist agenda as the synonym for good.[51] This is one among many reasons why it is fatuous for Michael Parenti to portray Hitler as an apostle of capitalism. Parenti misleads his readers when proclaiming that Hitler aimed to "abolish labor unions" and that he "eliminated worker benefits" and "ignored occupational safety standards."[52]

Pulitzer Prize-winning historian John Willard Toland edifies that the German dictator held faith in the "organized economy" while he nurtured "the bohemian's rather than the revolutionary's disdain for private property."[53] In a 1929 essay Dietrich Klagges, an official who assisted in mapping out the NSDAP's economic agenda, proclaimed, "Comprehensive social justice can be brought to bear only by withdrawing the decision making on interest rates, prices, and wages from the sphere of economic power and transferring it" to the authority of State regulation. Richard Steigmann-Gall surmises that Klagges "found inspiration for his ethical imperative in Christ's whipping of the usurers and the money changers,"[54] and notices that "the 'socialism' in National Socialism was of particular importance" to Joseph Goebbels, who described his political outlook as "Christ Socialism" but not "Christian Socialism."[55] Recall that American eugenicists Richard T. Ely and Edward Bellamy, whom we discussed in chapter 4, described their own ideology as *Christian Socialism.*[56] Steigmann-Gall apprehends the difference between Goebbels's "Christ Socialism" and the Christian Socialism of Ely and Bellamy. It is, in Steigmann-Gall's phrasing, that Goebbels regarded the former as an "unadulterated return to Christ's teachings, free of the temporal church . . ."[57] Niall Ferguson similarly confirms that the Nazis viewed their own ideology as having been inspired by Christianity.[58]

With such convictions, Goebbels wrote in his book *Michael*, "The idea of sacrifice first gained visible shape in Christ. Sacrifice is intrinsic to socialism. . . . The Jew, however, does not understand this at all."[59]

In a private meeting in 1930, Hitler discussed his "real socialism" with his advisers.[60] He expressed to everyone his concern that too many Germans were "imprisoned in the liberalistic attitude. Do you think that a confirmed industrialist is prepared suddenly to admit that his property is not a right . . .? That capital should no longer rule but be ruled? That it is not the life of the individual that matters but the totality? That the principle of the soldier's sacrificial death should be transformed into the readiness of every working person—whether he be active in the economy or elsewhere—to sacrifice himself for the community? . . . For almost two thousand years the Gospel of Christ has been preached, for two thousand years the sense of community has been taught: . . . care for one another . . .! But today, at the end of these two thousand years, economic liberalism flourishes as never before!" The Führer judges "the mechanization of manual labor and the industrialization of the economy" to be contradictory to Christian altruism.[61]

At a different private meeting with Wagener, Hitler prattled, "Socialism is a question of . . . the ethical outlook on life of all who live together in a common ethnic or national space." And the socialist philosophy is an ancient one. "Whenever I read the New Testament Gospels and the revelations of various of the prophets . . . I am astonished at all that has been made of the teachings of these divinely inspired men, especially Jesus Christ . . . *They* were the ones who created this new worldview which we now call socialism, . . . they taught it and they lived it! But the communities that called themselves Christian churches did not understand it! Or if they did, they denied Christ and betrayed him!" The modern church, Hitler maintains, does not consistently support the true altruism and socialism of Christ. Therefore, the Nazis *"are the first to exhume these teachings!"*[62] (emphases Wagener's). As early as one Yuletide dinner in the late 1920s, Hitler proclaimed to other Nazis present that he was destined to "translate the ideals of Christ into deeds." Naziism, the demagogue proclaimed, would implement "the work which Christ had begun but could not finish."[63] Years later, as dictator, Hitler encouraged Wagener with these words: "We are living in an age of . . . evolution from individualism to socialism, from self-interest to the public interest, from the 'I' to the 'we.'"

In complete contrast to Edwin Black's misleading portrait, Hitler was not an elitist social Darwinist who wanted to exterminate the financial underclass. Hitler plotted to wipe out the Jews as a result of his feeling that *they* were the elitist, classist, capitalist social Darwinists. The tyrant told Wagener, "Of course it is not appropriate for humankind to . . . eliminate what is sick and weak . . . Whatever has entered human society must somehow be placed in the service of this society and cared for."[64]

"Socialism, in Nazi terminology," summarizes George L. Mosse, "meant competition in the service of the Volk . . ."[65]

Having that same goal in mind, Goebbels commanded, "Be socialists of action."[66] Hitler seriously conducted himself according to this credo. To him, socialism was not all talk; he walked his talk by ministering a number of the same policies championed by America's so-called progressives. These are the same progressives whom Richard Hofstadter and other historians have credited with defeating social Darwinism on account of their support for such welfare-state policies.[67]

Anti-capitalism can be detected in the Nazi Party's treatment of department stores—the 1920s equivalent of the 1990s' big-box retailers. Hitler demonized them in his own day for the same reasons that big-box retailers are presently being denounced—allegedly for competing against smaller, mom-and-pop stores and putting them out of business.[68] Chew on an observation by Ian Buruma and Avishai Margalit. They write that "Berlin department stores . . . were vilified as symbols of 'Jewish materialism,' . . . depicted in Nazi publications as slimy octopuses strangling small German enterprises and honest German craftsmen . . ." Compounding the irrationality, these publications then added that the stores corrupted "German womanhood with decadent, 'cosmopolitan' products, such as cosmetics and cigarettes . . ."[69]

Plank 16 of the Nazi Party's 1920s platform, co-authored by Hitler himself, reads, "We demand . . . immediate communalization of the great warehouses"—meaning department stores housing their wares—"and their being leased at low cost to small firms, the utmost consideration of all small firms in contracts with the State, country or municipality."[70] Once they took absolute control over Germany, the Nazis did not end up fully nationalizing the department stores. But, as Jonah Goldberg (b. 1965) reveals, the Nazis did ban the stores "from entering a slew of businesses—much as today's critics would like to do with Wal-Mart."[71] There was a time, for instance, that Wal-Mart tried to set up its own automated teller machines. There have been legislative initiatives, for example, to stop this, barring Wal-Mart from the banking industry.[72]

Moreover, reveals Robert Gellately, when the Nazis began to institute a forcible boycott on businesses either owned by or stereotypically associated with Semites, "the boycott was directed particularly at department stores, many of them owned by Jews."[73]

In a predictable irony, it is these big-box retailers, the present-day analog to Jewish department stores—rather than the governists itching to shackle them—that are compared to totalitarians. New York *Daily News* columnist Neil Steinberg derides Wal-Mart as "an enormous fascist beast rising to its feet and searching for new worlds to conquer." Note the upcoming equivocation between the company's peaceful commerce with governmental violence. Steinberg opines, "Wal-Mart destroys the unique business and cultural identity of a city with the ruthless efficiency of Red Guards dynamiting a 1,000-year-old Tibetan monastery."[74]

As an additional measure, the NSDAP proposed that rich landholders' real estate be seized in Germany's then-equivalent of eminent domain, and then reapportioned toward poorer Aryans. In 1920, its program advocated "the expropriation of land for communal purposes without compensation, the abolition of ground rent, and the prohibition of speculation in land." Logically, there were observations that this plank sounded like something the Soviets would do. In reaction, Hitler decided to issue some governist double-speak. Hitler predictably rationalized the NSDAP's arrogation of the power to seize rightful private property at the behest of some collectivist rationale. On April, 1928, he prevaricated, "Since NSDAP stands on the basis of private property,"—now here comes to the contradiction of that claim—"it goes without saying that the phrase 'expropriation without compensation' refers simply to the creation of possible legal means for confiscation, when necessary, of land acquired illegally or not managed in the public good. It is, therefore, aimed primarily against Jewish companies that speculate in land."[75] But the Nazis' implementation of collectivist anti-capitalism goes farther. "Like [Franklin D.] Roosevelt," John Toland admits, Hitler campaigned to maintain "social security and old-age benefits." The German ruler also held up his oath to force labor regulations upon businesses for the ostensive benefits of employees. *This* is why the Nazi despot felt that private labor unions were no longer needed.

For the rest of part II, I shall expound upon how Hitler facilitated a welfare state, one that was, in astonishing respects, politically correct by twenty-first-century standards. I do not accuse contemporary progressives of condoning any of the atrocities the Nazis committed; almost no one in the freer world condones them. This is to stress, though, the disingenuousness of the claim that the Third Reich practiced any sort of "social Darwinism" consistent with what left-wing writers presume to be the social Darwinism of nineteenth-century laissez-faire proponents. To be sure, few of these welfare-state initiatives began, *in utero*, with the Third Reich. Under his own system, the Führer merely expanded upon the precedents that Bismarck had set in the nation earlier, under the Second Reich with which the expression *welfare state* originated. Hitler told one interviewer, "The bourgeois must no longer feel himself a kind of pensioner of...capital, separated from the worker . . ., but must aim to accommodate himself as a worker to the welfare of the community."[76]

That the German Labor Front embodied the government nationalization of traditional unions does not prove Hitler insincere in degrading the position of the entrepreneur or in promising regulations to boost the employees' ostensive well-being. While strikes were banned under the new regime, some wildcat strikes took place during its reign anyway. "The 'socialism' in the party title," elucidates George L. Mosse, "was given concrete expression through a paternalism which was supposed to end class differences on behalf of the unity of the Volk." The emphasis on solidarity manifested in the Strength Through Joy (*Kraft durch Freude*) program, which began in 1933 as a private initiative but soon became a State project. At the Labor Front, Robert Ley erected "an empire which . . . organized and looked after the social well-being of the salaried workers . . ."[77]

In John Toland's words, Hitler's laws demanded that workplaces had "more windows, less crowding, and better washrooms." For the employees of the government and private companies alike, the Strength Through Joy campaign expended tax money to subsidize their attendance of motion pictures,[78] operatic concerts, stage plays, adult education, lectures, and even luxury cruises and other tour vacations.[79] The State financed the employees' formation of sports teams as well.[80] And Toland observes that this "spirit of equality was even felt in the armed forces. There was far more camaraderie than formerly between officers and enlisted men in the regular service and the elite SS units were models of democracy. . . . Nowhere was egalitarianism more evident than in the Youth Labor Service where young men and women of all classes between the ages of seventeen and twenty-five were obliged to work for a period as farm hands and laborers for *Volk* and *Vaterland*"—fatherland, meaning the nation. "This service had been instituted to alleviate unemployment but went far beyond Roosevelt's Civilian Conservation Corps, which had a similar aim, to become the manifestation of socialism."[81] Recognizing that the Nazi regime facilitated a full-scale welfare state, Hitler announced to the Reichstag on January 30, 1937, "A radical transformation has...produced results which are democratic . . ."[82] Two years later, upon watching the growth of this same welfare state, Hitler boasted to a Munich assembly on February 26, "If positive Christianity means . . . the clothing of the poor, the feeding of the hungry . . . then it is we who are the more positive

Christian. For in these spheres the people's community of National Socialist Germany has accomplished prodigious work."[83]

Despite what he knows to be Hitler's evil, John Toland admits that these measures evince a "spirit of social democracy" and states, "In practice, this concept glorified the worker while underlining Hitler's theory of social equality." Not denying Hitler's evil, Toland remarks that among all economic classes of gentile Germans, "the nation was gaining in equality . . ."

In spite of their being at war with one another, Hitler felt some kinship with Josef Stalin for philosophic reasons. As Toland discerns, they both "operated under ideologies that were not essentially different. . . . Hitler had long admired Stalin, regarding him as 'one of the extraordinary figures in world history,' and once shocked a group of intimates by asserting that he and the Soviet leader had much in common since both had risen from the lower classes . . ." When one interlocutor pointed out to Hitler that, in his youth, Josef Stalin committed bank robbery, the Führer replied that the robbery was justified on account of Stalin committing it for the ultimate benefit of his own political movement—a gesture any Nazi ought to appreciate.

That should tell something to those who assert that communism and Naziism are opposites. More superficially outrageous, though, Hitler expressed strong approval for Franklin Roosevelt with regard to his domestic fiscal policies.[84] Hitler told a *New York Times* correspondent in July 1934, "I have sympathy for Mr. Roosevelt . . ."[85] On March 14, 1934, Hitler sent the following letter to then-U.S. Ambassador Thomas J. Dodd (1907–1971)[86] : "The Reich Chancellor is in accord with the President that the virtues of sense of duty" and "readiness for sacrifice . . . must be the supreme rule of the whole Nation. This moral demand, which the President is addressing to every single citizen, is also the quintessence of German philosophy of the State, expressed in its motto 'The public weal before private gain.'"[87]

John P. Diggins erroneously denies that New Deal adviser Rexford Tugwell was influenced by Italian fascism. Nonetheless, he does admit that New Deal advisors Hugh Johnson and James Farley expressed positive assessments toward Mussolini's programs. Diggins acknowledges that Farley "wrote Roosevelt a brief but glowing estimate of Mussolini's reclamation projects; Hugh Johnson, head of the NRA, carried a copy of Raffaello Vigone's *The Corporate State*, spoke respectfully of Fascism, and upon resigning invoked in a farewell speech what he called the 'shining name' of Mussolini."[88]

Franklin D. Roosevelt himself had some warm feelings toward Il Duce prior to declaring war on him. In June 1933, FDR wrote to Ambassador Henry Breckinridge Long about the Italian ruler, "There seems no question that he is really interested in what we are doing and I am much interested and deeply impressed by what he has accomplished and by his evidenced honest purpose of restoring Italy and seeking to prevent general European trouble."[89] On July 27, 1933, FDR wrote in a letter to John Lawrence, who also lauded Mussolini, "I don't mind telling you in confidence that I am keeping in fairly close touch with that admirable Italian gentleman."[90] John D. Diggins reveals that even "as late as 1939," when Italy had joined the Axis Powers, "Roosevelt could still look back on Il Duce's regime with some sympathy."[91]

Such observations belie the denials of FDR/Kennedy hagiographer Arthur M. Schlesinger that the New Deal enacted economic organization policies paralleling those of fascism.[92] And there were others, besides Hitler himself, who noticed these similarities in the outlook between him and FDR. Broadcast journalist and *Masterpiece Theatre* host Alistair Cooke (1908–2004)—hardly a right-wing ideologue—recollects that under FDR's administration, "America had a fling at National Socialism. Roosevelt was for all administration purposes a dictator," though, unlike the Führer, he was "a benevolent one."[93]

When reviewing FDR's book *Looking Forward,* Mussolini wrote, "The appeal . . . with which Roosevelt here calls his readers to battle, is reminiscent of the ways and means by which Fascism awakened the Italian people." He also was glad that FDR realized that the economy could not "be left to its own devices . . .,"[94] and therefore praised the New Deal for being "boldly . . . interventionist in the field of economics."[95] Although the Italian dictator felt that the New Deal could not be judged as full-fledged fascism, as it did not go far enough,[96] he still judged, "Without question, the mood accompanying this sea change resembles that of Fascism,"[97] as the New Deal's elevation of the collective above the individual can definitely "recall the bases of fascist Corporatism."[98] Recall from chapter 6 that Mussolini also wanted fascism to be called corporatism, and that he identified the New Deal's policies as "corporatism." At this juncture it appears that Mussolini's system of government more resembles the one that progressive Robert F. Kennedy, Jr., advocates than the laissez-faireist sort that he impugns.

Yale's Henry Ashby Turner, Jr., explicates that Hitler abided by a tradition of the Nazi Party that went back to the 1920s. The Nazis in the Reichstag introduced "several bills calling for the confiscation of 'the fortunes of the princes of bank and stock market'. . . Similarly, the Nazi deputies generally supported proposals by the parties of the left for increases in the level of expenditures on state welfare and social programs." The NSDAP indeed exhibited a "general tendency to side with the left on socioeconomic issues," but it went farther than that. In numerous circumstances in the Reichstag, NSDAP diehards were the sole loyalists to endorse radical bills that the communists introduced, including higher taxes on the rich. One instance of a Nazi/communist alliance pertains to an incident in Ruhr in 1928, when workers in the steel and iron industry went on strike against their employers. The communists introduced a bill in the Reichstag to provide financial relief to the striking workers—at a much a higher rate than what the other parties supported. This same bill demanded that the steel and iron companies being struck against be made to provide this monetary assistance. It happens that no other parliament members except those of the NSDAP supported that bill. [99]

Toland observes that the Reich chancellor additionally practiced a substantial amount of "city planning . . ."[100] The U.C. Berkeley political scientists Aaron Wildavsky and Carolyn Webber, too, write that the Third Reich "initiated a large public-works program to build waterways, railroads, public buildings, and superhighways, stipulating that when possible work was to be done by hand rather than machine"—which would ostensibly reduce unemployment— "and that new employees were to be hired from relief rolls."[101]

On this matter, governist apologist John Kenneth Galbraith acknowledges that Hitler initiated massive public works projects in line with Keynesian fiscal policy, and that this was prior to Lord Keynes's specific recommendations gaining favor among heads of state.[102] Later, when Keynesian fiscal policy had been brought to his attention, Mussolini conceded its similarities with his own central planning of the Italian economy. "Fascism," said Mussolini, "entirely agrees with Mr. Maynard Keynes, despite the latter's prominent position as a Liberal. In fact, Mr. Keynes' excellent little book, *The End of Laissez-Faire* might, so far as it goes, serve as a useful introduction to fascist economics."[103]

In the preface to the German-language edition of his classic *General Theory of Employment, Interest, and Money*, Lord Keynes himself conceded, ". . . the theory of output as a whole, which is what the following book purports to provide, is much more easily adapted to the conditions of a totalitarian state than is the theory of production and distribution of a given output product produced under conditions of free competition and a large measure of *laissez-faire*."[104] And we recall from chapter 4 that Lord Keynes took pronounced interest in government-enforced eugenics.

Consider another issue in which the Nazi program might be considered wise by the standards of the turn of the twenty-first century—gun control. The Third Reich definitely did not invent gun control—this policy was enacted in the Weimar Republic in the years subsequent to the First World War, and it probably would have remained in place had the Nazi regime never taken over. Nevertheless, contradicting what might be expected of a right-wing ideologue, Adolf Hitler felt comfortable with gun control measures on the books.[105] Given what he wanted to do to them, the Führer particularly found it advisable to deprive the Jews of their arms. Although there is some controversy about whether this interview is historically authentic, one alleged interview with Hitler quotes him saying, "The most foolish mistake we could possibly make would be to allow the subject races to possess arms. History shows that all conquerors who have allowed their subject races to carry arms have prepared their own downfall by so doing."[106] Supposing that that interview turns out inauthentic, it remains true that the Nazi regime maintained gun control and did not find that it conflicted with its political program.[107] Richard Grunberger is therefore incorrect to assert that the Nazis misidentified their ideology as "'Socialism' where 'Social Darwinism' would have been more appropriate" a label.[108] At this point, we recognize that there were no social Darwinists but that the Nazis were indeed socialists.

Moreover, Nazi treatment of the arts coincides with the modern popular opinion about government subsidies to artists.

Artists and the Nazis

Sometimes I hear a story, perhaps cherished among modern bohemian painters, musicians, and other artists, that is to the effect of the following: from 1920 to 1933, before Germany's Weimar Republic metamorphosed itself into Hitler's Third Reich, the artistic community immediately opposed the Nazi Party. This was the inevitable result of the culture clash between left-wing artists and Nazis. The Nazis, we are reminded, were uptight, anal-retentive, *censorious*, bourgeois bores. Conversely, left-wing artists are creative bohemians who oppose any and every variation of censorship.

An informative authority on this aspect of history is Alan E. Steinweis, the Hymen Rosenberg Professor of Modern European History and Judaic Studies at the University of Nebraska, Lincoln. Dr. Steinweis concurs that the "picture usually painted" in such a recounting of events "is one in which the regime acted and German artists *re*acted" (emphasis his). Once at the helm of government, the myth continues, the Nazis clamped down on the bohemian artists and ruined them. As Steinweis paraphrases the tale, the regime resorted "to monetary fines, professional bans, and even concentration camps when it became necessary to discipline wayward artists."[109] Afterward, art in Nazi Germany could only be created by commercialistic hacks who would not have become well-renowned masters if the Weimar Republic had not descended into tyranny.

This fable sounds so plausible because it speaks to archetypes formed among politically correct Western hippies in the late 1960s and early 1970s. Fascists represent the military-industrial complex and The Man. Contrariwise, the left-wing bohemian artists were supposedly the prototypical inspiration for the hippies who would rebel against such a repressive, suffocating military-industrial system. The historical data, however, shed light on some truths that are inconvenient for those who would like to think that artistic bohemianism is anathema to everything for which the Nazis stood.

It cannot be denied that, subsequent to the full establishment of this despotic reign, the Nazi regime ended up censoring any art of which it did not approve—particularly art critical of the Nazi regime itself. Nevertheless, the record also evinces that when the NSDAP campaigned for political influence in the late 1920s and early 1930s, it received vocal support from the vast majority of Germany's bohemian artists. Germany's bohemian artists shared Hitler's animus toward the bourgeoisie. Parisian social scientist Eric Michaud makes a telling observation about Walther Funk—Hitler's economics minister and the government-appointed chief executive of the Reichsbank. Funk was "seen in the company of artists more often than among businessmen and bankers."[110] Indeed, adds Joachim Fest, when Hitler was but a young man his landlady saw him as "a real bohemian."[111]

Germany's bohemian artists strongly urged the public to place NSDAP members into political offices. Complementarily, they spoke out against the laissez-faire liberals and other political parties. They performed these actions on account of the NSDAP vowing to exploit any position it might occupy in government to enlarge tax patronage for paintings, symphonies, and sculptures. Artists fearing for their income formed an association in 1928 called the Combat League for German Culture—*Kampfbund* for short—that advocated tax subsidies to arts through its periodical *Deutsche Kultur-Wacht* (*German Culture Watch*). Its articles, writes Alan Steinweis, "displayed sensitivity to the concrete socioeconomic impact of Weimar era developments (both real and imagined) on artists, calling attention to the ravages of capitalism on cultural life. For example, many articles sketched out ideas for rationalization of the German theater economy, emphasizing the theme of de-liberalization, which usually entailed a combination of professionalization measures, work creation programs, and audience mass-subscription arrangements. Similarly, in the field of music the Kampfbund called for greater official regulation of the progressive 'mechanization' of music through radio, film, and records . . . Cultural renewal and structural de-liberalization of artistic life were seen as two sides of the same coin."

Joseph Goebbels stayed true to his pledge to have taxes finance the arts, tasking the Reich Chamber of Culture with dispersal of these outlays. On November 27, 1936, he formally announced that the government agency "regulates" the artist's "material needs and "provides him, as far as possible, with security against old age and disease."

Steinweis goes on that the Nazi German cultural chambers "acted as agents of *völkisch* paternalism by promoting the expansion of the social insurance system to several categories of hitherto unprotected artists. . . . Finally, the chambers preserved, and in some cases expanded, the various social and professional services offered to artists by the *Berufsverbände* of the Weimar Republic." Therefore, Steinweis continues, we have every indication "that Goebbels—and Hitler too—sincerely sought the expansion of social insurance for German artists . . ." Consequently, between 1934 and 1936 the Nazis' agencies "acted together in issuing a series of wage orders and other regulations intended to create a framework of minimum wages, fees, and prices. The chief goal was to guarantee a reasonable level of income for artists in employment." For instance, in 1934 Germany's Theater Chamber "mandated minimum salaries for theater personnel employed by Berlin theaters." The Third Reich "sponsored or cosponsored concerts, theatrical performances, exhibitions, and other cultural programs. From time to time the chambers were also able to set aside funds for stipends to support artists in need of work. . . . Because theater was largely a state enterprise, the theater professions benefited directly from the Nazi regime's increase in direct subventions to the arts. According to the estimate of one historian, appropriations by the Reich government for theater alone climbed from just under RM 10 million in 1934 to almost RM 45 million in 1942."[112]

Consequently, an article in a 1934 issue of *Volkischer Beobachter* proclaimed, "German artists are grateful to the Führer for his great and warm interest. . . . In the Reich of Adolf Hitler there is not a single German artist who does not respond affirmatively, out of his deepest convictions, to the Führer's plan and spirit in both politics and art."[113]

Thus, analyzes Eric Michaud, it was through a tacit arrangement that a number of Germany's bohemian artists "transferred their liberty and strength to the person of the artist-Führer and in exchange received his protection: buyers for their works, exhibitions, allocated studios, decorations, and dispensation from frontline fighting in the war."[114] This came at taxpayers' expense, of course, not Hitler's.

Hitler's program of government subsidies to stage productions is mirrored in the New Deal's Federal Theater Program (FTP). Under the auspices of the Works Progress Administration (WPA), the FTP was founded under the usual governist presuppositions. Namely, the State was morally obligated to direct the course of culture by confiscating tax dollars from U.S. citizens and then funneling this capital to the creation of art that such U.S. citizens might not have agreed to bankroll voluntarily.[115] That the Nazis' approach to arts funding has become PC becomes clear by how vehemently an American politician is vilified whenever he suggests cutting tax funding to art. In 1999 American actor Tim Robbins (b. 1958) wrote, produced, and directed a semi-fictionalized motion picture entitled *Cradle Will Rock*, which focused on the FTP. In the movie, villainous advocates of the night watchman state object to the FTP's tax expenditures on plays with stories that amount to anti-capitalist propaganda. Robbins's basic point is that to yank tax funding from artwork is an abrogation of free speech and also an attempt to utilize government authority to manipulate public opinion.[116]

Not surprising is the political rhetoric of some recipients of grants from what has served as the FTP's successor since its creation in 1965 by President Lyndon B. Johnson[117]—the National Endowment for the Arts (NEA). NEA apologists have repeatedly accused the pro-privatization, free-market opponents of tax funding of being crypto-Nazis. John Frohnmayer, who served as the NEA's chairman from 1989 to 1992, announced to the National Press Club in the latter year that cutting his agency's funding would set in motion a chain reaction not unlike Hitler's conquests of various countries. "If the National Endowment for the Arts gets picked off . . . there will be no end to it," he cried. "It's the Sudetenland now, Czechoslovakia next week, and, after that, Poland."[118] Robert Brustein, whose theater has received a multitude of NEA grants, self-righteously asserts, "The distinction between censorship and determining the distribution of taxpayers' dollars . . . derives from the pernicious American tradition of letting the marketplace rather than a Commissioner of Culture or a Minister of Propaganda function as the censor of the arts. Everyone has a First Amendment right to subsidy."[119] U.S. Rep. Peter Kostmayer of Pennsylvania likened any criticism of NEA funding to "book burning in America, 1990."[120]

Here I should explain why such insinuations prove to be at variance with reality. I am tired of this assumption that for the State to avoid taxpayer financing of art is the same as violating artists' freedom of speech. We shall examine Frohnmayer's proclamation that tax funding for art equals "freedom of expression and the right of its citizens to create."[121] New York University Law School professor and former ACLU president Nadine Strossen actually concedes that "no artist has a First Amendment right to government funding, so the government's denial of an NEA grant would not ordinarily implicate constitutional rights."[122] Yet that is actually not the ACLU's party line in practice, as other prominent ACLU attorneys do argue that an artist has a First Amendment right to government funding. Exhibit *A* of this is someone who has worked directly under Dr. Strossen—the founding director of the ACLU's Art Censorship Project, Marjorie Heins. Despite Strossen's aforementioned concession, Strossen and Heins go on to argue that whenever some conservative politician tries to cut government funding for the arts in general, it is often in response to public outcry from some taxpayer-funded artwork being considered lewd or vulgar.[123] Strossen notes that it is hypocritical for many of those participating in the outcry to want to pull funding for offensive taxpayer-funded artwork but to continue having tax dollars fund artwork that does not offend them. However, Strossen concludes that for a politician to yank tax funding for all art in response to some art being offensive is somehow evil and discriminatory, and therefore an attack on civil liberties and equal treatment of every citizen under the law. In other words, Strossen conjures up a pedantic pretext for invoking the law to fight against multiple attempts by free-marketers to privatize arts funding to the fullest extent.[124] Following in the pedantry, Heins pronounces conservative Republican opponents of NEA funding to be guilty of yet another double standard as well. Religious conservatives call for the de-funding of offensive artwork on the grounds that taxpayers should not be forced to subsidize speech they find offensive, to be sure. Yet, argues Heins, taxes simultaneously finance the creation of public parks, "public" schools, "public" libraries full of offensive books, and various other public works projects that some taxpayers may find offensive.[125] Because Heins takes it for granted that these enterprises deserve tax funding, Heins proclaims that there is nothing inherently wrong in forcing people, at gunpoint, to finance enterprises that disgust them. According to Heins's argument, if you object to financing some spiteful artwork or some wasteful public-works project, you ought to suck it up; it is for the public interest anyway. This argument very successfully shuts up most conservatives, as they fear the prospect of stating publicly that these other socialist projects should likewise have their tax funding discontinued. This intimidating debate tactic does not neutralize *me*, however. No public works project should be financed through tax extortion—period.

Contrary to Frohnmayer, Brustein, Strossen, and Heins, free speech simply means one can employ his own private property—as well as the property of other consenting adults—to transmit whatever messages he wants to other consenting parties. Free speech does not mandate that if Cassie is not willing to buy my paintings, then I am right to have government officials point guns at her and threaten, "Buy these paintings or eat lead!" This gives the lie to George Orwell's importation, "the fact that most of the press is owned by a few people operates in much the same way as state censorship."[126]

Further, as Alan Steinweis's research on the Third Reich substantiates, government financing of artwork is itself an effort by the State to manipulate the public's opinions and tastes. When the State mulcts money from Cassie to finance the production of art that repulses Cassie, that leaves her with much less cash to spend on the art that she loves. This depletion of funding for the art that Cassie favors, in turn, reduces the opportunities of Cassie's favorite artists to create art that Cassie would actually appreciate enough to finance consensually. Likewise, the Third Reich's ability to apportion tax money to fund artwork meant that it could sap investment capital away from forms of art it considered too "Jewish"—thereby eradicating the market for such "Jewish art"—while it invested more money into the creation of art that was considered "non-Jewish."

There is an eerie parallel between the Third Reich's art patronage and the FTP's. The Third Reich financed propagandistic art that glorified national socialism and vilified Jewish industrialists. Likewise, The FTP financed a propagandistic stage production that glorified international socialism and vilified industrialists in general. Remember from Book One of our trilogy that the Third Reich employed the system of tax-funded public libraries to promote the ideas it preferred and demote the ideas it disliked; the same sort of manipulation is conducted by governments assigned with financing artists through tax money. Harvard University reference librarian Margaret Stieg Dalton mentions that under the Third Reich, the "political character of the public library was emphasized; the idea of the Volk was integrated."[127] Under the Nazis, writes Matthew Battles, government-employed librarians "joined in the great and mystical task of fashioning the German people into a *Volk*. . . . As defined by the Reich, the job of the library would be to control the dangerous, the bourgeois, the effete and dissipating energies of reading, to help the *Volk* find useful information without degrading its 'spirit.'" The librarians of the government-controlled libraries put together were keen on "*eliminating their chief rivals in the small-scale commercial lending libraries* . . ." (emphasis added). Matthew Battles observes the ideological origin of this program—it "gave a romantic slant to the progressive ideals" that demeaned classical liberalism.[128]

Extorting Money versus Keeping What Is Yours: Which Is More Censorious?

Political journalist James Bovard ascertains that since a government agency for arts subsidization "cannot finance all artists, political decisions must be made on who will receive subsidies."[129] Former NEA deputy chairman Michael Straight admitted in 1991 that "when 900,000 people call themselves artists in the census, it's preposterous for the federal government to pick out 500 of them and say, 'These are the ones who deserve support.'" Straight further confesses that this agency has, for quite a number of years, enjoyed "playing God" when choosing which artists receive tax support.[130]

Apologists for NEA funding proclaim that the U.S. agency distinguishes itself from its Nazi counterpart on account of NEA funding being indiscriminate, not motivated by the agency's political biases. This argument is strongly vocalized by Marjorie Heins. She claims to support "government funding without strings attached."[131]

When it comes to arguing that it is just for the State to coerce people into financing activities of which they do not approve, Heins finds an unlikely ally—cultural conservative legal scholar Robert Bork, who openly approves of censorship. Granted, Bork denounces a variety of NEA-funded artistic works on the grounds of taste. When an artist took a photograph of a crucifix soaked in his own urine, Bork reviled that. Still, Bork maintains that this artwork, *Piss Christ,* should be denounced solely on account of its tastelessness, and not on account of its having been financed though governmental extortion. Bork ridicules everyone he has caught "complaining that works like these should not be subsidized with 'taxpayers' dollars,' as if taxpayers should never be required to subsidize things they don't like."[132]

John Frohnmayer comparably congratulates himself for ensuring that the NEA doles out grants "with a level playing field, no blacklists, and no ideological preconceptions. . . . You and I don't have to like everything that the Endowment supports, because your government is not a sponsor of those ideas. The government is merely an enabler."[133] He is right to choose that last word for reasons other than those intended. The government is an "enabler" of poorly chosen expenditures the same way that someone makes himself an "enabler" by helping an alcoholic procure more booze.

The facts belie the assertions that politics plays no role in deciding which projects the NEA does or does not fund; James Bovard has documented examples of the NEA's imposition of a PC agenda. *The Hudson Review,* one of America's leading literary quarterlies, had received several NEA grants until 1993. A May 27, 1993, letter from the NEA gave the reason for this—"writers of color were significantly underrepresented in the *Hudson Review."* Another reason the NEA delivered for rejecting the grant application was that this journal published in its summer 1992 issue an article about black novelist Richard Wright that the NEA panelists deemed "isolating and condescending. . . . This concern was exacerbated . . . when this essay was compared with the fulsome essay about [Emile] Zola in the same issue."[134] Roger Kimball, *The New Criterion*'s managing editor, opined, "The message from the NEA's panel is clear: Only institutions waving the banner of political correctness need apply. . . . Even if you have published articles about, say, black authors, were the articles sufficiently—that is, unequivocally and unreservedly—enthusiastic?"[135]

There are other examples of government funding amounting to the manipulation of the free expression of political thoughts. One can be found in the case of the National Parks Service's (NPS) magazine. In the publication's February 1990 issue, a column quipped about U.S. Congressmen in general, "Having assured themselves a significant pay increase while retaining many of the 'perks' attendant to being a member of Congress, they made the nation safe again by recessing for the Christmas holidays in late December." Some months later, the House & Senate Appropriations Committee withheld the magazine's annual 75,000 dollars publishing cost from the NPS's budget, thereby ceasing its publication. This was a reprisal for the magazine's wisecrack.[136]

Some members of the profession speak out about how NEA funding stacks the deck against the private arts market. Consider this case: in 1991, the Arts Endowment dispersed 174 million dollars on various projects it deemed worthy, and such grants were matched by city and state governments. "The funds have, paradoxically, undermined serious theater," observes playwright John Chodes. On account of the taxpayer subsidies, the price of renting theaters went up drastically, particularly for consensually financed theatre troupes. A New York City theatre that would have rented for 400 dollars per week in 1981 correspondingly could not be rented for less than 2,000 dollars per week in 1991. Moreover, tax-financed troupes were able to rent out some stages permanently. "These favored companies monopolize theater access. The endowment has a penchant for financing 'social issue' plays, and the lure of subsidies" motivates various troupes into staging them. Note the common refrain that artists should be subsidized by taxes because, without such subsidies, these artists would have to cater to the vulgar tastes of the market. Here we witness the reality—when taxpayer dollars finance art, artists end up catering to the hobbyhorses of government bureaucrats. Such "hewing to the NEA priorities," Chodes conclude sadly, "is the antithesis of a personal artistic perspective."[137]

Many critics classify the *cubist* and *Dadaist* painting genres as bohemian artistry. To the degree that they would prefer to see bohemian artists as the antithesis of Nazism, adherents to the conventional "bohemianism-is-the-opposite-of-Nazism" narrative can take comfort. Hitler despised these particular (post)modernistic genres and had them outlawed.[138] I dislike those genres, too, though I would never pass any legislation regarding them. Despite Hitler's distaste for avant-garde styles, we must still confront a still more glaring consideration. It is that, rather than stand up to the NSDAP in the early 1930s on behalf of free speech, the majority of Germany's bohemian artists wholeheartedly supported Hitler. And this was on the order that the Nazi Party offered the exact same tax patronage for the arts that has become so politically correct in the modern USA.

George Mason University economist Tyler Cowen stresses that, per capita, NEA funding costs the average taxpayer a mere 70 cents per year.[139] Still, I consider this an argument for cutting tax funding rather than retaining it. The United States would not suffer from a dearth of great artwork if each American were allowed to keep the 70 cents that would otherwise be extracted from him at gunpoint. Millions of U.S. dollars are consensually endowed to artists every year.

And tax patronage to art is just one among an array of plans in the Nazi political program that have become PC. American progresssives might have to confront the possibility that there is much in the Third Reich's public-health-related- and environmental measures that they would appreciate.

NOTES

1. Hitler 2001, 391, 233–34, 340, 45, 317–18.
2. Hitler, qtd. by Wagener 1985, 148–49.
3. Kaczynski 1995, accessed online Tuesday, November 15, 2011.
4. Ed Marston, "Dear Friends," *High Country News*, April 28, 1997, http://www.hcn.org/issues/108/3399, accessed Tuesday, August 26, 2014.
5. Hitler 2003, 231–32.
6. Qtd. by MacFarquhar 2004, http://tinyurl.com/2nz66j, accessed Tuesday, June 19, 2007.
7. M. Moore 2003, 137–148.

8. H. A. Turner 1985, 74–75.
9. Hitler 2001, 639.
10. Hitler, qtd. by Wagener 1985, 173.
11. Hitler, qtd. by B. Rubin and Rubin 2004, 97. They cite Bormann 1953, 155.
12. Hitler, qtd. by B. Rubin and Rubin 2004, 96. They cite Compton 1968, 17, 35.
13. Qtd. by B. Rubin and Rubin 2004, 137. Nazi philosopher Martin Heidegger, too, hated the cultural influence of the United States. Writes Buruma and Margalit 2005, softcover, 8, "Heidegger was a sworn enemy of what he called *Amerikanismus*, which in his view sapped the European soul."
14. Hitler, qtd. by Wagener 1985, 160–61.
15. Wagener 1985, 161.
16. Hitler, qtd. by Wagener 1985, 161.
17. Michael Moore, interviewed by Ian Hodder, "Michael Moore's Ongoing Crusade Against Corporate Greed Won Him an Audience With Big, Bad Nike," *Industry Central*, 1997, http://tinyurl.com/25mkhg, accessed Friday, June 15, 2007.
18. Brian Reade, interview with Michael Moore, "The Awkward Conscience of a Nation," *The Mirror*, March 11, 2003, http://tinyurl.com/2na6u3, accessed Tuesday, June 19, 2007.
19. Michael Moore, "Not All of America Is Crazy," *Die Zeit*, November 6, 2003, http://www.zeit.de/2003/ 46/AbdruckMoor, translated from German into English by Ray Drake for David Kaspar, "Michael Moore: Americans Are Dumb, Germans Educated," *Davids Medienkritic-Online*, http://tinyurl.com/37bqed, both websites accessed Tuesday, June 19, 2007.
20. Andreas Tzortzis, "Michael Moore at Home in 'Old Europe,'" *DW [Deutsche Welle] World*, November 17, 2003, http://www.dw-world.de/dw/article/0,,1035379,00.html, accessed Tuesday, June 19, 2007.
21. Qtd. by MacFarquhar 2004, http://tinyurl.com/2nz66j, accessed Tuesday, June 19, 2007.
22. Hitler 2001, 214.
23. Bin Laden 2002, http://tinyurl.com/kva1, accessed Tuesday, June 19, 2007.
24. M. Moore 2003, 148–155.
25. Steinweis 1993, 159.
26. Mikela J. Mikael narrating in Achbar et al., prods. 2003, transcript online in the PDF at http://hellocoolworld.com/files/TheCorporation/Transcript_finalpt2copy.pdf, accessed Thursday, May 17, 2007.
27. Hitler 2003, 30–31.
28. Milward 1977, 6.
29. Hitler 2001, 406.
30. M. Moore 2003, 154–55.
31. Hitler, qtd. by Wagener 1985, 16–17.
32. Hitler 2001, 151, 227, 298, 336, 423.
33. H. A. Turner 1985, 75.
34. Fest 1974 U.S. version, 146, 167.
35. Qtd. by Fest 1974 U.S. version, 167.
36. Fest 1974 U.S. version, 145.
37. For Hitler's mention of Karl Lueger, see Hitler 2001, 56, 93–95, 111–13. Zakaria 2003, 60, brought to my attention the fact that Hitler mentioned Lueger by name in *Mein Kampf*. Zakaria 2003, 59–60, pointed out to me that Hitler modeled himself after Lueger.
38. For Hitler's mentions of Lueger's Christian Socialism, see Hitler 2001, 56, 93–94, 111–13. Corroboration of Lueger being a Christian Socialist, and candidate for the Christian Socialist Party, is in G. L. Mosse 1975, 118; G. L. Mosse 1985, trade paperback, 146–47; and Zakaria 2003, 59–60.
39. G. L. Mosse 1985, trade paperback, 146–47.
40. Shirer 1990, 35–36.
41. Fest 1974 U.S. version, 117–18, 123.
42. Shirer 1990, 35–37, 40–41, 260, 759.
43. Tom Perkins, "Progressive Kristallnacht Coming?", *Wall Street Journal*, January 24, 2014, http://online.wsj.com/news/articles/SB10001424052702304549504579316913982034286, accessed Tuesday, August 26, 2014.
44. David Christopher Bell, "4 Cringeworthy Justifications People Used for Being Assholes," *Cracked.Com*, March 8, 2014, http://www.cracked.com/quick-fixes/4–cringeworthy-justifications-people-used-being-assholes/, accessed Tuesday, August 26, 2014.
45. Hitler 2001, 436–447, 599–600.
46. H. A. Turner 1985, 181.
47. Shirer 1990, 202.
48. To some degree, E. Faye 2009 U.S. version, 141, does this.
49. G. L. Mosse ed. 1955, 341–42.
50. Shirer 1990, 202–03.
51. H. A. Turner 1985, 77.
52. M. Parenti 2004, 117–18.
53. J. Toland 1976, 314.

54. Dietrich Klagges qtd. by Steigmann-Gall 2003, 41–42.
55. Steigmann-Gall 2003, 44.
56. J. Cort 1988, 229; and R. J. Ellis 2005, 27–28.
57. Steigmann-Gall 2003, 44.
58. N. Ferguson 2006, 245.
59. Steigmann-Gall 2003, quoting from Goebbels 1987, 66.
60. Steigmann-Gall 2003, 46, citing Wagener 1985, 56.
61. Hitler, qtd. by Wagener 1985, 56–57. These words are alluded to in Steigmann-Gall 2003, 46.
62. Hitler, qtd. by Wagener 1985, 139–140. A part of this quotation appears in Steigmann-Gall 2003, 46–47.
63. J. C. Fest 1974 U.S. version, 252.
64. Hitler, qtd. by Wagener 1985, 23, 145.
65. G. L. Mosse ed. 1966, 342–43.
66. Goebbels, qtd. by Steigmann-Gall, 44.
67. Milward 1977, 8.
68. J. Goldberg 2007, 304. Gellately 2007, 96, states that in Germany at the time, department stores "were identified mainly with Jewish firms . . ."
69. Buruma and Margalit 2005 paperback, 28–29.
70. This is republished in *Nazi Conspiracy and Aggression* vol. 4, (Washington, D.C.: United States Government Printing Office, 1946) republished online at Ley ed. 2007, http://www.yale.edu/lawweb/avalon/imt/document/nca_vol4/1708–ps.htm, accessed Monday, March 24, 2008.
71. J. Goldberg 2007, 304.
72. Bernard Wysocki, Jr., "How a Broad Coalition Stymied Wal-Mart's Bid to Own a Bank," *Wall Street Journal*, October 23, 2006, http://online.wsj.com/news/articles/SB116118495912296504, accessed Sunday, June 9, 2014.
73. Gellately 2001, 26.
74. N. Steinberg 2005, http://tinyurl.com/ysz5gb, accessed Monday, March 24, 2008. I learned about Neil Steinberg saying this from J. Goldberg 2007, 304.
75. Qtd. by Gellately 2007, 189.
76. J. Toland 1976, 403, 406.
77. G. L. Mosse ed. 1966, 341–43.
78. J. Toland 1976, 405–06.
79. G. L. Mosse ed. 1966, 342; and J. Toland 1976, 405–06.
80. G. L. Mosse ed. 1966, 342.
81. J. Toland 1976, 405–406.
82. Qtd. by J. Toland 1976, 405–06.
83. Steigmann-Gall 2003, 190. Hitler spoke these words at the assembly of *Alte Kämpfer* in Munich on February 26, 1939.
84. J. Toland 1976, 312, 405–07, 542.
85. Hitler qtd. by J. Toland 1976, 312.
86. It is not very relevant to this discussion, but I found it interesting that Ambassador Dodd was the father of U.S. Sen. Christopher Dodd (D-CT; b. 1944).
87. Adolf Hitler, letter to President Franklin D. Roosevelt and U.S. Ambassador Thomas Dodd, March 14, 1934, republished in "Message from Hitler to Roosevelt," *History-of-the-Holocaust.Org*, http://tinyurl.com/3chujh, accessed Wednesday, May 23, 2007.
88. J. P. Diggins 1972, 280.
89. FDR to Henry Breckinridge Long, letter, June 1933, qtd. by J. P. Diggins 1972, 279.
90. FDR to John Lawrence, letter, July 27, 1933, qtd. by J. P. Diggins 1972, 279.
91. J. P. Diggins 1972, 279.
92. Arthur M. Schlesinger makes this denial in his essay "Sources of the New Deal." I learned about this essay from J. P. Diggins 1972, 232.
93. A. Cooke 1973, 329.
94. J. Goldberg 2007, 148, quoting Mussolini 1933.
95. J. P. Diggins 1972, 281, quoting Mussolini 1933.
96. J. P. Diggins 1972, 281.
97. J. Goldberg 2007, 148, quoting Mussolini 1933.
98. J. P. Diggins 1972, 281, quoting Mussolini 1933.
99. H. A. Turner 1985, 66–67.
100. J. Toland 1976, 404.
101. C. Webber and A. Wildavsky 1986, 461.
102. J. K. Galbraith 1987, 222.
103. Qtd. by J. S. Barnes 1928, 113–14.
104. J. M. Keynes 2013, 3.
105. Kopel and Griffiths 2003, http://tinyurl.com/cgkk, accessed Wednesday, May 30, 2007.

106. Adolf Hitler, April 11, 1942, interview, in Hitler 2000b, 321.
107. "Did Hitler Ban Gun Ownership?", *The Straight Dope*, June 16, 2000, http://www.straightdope.com/columns/read/1791/did-hitler-ban-gun-ownership, accessed Saturday, September 28, 2013.
108. R. Grunberger 1971, 199.
109. Steinweis 1993, 2.
110. E. Michaud 2004, 7, 32.
111. Fest 1974 U.S. version, 124.
112. Steinweis 1993, 3, 19–20, 23, 26–27, 73–75, 80, 175.
113. Qtd. by E. Michaud 2004, 50.
114. E. Michaud 2004, 50.
115. Joan Blumenthal, "Art for Power's Sake, Part 1," *The Objectivist* vol. 7 (no. 11, November), republished in Rand 1982, 550; A. Cooke 1973, 331; Leuchtenburg 1963, 126–27; and Valk 2003/2004, http://tinyurl.com/yp3hh3, accessed Sunday, June 3, 2007.
116. T. Robbins et al., prods. 1999. That the basic plot of the film is true—but not Robbins's interpretation of those events—can be corroborated in Joan Blumenthal, "Art for Power's Sake, Part 1," *The Objectivist* vol. 7 (no. 11, November), republished in Rand 1982, 550.
117. That the NEA was created in 1965 by the Johnson administration is noted in Joan Blumenthal, "Art for Power's Sake, Part 1," *The Objectivist* vol. 7 (no. 11, November), republished in Rand 1982, 549.
118. "National Press Club Luncheon Speaker John Frohnmayer, Chairman, National Endowment for the Arts," Federal News Service, March 23, 1992, qtd. by Bovard 1994, 159.
119. Robert Brustein, "The First Amendment and the NEA," *The New Republic*, September 11, 1989, p. 27, qtd. by Bovard 1994, 159.
120. *Congressional Record*, October 11, 1990, p. H9406, qtd. by Bovard 1994, 159.
121. John Frohnmayer, qtd. by Bovard 1994, 159.
122. N. Strossen 2000, paperback, 99.
123. M. Heins 1993, 128–130; and N. Strossen 2000 paperback, 99–100.
124. N. Strossen 2000, paperback, 99–100.
125. M. Heins 1993, 128–130.
126. Orwell 1968b, 39–40. Huber 1994, 56, brought this statement of Orwell's to my attention.
127. M. Battles 2004, paperback, 172, quoting M. F. Stieg 1992.
128. M. Battles 2004, paperback, 170–72.
129. Bovard 1994, 159.
130. Qtd. by Bovard 1994, 159, citing Mark Laswell, "How the NEA Really Works," *Spy*, November 1990, p. 58.
131. M. Heins 1993, 127.
132. Bork 2003a, 150.
133. "National Press Club Luncheon Speaker John Frohnmayer, Chairman, National Endowment for the Arts," Federal News Service, March 23, 1992, qtd. by Bovard 1994, 159.
134. Qtd. by Bovard 1994, 160, citing Roger Kimball, "Diversity Quotas in NEA Skewer Magazine," *Wall Street Journal*, June 24, 1993.
135. Roger Kimball, "Diversity Quotas in NEA Skewer Magazine," *Wall Street Journal*, June 24, 1993, qtd. by Bovard 1994, 160.
136. Bovard 1994, 160, citing John Lancaster, "Payback Time for Parks' Poke at 'Perks,'" *Washington Post*, November 20, 1990.
137. John Chodes, "The S.E.C.'s War on the Theater," *New York Times*, January 11, 1992, qtd. by Bovard 1994, 160.
138. N. Ferguson 2006, 425; G. L. Mosse 1975, 191; and Steinweis 1993, 22–23, 27.
139. T. Cowen 2000, paperback, 38.

NINE

They Loved Blood and Soil but Not the Mind

Your Personal Health Is Everyone's Business?

Appropriately, the ideological position for which the Nazis are most famous—white supremacism—became politically incorrect decades ago. I definitely do not dispute political correctness on that count. Sadly, we discern that in various other areas aside from support for tax patronage to the arts, the Nazis advanced an agenda consistent with modern PC notions. The center-left political orthodoxy of Western Europe, Canada, the United States, and Australia is replete with governist ideas that were popular under both the Second and Third Reichs. One governist academician both knowledgeable and supportive of those PC ideas is Pennsylvania State University science history professor Robert N. Proctor. During his research on the atrocities that the Nazis committed in their experiments on human health, Proctor came to discover something that astonished him. It was that the Nazis implemented a plethora of tropes of Proctor's own modern environmental ideology before the New Left made them popular in the United States. More astonishing yet, Proctor found himself strongly appreciating this and wishing that the current U.S. government would actually adopt more of what he considers to be the better aspects of the Nazis' public health programs.

Proctor writes that "fascist ideals fostered research directions and lifestyle fashions that look strikingly similar to those we"—by *we*, he means his center-left environmentalist allies and himself—"might embrace." The Nazi's tax-funded public health program was "pursued not just *in spite of* fascism, but also *in consequence of* fascism" (emphases his). "This is the 'flip side' of fascism—the side that gave us struggles against smoking, campaigns for cleaner food and water, for exercise and preventive medicine." It is "disturbing," he mutters, that "Nazi doctors and public health activists" were "involved in work that we . . . might regard as 'progressive' or even socially responsible—and that some of that work was a direct outgrowth of Nazi ideology. Nazi nutritionists stressed the importance of a diet free of petrochemical dyes and preservatives… Many Nazis were environmentalists; many were vegetarians. Species protection was a going concern, as was animal welfare."[1]

Among scholars who study the Third Reich's history, Proctor is not solitary in this assessment. According to John Toland, "the problem of pollution" substantially worried the Reich chancellor. "Anti-pollution contrivances were already installed in some factories in the Ruhr basin, and new plants were required to construct preventive devices to avoid pollution of the waters." Moreover, under Hitler's reign, "welfare and training of the youth of the nation were also given priority." The Führer put a new emphasis on policing the diets and physical activity of children. Kids were forced to exercise for five hours each school day.[2] The Nazis emphasized team sports in their physical education for two reasons. First, they believed it would cultivate physically healthy Aryans who would disseminate their healthfulness to subsequent generations, thereby contributing to the collective benefit of the race. Secondly, having children work in concert in teams would instill in them a feeling of loyalty to social collectives in general. That would be a form of training for serving the community-state in adulthood. A 1937 Nazi Party document entitled *Guidelines for Boys' Schools* instructs,

> 1. Physical education is a fundamental and inseparable part of National Socialist education. . . . **National Socialist education . . . grasps man in his totality in order to make him able and ready to serve the community of the people . . .**
>
> 1. In the training of the youth in the schools, physical education, within the framework of education as a whole, is of the greatest importance. . . . a) Physical education is education in community. By demanding obedience...it trains them in those virtues which constitute the foundations of the Volk community [boldface added].[3]

Such health policing and health-related indoctrination was imposed upon adults as well, and for similar reasons.

The Nazis, Proctor proclaims, were cognizant of the harmful health effects of cigarettes, anticipating the U.S. federal government by decades. Hitler foresaw the late-1900s movement to prevent private tavern owners from deciding whether smoking may or may not be allowed within the walls of their own private establishments. Taxes paid for the dissemination of antismoking propaganda in schools. Proctor additionally details the many state-imposed limits on cigarette marketing.[4] This Pennsylvania State University science historian assumes that any government—whether that of the USA or the Nazis—is morally right to spoliate business owners if it is with the ostensive intent of preserving the health of the commonweal. Consequently, Proctor does not express offense that the Nazis threatened force upon a restaurateur who permitted smoking on his own private premises. Nor does Proctor judge anything terrible about tax money being confiscated to pay for this. He surmises that in the final analysis the Nazi campaign proves the need for tax funding of public health efforts. Buttressing this is his claim that the Nazis discovered tobacco's toxicity before the U.S. did, and that this was on account of the Nazis having committed more tax money to public health research. However, from the 1910s to early 1930s, privately funded scientific research in the United States already demonstrated the carcinogenic properties of smoking. In 1928, the medical doctors Herbert L. Lombard and Carl R. Doering tested for cancers in tobacco-users and non-users alike—the users including pipe smokers, cigarette smokers, cigar smokers, and tobacco-chewers—and discerned a statistically significant positive correlation between tobacco use and incidence of cancer.[5] In 1915, the medical doctor Frederick L. Hoffman (1865–1946) made note of the link between tobacco use and cancer incidence. He cited data from experiments conducted in Germany, but those experiments were conducted in 1906 and not financed by the Nazi regime.[6]

It is easy for someone to conclude that Proctor's proposal for further U.S. emulation of the Third Reich's public health program sounds like tacit approval, on Proctor's part, for specific aspects of Naziism. Aware of this, Proctor takes a preemptive swing at those who would openly observe as much. "My own intention is not to argue that today's antitobacco efforts have fascist roots, or that public health measures are in principle totalitarian—as some libertarians seem to want us to believe."[7] Proctor's endnotes section evinces that the "libertarians" to whom he alludes include Jacob Sullum (b. 1965) of the Reason Foundation.[8] In 1999 Sullum responded by noting that of course some institution is not automatically objectionable solely as a result of Hitler or other Nazis having approved of it; mustaches do not become innately evil just because Hitler had one. The larger point of Sullum's, to which Proctor gives short shrift, is that the "public-health movement would be worrisome even if the Nazis had taken no interest in it . . ."[9] The issue is not that the Nazis had inspired America's public health regulations. Nay, the issue is that every government regulation over what private individuals peaceably do with their own bodies originates from the same governist assumptions.

Each of these regulations, recall, is maintained by the threat of State force against those who fail to comply with the regulations. That is the reason why there are several ominous parallels between the Nazi public health practices that Proctor praises, and the movement for governist public health measures presently existing in the West. For instance, Proctor admits that the Nazis adopted "health as a duty" as a national slogan in 1939. In the "Nazi ideology," he explains, the laissez-faire "liberal distinction between public and private sphere was abandoned. As one Hitler Youth health manual put it: 'Nutrition is not a private matter!'. . . The state thereby acquired a stake in the maintenance of the individual body—whether it was healthy or poisoned, exercised or abused, and so forth."[10]

Sullum then exposes the parallels between such a mentality and those of modern governist public-health activists who employ a specific rationalization. These public-health activists proclaim that on account of how tax money finances the health care of everyone, it follows that people who experience worse health complications drain the exchequer more than those who contract fewer such difficulties. Therefore, says the argument, it stands to reason that, everything else being equal, a fatter person who smokes more will ultimately cost the public treasury more in health spending than will a thinner or more athletic nonsmoker. The argument thus concludes that to reduce costs for everyone, the State is justified in trying to micromanage everyone's health decisions. That argument was uttered in the Third Reich, wherein tax money financed most health care, and it is uttered in the USA, where—despite the USA's reputation for holding out on adopting fully socialist single-payer health care—tax funding still accounts for more than a quarter of spending on health-care provision. As an example, Sullum quotes Dr. Faith T. Fitzgerald, a U.C.-Davis Medical Center instructor. In the *New England Journal of Medicine*, Dr. Fitzgerald wrote, "Both health care providers and the commonweal now have a vested interest in certain forms of behavior, previously considered a person's private business, if the behavior impairs a person's 'health.' Certain failures of self-care have become, in a sense, crimes against society, because society has to pay for their consequences." The Reason Foundation libertarian chews out the Penn State scholar for failing to scrutinize his own premise "that the government should protect 'the public health' from individuals who choose to trade longevity for pleasure."[11]

Throughout this trilogy I have argued that life is the standard of value, and that freedom's justification is that it allows for the maximization of life. For those reasons, one might wonder how I can justify that right that Sullum defends, the right to "choose to trade longevity for pleasure." Recall that when I argue that the goal of an individual's life is to maximize her life, I mean that her goal is to maximize her eudaemonia—that is, the level of happiness within her life. One's continued physical survival is merely the prerequisite to any future state of eudaemonia. To review, liberty's justification is that liberty best allows one to find eudaemonia. That eudaemonic self-interest justifies liberty, however, does not place upon anyone some duty to extend her own physical survival as lengthily as possible. One especially possesses no such duty if the actions necessary for extending her physical survival really long, such as denying various delicious foods to herself, comes at the expense of years' worth of happiness. My doctrine of eudaemonia-maximization takes into account the fact that someone can maximize her own eudaemonia in specific pursuits—such as eating wonderful meat—cognizant that such choices might come at the cost of some years to her life, additional years that she otherwise would have experienced. When it comes to such tradeoffs in the pursuit of eudaemonia, every individual must engage in introspection and rational study to render her own judgment calls on what she does.

The Nazis would not countenance my idea—leaving people free to act on their own choices concerning their own health. Proctor states that in the Third Reich the ideology of illness "prevention merged with the ideology of 'one for all and all for one' (*Gemeinnutz geht vor Eigennutz*) that was yet another hallmark of Nazi thought: as one anti-tobacco activist put it, nicotine damages not just the individual but the population as a whole."[12]

Sullum points out that this idea sounds similar to a sentiment expressed by C. Everett Koop (1916–2013), the U.S. surgeon general under the Clinton administration. He quotes this statement of Koop's: "The government has a perfect right to influence personal behavior to the best of its ability if it is for the welfare of the individual and the community as a whole." Both Koop and the German governists prioritize the collective above the individual's well-being.[13]

The Malignancy of Misconceptions About Cancer

Nazi practices that are still yet more PC by the standards of contemporary environmentalists are the dietary habits that the Third Reich forced upon the population. Despite his own support for left-wing environmentalist governism, George Washington University geographer Martin W. Lewis admits, ". . . German national socialism was heavily laden with a romantic, ecologically imbued 'blood and soil' ideology."[14] Returning to the subject of Germany's past, Robert Proctor illuminates, "One common theme of Nazi food rhetoric was the need to return to a more natural diet free of artificial colorings and preservatives. . . . Meat eating was to be minimized, and fresh foods were to be chosen over preserved foods in tins." Through it, "Vegetarianism got a boost." Meanwhile, "Efforts were made to control food additives and to limit the oversalting of prepared foods."[15]

Airing the same arbitrary assumptions that hundreds of anti-capitalist environmentalists air in the era of this writing, Nazi officials alleged that the synthetic fertilizers, pesticides, and antibiotics utilized by agribusiness and factory farms had rendered the food supply unsafe and significantly more carcinogenic than it otherwise would be. Just as New Age food faddists currently do, Nazis assimilated this fetish about eating "organic" produce grown in the absence of manmade chemicals. As usual, though, this famed Nazi practice did not begin under the Third Reich but under the influence of the Second. It was in Bismarck's Germany, it will be recalled, that governism and anti-reason, anti-industrialist Romanticism held sway. Organic farming was first preached in 1924 by the doctor and mystic Rudolf Steiner (1861–1925), who built his entire lifestyle around belief in the supernatural, participating in the New Age movement "theosophy"—until his dispute with fellow mystic Jiddu Krishnamurti (1895–1986)[16] —and then founding "anthroposophy."[17] Although Bismarck's rule had ended by then, the philosophic climate he engendered had remained strong in 1920s Germany when Steiner's teachings first caught on. Nor is it a surprise that Steiner advocated a controlled economy. Steiner's ideal social system, writes Colin Wilson, was what Steiner "called the Threefold Commonwealth—in effect, a society in which thinkers and artists would make the decisions, while the politicians and businessmen would ensure the smooth running of the state."[18]

Steiner did not call non-chemical farming "organic," though; the name he used was *biodynamic farming*.[19] Steiner's preaching heavily influenced royal family members in England, particularly Sir Albert Howard (1873–1947) and Lady Eve Balfour (1899–1990) in the 1930s and 1940s.[20] In turn, Sir Albert's writings struck a chord with American-born accountant Jerome Irving Rodale (1898–1971). It was Rodale who first termed this non-chemical agriculture "organic," and, from the 1930s to the 1950s, he started various organizations and enterprises, including *Prevention* magazine, to propagate his organic-vegetable-diet lifestyle.[21] Some years before Sir Albert and Rodale—also in the 1930s—Nazi leader Rudolph Hess absorbed German mystic Rudolf Steiner's teachings about the need to eat nothing but "organic" produce grown in the absence of those manmade chemicals.[22] Heinrich Himmler then followed in turn.[23] Other prominent Nazi supporters of biodynamic farming include inner security department official Alfred Bäumler (1887-1968), head of the Department of Inner Security; Gestapo leader Hermann Wilhelm Göring (1893-1946), propagandist Alfred Rosenberg (1893–1946), wilderness protection officer Walter Schoenichen, and SS general Otto Ohlendorf.[24]

Organic vegetables became popular in Weimar Germany—and by the 1990s, also in America and the rest of Europe—as a partial result of Germans arbitrarily stating in the 1930s that the microscopic particles of synthetic pesticides residue on produce had started a cancer epidemic. *The Corporation* repeats such allegations in its interview with University of Illinois environmental medicine professor emeritus Samuel Epstein. In the film, Epstein laments that the Industrial Revolution ignited an "era of the ability to create and synthesize, on an unlimited scale, new chemicals that had never before existed in the world. . . . As the petrochemical era grew and grew, warning signs emerged that some of these chemicals could pose hazards . . . a body of data started accumulating to the extent that we now know that the synthetic chemicals which have permeated our workplace, our consumer products, our air, our water, produced cancer, and also birth defects and some other toxic effects. . . . We are now in the midst of a major cancer epidemic—and I have no doubt, and I have documented the basis for this, that industry is *largely* responsible for this overwhelming epidemic of cancer, in which one in every two men gets cancer in their lifetimes, and one in every 3 women gets cancer in their lifetimes"[25] (emphasis Epstein's).

Cal State-Northridge political scientist Michael Parenti agrees—"Chemicalization generates the conditions for further environmental degradation." And he touts what he judges to be the solution: "Thousands of farmers have abandoned chemical farming and have turned to organic farming (now called 'low input' farming)."[26]

Osama bin Laden buys into the scare stories about synthetic chemicals, citing it as one of his reasons for declaring war against the U.S. in his Open Letter to America— "You have destroyed nature with your industrial waste and gases more than any other nation in history."[27] University of Hawaii anthropologist Leslie E. Sponsel agrees, shouting, "Pesticides are actually biocides because chemicals applied to crops may harm more than their targets. . . . Collectively and cumulatively through time, individual lifestyles as well as corporations and others contribute to the poisoning of the planet and other environmental problems."[28]

But Samuel Epstein's accusations exaggerate the dangers. Epstein neglects to mention the true reasons why the number of cases of Americans with cancer has increased since the 1900s. (1) The population is larger in the present than it was a hundred years ago. As we addressed in Book Two, the larger aggregate population resulted from liberalization-encouraged industrialization reducing the death rate, which actually indicates great progress in man's standard of living. (2) Cancer is a disease that more commonly afflicts people in their old age.[29] (3) More people are living long enough to get cancer in the present age than in past decades, which, again, is an indication of progress.[30]

You can think of it in this manner. Prior to the Industrial Revolution, the reason why such a small percentage of the population died of cancer was that most people died of every other possible cause. That is, most people died on account of many other hazardous health conditions. As liberalization and the consequent industrialization gradually removed these other causes of death from the equation, heart disease and cancer were left among the few remaining major causes of death.

Further, most of the increase in cancer from 1900 to 2007 had been due to many Americans taking up smoking in the 1950s.[31] Epstein's insinuations that industrial pollution instigates most cancers are pulverized by the renowned team of physiologist Richard Doll (1912–2005) and epidemiologist Richard Peto (b. 1943). The two Richards deciphered that manmade chemicals, including industrial pollution, account for 2 percent of cancer cases.[32] Microscopic particles of synthetic pesticides on food are responsible for less than 1 percent of the cases.[33] By contrast, of the yearly cancer deaths, tobacco and dietary choices, respectively, explain 30 percent and 35 percent of cancer cases.[34] A significant share of the remainder of cancer cases result from genetically inheritable conditions.

When one factors out smoking-induced lung cancer, one finds that the per-capita, age-adjusted rate of cancer incidence has overall declined since 1990,[35] particularly stomach-, uterine, and colorectal cancer.[36] To some degree, cancer is a risk that comes from being a multi-cellular organism. As a multi-cellular organism's cells divide and its DNA is replicated, there is always the risk that some mutation will occur that obstructs the normal process of DNA replication. Such mutations can result in cells becoming cancerous. This is why cancer is found not merely in human beings but in other multi-cellular organisms—traces of it are discovered in shark specimens and in fossilized dinosaur remains.[37] There might not be a long-term solution to cancer until scientists achieve major breakthroughs in genetic engineering. That would involve employing some method of "reprogramming" the DNA of a cancer sufferer in order to correct the errors in the genetic code and thereby suppress the tumors. Other forms of biotechnology, by the way, shall be addressed in the next chapter.

In any case, Bruce N. Ames (b. 1928), whom I briefly quoted in Book Two, wrote in 1995 that "once one has adjusted the rates for age and smoking," cancer death rates "have decreased 14 percent since 1950. The types of cancer deaths that have been decreasing since 1950 are primarily stomach, cervical, uterine, and rectal."[38]

Ames, a professor of biochemistry and molecular biology at the University of California at Berkeley, at one time inadvertently assisted in producing the late-twentieth-century cancer hysteria. In 1973 this biochemist devised what is famously known as the Ames Test. This test measures the degree to which a substance is carcinogenic. It involves a scientist exposing that substance to a bacteria culture in a Petri dish and examining whether the bacteria become cancerous. Whenever the bacteria developed malignant growths, Ames inferred that the substance in question was deathly carcinogenic to humans. Upon discovering such cancerous growths developing from the bacteria's exposure to synthetic hair dyes and flame-retardants in children's pajamas, Ames aided public health officials in their efforts to ban these synthetic materials.

However, Ames later noticed that his eponymous Test began to "prove" that everything was carcinogenic. He began to suspect his own test of being too sensitive. He changed his mind, explaining that most substances on Earth can become a carcinogen in too high a dosage. It is the dosage level that determines whether too much contact with a substance can render it toxic. Moreover, contrary to the vociferations of organic vegetable faddism, Dr. Ames ascertains that we have yet to encounter concrete evidence of microscopic particles of synthetic pesticides, and fertilizers, causing any long-term negative health effects, let alone cancer, in the industrialized countries.[39] We are exposed to dosages too small to inflict significant long-range harm. In Ames's words, the "very low levels of chemicals to which humans are exposed through . . . synthetic pesticide residues are likely to pose no or minimal cancer risks."[40] For you to ride a bicycle without a helmet would be actuarially likelier to kill you than for you to eat, on a daily basis, an apple that has microscopic synthetic pesticide residue on it.[41]

University of Aarhus statistician Bjørn Lomborg warns that if the entire industrialized realm switched to organic farming and ceased the application of synthetic pesticides, fertilizers, and agricultural antibiotics, the consequence would be an increase in the cancer rate. This is because these agricultural chemicals so heavily derided by environmentalists have, beginning with the early 1900s, increased crop yields while providing a statistically negligible increase in the risk of spreading carcinogens. To avoid use of these chemicals would thereby shrink crop yields, consequently reducing the supply of industry-created fruits and vegetables that do more to fight cancer than cause it. This would raise the price of such produce on the market, which would lead to consumers buying them in smaller quantities and consuming less of them.[42]

Bruce Ames agrees, "The effort to eliminate synthetic pesticides because of unsubstantiated fears about residues in food will make fruits and vegetables more expensive, decrease consumption, and thus increase cancer rates."[43] The American Cancer Society's (ACS) website avows firm agreement: "Pesticides and herbicides can be toxic when used improperly in industrial, agricultural, or other occupational settings. Although vegetables and fruits sometimes contain low levels of these chemicals, overwhelming scientific evidence supports the overall health benefits and cancer-protective effects of eating vegetables and fruits. At present there is no evidence that residues of pesticides and herbicides at the low doses found in foods increase the risk of cancer, but produce should be washed thoroughly before eating."[44]

In fact, most plants we eat naturally produce their own pesticides as a consequence of natural selection. Resultantly, we ingest 10,000 times more (by weight) natural pesticides than we do the synthetic pesticides that farmers spray.[45] To quotes Bruce Ames, the synthetic chemical residues on foods are "likely to be a minimal carcinogenic hazard relative to the background of natural carcinogens."[46] Dr. Robert J. Scheuplein, who was a director of the FDA's Office of Toxicological Sciences, finds that more than 99 percent of food-related cancer risks come from these naturally occurring substances. Synthetic pesticide residues and the antibiotics administered to barn animals each account for 0.01 percent of the risks.[47] Bruce Ames additionally brings our attention to the fact that a person subjects himself to greater exposure to carcinogens by consuming a single cup of coffee—including coffee made from organically grown Fair Trade coffee, mind you—than he does in an entire month of having three meals per day, wherein each meal includes produce containing synthetic pesticide residue.[48] In 1989 the FDA estimated that the risk of someone dying from building up a collection of microscopic synthetic pesticide particles in his body was 0.0015 in a million. By contrast, the risk of dying from insect stings was 0.02 in a million; from a lighting strike, 0.3 in a million; from electrocution, 3 in a million; from drowning, 16 in a million, from an accidental fatality at home, 90 in a million; and, from a traffic accident, 192 in a million.[49] A two-decade study concluded in 2014, headed by University of Oxford nutritionist Kathryn E. Bradbury, examined over 600,000 women in the United Kingdom. It classified *organic* food as food grown in the absence of synthetic pesticides and synthetic fertilizers and which were not provided antibiotics. Conversely, it classified *conventional* food as food that might have received these treatments. This study concluded that except for the case of non-Hodgkin lymphoma, women who consumed the organic food were no less likely to develop cancer than those who consumed conventional food.[50]

For these reasons, Ames sighs that the fable of pesticides "bringing disease," rather than "freeing us" from it, is nothing less than "twisted . . . Various misconceptions about the relationship between environmental pollution and human disease, particularly cancer, both stem from and drive the maligning of pesticides."[51]

"There is no evidence . . ." Science Journalism Award winner Michael Specter concurs, ". . . that a single person has died or become seriously ill as a result of the accumulated residue of pesticides in their food. The same cannot be said of the toxins contained in 'natural' food—as any number of…raw milk poisonings in the United States continually demonstrate." Specter curses this popular belief that "Civilization causes cancer," as this belief increasingly motivates hundreds of thousands of people in the West to judge civilization as being something for which they ought to feel guilty.[52]

Cancer Hysteria Wrong in Both Nazi Germany and Modern USA

In 1993 the team of George Mason University communications professor Samuel Robert Lichter and former Smith College political scientist Stanley Rothman performed two surveys, one of environmental activists and the other of a random sample of 260 members of the American Association of Cancer Researchers (AACR), specialists in the scientific study of cancer, known as oncology. Sixty-three percent of respondents in the AACR survey were academicians, 46 percent taught in medical schools, 17 percent were of other academic disciplines, 13 percent worked in government agencies, 1 percent were medical doctors in private practice, 3 percent were hospital employees, 4 percent were employed in industrial medical labs, and but 7 percent were in private industry. Seventy-three percent never consulted for private industry; of those who did, but 8 percent did so on more than three occasions. Ninety-two percent were involved in research on cancer causation or prevention at the time, and 55 percent had published 40 or more professional journal articles. Overwhelmingly, the AACR sample was well-qualified to assess the risk of developing cancer in exposure to various substances.

Both groups surveyed pronounced themselves politically left-wing and supportive of government regulation. Of the AACR sample, 48 percent considered themselves welfare-state "liberals"; only 17 percent called themselves conservative, 28 percent described themselves as moderate, and 8 percent were undecided on political ideology. Sixty-two percent claimed to lean toward the Democratic Party, 18 percent toward Republicans, 13 percent were independent, and 7 percent said it did not know its own affiliation.

Despite the AACR sample's governist biases, the group's cancer experts flatly contradicted the popular governist claims about synthetic chemicals producing a cancer epidemic. Lichter and Rothman compared the AACR's opinions about the issue to those of media reports. In the media survey, Rothman and Lichter went over 1206 news items from 1972 to 1992, including stories from ABC, CBS, and NBC evening newscasts, stories on the front page of any section of the *New York Times*, *Washington Post*, and the *Wall Street Journal*, plus stories in *Time*, *Newsweek*, and *U.S. News & World Report*.

Whereas 85 percent of the media sources pronounced that the U.S. faced a cancer epidemic, 69 percent of the AACR sample disagreed. Though 66 percent of the media sample believed that cancer-causing agents are unsafe at any dosage, the same ratio of AACR members denied the veracity of such an assumption. When asked whether they

Table 9.1. Cancer Scientists whom Rothman and Lichter Surveyed in 1993

Ideology	Official Party Affiliation	Party Leaning
Welfare-State Liberal 48 percent	Democrat 46 percent	Democrat 62 percent
Moderate 28 percent	Republican 11 percent	Republican 18 percent
Conservative 17 percent	Independent 37 percent	Independent 13 percent
Don't Know 8 percent	Don't Know 6 percent	Don't Know 7 percent
Total 100 percent	Total 100 percent	Total 100 percent

Source: S. Rothman and S. R. Lichter 1996, 233.

found various news sources reliable in providing accurate information about cancer risks, a paltry 22 percent of the AACR's members found the *New York Times* reliable, while a pitiable 9 percent and 6 percent, respectively, evaluated the weekly news magazines and TV network newscasts as reliable.

The Ralph Nader-employed scientist-activist Sidney Wolfe, along with Samuel Epstein, were the two cancer experts most frequently cited by those news media. Simultaneously, environmental activists are apt to write Bruce Ames off as a dishonest industry shill. Considering these reputations, Lichter and Rothman decided to poll the anti-capitalist AACR members on how reliable they judged each of a whole slew of cancer experts, which included Wolfe, Epstein, and Ames. Of the 10 cancer experts given, the AACR rated Ames the most reliable—67 percent rating him of high reliability and 19 percent in the medium area. No more than 6 percent rated him low, and 8 percent abstained from answering. In the case of Epstein and Wolfe, they were respectively the lowest and second-lowest rated. A pitiful 24 percent of the AACR members rated each of them as being highly reliable.

Finally Rothman and Lichter had both the AACR members and the environmental activists rate the cancer risks of various activities and substances, on a scale of 0 to 10. Seven through ten would be major risks, 4–6 would be moderate, and 0–3 minor. Whereas environmental activists gave the chemical dioxin a mean rating of 8.1, AACR members rated it a much smaller 4.59. Though environmentalists gave nuclear power plants a 4.6 rating, AACR judged them a 2.46. Environmentalists assigned the synthetic pesticide DDT a 7.3 rating; among AACR, DDT's rating was 3.83. And whereas the apple additive Alar got 4.1 from environmentalists, it received a mere 2.18 from AACR. By contrast, the AACR members rated smoking at 9.19, second-hand smoke at 5.88, fat in one's diet as 5.39, a low-fiber diet at 4.83, and alcohol consumption 4.22. It is ironic, then, that a significant number of Americans are more frightened of nuclear energy than they are of drinking beer. Rothman and Lichter comment on their two surveys that while "using the

Table 9.2. How Much Confidence Do the AACR Experts Have in Various Cancer Experts Cited in Media?

Individual Rated	High Level of Confidence (percent)	Medium of Level of Confidence (percent)	Low Level of Confidence (percent)	Don't Know (percent)
Bruce Ames	67	19	6	8
Richard Peto	57	15	3	24
Richard Doll	53	8	2	37
Samuel Broder	49	18	7	26
Robert Weinberg	47	18	11	24
Joseph Fraumeni	40	15	4	41
Irving Selikoff	30	15	4	51
John Higginson	29	12	4	56
Sidney Wolfe	24	15	11	50
Samuel Epstein	24	20	17	40

Source: S. Rothman and S. R. Lichter 1996, 233.

mean score of carcinogenicity," approximately double the number of AACR researchers "as environmentalists rated Alar and nuclear power as minor carcinogens; three times as many researchers regarded food additives and dioxin as minor threats; and a whopping seventeen times as many cancer specialists as environmental leaders (by thirty-four to two percent) saw pollution as a minor contributor to cancer rates."

Perversely, of the news items in Rothman and Lichter's media survey, there was a greater focus on manmade chemicals and industrialization as being the major causes of cancer, rather than the far-greater cancer risks (such as low-fiber diets). Of the 1,206 news stories, 498 (41 percent) focused on man-made chemicals, whereas a significantly smaller 292 (24 percent) were on tobacco. Whereas pollution received 222 stories (18 percent) and synthetic pesticides 194 (16 percent), the much greater cancer risks of dietary choices, alcohol, lifestyle, and aging, respectively, received 136 stories (12 percent), 79 stories (7 percent), 70 (6 percent), and 9 (1 percent).[53] Indeed, to the extent that reliable results come from testing the toxicity of various substances on lab rodents, it has been established that the risk of consuming microscopic pesticide particles on foods is much lower than is exposure to ordinary foods and drinks. Drinking a single liter of water holds twice the cancer risk of pesticide particles, a single peanut butter sandwich is 60 times as deadly, eating a single raw mushroom is 200 times riskier, two slices of bread are 800 times worse, and guzzling 12 ounces of beer boasts 5,600 times the cancer risk.[54] The chemical pesticide DDT, which the Environmental Protection Agency banned for U.S. use in 1970, is less carcinogenic than the natural pesticides found in organic vegetables.[55]

In *The Corporation* and in other media, Samuel Epstein promulgates still other horrifying charges. He specifically charges that agribusinesses' injections of its milk-producing cows with genetically engineered growth hormones (called "bovine somatotropin," or BSTs for short[56]) has precipitated an increase in their levels of the hormone IGF-1, some of which might get in their milk. Epstein argues that this practice has driven a rise in the incidence of prostate cancer in men who have consumed this milk.[57] But rigorous testing reflects that cows receiving the BST supplement do not have higher BST levels in their milk than do cows that do not receive it.[58] And as the American Council on Science & Health (ACSH) clarifies, both cows and humans already produce IGF-1. More importantly, the "real questions are: 1. whether or not this IGF-1 substantially increases the concentration of the hormone in milk and 2. whether or not the hormone would be absorbed in sufficient amounts to raise blood levels in humans. And the answer, in each case is 'No.' Research has shown that the concentration of IGF-1 in milk from rBST-treated cows is not elevated over that in untreated cows. Furthermore, even if all the IGF-1 present in milk were absorbed into a person's blood, there is so much naturally there that the additional hormone would hardly be detectable."[59]

ACSH provides a plethora of quotations from far more reliable sources about whether consuming milk produced by rBST-injected cows will largely contribute to the incidence of prostate cancer. The American Cancer Society reports, "There are no valid findings to indicate a risk of human carcinogenesis." Likewise, the American Medical Association states, "BST is a protein hormone that is produced naturally by cows to help them make milk. Supplementing cows with small amounts of BST has been shown to increase their milk production by 10–40 percent per cow without harming the animal or altering the nutritional value of their milk." In the *Journal of the American Medical Association*, the National Institutes of Health let us know, "rBST-treated cows experience no greater health problems than untreated cows."[60] Additionally, Dr. John Ausberg of the Food & Drug Administration certifies that BSTs are "100 percent safe."[61]

Proctor takes it for granted that the chemical phobias propagated by Epstein and his allies are fundamentally rational. Then Proctor credits the Nazis for pioneering in such insightfulness in related matters. Recall that when farmers raise crops, those crops must compete against weeds for nutrients in the soil. Should the weeds usurp every last nutrient, they leave nothing for the crops, which die. Therefore, farmers have had to develop methods for killing the weeds. From the late nineteenth century to early 1900s, farmers applied arsenic and other toxic metals to the soil to kill the weeds. Proctor applauds the Nazis for enacting measures to prohibit the application of these toxic metals as herbicides.[62] Yet University of Houston economist Thomas R. DeGregori notices that the Nazis were really not trailblazers in this regard. "By the 1930s," he notes, "arsenic's harmfulness was obvious. The Nazi anti-arsenic movement was by no means special." Indeed, the synthetic liquid herbicides applied from the 1950s to the present were introduced precisely to replace the toxic metals that had previously been applied as herbicides—metals far more toxic than the liquid herbicides presently utilized. Therefore, says DeGregori, "to liken, as Proctor has tacitly done, the Nazi opposition to the agricultural use of arsenic to the present-day opposition to the agricultural use of modern pesticides—which have obviated the use of their highly toxic forerunners—is ridiculous."[63]

Indeed, just as the Soviet Union did in its governmental support of Trofim Lysenko's illogical eugenics, the Third Reich disbursed tax funding to outright pseudoscience, including faith healing. The Nazi government provided expenditures to Hans Löhr, the medical director of the University Hospital of the University of Kiel, to remunerate him for spouting this sort of romanticist nonsense: "The concept of an 'unbiased and objective' science, aiming at 'absolute truth' based on pure reason, which arose in the [less-regulated] liberalistic period" of the eighteenth and nineteenth centuries, has "entirely lost its reason and justification for existing. . . . Since the ultimate process of life can never be fully explained through causal-mechanical analyses, the question arises whether the physician...may not also have at his disposal some other means of knowledge. At once such concepts as . . . intuition spring to mind." Data gathered through sensory "experience is deceptive." And another "question is whether intuition is really a new kind of knowledge . . ." Löhr states this position in an official Nazi government document, *About the Position and Role of Medicine in the National Socialist State*. In this same document, Löhr concedes that supernatural faith healing is a legitimate medical practice, and that for medical doctors to execute life-and-death decisions according to "intuition" is no less sound that going by empirically demonstrable facts.[64]

Along with their colleagues, Christian Pross—medical director of the Berlin Center for the Treatment of Torture Victims—and historian Götz Aly point out "the preference of early Nazi health policy" over unscientific, romanticist alternative "holistic medicine . . . over decadent, Jewish"—translation: empirically substantiated—"medicine."[65] Among these alternative healing techniques were those inspired by Rudolf Steiner's anthroposophy. "Anthroposophical medicine," writes Marquette University historian Peter Staudenmaier, ". . . was sponsored by the Nazi Party's Main Office for Public Health, and the anthroposophist physicians' organization, the League for Biodynamic Healing, was a central member of the officially sanctioned Reich Committee for a New German Art of Healing. Biodynamic treatments and Weleda products were avidly promoted in various Nazi contexts."[66]

Not every Nazi officer and not every Nazi environmentalist endorsed Steiner's farming methods. Some of them approached it guardedly, as they knew of Steiner's connection to occultist theosophy. These Nazis distrusted anything foreign, and therefore they looked askance at the Asian influence over theosophy. Nonetheless, German followers of Steiner's biodynamic agriculture movement convinced a large enough number of prominent Nazis to ensure that entire government agencies of the Third Reich applied such methods in their farming projects. Throughout the 1920s, anthroposophist Erhard Bartsch advised other Germans on biodynamic farming, and in 1934 the Third Reich's interior minister, Wilhelm Frick, sought Bartsch's counsel. Bartsch said he reviled the "Americanization and mechanization of agriculture . . ."[67] Interior minister Walther Darré started out a bit doubtful of anthroposophy, but he had been won over by biodynamic techniques by 1940, when welcomed on Bartsch's estate as an honored guest.[68] It was not until 1941 that the regime scaled back on promoting these methods, though the Third Reich continued to employ them in some limited contexts. That year, the State dismantled the Reich League for Biodynamic Farming and went as far as temporarily jailing Bartsch. Despite the Third Reich no longer supporting biodynamics so publicly though, writes Staudenmaier, the agricultural measures consistent with biodynamic strictures "continued apace under the unlikely protection of Himmler and the SS . . ."[69]

One might retort, though, that if the Nazis revered the wilderness above technology and industrialization, such wilderness-worship would not comport with the Third Reich's construction of the Autobahn—the impressive superhighway that predated Dwight D. Eisenhower's federal U.S. interstate system. Although the construction of the Autobahn proved to be an implicit concession to the need for industrialization, it does not mean that this construction paid no obeisance to environmentalist ideology. The Third Reich had a group of environmentalists led by Alwin Seifert, called Advocates of the Landscape, advise the engineers who designed the Autobahn. They ensured that in the construction of the highway, swamps that otherwise would have been dredged would remain unaffected.[70]

Earth First, Jews Last?

The Nazis' Foreign Press Chief, Ernst "Putzi" Hanfstaengl (1887–1975)—who also worked for FDR—commented that the Fuhrer had "made a dogma" out of his own vegetarianism.[71] The Nazis promoted such vegetarianism for much of the same reasons as the early-twenty-first-century's food faddists. One of the reasons cited by Nazis and non-Nazis alike is that they consider vegetarianism to be healthier than eating meat.

But the Nazis also contended that nonhuman animals possessed rights, and that it would therefore be heinous for a man to kill an animal for his own gain. Here, the Nazis obviously gloss over their own double standard about murdering other human beings.[72] The practice of vegetarianism for reasons of "animal rights" ideology—called *veganism* at the date of this writing—found vocal spokesmen not just in the Führer but also in Rudolf Hess and Heinrich Himmler.[73] Hess and Himmler, doing whatever they could to advance animal activism, environmentalism, and organic farming, vociferously lent their support to Nazi agriculture minister Walther Darré, who made each of these ideas a plank of his "Blood and Soil" economic program.[74] It was not uncommon, writes Ian Kershaw, for Adolf Hitler to burst "into a lengthy discourse on vegetarianism" in which he "aimed at overpowering, not persuading, the listener."[75] Following suit, the Nazi regime overall discouraged the consumption of beef.[76]

"A bizarre moral inversion," writes psychologist Hal Herzog, "occurred in prewar Germany that enabled large numbers of reasonable people to be more concerned with the suffering of lobsters in Berlin restaurants than with genocide."[77] In August of 1933, Hermann Wilhelm Göring (1893–1946) announced his plan to abolish what he called the "unbearable torture and suffering in animal experiments . . ."[78] He reminds others that nonhuman mammals "feel pain and experience joy and prove to be faithful and attached."[79] Thus, he threatened, "I will commit to concentration camps those who still think they can treat animals as inanimate property."[80]

On November 24 of that same year, the German government enacted the world's most comprehensive animal protection statute. According to historian Jonathan Peter Spiro, that Law for the Protection of Animals "specifically prohibited experiments on animals involving the use of cold, heat, or infection. (It need hardly be said that it was precisely such experiments that the SS doctors under Himmler's command conducted on the Jews in the extermination camps.)"[81] The NSDAP then issued a press release bragging, "The New Germany not only frees people from the curse of materialism" and "egotism . . . but also gives rights to the tortured, tormented and until now, completely unprotected animals." What "Reich Chancellor Adolf Hitler and Prime Minister Göring did and will do for the protection of animals, stands as a guideline to the leaders of all civilized states."[82] Similar legislation followed. In 1936, the government dictated that restaurateurs could not feed their patrons lobsters unless they could kill these crustaceans as quickly possible, and that fish would have to be anesthetized prior to being killed.[83]

As propaganda minister, Joseph Goebbels provides some supplemental insight into his leader's animal-rights creed. In his words, "The Führer is deeply religious . . ." The dominant Western religions and denominations "have no point of contact to the animal element, and thus, in the end, will be destroyed. The Führer is a convinced vegetarian, on principle. His arguments cannot be refuted on any serious basis. They are totally unanswerable."[84]

Much of this thought can be traced to the German biologist and eugenicist Ernst Haeckel, who, again, coined *ecology*. According to Arthur Herman, Haeckel himself "argued that . . . animals should enjoy equal status to men as fellow sentient, social, and . . . (in the case of the higher vertebrates) rational creatures. Haeckel's early version of 'animal rights' formed part of his campaign against the anthropocentric view of nature, that 'boundless presumption of conceited man' that 'has misled him into making himself "the image of God."'"[85]

"As you might expect," adds Hal Herzog, "contemporary animal activists don't relish the idea that Adolf Hitler was a fellow traveler, and some activists adamantly deny that he was either a vegetarian or an animal lover."[86] As an example, Herzog cites the book *Hitler: Neither Vegetarian Nor Animal Lover* by animal activist R. Berry.[87] Nonetheless, Herzog concludes, "anthrozoologist Boria Sax has carefully documented the evidence that many leading Nazis, including Hitler, were genuinely concerned about the treatment of animals."[88]

It is therefore ironic, Thomas DeGregori opines, that gangs of activist groups that Robert Procter "apparently would regard as 'socially responsible' or as promoting 'progressive' ideals, themselves equate their opponents and Nazis. Animal-rights advocates and vegans, for example, often represent meat-eating and the use of animals in medical research as a 'holocaust.'"[89] An example of this is that of a roving exhibition that the organization PETA (People for the Ethical Treatment of Animals) put on in 2003. Titled "A Holocaust on Your Plate," it juxtaposed photographs of Nazi concentration camps with those of animals raised on "factory farms."[90] When it was pointed out that one of the young Jewish prisoners in PETA's photographs was none other than Nobel Peace Prize-winning author Elie Wiesel (b. 1928), and that the entire display was in poor taste, PETA's cofounder-president Ingrid Newkirk (b. 1949) retorted, "Six million people died in concentration camps, but six *billion* broiler chickens will die this year in slaughterhouses"[91] (emphasis hers). As we have recognized in chapter 3, Anti-Defamation League's executive director and Holocaust survivor Abraham Foxman unfortunately swallows the untruths about Enlightenment-era free-marketers inventing social Darwinism. Fortunately, on the matter of PETA's Holocaust campaign, Foxman exhibited wiser judgment, issuing this statement: "The effort by PETA to compare the deliberate systematic murder of millions of Jews to the issue of animal rights is abhorrent. . . . Abusive treatment of animals should be opposed, but cannot and must not be compared to the Holocaust. The uniqueness of human life is the moral underpinning for those who resisted the hatred of Nazis and others ready to commit genocide even today."[92] PETA delivered an apology to the Jewish community eighteen months later. Yet that did not stop other animal-rights groups from attempting the same Holocaust comparisons again by August 2005.[93]

While PETA claims to support the humane treatment of animals, it sometimes promotes what behavioral psychologist Edwin A. Locke III, retired from the University of Maryland, accurately characterizes as "the animalistic treatment of human beings."[94] In 1992 there was a man suffering from liver failure wrought from his hepatitis B. A transplant of a human liver was not enough to save him. His doctor, Thomas E. Starzl, judged that since the hepatitis virus does not damage baboon livers, it made sense to provide this man with a baboon liver.[95] PETA and other animal rights groups responded by picketing outside the hospital. When the liver patient died, a spokesman from the Fund for Animals stated, "Animal rights groups should feel vindicated, though we don't applaud the death of any person."[96] Apparently the spokesman meant that this demonstrated that he was correct that the liver transplant would not save the patient anyway. The gloating, though, is still dubious. And it has not stopped. TV host Bill Maher, who sits on the board of PETA, has gone as far as proclaiming that the saving of human lives is not sufficient to justify medical research on animals.[97] Maher can disapprove as much as he chooses, but Hal Herzog provides an impressive list of medical treatments that resulted from animal research. In the absence of such scientific experimentation, he writes, "we would not have immunizations for polio, mumps, measles, rubella, or hepatitis. Nor would there be antibiotics, anesthetics, blood transfusions, radiation therapy, open-heart surgery, organ transplants, insulin, cataract surgery, and medications for epilepsy, ulcers, schizophrenia, depression, bipolar disorder, or hypertension."[98] Frederick Banting relied on research on animals in testing the safety of administering insulin. So did Jonas Salk in examining the effects of his polio vaccine. Penicillin, too, saved millions of lives. Ernst Chain demonstrated the efficacy of penicillin in his studies on mice. Alexis Carrel applied organ surgery to animals prior to applying his technique to human patients.[99] Herzog goes on that your pets also would face hardship in the absence of such research. "We would not have immunizations against rabies, distemper, parvo, or feline leukemia, nor treatments for heartworm, brucellosis, cancer, or canine arthritis."[100]

In keeping with their tenet that, for the ostensive welfare of non-sapient species, we renounce our own eudaemonia, the Nazis demanded that Western civilization subordinate its own interests to environmentalism. Proctor writes about Walter Schoenichen's 1942 treatise about the conservation of the wilderness, which is called *restituiernder Naturschutz*, and of indigenous cultures. It emphatically rejects the "liberalistic worldview," which is that indigenous peoples can abandon their old illiberal collectivism and be assimilated into a laissez-faire liberal republican body of law. Proctor rhapsodizes about Schoenichen's paean to the bucolic lifestyle, as if it is humane to expect an impoverished people never industrialize.[101] Thomas DeGregori exposes some truths about this that Proctor omits. Preservation of indigenous culture, wherein the culture is exempt from liberalization, results in poverty being "maintained, according to what others define as 'traditional'—all for the benefit" of paternalists who care more about preserving these customs for posterity than about the well-being of those who are expected to practice such customs.[102]

One may notice parallels between this outlook and the modern philosophical movement known as *deep ecology*, which promotes a radical de-industrialization in favor of a return to hunter-gatherer subsistence. Norwegian philosophers Arne Naess (1912–2009) and Johan Galtung (b. 1930) coined this particular term,[103] but the ideas behind it are not original to them. It is no accident that Earth First publicist Christopher Manes admits that the deep ecology philosophy's roots can be traced back to Nazi philosopher and propagandist Martin Heidegger.[104]

That environmental regulation and governist eugenics received support from the same early-twentieth-century figures is noteworthy. Recall Ernst Haeckel, the German whose writings deeply influenced the Third Reich's policies on both eugenics and environmentalism. Jonathan Peter Spiro points out that it was Haeckel who coined the word *ecology*.[105] Likewise remember another eugenics advocate, Madison Grant, who was the one American to receive fan letters from both Theodore Roosevelt and Adolf Hitler. Madison Grant, too, pioneered in environmental conservation.[106] The main aspects of this wilderness-exalting dogma—shared by Nazis and much-less-murderous modern environmentalists alike—arose from a rejection of the ethical notion that a human's life, particularly one's own life, is the standard whereby every value is generated. Instead Nazis and environmentalists cling to the supposition that man is not a part of nature, but an alien blight on it.

This comes from the environmentalists' acceptance of the bromide that morality arises not from eudaemonic self-interest, but from the ethical duty to sacrifice one's own well-being for others. The *other*, in this situation, alludes to every aspect of the natural realm that is nonhuman. University of Wisconsin political science professor Jonathan Olsen explains, "According to the national socialist ideology, an anthropocentric view of nature—that man stands above" the non-sapient wilderness "was to be decisively rejected."[107]

Himmler summarized the Nazi outlook as "man is nothing special . . ."[108] The Nazis consequently passed laws violating the actual private property rights of human beings in order to protect, allegedly, some animal or wilderness area. Speaking for them, Nazi environmentalist Oskar Karpa pronounced, ". . . the right of private property only has validity insofar as it does not violate the well-being of the general public," with non-human entities counting as part of the general public.[109] Likewise, in 1940 the anthroposophist Armin Süßenguth proclaimed, "Ordinary materialism"—that is, observation-based rationality—"is digging its own grave: the cow is not a milk factory, the hen is not an egg-laying machine, the soil is not a chemical laboratory, as the Jew-professors would have us believe."[110] Environmentalism was not an idea to which Nazis simply gave lip service, Peter Staudenmaier reminds us. Rather, the Nazis genuinely believed in the doctrine: ". . . Nazi initiatives around environmentally sensitive public works, organic agriculture, habitat protection, and related matters are perhaps better seen not as mere camouflage or peculiar deviations from the destructive path of the Nazi juggernaut."[111]

I hope that at this point it should be easy to understand why no non-sapient organism has actual Lockean rights. Rights are not intrinsicist Platonic absolutes that are "just there." Nay, they are contextual absolutes that apply to the day-to-day dealings that sapient beings have with one another, applying to sapient organisms alone due to the nature of their sapience. Rights can belong to—and be respected by—no sort of organism other than that which belongs to a species biologically capable of volitional conceptual consciousness. Nonhuman mammals evince perceptual consciousness, but not conceptual consciousness. The absence of conceptual consciousness is apparent in as sophisticated a family as the great apes. That is notwithstanding the fact that primates' memories, symbol recognition ability, and peer-mimicking abilities were necessary preconditions to the eventual evolution of man's conceptual consciousness.

As Lockean rights are a complex concept, they cannot be consistently applied—either practiced by or practiced upon—any organisms except in the context of one concept-holding being with another concept-holding being. Lions hunt after zebras and eat them, and neither lion nor zebra have any conceptual volitional choice in the matter. Were we to engage in serious efforts to apply the concepts of Lockean rights to non-sapient animals, we would have to imprison lions for murdering the zebras. Yet that mass-scale captivity for animals would be contrary to what PETA actually favors.[112] Though Book One cites primatologist Jane Goodall on how the structure of chimpanzee bands provides us a glimpse on how human social institutions evolved, I must fault her on the issue of animal rights. She advocates the "rights" of chimpanzees when her own research demonstrates that chimpanzees cannot have rights—at least not the Lockean ones this trilogy upholds. A sapient being possesses rights insofar as he or she refrains from spoliating any other innocent, non-spoliating sapient being with rights. Dr. Goodall discovered that it is an overwhelmingly statistical commonality for a male chimpanzee to obtain sex from females by means of assault.[113] Were we to apply human concepts of social interaction to those of chimpanzees, we would have to characterize most male chimps as rapists and concede that their propensity to rape is congenital. Chimps are also violent with one another to an extent that has not been tolerated among humans in the modern industrialized world since the Industrial Revolution.

Were it true that chimpanzees possessed Lockean rights, a night watchman state would have to imprison most chimpanzees on the grounds that they abrogated the "rights" of other chimpanzees through spoliation. As we recall from Book Two, it is not unusual for Male Chimp 1 to "murder" the babies of Male Chimp 2 so that Male Chimp 1's babies will not have to share food or living space with Male Chimp 2's babies.[114] The same goes for orangutans.[115] It is proper, of course, for someone who aesthetically values animals to campaign peaceably against wanton cruelty toward them. An appropriate name for such a campaign is *animal welfare*, not animal "rights."

Himmler's notion that man is nothing special has continued with modern environmentalists in the more-developed countries (MDCs). In 1974 the Club of Rome proclaimed, "The World Has Cancer and the Cancer Is Man."[116] Former Rockefeller Foundation president Alan Gregg, whom we quoted in Book Two, pronounced in a 1955 issue of the journal *Science*, "that the world has cancer, and that the cancer cell is man . . ."[117] Paul R. Ehrlich asserts, "A cancer is an uncontrolled multiplication of cells; the population explosion is an uncontrolled multiplication of people. . . . We must shift our efforts from the treatment of the symptoms to cutting out the cancer."[118]

What Is the Source of Similarity Between the U.S. Left and Early-1900s Germany?

None of this is to convey that somebody is a Nazi on account of professing that animals have rights or that we human beings should sacrifice our own interests to those of an unmolested wilderness. What proved the Nazis deserving of condemnation was not that they held such beliefs, but that they inflicted violence on innocent people in order to impose these beliefs about race, eugenics, religion, and environmentalism. Further, what Jacob Sullum said about public health regulations also applies here—the enactment of spoliative environmental and "animal activism" regulations would be just as deleterious as they are in the modern context were the Nazis never to advocate them.

Some readers might regard it as inappropriate that I have placed so much emphasis on the Third Reich's politically correct regulatory-entitlement-state policies, as if that emphasis trivializes the horrors of the Holocaust. Such readers may believe that the welfare-state policies are kind and humanitarian and therefore contradict the obvious evils of Auschwitz and the gas chambers. Yet, despite their differences in severity and magnitude, each of the regulatory-entitlement governist policies shares something in common with the Nazis' persecution of Jews. Each of these programs was enforced through violence against sapient human beings possessing actual rights. Recall that the German government would issue violence against anyone who ceaselessly resisted the compulsory taxation that financed the Nazi welfare state. The same principle applies to anyone trying to disobey the Third Reich's economic regulations. I am mindful that the Nazis' taxation and regulation measures were minor in comparison to the pure savagery of the Holocaust. But the threat of government-imposed violence remains a common thread in each of these measures. No matter the degree of severity, spoliation runs rampant through every aspect. Yes, Nazi-imposed taxation was a small nuisance relative to the actions for which the Nazis are most infamously remembered. But the mass abductions, enslavements, murders, and other atrocities the Nazis committed must be recognized as Nazi governist spoliation taken to its final extreme. And that other forms of governmental spoliation may be minor in comparison to others does not render the tamer spoliation wholly unimportant. No form of spoliation is ultimately acceptable.

Definitely, there are numerable ominous parallels between Nazi's domestic economic policies and the modern U.S. progressive ideology. This is not on account of U.S. progressives reading up on Nazi history and then self-consciously copying the Nazis' policies. Rather, both Nazi policy and modern U.S. progressivism are logical extensions of the animating philosophy of the Second Reich—of nineteenth-century Germany under Otto von Bismarck. Bismarck's regime, notes Fareed Zakaria, was typified by "government subsidies, regulations, and tariffs."[119]

Nineteenth-century Germany, to be sure, was not completely devoid of economic freedom. In spite of the massive controls in place in this regulatory-entitlement state, there was enough wiggle room for entrepreneurs to innovate. Hence, Gottlieb Daimler and Carl Benz pioneered in automobile technology in this era.[120] The relative liberty also allowed for the breakthroughs of Fritz Haber and Carl Bosch in synthesizing nitrogen fertilizer.[121] The zeal for freethinking motivating these scientists, engineers, and entrepreneurs, sadly, was being eclipsed by what became Germany's two most prominent philosophies.

The first of these two philosophies was governism. In the Enlightenment period there were indeed some German philosophers sympathetic to laissez-faire liberalism, such as Friedrich Schiller and Wilhelm von Humboldt.[122] Yet the governists, who held much greater influence over the intelligentsia, greatly outnumbered them. Much more renowned German philosophers, exemplified by Georg Hegel, heralded the transmutation of generally less-regulated capitalism into a new governist regime where the State controlled minute aspects of everyone's doings. The crude consensus was that the individual exists for no purpose higher than serving the social collective, often embodied in the State. Marxism was merely a highly pedantic and specialized doctrine that branched off from this more general governist collectivism.

At this moment, most governists still nursed an overall positive attitude toward industrialization and technology. It was not their desire to prohibit particular technologies or push any legislation to discourage the usage of some technologies. Karl Marx merely wanted ownership of existing industrial technologies to change hands, being forcibly transferred from control by private industrialists to collective worker committees. Moreover, despite their irrational appeal to force, the Marxists and the pioneering nineteenth-century socialists branded themselves the foremost advocates of reason. They would not acknowledge that fideism was integral to their philosophy.

Also reigning in this era was the philosophy of Romanticism,[123] which I discussed in chapter 3. Johann Georg Hamann (1730–1788), Friedrich Heinrich Jacobi (1743–1819), Johann Gottfried Herder (1744–1803), and Johann Gottlieb Fichte (1762–1814) were among the earliest of the German romantics. In spite of their criticisms of Immanuel Kant—they believed he was too individualist in such writings as "What Is Enlightenment?"—Hamann, Herder, Fichte, and Jacobi were nevertheless influenced by Kant's criticisms of sensory perception and inductive reasoning. These four realized that insofar as Kant continued to accept the idea of each person being an individual, he had contradicted himself, for individualism could be premised on nothing but confidence in the individual's inductive reasoning abilities.[124] As Fichte and Jacobi argued, it is through observational reason that you determine that your consciousness is an individual one—it is through perceptual observation that you recognize that you are a being separate from other sapient beings, thus establishing you as an individual and separate unit. Jacobi drew from David Hume's attack on observational reasoning,[125] whereas Fichte was inspired by Kant's.[126] Both Jacobi and Fichte surmised that insofar as their respective philosophic mentor was correct that sensory evidence provides no objective confirmation of any proposition, it follows that there is no proof that anyone is an individual separate from anything else. Insofar as Kant is right, declared Jacobi and Fichte, it follows that it is theoretically possible that humans are not individual and autonomous units, but a single, collective entity.[127] Taking that argument farther, Herder proposed that Germany was comprised not of individuals but just one entity. This served as the precursor to the extolling of the *Volk*.[128]

And the romanticist movement was not limited to Germany;[129] the movement concurrently existed in England and France. The English and French versions of romanticism avoided adopting the explicit political collectivism of their German counterpart, though they maintained the hostility to political collectivism's implicit opposite—individualistic profit. Such Romanticists as Mary Shelley's husband, Percy Bysshe Shelley, explicated their aversion to capitalism, denouncing the factory system.[130] As Percy framed it, "'Poetry," which represents everything that is good, versus "the principle of Self, of which money is the visible incarnation, are" respectively, "the God and the Mammon of the world."[131]

There were three areas, though, where the Romantics proved to be unlike Marx and the early nineteenth-century's utopian socialists. First, the Romanticists did not consider it a priority to effect a change in ownership over factories or mills; they wanted these mills to cease to exist altogether, or at least to dwindle in number. It was in the Romanticist movement that we notice the first ripples of the anti-industrialism current that has ever since been part of the environmentalism movement.[132] Secondly, the Romantics admitted their distrust of inductive reason; they announced their fideism. Thirdly, with the exception of Jean-Jacques Rousseau, few of the Romantics peddled a complex political platform that they strived to implement or ratify. On the whole, the Romantics did not present plans *for* a new social system; they merely knew what they were against—capitalism and industrialism—and blindly rebelled as a result.

As noted by U.C. Berkeley political scientists Aaron Wildavsky and Carolyn Webber, "Celebrating intuition and emotion instead of cold reason, the Romantic movement was certainly antimarket; it condemned the 'cash nexus' as an inappropriate basis for personal relationships. In the impersonal industrial system, leftists and rightists alike saw only dehumanized work under abhorrent conditions. . . . If life was harsh for laborers in the new order, it was not only because businessmen were greedy, but because structural instability was built into the system . . . the industrial establishment's rape of unspoiled nature would lead inexorably toward disaster. A lament for loss of rural virtue was the Romantic poets' response to industrialization; their metaphors rejected the present and idealized the past."[133]

Joachim Fest, biographer of Hitler, elaborates on Romanticism overtaking Germany from the 1800s to the beginning few decades of the 1900s: "Such writers as Paul de Lagarde, Julius Langbehn, and Eugen Duhring became spokesmen for a widespread mood hostile to modern civilization. . . . there was a reaction against the . . . life-affirming optimism of the age . . ." The more-civilized "present was fiercely condemned. . . . In Germany the spokesmen for this attitude despised progress . . . they opposed the stock exchange, urbanization, . . . the global economy and positivistic science. . . . In brief, they were against the whole concept of modern improvement . . ." Rejecting industry, they revered the wilderness, everything feral and nonhuman.[134]

George L. Mosse elucidates on how the environmentalism of twentieth-century Westerners and Nazis, alike, originated with the nineteenth-century Volkish movement's fetish over the undeveloped wild habitat—"Man was seen not as a vanquisher of nature, nor was he credited with the ability to penetrate the meaning of nature by applying the tools of reason; instead he was glorified as living in accordance with . . . mystical forces. In this way, instead of being encouraged" to contemplate the properties of his own relationship with "urbanization and industrialization, man was enticed to retreat into a rural nostalgia. Not within the city, but in the landscape, the countryside native to him, was man fated to merge with and become rooted in nature and the Volk. And only in this process, taking place in the native environment, would every man" be able to attach some meaning to his existence.[135]

Are You for Riehl?

One nineteenth-century Munich professor, Wilhelm Heinrich Rihel, would provide inspiration both to the Volkish movement and to the twentieth-century German Green Party. He did so through his treatise *Land und Leute* (*Places and People*), penned between 1857 and 1863. Mosse tells us that Riehl "discussed the organic nature of a Volk which he claimed could be attained only if it fused with the native landscape. . . . The culture of a Volk," which was to worship the untrammeled wilderness, "was posited as the very opposite of a mechanical and materialistic civilization. Riehl rejected all artificiality and defined modernity as a nature contrived by man and thus devoid of that genuineness to which living nature alone gives meaning. Such a contrivance—a city and its factories, for example—was seen as lacking genuineness. . . . For Riehl, as well as for subsequent Volkish thinkers," nothing but the inhuman wilderness was "genuine, since it was infused with both the life force and historical meaning for the Volk. Any merely man-made improvement upon it would destroy the ordained meaning of nature . . ." And Mosse continues, "Starting once more with the ideal of nature, Riehl held up the unspoiled countryside as the model for the social structure he desired." Mosse writes that Riehl estimated the bourgeoisie to be "a disruptive element" that was "composed mainly of merchants and industrialists who had no close connection" to the divine wilderness.

To Riehl and other Volkish thinkers "the city came to symbolize the industrial progress and modernity that all adherents to the Volkish ideology rejected. . . . Riehl, who was credited with this apocalyptic vision, criticized big cities for wanting to become international urban centers . . ." Continuing the summary, Mosse writes that Riehl feared that the cosmopolitan bourgeoisie would "exercise a suzerainty over a world in which all that was natural had been destroyed . . ." Reviling individualism, Riehl held in high regard the utopian collectivist communities started by the likes of the wealthy Robert Owen—the precursors to our own era's hippie communes. As Mosse paraphrases Riehl, an individual should aspire to having his economic output "shared by all his fellows," as that would result in cultural "simplicity, removal of the frustrations inflicted by an artificial urban way of life, genuineness of emotions and their uninhibited expression, and the rewarding sense of belonging within both a historically and cosmically sanctioned whole." As Mosse describes the Volkish collectivist communes, "The utopias were effectively removed from the real world, rural islands in a sea of industrialization. Basic to the utopian movement was the urge to return to the land. It embodied an effort to root the natural rhythm, that soothed the discord of urban life. The appeal was leveled at city dwellers, at disgruntled intellectuals and educators, people who had ideologically declined positive identification in urban life precisely because they rejected the values of modernity. It was spurred on by a double purpose: to escape the debilitating effects of city and industry and to reconstitute the Volk and the race. . . . the utopian settlements were to be not only compelling examples for the Volk, but also living indictments of modern capitalism. . . . To them the migration to the city and the commercial exploitation of the peasants represented the destruction of the organic base of the Volk." The communes swore that the peasant estate should never be "replaced by a factory." Mosse goes on that "there was a distinct element of anti-capitalism" in the literature of the Volkish commune movement, as "presented . . . in the famous book written by Ferdinand Fried (Friedrich Zimmerman), *Das Ende des Kapitalismus* (*The End of Capitalism*) (1931), and published by Eugen Diederichs," the mystic who also published Paul de Lagarde. Fried "decried capitalism in its modern form, labeling it 'Manchesterism.'" The book's "argument culminated in the defense of socialist communal arrangements insofar as they facilitated the goal of complete nationalization of the land." You may recall from our discussion of intellectual property rights from Book Two that Henry George advocated nationalization of real estate. He was actually more popular in Europe, particularly among the Volkish romantics, than he was in his native United States.[136] Among other environmentalist romantics was Rudolf Steiner, academically trained as a scientist but most famous for espousing his religious views in the twentieth century's three earliest decades, first in the Theosophical Society and then in his own Anthroposophical Society.[137] A volume published by Rudolf Steiner's own press firm announces proudly that Steiner's "main endeavour is to counter the narrow cause-and-effect conceptions which dominate modern outlook and research. He sees the latent possibilities in man of advancing beyond the present-day limitations of cognition to include 'knowledge of higher worlds.'"[138] Anti-realist epistemology and radical environmentalism are more closely linked than many of us would prefer to think.

Richard Grunberger admits that the Nazis' explicit philosophy played upon a desire to return to a more primitive age, one that preceded the Industrial Revolution—"Pre-capitalist yearnings, i.e. the desire for . . . a static traditional order, created an undertow within the general anti-capitalist current of the period."[139] It was from Riehl's teachings that the Volkish movement formed the idea that a good German's life must prioritize *Blood and Soil*. As we recall from chapter 3, the *blood* of *Blood and Soil* refers to supposedly unchosen obligations to one's bloodline—one's family, ethnicity, and tribe. Embedded into the *soil* is a twofold meaning. As numerous left-wing scholars confirm, the *soil* alludes to the nation and fatherland—the Germans' nationalistic devotion to their country. Yet there is a second meaning that left-wing scholars mention less often. It is that one must be loyal to the *soil* more literally—that is, willing to preserve the landscape in its wild condition, to be forever undeveloped and untrammeled by industrial technology.

The Students Becoming the Masters

Hitler Youth patterned itself after the turn-of-the-twentieth-century environmentalist German Youth Movement, which found inspiration in Riehl's *Places and People*, as well as the anti-capitalism of Paul de Lagarde and Julius Langbehn. This movement, writes George L. Mosse, "immersed youth in a romanticism based upon the native landscape" and nostalgia for the alleged lost paradise of Germany's preindustrial past. Despite the youth movement's white supremacism, one if its leaders, Karl Fischer, taught adolescent members to hold the American Indians in awe, as they were seen as wise on account of their reputation for being socially collectivist, preindustrial, and strongly believing in wilderness deities.

Much as with the Hitler Youth who succeeded them, the German Youth sought a Third Way as a middle path between political-economic liberalization and communism, a collectivist concept they called the *Bund* (group or bundle). Mosse explains that "the idea of the *Bund* . . . was elevated to a general principle not only of life but of government as well. . . . The *Bund* structure . . . appeared as an alternative to both Marxism and capitalist class society . . . basing itself on . . . Volk . . . and community. The Youth Movement's concept of the *Bund* was to remain a constant inspiration . . ." As Mosse phrases it, Hans Breuer, another leader of the German Youth Movement, believed that "the age of science and reason . . . had done Germany serious injury . . ." Resultantly, Breuer turned, as his guide, to a writer who solidified the philosophic mainstream's revolt against inductive reason. Breuer, writes Mosse, "cited Kant . . . in his appeal for a Volkish commitment. The philosopher's principles of civic duty were read as Volkish cultural imperatives."

This same book of Riehl's that George L. Mosse describes—*Places and People*—was held, by the German Green Party at the time of its founding in 1980, as one of the main inspiration's for the party's political philosophy. On account of their reverence for the undeveloped landscape, the Volkish writers—as with late-twentieth-century environmentalists—appropriated the tree as the symbol for their movement.[140] In Mosse's words, Paul de Lagarde believed that man "should listen to the trees of the wood . . ."[141] Riehl, Julius Langbehn, and Lagarde each preached that humanity could find no fulfillment except for emotionally subordinating their own interests in modern conveniences to their appreciation of the untouched wilderness.

Recall that Lagarde's publisher, Eugen Diederichs, also influenced the Volkish movement. Diederichs's volkish idealism, writes George L. Mosse, would not "tolerate increasing industrialization. . . . The soil, the earth, the Volk were the component parts" of the grander whole. "The soil was once more equated with the Volk."[142]

"I have," Diederichs explained, "always felt close to mother nature."[143] Likewise, Arthur Moeller van den Bruck's political views informed those of Otto Strasser, the ideological ally of Joseph Goebbels during his early days at *National Socialist Briefings*.[144] Rudolf Steiner's anthroposophy and biodynamic environmentalism also played a role, ultimately influencing the Nazis' agriculture policies as it one would day later inform the protocols of organic farming in the late-twentieth-century USA. There was, notes Peter Staudenmaier, "convergence between biodynamic philosophy and the tenets of blood and soil, some of them stemming from common roots in pre-Nazi culture."[145]

It was also the Romantic movement that spawned the nineteenth century's first animal activists, who called themselves anti-vivisectionists. Then, as in the present, they protested scientific experiments on animals. Despite their differences, where the governist movements and Romantic movements overlapped was their opposition to capitalism and middle-class businesspeople. Ultimately, from the 1800s to the turn of the twentieth century, the two movements fused into one. The new, unified movement embraced the governists' prescriptions for governmental control of industry. Likewise, it adopted the Romanticist notion that the non-sapient wilderness was morally superior (1) to inductive reason and (2) to inductive reason's practical consequence, industrial technology. At the start of the 1800s, Hegel wedded together fideism, collectivism, and governism. Still, it took some time for the fideism-collectivism-governism trifecta to catch on. This trifecta never lost popularity in the Weimar years, pursuant to Germany's defeat in the First World War. Hitler and scores of other Germans remained loyal to this philosophy in the Weimar years, and thus implemented the trifecta under the Nazi regime. In effect, the Third Reich inherited its fideism, environmentalism, animal activism, societal collectivism, and governism from the Second Reich. Therefore, notes Joachim Fest, the Reich chancellor came to fear "American technology, . . . big cities, 'industrialization as unrestricted as it is harmful,' the 'economization of the nation,' corporations," and "the 'morass of metropolitan amusement culture' . . . "[146] Much of the German romantics' aversion to capitalism originated from their aversion to the sources of the capitalist profit motive: individualism and peaceable self-interest. These German romantics faulted the Enlightenment, writes Anthony Padgen, based on their belief that the Enlightenment philosophers "had made the 'I' the center of all inquiries into the human condition."[147]

In Book Two we touched upon the benefits of international trade in material goods and financial services. Sadly, whereas most imports benefit those who purchase them, one import that the USA and United Kingdom brought in from Germany proved harmful to every nation that housed it. That harmful import was the philosophy that developed in Germany during the reign of the Second Reich. Insofar as Germany economically prospered in the 1800s, it was in spite of, rather than because of, the Romantic and governist philosophies brewing there. Under Bismarck's governist regime, damage was already being done. Although George Orwell does not agree with his assessment fully, Orwell splendidly paraphrases F. A. Hayek's argument from *The Road to Serfdom*: ". . . in Germany the Nazis were able to succeed because most of the Socialists," Bismarck, and other governists "had already done most of their work for them, especially the intellectual work of weakening the desire for liberty."[148] Various intellectuals from France, England, and the USA visited Bismarck's Germany and, in their poor judgment, found themselves impressed by the Romantic philosophy and governist policies.[149]

"Around 1900," records Fareed Zakaria, "Germany was seen by many serious observers as the most progressive state in the world." They admired the massive "welfare state."[150] American governists could not help but be intrigued by the manner in which Prussia's government mandated that children attend tax-funded schools,[151] as well by various other aspects of the welfare state, such as Social Security.[152] They relished that in the industry of public libraries, the user-fee-funded libraries were driven out of business by the state-owned, tax-funded libraries. Such visiting American professors as John Dewey—who absorbed Hegel's ideas on government[153] —and Richard T. Ely and Simon Patten returned to their respective universities and extolled the greatness of governism and the purported failure of the Enlightenment liberal philosophy.[154]

You may remember, from Book One, quotations from various twentieth-century intellectuals to the effect that it was wonderful when the USA finally adopted regulations that compelled children to attend schools. Yes, the USA here was modeling itself after the system of government schooling established in Germany. The Prussian government instituted full tax funding for schools in 1806, and in 1826 enacted statues mandating that every child attend one.[155] The Prussian school system, of course, was not as competent as those governists made it out to be. The pressures the child attendees faced were excruciating and unwarranted. George L. Mosse reports that from 1883 to 1888 in Prussia, 289 children took their own lives. One-hundred and ten of them were high school students.[156]

A German governist named Friedrich List (1789–1846) has been cited as influencing the Germans of Bismarck's tenure,[157] many of whose programs the Nazis adopted. List's teachings were ingrained in the mind of Paul Bang, the economic expert for the German National Volk Party (Deutschnationale Volkspartei, DNVP). This DNVP was a role model for the same DAP of Anton Drexler, which Hitler made into the NS-DAP. Mosse remarks that Bang "demanded a return to the mercantilist ideas of Friedrich List." Moreover, the "Jews were free traders, and could thus only oppose the true spirituality of the Volk."[158] Drawing from both List's governism and the prevailing racism, the DNVP mouthed these words—"Mammonism does not reside in the purse but in the heart. . . . Man must accept the task which God and destiny have given him and swear a holy loyalty to the Volk."[159] The same Friedrich List who inspired Paul Bang became an icon among the same American progressives[160] who would eventually prescribe compulsory sterilization and other eugenicist measures.

Among the influential Americans to take in the Germanic worldview was jurist and political philosopher Francis Lieber (1798–1872). Educated in Germany and influenced by Immanuel Kant's ethics, Lieber sought to have the State enforce collective social duties.

As Johns Hopkins University historian Dorothy Ross paraphrases Lieber's creed, "Natural rights did not inhere in the lone individual . . .; people were social beings who derived primordial, natural rights from their (social) humanity."[161]

As significant is progressive eugenicist Woodrow Wilson. Although he would ultimately lead the USA in battle against Germany in World War One, it happens that in 1887, as a Princeton University political science professor, Woodrow Wilson considered Prussia's system a role model for the USA to mimic. At the time, Wilson judged that Prussia ran an "admirable system . . . the most studied and most nearly perfected." It "transformed arrogant and perfunctory" bureaucracies "into public-spirited instruments of just government."[162]

Convergence

Thus the Romanticism-collectivism-governism axis gained prominence in U.S. universities by the former half of the twentieth century. Resultantly, writes Dorothy Ross, "the language of 'control,' with and without the 'social' prefix, became pervasive" among the progressives of early-1900s "American social science."[163] Still yet, prior to the Vietnam War, the governist Progressives, following Germany's nineteenth-century socialists, still cast glances of approval upon technology and industrialization. As the 1960s ended, though, it became increasingly obvious to American leftists that the nation's stockholders would not be overthrown by their employees. The U.S. Left began looking to other philosophies for an explanation. They sought a rationalization in the early twentieth-century's Frankfurt school of neo-Marxist cultural thought (which was, unsurprisingly, German) and discovered its critique against Aristotelian rationality and the Enlightenment.[164] In the 1970s, American leftists took the Frankfurt school's synthesis of Romanticism and anti-capitalism and welded it to their own governism. Among the Frankfurt school members who epitomized the New Left thought of the 1970s was philosopher Herbert Marcuse, himself heavily influenced by romanticist philosopher and Nazi eugenicist Martin Heidegger.[165] Hence there was a philosophic synthesis in the USA similar to the one that took place decades earlier in Germany. The American synthesis of Romanticism and governism gave birth to the New Left movement peopled by hippies, yippies, environmentalists, and animal activists.[166] On account of this synthesis, writes UCLA political scientist and historian Anthony Padgen, "the more radical opponents of 'globalization' have cast the Enlightenment, as Adorno and Horkheimer cast it, as the apotheosis of . . . science," which "devastates the environment; strips indigenous peoples of their moral dignity . . . and possibly their cultural identities; and condemns more than two-thirds of the population of the world to penury and misery in order to supply the excessive wants of the other third."[167] John Gray disparages that development as "western cultural imperialism."[168]

Consequently, Theodore Kaczynski admits that contemporary "leftish philosophers" influenced by Romanticism "tend to dismiss reason, science," and "objective reality and to insist that everything is culturally relative." The domestic terrorist thinks it "obvious that modern leftish philosophers are not simply cool-headed logicians systematically analyzing the foundations of knowledge. They are deeply involved emotionally in their attack on truth and reality."[169]

To this day, University of Cambridge chemist Terence Kealey reminds us, the reputation of the human mind has "never fully recovered from the success of Rousseau and the Romantic Movement. . . . The Romantics made it fashionable to despise scientists as overrational, megalomaniac, emotional cripples, in hock to materialism, vainglory and greed. That view persists; a 1962 survey of American college students' attitudes" showed that they judged scientists to be "intelligent, but socially withdrawn and possessed of a 'cold intellectualism.'"[170] Kealey is right—the Romantic movement's distrust of epagoge and its practical result, liberated wealth creation, remains fashionable throughout the present era. As hostess for the May 1994 television documentary *The Nobel Legacy*, McGill University classics professor Anne Carson brazenly spouts, "What an irrational idea. . . . The happy delusion that there are such things as facts, and they do not deceive us, underlies the whole progress of science and chemistry to this day." To her, that is not a benign development. ". . . I don't believe in progress, and I am skeptical of how chemistry is contributing to my humanity." She has evidently overlooked the importance of chemistry in the development of the polio vaccine. "Now that we've filled the world with Styrofoam cups, carbon monoxide, and holes in the ozone, maybe it's time . . . to stand still and pay attention to the real relation between our humanity and our progress."[171] Given the low opinion of self-interested industrialism, it comes as little surprise when oceanographer Paul R. Gross, mathematician Norman Levitt, and geographer Martin W. Lewis air the prevailing interpretation—"Only ignorance or duplicity permits the conclusion that, in the state of the biosphere and its physical environment, all is well."[172]

The American journalist Michael Specter relates that in 1999 he ran afoul of a U.S. group calling itself Progressives Against Scientism, an activist cell dedicated to advocating vandalism and government regulations—both forms of spoliation—to curb what would otherwise be unregulated innovation in technology. Here in this group we detect the confluence of pro-governist Progressivism and anti-science Romanticism. We observe such influences in the radical environmentalist effort to employ eco-terrorism or government control to destroy the genetically modified foods and other biotechnologies employed by the agribusiness corporation Monsanto, which they decry as "Monsatan."[173] One member of this environmentalist campaign, Lord Peter Melchett of the United Kingdom, actually descended from a founder of a chemical company. That lineage does not deter him from pronouncing on behalf of the British Soil Association, "We say it is time to stop assuming that discoveries only move us forward. The war against nature has to end. And we are going to stop it."[174] Kent A. MacDougall, a former *Wall Street Journal* reporter and U.C. Berkeley journalism professor, takes the ideology to another extreme. You may recall from Book Two that I mentioned that Garrett Hardin compared Asians to a tumor on the Earth. Kent MacDougall is less discriminatory about ethnicity than Hardin, for MacDougall proclaims in the academic journal *Wild Earth* that humanity, as such, is a cancer on the biosphere.[175] Warren M. Hern, a medical doctor based in Boulder, Colorado, goes with that analogy as well in the journals *Population & Environment* and *BioScience*.[176] Despite his own sympathies for the environmental movement, *New Scientist* magazine environmental consultant Fred Pearce admits that the environmentalist movement takes an "anti-science" position with respect to its opposition to genetic engineering, nuclear energy, and shale gas.[177] Once in a while, a leading environmentalist will himself acknowledge that his ideology does not hold science to be paramount. Back in the year 2000, acting as the head of Greenpeace's United Kingdom chapter, Stephen Tindale said of his own organization, "Greenpeace has never based its campaigns solely on science. Cartesian science strips everything down to cold logic: there is no room for ethics or emotion. We believe, in contrast, that there is a moral basis for our defense of the natural world," *moral* referring to the antirational doctrine of self-denial.[178]

That integration of anti-science and pro-governism did bear one major Western precedent—in the philosophy of Rousseau, one of the few Romantics to espouse a pogrom for political change. As Terence Kealey paraphrases Rousseau's argument: ". . . under the distortions induced by reason (which is emotionally sterile), by capitalism (which inflames greed) and by private property (which isolates and destroys sympathy) a man's personality degrades, and he becomes psychologically alienated, both from himself and from other people. His selfishness grows into a devouring monster of vanity and egotism, *amour-propre.*"[179] Consistent with the convergence of Romanticism and anti-capitalism, in 1980 the former SS officer August Haussleiter helped cofound the German Green Party.[180]

These two factors—mysticism and governism—were each systematically advocated by Plato. As Plato argued for mysticism and governism, Ayn Rand characterizes a society and era dominated by them as one of "Platonism." And in a lecture series on Objectivism he delivered with Ayn Rand's approval, California-based psychotherapist Nathaniel Branden holds up Aristotelianism as the model for progress. As we discerned in *The Freedom of Peaceful Action*, Aristotle was not a consistent advocate of liberalization but he certainly did better than Plato on this front. Moreover, as Branden notes in this lecture under Rand's auspices, Aristotle's emphasis on objective reality, practical reason (phronesis), and eudaemonia-centered metaethics laid the rational groundwork whereby other philosophers, notably John Locke and Ayn Rand, would logically induce a consistent and objective politics of natural liberty. For that reason, a culture and time period conducive to a greater emphasis on phronesis and its consequences—rational self-interest, liberalization, and individualism—is one that Rand christens Aristotelian.

Through such terms, Branden explains how the nineteenth century could produce such achievements in liberalization, commerce, and scientific and technological innovation in spite of the philosophical zeitgeist at the university level simultaneously taking a turn for the worse. "The tragic paradox of the 19th century," he sighs, "was that it was Aristotelian in its actions and Platonist in its thinking. . . . While the Aristotelian trend was achieving political freedom for men, the Platonists were laying the philosophical foundations for the cult of the state."[181] Nowhere was this more apparent in nineteenth-century Germany, where the remnants of liberty were enough to ignite the advances in industry and technology. By contrast, as this went on, the governist paradigm reemerged as a cultural force.[182] The reason why these phenomena could commence concurrently is that there is usually a time lag. Ideas—whether they foster reason and eudaemonia and liberty, or do the opposite—start within the intelligentsia at the university level, and they gradually filter on down to other levels of intellectuals, such as to social scientists, journalists, and artists. It takes several decades before such ideas become entrenched at the political level and result in unprecedented legislations. That is why, as German philosophers and the American thinkers they had influenced had come to embrace unreason and governism in the 1800s, it would be decades before the practical results of these ideas would be seen in the USA in the realm of politics.[183]

Richard Grunberger sums this up well—the Nazis hated the Jews on account of their judging the Jews to be consistent practitioners of secular practical reason, peaceable egoism, and laissez faire. The anti-Semitism of the Volkish movement, writes Grunberger, "was fed by two currents, one economic" and the other being epistemological. "In its economic aspect, anti-Semitism . . . was a defensive reaction . . . to the advent of full-blown capitalism. Jews were agents of change: promoting free trade, commercial publicity, installment payments and the sale of ready-made goods, they intruded between producers and customers and breached the monopoly of specialized shops ranging from second-hand stalls to department stores. In fact, they prefigured the twentieth-century trend toward urbanization, commercialization and white-collar specialization . . ." On other philosophical matters, anti-Semitism amounted to "a reaction against such features of modernity as parliamentarism," *phronesis*, and "enlightened *self-interest*—'alien' notions" whose influence such Volkish commentators as Langbehn and Lagarde juxtaposed against what they proposed as the morally supreme alternative: their own "primitive...*völkische* ideology"[184] (emphases added).

The Nazis therefore opposed everything held valuable by Rand and other rational enterprisers. In contrast to Rand's individualism, reason, egoism, and capitalism, the Nazis extolled social collectivism, fideism, sacrificial altruism, and governism. Insofar as

Randian capitalism is social Darwinism, the Nazi ideology cannot be social Darwinism. The converse is also true—inasmuch as the Nazi ideology counts as social Darwinism, Randian Objectivism cannot be that.

Yet, despite such considerations, some governists interject, "Not so fast!" They tell me that it is unfair to classify Nazis as mystics who consistently refused to exercise the faculty for practical reasons. These governist interlocutors of mine insist, for instance, that the Nazis did put their *phronesis* into play to manage the German economy successfully—much more successfully than did the Inventive Period's market economy or the Soviet Union's central planners. That claim must not go unchallenged. Also to be subjected to scrutiny is the very idea that there is any legitimacy in criticizing anyone for being a social Darwinist, whatever that is.

NOTES

1. Proctor 1999, 5, 248–49, 11.
2. J. Toland 1976, 403–04.
3. Nazi Party 1966, 281–82, adapted from Nazi Party 1937, 7–8.
4. Proctor 1999, 201–05.
5. H. L. Lombard and C. R. Doering 1928, http://www.epidemiology.ch/history/PDF bg/Lombard HL and Doering CR 1928 cancer studies in mass II habits, characteristics.pdf, accessed Monday, September 23, 2014.
6. F L. Hoffman 1915, 186.
7. Proctor 1999, 277.
8. Proctor 1999, 277 says "as some libertarians seem to want us to believe" and then places endnote 72 at the end of this sentence. On page 350 n. 72, Proctor cites Sullum 1998.
9. Sullum 1999, accessed online Thursday, May 31, 2007.
10. Proctor 1999, 124.
11. Sullum 1999, accessed online Thursday, May 31, 2007.
12. Proctor 1999, 25–26.
13. Sullum 1999, accessed online Thursday, May 31, 2007.
14. M. W. Lewis 1992, 38.
15. Proctor 1999, 124–25, 26.
16. Steiner's involvement in theosophy—and resultant dispute with theosophy leader J. Krishnamurti—is documented in C. Wilson 2000, 141–46.
17. That it was a German mystic named Rudolph Steiner who started the "organic" food craze is discussed in L. M. Silver 2006, 225–231. Steiner's involvement in theosophy is discussed in C. Wilson 2000, 141–46.
18. C. Wilson 2000, 142.
19. Paarlberg 2010, 140; and L. M. Silver 2006, 226–28.
20. Paarlberg 2010, 140.
21. Paarlberg 2010, 141; and L. M. Silver 2006, 230.
22. DeGregori 1999, accessed online Wednesday, May 30, 2007; and Paarlberg 2010, 140.
23. Paarlberg 2010, 140.
24. DeGregori 2001, https://web.archive.org/web/20030802013328/http://www.acsh.org/publications/priorities/1302/coverstory.html, accessed Wednesday, October 1, 2014,
25. Samuel Epstein, interviewed in Achbar et al., prod. 2003, transcript online in the PDF at http://hellocoolworld.com/files/TheCorporation/Transcript_finalpt2copy.pdf, accessed Thursday, May 17, 2007.
26. M. Parenti 1995 6th ed., 112, 119.
27. Bin Laden 2002, http://tinyurl.com/kva1, accessed Tuesday, June 19, 2007.
28. L. E. Sponsel 2013, A11.
29. B. Ames 1995, 589.
30. Jeff Diamond, prod. 1994, provides these facts. They are also provided by M. W. Lewis 1992, 128.
31. B. Ames 1995, 589. On this same page Bruce Ames adds that 30 percent of cancer deaths in the United States can be attributed to smoking.
32. Doll and Peto 1981, 1197, 1256, cit. by Logomasini 2002, 159, and also cit. by Bast et al. 1994, 41–42.
33. Bast et al. 1994, 41–42, citing Doll and Peto 1981, 1256.
34. Doll and Peto 1981, 1197, 1256, cit. by Logomasini 2002, 159, and also cit. by Bast et al. 1994, 41–42.
35. Jeff Diamond, prod. 1994; and Fumento 1999, accessed online Thursday, May 31, 2007, provide these facts.
36. Stossel 2006, 7.

37. G. Johnson 2013.
38. B. Ames 1995, 589.
39. This information about Bruce Ames comes from L M. Silver 2006, 233; and Stossel 2006a, 2–3. Also read these important papers of his: B. N. Ames et al. 1987; and B. N. Ames et al. 1990.
40. B. Ames 1995, 590.
41. M. Specter 2009, 53.
42. Lomborg 2001 paperback, 9–10. That these chemicals are responsible for larger crop yields is also noted in R. Bailey 2002, accessed online Thursday, May 31, 2007.
43. B. Ames 1995, 594, in J. L. Simon ed. 1995 trade paperback, 594.
44. American Cancer Society (ACS) 2006, http://tinyurl.com/5tmsw, accessed Thursday, May 31, 2007.
45. Bast et al. 1994, 20–21, citing Bruce N. Ames.
46. Ames, qtd. by Bast et al. 1994, 21.
47. Bast et al. 1994, 21–22.
48. M. Specter 2009, 134.
49. Bast et al. 1994, 147. According to U.S. Census Bureau 2008, Table 111, PDF http://www.census.gov/compendia/statab/tables/08s0111.pdf, accessed Thursday, July 24, 2008, the chances of an American dying from a traffic accident was 153 in a million in 2004, while the chances of dying from slipping and falling was 64 in a million.
50. K. E. Bradbury et al. 2014.
51. B. Ames 1995, 589.
52. M. Specter 2009, 106, 128.
53. Rothman and Lichter 1996, 231–240.
54. Bast et al. 1994, 21, citing B. N. Ames et al. 1987, 271.
55. L. M. Silver 2006, 234.
56. R. Bailey 1994, paperback, 104.
57. One can find Epstein making such accusations in his interview in Achbar et al., prod. 2003, transcript online in the PDF at http://tinyurl.com/2h27kx, accessed Thursday, May 31, 2007. These accusations are also summarized in American Council on Science & Health (ACSH) 1998, http://www.acsh.org/news/newsID.445/news_detail.asp; and ACSH 2002, http://tinyurl.com/67sfs, both accessed Thursday, May 31, 2007. Irate that a number of individuals and organizations use facts to disprove their propaganda, a number of anti-capitalist activists issue an *ad-hominem* smear. They arbitrarily insinuate that any expert or organization who disagrees with them, must have—at some time or another—received some kind of funding from the big corporations that the anti-capitalists hate. In the anti-capitalist's book, should it be the case that Dr. X has ever received corporate funding, must be some paid shill, then any information that Dr. X can cite should be rejected as categorically unreliable.

That attitude is utterly asinine. Facts are facts, regardless of who cites them or for what purpose. Were an expert to tell you that gravitation exists, should you automatically disbelieve in gravitation if Exxon paid the expert to divulge this fact to you? The facts can be assessed by themselves. A fact cannot be reasonably discounted on the sole basis that the fact's divulger received corporate funding at some time.

The anti-capitalist's smear relies on these assumptions: (1) anything said by a not-for-profit group that rails against selfishness, is necessarily honest; and (2) any fact cited by a for-profit organization, for self-serving purposes, is necessarily untrustworthy. Since I want to put this demagoguery to rest, something else must be mentioned. It is utterly self-defeating for somebody to say that anti-business groups have more credibility than pro-market groups on account of there being a direct positive correlation between how much corporate funding a think tank receives and how dishonest it is. The reason is that business-hating advocacy groups receive more corporate funding than do free-market think tanks.

Bidinotto 2006, 40, http://www.atlassociety.org/ct-1814-cash_left.aspx, accessed Sunday, June 1, 2008, notes this. David Hogberg and Sarah Haney looked at the 2003–2004 IRS tax filings of the 53 *Fortune* 100 companies that have not-for-profit charitable foundations for donating to political groups. They divided political groups between "rightwing" and "leftwing." In this context, "leftwing" means those that favor of higher taxes, more regulation, and more social welfare programs. Concordantly, "rightwing" groups favor the opposite. Hogberg and Haney found, "The political left received nearly 59 million dollars, while the political right received only about 4 million dollars, a ratio of 14.5 to 1." When one subtracts the Goldman Sachs Foundation's enormous grant to the Wildlife Conservation Society, donations to the political Left are still greater than to the Right by 5.8 to 1. The J. P. Morgan Chase Foundation donated almost 1.2 million dollars to leftwing groups and nothing to any rightwing group. Citigroup donated 55,000 dollars to rightwing groups and over 1 million dollars to the Left. The timber company Weyerhaeuser, regularly denounced by environmentalists, donated 53,500 dollars to free-market think tanks whereas it donated 673,300 dollars to leftwing groups—including environmental groups. Were one to assume that this was hush money, then it did not work, as the anti-capitalists' denunciations continue. The entire report is Hogberg and Haney 2006, 1–8, http://www.capitalresearch.org/pubs/pdf/v1185995021.pdf or http://www.humanevents.com/article.php?id=16588, accessed Sunday, June 1, 2008.

An anti-capitalist who tells you to reject any information from a corporation-funded organization, cannot have it both ways. He contradicts himself if he proclaims:

(1) You should believe leftwing anti-business groups over free-market think tanks for the reason that
(2) there is a direct positive correlation between how much corporate funding a nongovernmental organization (NGO) receives and how dishonest it is.
Were it true that an NGO's dishonesty is measured in dollars it gets from controversial corporations, then the anti-business NGOs are, by that criterion, inexorably more dishonest than free-market organizations. Which is it?

58. E. M. Whelan 1995, 597.
59. American Council on Science & Health (ACSH) 1998, accessed online Wednesday, October 1, 2014.
60. These organizations quoted by American Council on Science & Health (ACSH) 2002, accessed online Wednesday, October 1, 2014.
61. John Ausberg, qtd. by R. Bailey 1994, paperback, 105.
62. Proctor 1999, 104–105.
63. DeGregori 1999, accessed online Wednesday, October 1, 2014.
64. H. Löhr 1966, 231, adapted from H. Löhr 1935, 19–23, 26–29, 32–35.
65. G. Aly et al. 1994, 9. G. Weissman 1996b, 185, brought this statement to my attention.
66. P. Staudenmaier 2013.
67. P. Staudenmaier 2013.
68. P. Staudenmaier 2013.
69. P. Staudenmaier 2013.
70. P. Staudenmaier 2013.
71. Hanfstaengl, qtd. by Kershaw 2000, paperback 262.
72. DeGregori 1999, accessed online Wednesday, October 1, 2014; and Proctor 1999, 134–141.
73. A. Herman 1997, 418; and Proctor 1999, 127, 137-39. Spiro 2009, 379, likewise points out Himmler's support for animal "rights."
74. A. Herman 1997, 418.
75. Kershaw 2000, paperback, 345.
76. Proctor 1999, 127.
77. Herzog 2010, 58-59, citing A. Arluke and B. Sax 1992; and B. Sax 2000.
78. Qtd. by Proctor 1999, 129.
79. Qtd. by Herzog 2010, 58–59.
80. Qtd. by Herzog 2010, 58–59; and Proctor 1999, 129.
81. Spiro 2009, 379.
82. Qtd. by DeGregori 2001, accessed online Wednesday, May 30, 2007.
83. Herzog 2010, 58–59, citing A. Arluke and B. Sax 1992; and B. Sax 2000.
84. Qtd. by Proctor 1999, 136.
85. A. Herman 1997, 416. His quotations of Haeckel are from Haeckel 1900, 363.
86. Herzog 2010, 58–59.
87. The book is R. Berry 2004; Herzog 2010, 58–59 cites it.
88. Herzog 2010, 58–59, citing B. Sax 2000.
89. DeGregori 1999, accessed online Wednesday, May 30, 2007.
90. "Group Blasts PETA 'Holocaust' Project" 2003, http://www.cnn.com/2003/US/Northeast/02/28/peta.holocaust/, accessed Thursday, May 31, 2007.
91. Center for Consumer Freedom (CCF), "People for the Ethical Treatment of Animals," *ActivistCash.Com*, 2007, http://www.activistcash.com/organization_overview.cfm/oid/21, accessed Thursday, May 31, 2007.
92. Anti-Defamation League (ADL) 2003, accessed online Thursday, May 31, 2007.
93. Anti-Defamation League (ADL) 2005, accessed online Thursday, May 31, 2007.
94. E. A. Locke 1995, https://web.archive.org/web/19961018083137/http://www.aynrand.org/animals.htm accessed Tuesday, October 21, 2014.
95. Dr. Ken Walker, M.D., "Why a Baboon Liver?", October 11, 1992, republished on *Canada Free Press*, http://www.canadafreepress.com/medical/surgery101192.htm, accessed Sunday, May 18, 2008.
96. J. L. Simon 1998, trade paperback, 567, citing the *Washington Post*, September 3, 1992, p. A2.
97. J. Hirsen 2003, 123.
98. Herzog 2010, 216.
99. I. E. Levine 1959, 92–99; D. Durrett 2002, 28; M. A. Meyers 2007, 72–73; M. Wainwright 1990, 56; and N. L. Tilney 2003, 40, 123.
100. Herzog 2010, 216.
101. Proctor 1999, 266–67.
102. DeGregori 1999, accessed online Thursday, May 31, 2007.
103. A. Herman 1997, 423, mentions Arne Naess coining the term.
104. C. Manes 1990, 143, 226.
105. Spiro 2009, 378, specifying that the word first appeared in Haeckel 1900, 11ff, 244, and ch. 14. Bova 1998, 61, likewise points out that Haeckel coined *ecology*.
106. See Spiro 2009.

107. J. Olsen 1999, qtd. by DeGregori 2001, accessed online Wednesday, May 30, 2007.
108. Qtd. by T. DeGregori 2001, accessed online Wednesday, May 30, 2007, citing J. Olsen 1999.
109. Spiro 2009, 379.
110. Qtd. by P. Staudenmaier 2013.
111. P. Staudenmaier 2013.
112. J. L. Simon 1998, trade paperback, 458, wisely disputes the notion that animals have rights.
113. To see the extent to which chimpanzees spoliate one another, consult Goodall 1971; Goodall 1986, 283, 514, 530; and F. Waal 2005, 26–27, 127. For information on chimps committing "rape" on other chimps, see R. Wright 2000, trade paperback, 50. Orangutans "rape" one another too, according to van Schaik 2004.
114. F. Waal 2005, 27. I argue that for an organism to have Lockean rights, it must be a sapient being with a rational faculty, as a rational faculty is a prerequisite to having a concept of rights and being able to adhere to those rights in others. Some might wonder if chimpanzees, bonobos, and other great apes might fit into this category, as some of them have allegedly been taught sign language. It is true that great apes are more adept than most species in recognizing and memorizing human symbols, such as the Arabic numerals. That is seen in J. Rubin, prod. 2008. Great apes do not actually understand American Sign Language (ASL), however. The psychologists who claimed to have taught ASL to great apes like Koko the gorilla and Washoe the chimpanzee were conveniently interpreting the apes' gestures as ASL. When deaf people who actually knew ASL tried to communicate with the great apes through ASL, they found that the apes were not using ASL and yet the psychologists demanded that the great apes' random gestures be interpreted as official ASL signs. For information on this, see A. Neisser 1983, 202–233; S. Pinker 2000, paperback, 342–350; and J. Trefil 1997, 57–60. As observed by J. Trefil 1997, 59–60, 350, one truly impressive case where it appears that a great ape might understand some human symbols is that of Kanzi the bonobo. Even in this case, Kanzi's vocabulary is too limited to consider Kanzi on par with an adult sapient human being. J. Trefil 1997, 60, observes that Kanzi was once discovered to have "about the same linguistic ability as a two-and-a-half-year-old child. . . . In the years since those initial findings, the length of Kanzi's sentences has not grown past a couple of words . . . Based on this finding, it seems safe to say that human language, as it is now understood, can be numbered among the unique adaptations of our species, and one that is not shared with any of the rest of the animal kingdom."
115. van Schaik 2004, 86.
116. Msarovic and Pestel 1974, 1.
117. A. Gregg 1955, 681, accessed online Wednesday, April 23, 2014. Zubrin 2012, 1, brought this statement to my attention.
118. P. R. Ehrlich 1968, 152.
119. Zakaria 2003, 63–64.
120. R. B. Jackson 1968, 15; and Montagu and Burgess-Wise 1995.
121. M. Goran 1967; and T. Hager 2008.
122. See Humboldt 1854, ch. 3, para. 1, accessed online Monday, March 5, 2012. I learned about Humboldt's pro-liberty arguments from G. H. Smith 1991-D, 207.
123. Romanticism being the preeminent philosophy of nineteenth-century Germany is noted in I. Berlin 1980b, 19.
124. A. Padgen 2013, 382–86, points out that Hamann and Herder were influenced by Kant but believed that Kant contradicted himself in believing humans were individuals, and that they also believed that by denying that sensory evidence confirmed the individual separateness of each person, they were taking Kant's epistemology to its logical conclusion. *Ibid.* also points out that Hamann and Herder pioneered in the German romanticist movement. I. Berlin 1998a, 242–44, points out that Fichte similarly believed that he was taking Kant's epistemology to its logical conclusion by saying that since sensory evidence provides no objective knowledge, it follows that there is no objective knowledge that each person is an individual. Consequently, points out *ibid.*, Fichte believed he could justly argue that it was theoretically possible that all human beings share the same collective consciousness; sensory evidence to the contrary be damned.
125. I. Berlin 1980b, 184–87.
126. I. Berlin 1998a, 242–44.
127. I. Berlin 1980b, 184–87 with respect to Hume's anti-induction argument influencing Jacobi's collectivism; and I. Berlin 1998a, 242–44, with respect to Kant's anti-induction argument influencing Fichte's collectivism.
128. A. Padgen 2013, 386.
129. Fest 1974 U.S. version, 96.
130. A. Herman 1997, 41.
131. Qtd. by T. Kealey 1996, 314.
132. A. Herman 1997, 41.
133. C. Webber and A. Wildavsky 1986, 314–15.
134. Fest 1974 U.S. version, 96.

135. G. L. Mosse 1964, 15. T. P. Hughes 1989, 293, also remarks upon the anti-industrialism popular in early-twentieth-century Germany.
136. G. L. Mosse 1964, 19–23,108–09, 121.
137. A. P. Shepherd 1983 U.S. version, 70–73, 200–03.
138. L. F. Edmunds 1982 2d ed., 15.
139. R. Grunberger 1971, 44–45.
140. G. L. Mosse 1964, 44–45, 172–74, 180, 26.
141. G. L. Mosse 1985 trade paperback, 100.
142. G. L. Mosse 1964, 43, 55.
143. E. Diederichs 1936, 84, translated into English and quoted by G. L. Mosse 1964, 56.
144. G. L. Mosse 1964, 288. This influence can be seen in Otto Strasser's *Fourteen Theses of the German Revolution* from 1929.
145. P. Staudenmaier 2013.
146. Fest 1974 U.S. version, 101. Galbraith 1987, 92, admits that Adam Müller (1779–1829) was a German romanticist and advocate of governism who had an influence on Nazi ideology and Adolf Hitler. *Ibid.* also observes the influence of Friedrich List on Germany's nineteenth-century governist ideology.
147. A. Padgen 2013, 12.
148. Orwell 1968f, 118.
149. Mowry 1958, 22; and P. Watson 2010, 318–326. This is also admitted in Galbraith 1987, 165, 292.
150. Zakaria 2003, 66.
151. E. P. Cubberley 1934 rev. expanded ed., 449–454; and P. Watson 2010, 316–18.
152. Ekirch 1974, 24-25. Mowry 1958, 22, points out the nineteenth-century German influence over Richard Ely and Simon Patten.
153. D. Ross 1991, 165, points out Hegel's influence over Dewey.
154. For information on Dewey, see D. Ross 1991, 155–165; and P. Watson 2010, 539–541. Mowry 1958, 22, and D. Ross 1991, 105, point out that Richard T. Ely considered nineteenth-century Germany his model for the ideal sort of political system for the USA. Edward A. Ross himself studied in Germany and was influenced by its dominating turn-of-the-century political philosophy. The information on Simon Patten is from T. Russell 2010, 248. Ekirch 1974, 24–27, also admits that these Progressives were influenced by nineteenth-century big-government thought in Germany.
155. T. Kealey 1996, 349.
156. G. L. Mosse 1964, 157, citing L. Gurlitt 1906, 98.
157. See W. O. Henderson 1983. Galbraith 1987, 92, observes the influence of Friedrich List on Germany's nineteenth-century governist ideology.
158. G. L. Mosse 1964, 247.
159. The DNVP qtd. by G. L. Mosse 1964, 248.
160. M. Lind 2005, 244.
161. D. Ross 1991, 38–39.
162. W. Wilson 1968b, 365–67. Zakaria 2003, 66, brought this statement of Woodrow Wilson's to my attention.
163. D. Ross 1991, 236.
164. A. Herman 1997, 306–09. The rest of the book is how these ideas filtered into leftist thought in the USA and France throughout the late 1960s onward.
165. A. Herman 1997, 420.
166. The political Left's transition from that of the pro-industrialization, pro-technology Old Left to the anti-industrialization, anti-technology New Left of the 1970s is documented in A. Herman 1997.
167. A. Padgen 2013, 19–20.
168. A. Padgen 2013, 19–20, quoting J. Gray 1997.
169. Kaczynski 1995, http://www.washingtonpost.com/wp-srv/national/longterm/unabomber/manifesto.text.htm, accessed Monday, March 25, 2013.
170. T. Kealey 1996, 316.
171. McGill University classics professor Anne Carson in TV documentary *The Nobel Legacy*, aired May 1994, qtd. in D. R. Herschbach 1996, 12.
172. P. R. Gross et al., eds. 1996, trade paperback, 207.
173. M. Specter 2009, 1–2.
174. Qtd. by M. Specter 2009, 2.
175. K. A. MacDougall 1996, accessed online Monday, June 9, 2014.
176. W. M. Hern 1993, accessed online Monday, June 9, 2014. The papers of Kent MacDougall and Warren Hern came to my attention through Diana Hsieh, "Do You Want to Reduce Your Carbon Footprint to Zero?", *Noodle Food*, July 5, 2012, http://www.philosophyinaction.com/blog/?p=6721, accessed Sunday, July 9, 2014.
177. Fred Pearce, "Why Are Environmentalists Taking Anti-Science Positions?", *Yale Environment 360*, October 22, 2012, http://e360.yale.edu/feature/why_are_environmentalists_taking_anti-science_positions/2584/, accessed Monday, June 9, 2014.

178. Stephen Tindale, "Who Is Anti-Science?", *Prospect*, May 20, 2000, https://www.prospectmagazine.co.uk/magazine/whoisantiscience/, accessed Monday, June 9, 2014.
179. T. Kealey 1996, 312.
180. Zubrin 2012, 196–97.
181. N. Branden 2009, 20–21.
182. Other Objectivist thinkers notice that anti-reason, and anti-individual, anti-liberty philosophies gained renown in the West throughout the 1800s, particularly in Germany, even as relatively unprecedented liberalization, wealth creation, and advancement in knowledge took place in the West this same duration. See Peikoff 1993a, 100–03; Peikoff 1993b, 454–57; Rand 1961c, 23–44; Rand 1963e, 18; and Rand 1990b, 6.
183. The existence of this time lag is indirectly acknowledged in the discussion in Peikoff 1993a, 298–302.
184. R. Grunberger 1971, 454–55.

TEN

Extinction of the Social Darwinism Canard

Did the Nazis Vindicate the Controlled Economy?

Some critics may warn that we should not be too hasty in writing off Robert Proctor's thesis that some of the Nazis' welfare-state policies could be worthy of imitation in the USA. Some progressive academicians contend that, despite the evils of the Nazis' racism and genocide, Hitler's totalitarian shackles on the economy nonetheless drastically improved living standards for every Aryan in Deutschland.[1]

Such an apologia for the Nazis' command economy escapes the mouth of W. E. B. Du Bois. In spite of his censure of the Nazi persecution against the Jews, he nevertheless paid these compliments to the regime on December 5, 1936. "Germany has food and housing, and is, on the whole, contented and prosperous. Unemployment in four years has been reduced from seven to two million or less. The whole nation is dotted with new homes for the common people, . . . and new public works of all kinds. Food is good, pure and cheap. Public order is perfect, and there is almost no visible crime."[2]

George Orwell likewise bought into such nonsense. Although Orwell loved democratic socialism and hated fascism, he found in fascism the benefit that it "borrows from Socialism just such features as will make it efficient." To him a fascist state "can solve the problems of production and consumption. . . . The State simply calculates what goods will be needed and does its best to produce them. The mere efficiency of such a system, the elimination of waste and obstruction, is obvious." By contrast, "private capitalism—that is, an economic system in which land, factories, mines and transport are owned privately and operated solely for profit—does not work. It cannot deliver the goods." The reason why fascism is more efficient that laissez faire is that a "planned economy is stronger than a planless one." In the end, "However horrible" Naziism "may seem to us, it works. . . . British capitalism does not work..."[3] Michael Parenti, too, proclaimed that the Nazi economy was one of "low business taxes" and "high profits."[4]

In what appears to be a misguided effort to rescue the reputation of central planning of the national economy by the government, a number of progressive scholars continue delivering such appeals. As the vast majority of governist professors presently concede that the Soviet economy was a failure, they say that this was not due to the impracticability of central planning *per se*. Rather, they pontificate, the mismanagement of the economy under the Soviets and the North Korean government are isolated cases, and the Nazi example confirms that a State can still achieve prosperity through governism. A triumph in such social engineering, they pronounce, involves following the correct procedures that the Soviets did not abide. But some people know better. Paul du Gay of the Copenhagen

Business School admits, "The image of Nazi Germany as an efficient, bureaucratic machine . . . has been superseded in historical researches . . ."[5]

The writers who proclaim the Nazis' efficiency often give life to the misleading assertion that the Third Reich achieved a low unemployment rate. Germany's unemployment rate dropped from 34 percent in January 1933 to a still-high 14 percent in January 1936.[6] And that 14 percent rate is understated. University of Sydney economic historian Stephen Roberts, who resided in Germany in that period, elucidates that, back then, the Nazis' official unemployment statistics did not count persons whom the Nazi State targeted for persecution and official discrimination. These parties "lost their jobs" by State decree and were "cut off from relief; such persons do not appear in the official figures of unemployment. The refugees are ignored." The Third Reich actually cheated in other ways to have it appear as though unemployment had dwindled. The Nazis conscripted young adults, forcing most men into the army, and also coercing them into various civilian occupations through the Reich Labor Service, which undertook tasks similar to FDR's Civilian Conservation Corps. By 1939 the Third Reich conscripted laborers to attend to jobs where the State believed there to be a shortage of needed workers.[7] Other people were hired for public-works projects not on account of their qualifications but simply to provide Hitler with an excuse to boast about the high employment rate.[8] Stephen Roberts, adds, "Half a million women have been taken off the labour market in the last four years by means of the marriage allowance paid by the Government to entice them away . . ."[9] In actuality, the German economy began to recover from the Great Depression *before* Hitler came into power,[10] and, after gaining dictatorial status, the Führer wrecked the economy and sent it back into turmoil.

Ian Kershaw slays—forevermore—the myth that the Deutsch despot was competent at assisting in the financial standing of his Aryan subjects. "A summary of price and wage levels prepared for Hitler on 4 September 1935 showed almost half of the German workforce earning gross wages of eighteen Reich Marks or less per week. This was substantially below the poverty line. The statistics then went on to illustrate that a family of five—including three children of school age—existing on the low wage of even twenty-five Reich Marks a week earned by a typical urban worker and living on an exceedingly frugal diet could scarcely be expected to make ends meet. Wages, then, remained at the 1932 level—substantially lower than the last pre-Depression year of 1928 in the much-maligned Weimar Republic. . . . Overall living costs were higher by 5.4 per cent."[11] Richard Grunberger also describes the nightmarish regulations: the State "froze all chain stores at their existing size, forbade the establishment of new ones, and prohibited them from providing such services as shoe-repairing, barbering, baking and catering." Furthermore, the regulations "combated illicit cut-rate competition (*Schwarzarbeiten*) . . ., by police raids on workshops. Unemployed 'black workers' caught in this way were debarred. The establishment of all new shops and artisan enterprises required official permission, which was made dependent on local need and the applicant's professional, personal and political suitability."[12]

At the same time, Hitler centrally organized German industry in a manner similar to that of FDR's National Industrial Recovery Act. By July 15, 1933, the Nazi Ministry of Economics enacted laws compelling Germany's larger business firms to consolidate into cartels, either forming new cartels or joining already-established ones.[13] Moreover, by February 27, 1934, these cartels were, according to William Shirer, "put under the control of the State."[14] The cartelization was exacerbated by steep licensing rules written to make the remaining small businesses lose their balance. Grunberger adds, "From 1935 onwards it became compulsory for everyone wanting to open a new artisan enterprise to pass the Master's Examination in his craft. This exam had previously been optional. In 1931 less than a third of all practicing artisans held master's certificates—but after 1935 all younger nonqualified owners of craft enterprises were obliged to qualify for a certificate of employment." Owing to the regulations, the number of wholesalers in the nation fell by 33 percent. From 1935 to 1939, the number of radio retailers declined by more than half. Also in 1935, the State compounded the burden by insisting that the small businesses report every transaction to the regime. "Ledgers for incoming and outgoing goods ('supplies books' and 'customers' books') now enabled officialdom to subject all enterprises to searching tax control and to investigations of economic viability. This process was intensified by the ruthlessness of the fiscal authorities in collecting tax debts and refusal of the licensing authorities to grant new concessions to small shops wishing to diversify their selling lines."[15]

William L. Shirer writes that small businessmen witnessed the destruction of their enterprises, which plunged them "back into the ranks of wage earners. Laws decreed in October 1937 simply dissolved all corporations with a capital under 40,000 dollars and forbade the establishment of new ones with a capital less than 200,000 dollars. This quickly disposed of one fifth of all small business firms." Shirer continues, "At the head of an incredibly complex structure was the Reich Economic Chamber, whose leader was appointed by the State, and which controlled seven national economic groups, twenty-three economic chambers, one hundred chambers of industry and commerce and seventy chambers of handicrafts. Amidst this labyrinthine organization and all the multitude of offices and agencies of the Ministry of Economics and the Four-Year Plan and the Niagara of thousands of special decrees and laws even the most astute businessman was often lost, and special lawyers had to be employed to help a firm function."[16] From 1876 to the 1920s, Bismarck's irrational influence had already wrought a trend of massive government-backed cartelization of industry.[17] Still, Hitler further encouraged and ultimately exacerbated this trend. Just as Franklin D. Roosevelt had forcibly cartelized farming under the Agricultural Adjustment Act, Hitler likewise cartelized farming in Germany to a far severer degree.[18] This resulted in Germany's catastrophic food shortage.[19] In fact, notes Ian Kershaw, by late summer in 1935 "the terms 'food crisis' and provision crisis' were in common use."[20] The price of wheat jumped by 15 percent, butter 40 percent, eggs 50 percent, meat 50 percent, and potatoes 75 percent.[21] Kershaw reports that increases of "33, 50, and even 150 per cent had been reported" for some other foodstuffs. Seeking to downplay the crisis, Germany's official statements reported that food prices had risen a mere 8 percent between 1933 and 1935.[22] That Du Bois preposterously cheered that Germans had enough "food" to be "contented and prosperous"—and that it was "cheap"—signifies Du Bois's ignorance of their actual situation. A number of factors contributed to the food shortage, each of which relates to governism. First, as FDR encouraged the cartelization of the USA's agribusiness with the Agricultural Adjustment Act, Hitler likewise instituted laws to strengthen Germany's agribusiness cartel. Hitler consequently ratified government regulations against the importation of food, thereby prohibiting foreign competition against German food producers. Thus German consumers paid higher prices for nourishment than they otherwise would have.

The food shortage may still seem strange, in light of the fact that, prior to the war, Germans Fritz Haber and Carl Bosch had developed the Haber-Bosch method for cultivating more crops. But this shortage is less surprising when one considers the myopia and misplaced priorities of those in charge of government-managed economies. The ammonium nitrate synthesized in the Haber-Bosch process is useful not just in fertilizer, but also in munitions. Tom Standage of *The Economist* magazine reports that the majority of synthetic ammonium nitrate produced was used for weaponry, and there was hardly any left over for German farming. Hence, the results at harvest time were pitiful.[23] As nonproductive as this absence of ammonium-nitrate fertilizer was, it was just fine by Nazi officers Rudolf Hess and Heinrich Himmler. As adherents to Rudolf Steiner's ideology of biodynamic (organic) farming, Hess and Himmler tried to institute restrictions on the usage of this and other synthetic fertilizers.[24] The two prioritized their own environmentalist dogma above any concerns about whether the population was fed.

Accordingly, the German Labor Front implemented government regulations to micromanage the terms of how and when an employee would sell his time and labor. "The salaried worker," comments George Mosse, "could not escape the tentacles of the Labor Front, for it controlled hiring and firing, workmen's compensation and insurance, as well as care for the elderly and disabled workers." Despite the alleged benefits provided by the welfare state, the rules were insufferable. To elucidate, "The wages of salaried workers were frozen and their ability to move from one job to another was rigidly controlled."[25] Aaron Wildavsky and Carolyn Webber concur, "Government-operated labor exchanges in effect dictated to workers where, for whom, and how long they could work."[26] Mosse adds that "there were other financial sacrifices, as the examination of the Nazi taxation system shows. . . . The rise in the cost of living has to be considered in relation to both the wage freeze for salaried employees and the tax structure."[27]

In observance of the data, Raymond Fletcher—a British Member of Parliament who was educated in Germany—disputes those who "think of Nazi Germany as a model of efficiency. . . . The Third Reich as an example of military or industrial efficiency is a ludicrous myth." For instance, in the Ruhr, "the Nazis continued to turn out tanks and armored personnel carriers well after they no longer could find rail transport to take them away. They used their scientists very poorly. Of 16,000 inventions of military significance made during the war, few ever actually got into production because of the prevailing inefficiency."[28]

Despite the casuistry from his best-selling textbook—which we criticized in Books One and Two—governist economist Paul A. Samuelson confronts the data concerning the Axis Powers. Contradicting Orwell, Samuelson says plainly, "Fascism turns out in history not to deliver the goods."[29] It is the case that the Nazis were no more successful than the Soviets in achieving economic efficiency through the elimination of the free market. Recall my point from Book One that a nation achieves self-destruction to the extent that its government steadfastly assumes itself to have statutory ownership over its own citizens. That principle applies just as accurately to the National Socialist dictatorship as it does to the United Soviet Socialist Republics.

Here we return to the topic of whether Hitler was a stooge of big business. Recall that William Shirer's *The Rise and Fall of the Third Reich* helped give currency to the idea that industrialists looked favorably upon the Nazi Party, supposedly on account of the Nazis being pro-capitalist. This very same Shirer, in the very same *Rise and Fall*, reports that once Hitler gained power in 1933, businessmen became "mere cogs in a war machine . . ." Their work was "circumscribed by . . . restrictions" and there were too many "forms to fill out . . ."[30] Aaron Wildavsky and Carolyn Webber report, "A business reluctant to comply with government preferences could be quickly brought in line through cuts in imported raw materials and reduced labor allocations."[31]

In 1939, Nazi economics minister Walter Funk ashamedly conceded that "official communications" with the national government's regulatory agencies "now make up more than one half of a German manufacturer's entire correspondence . . ." He also reported that "Germany's export trade involves 40,000 separate transactions daily; yet for a single transaction as many as forty different forms must be filled out."[32]

Thus Shirer reports, "Buried under mountains of red tape, directed by the State as to what they could produce, how much and at what price, burdened by increasing taxation and milked by steep and never ending 'special contributions'" to the Nazi Party, Germany's industrialists grew frustrated.[33]

This helps explain much of the Nazis' ineptitude. The steel industrialist Fritz Thyssen, who was once among the Nazi Party's most avid supporters in the business community, fled the country at the onset of World War Two, realizing, "The Nazi regime has ruined German industry."[34]

We must likewise dispense with the myth that at least Benito Mussolini's fascism "made the trains run on time." Recall from chapter 7 our quotation of the passages from Ida Tarbell's 1939 memoir that gush about Mussolini getting the whole nation back to productive work. British historian Denis Mack Smith exposes what really happened. Italy's rail system had fallen into disrepair throughout World War One. Throughout the 1920s the country oversaw a rebuilding of its train system, but this rebuilding began prior to Mussolini's 1922 coup. Moreover, when Italians who had lived under Il Duce's rule were interviewed decades later, they distinctly remember the trains consistently running late.[35]

Additionally, a note must be made about the Nazis' State-managed breeding facilities where women were kept for eugenicist purposes. Far from being pinnacles of efficiency, these facilities were poorly managed. As a consequence, their infant mortality rate was twice that of the general population where people still bred as they chose.[36]

Some German newspapers acknowledged that the Third Reich's price controls instigated the misallocation of resources. The Nazi State's control of the economy resulted in retail stores being loaded down with a surplus of goods that consumers did not want, while, conversely, there were shortages of goods that consumers greatly desired. On July 27, 1939, *Das Schwarze Korps* reported that the State was "Throttling the Retail Trade."[37]

That same year some German merchants gained the gumption to defy these edicts openly. These merchants, reports Richard Grunberger, "circumvented the price freeze by offering lower-quality goods at the prescribed charge, or coupled the non-profitable sale of essentials with the advantageous one of nonessential items—and thus fell foul of the law."[38]

"However," George L. Mosse adds, "punishment for violations of the price-control regulations was instantaneous and severe, as the butchers' guild in Bockum-Höve (a small town near Hamm in Westphalia) discovered. Such united resistance is a measure of desperation of a whole section of the retail trade, coming as it did at a time when the Third Reich was already six years old and its terror and cultural control in full swing."[39] As one example, Grunberger discloses, "A fruit wholesaler was fined 10,000 marks for coupling transactions, with the *Schwarzes Korps* clamorously demanding that he be 'shortened by a head' if the offense should recur." Thus, the Nazi government's threat to destroy noncompliant businesses "was not an empty one. By 1939 the 350,000 supernumerary enterprises—as estimated in 1936—had been reduced by roughly one half" as a consequence of the government shutting down the offenders.[40]

The butchers and other merchants tried to circumvent the regulations through other methods. They printed two different prices on the tag of every item. "On one side," reported the January 11, 1939, *Frankfurter Zeitung*, the tags "carry the correct price, and on the other side the illegal higher price." During inspections by Nazi officials, the merchants "simply turn the tags around . . ." Hence the inspectors would catch sight of nothing but the legally approved price, no one being the wiser. The newspaper therefore recommended to officials, " . . .Special care should be taken in the examination of bills and bookkeeping methods generally."[41] Sometimes, reports Grunberger, the merchants violated the law against their own intentions, being unable to adhere to so many edicts at once. These retailers "circumvented price regulations quite involuntarily since they were hard put to it to orientate themselves amidst a welter of complicated directives and constant variations in the quality of the goods supplied to them."[42]

In the end, the Third Reich successfully squashed the insubordination.[43] Here is the text from *Frankfurter Zeitung* in that same year, October 15: "members of the butchers' guild refused to abide by the prices for meats as fixed by the county administrator"—this county official is called the *Landrat*—"at the behest of the Government Price Control Office. The spokesman for the butcher told the police: 'We won't let the county administrator set the prices for us.' He was thereupon ordered by the chief administrative officer of the region," the *Regierungspräsident*, "to be taken into custody and was lodged in the police jail at Recklinghausen."[44] At one point, one grocer told the Nazis outright, "I have myself long contravened price regulations and cannot report any colleague for the same offense."[45]

Recall my previous citation of historian Götz Aly about how the Nazis favored their holistic healing methods over scientific "Jewish" medicine. This same Götz Aly presents evidence that the Nazi regime was temporarily able to sustain itself by living off of the assets it had expropriated from wealthy Jews in the 1930s. Such assets would run out eventually. The Nazi economy was ultimately unsustainable, as it allocated its most important resources toward the war effort, leaving hardly anything for day-to-day civilian sustenance. Had there been no world war, the government regulations would have brought German businesses to stagnation at best and collapse at worst.

We therefore find it disturbing that Californian political scientist and anti-capitalist Michael Parenti asserts the exact opposite of the truth: "After Adolf Hitler took state power in Germany in 1933, he set about establishing a . . . reactionary government that . . . drastically reduced wages, eliminated worker benefits," and "ignored occupational safety" regulations. Parenti would have us believe that Hitler's mission was to "cut taxes for the very rich" as he "privatized various state enterprises."[46] As I mentioned in Book One, there was a brief point in the Weimar Republic's history, prior to the ascension of Hitler, where, during the Great Depression, the Weimar Republic nationalized various industries. And, as I also mentioned in Book One, when Hitler converted those enterprises back into private property, those assets were privately owned in name only. In practice they remained controlled by the State.

Historian Joachim Fest thus concludes that Hitler "was not the last desperate gasp of dying capitalism, as a good many ideologists"—Michael Parenti implicitly included among them—"have described him."[47] No longer let business-haters tarnish capitalism's reputation through such a scurrilous association. Likewise, it should finally be comprehended that Naziesque eugenics has nothing to do with free enterprise. Shamefully, that does not stop Edwin Black and other governists from crying that free-market economics will indeed revive a Nazi-styled, Aryan-centered eugenics. They say this will happen when Americans are finally able to utilize prenatal genetic engineering to provide our future children enhanced immune systems and upgraded aptitudes.

Is Biotechnology the New Eugenics?

We humans opt for specific complex actions as a result of our own willful choices. Book Two outlined how those actions affect a person's own likelihood of being able to propagate her own genetic legacy beyond the next two generations. It then stands to reason that conditioning and individual human choices actually influence the biological evolution of the human species. Compounding that, in the near future, improvements in biotechnology may bestow the citizens of industrialized countries with the opportunity to play a more direct role in affecting the human race's biological evolution. It will be through a process that has been dubbed "germline genetic engineering" or "reprogenetics."[48] Alternately, for this same process I will also employ the label *human enhancement*—simply "enhancement" for short. Scientifically, the procedure would go as follows.

1. A husband and wife decide to apply *in-vitro* fertilization, in which some of the wife's eggs are removed from her body. The removed egg is artificially fertilized by her husband's sperm.
2. While it is still in a Petri dish, scientists genetically alter the fertilized egg to immunize it, prenatally, from developing genetic diseases in the future. They can also provide this embryo with the potential to develop specific aptitudes as it matures as an adult human being.
3. At least one of the altered embryos is then implanted back into the wife or into a volunteering surrogate mother.[49]

There are other technological methods of human enhancement. An adult human being might decide to merge his flesh with some machinery to enhance himself, and become a cybernetic organism ("cyborg" for short).[50] Nanotechnology and the aforementioned biotechnology may also be used. Noticing that eugenics had come to be associated with the coercive government policies that he disavowed by the end of World War Two, World Wildlife Fund founder Julian Huxley—brother to Aldous Huxley—came up with a different term for peaceable human enhancement. He called it trans-humanism.[51]

In one important respect, I deem the term *transhumanism* misleading. Those, such as Californian social activist Max More, who proclaim themselves "transhumanists," obsess over how they wish for people to employ technological enhancement to such a degree that they evolve into an entity that is longer human, but somehow better than human. To me, this reflects a misunderstanding of what it means to be human. As we recall from Books One and Two, technology is merely the exercise of one's rational faculty to reshape the natural environment to improve one's own well-being. Thus, to employ technology peaceably in order to engineer a better life for oneself, one's children, and one's descendents is not a feat that transcends human nature. No matter what biochemical or physiological changes technology may administer to our descendents, to the degree that they continue to be sapient they commensurately continue to be human. That our descendents may look different from us does not render them nonhuman; it simply renders them a new sort of human, and there is nothing wrong with either they or us being considered fundamentally human *qua* human. Anyhow, I support the right of any grown-up to participate in enhancement to the extent that the participation abides by the principles of Lockean freedom.

Should a couple choose to produce changes in the genetic code of their children, it will not follow that such changes will be a permanent fixture in the germlines of their descendents. Biologist Mario Capecchi proposes a private method of caution. As the default position is for humans to inherit 43 chromosomes, Capecchi suggests that as an embryo is being genetically modified, an extra chromosome be added. Within this chromosome is the genetic coding for the attributes that the parents desire to have engineered into their child. As Capecchi understands that that child may find these features to be unsatisfactory, he has the option of having these same genes "switched off" when he himself chooses to engineer his own child's embryo. That is, although he carries the genes for the attributes that displease him, he can ensure those same attributes will not appear in his own child.[52]

Peaceful human enhancement has enormous ramifications as far as the concept of biological evolution is concerned. The term *natural selection* has always been problematic in one respect. It is inappropriately anthropomorphic to say that ontological nature "selects" particular genes for survival in the wilderness, given that the word *select* refers to a conscious decision rendered by an individual's volitional consciousness. A private volitional consciousness is something that the Earth or wilderness, as a single unit, has not been proven to possess. For that reason, Gabriel Dover suggests a more accurate term for what happens in the wilderness than *natural selection—natural sorting*.[53] But once reprogenetics establishes itself, a form of natural selection that is literal "selection" will then direct evolution. This is because romantic couples will exercise their volition to select, most consciously, which genetic traits of theirs are replicated and which are not. Such a technology would be a boon to couples that worry about passing on their hereditary health ailments. The recombination of an embryo's DNA can help that embryo develop into a child who is immune from cystic fibrosis, sickle-cell anemia, diabetes, heart disease, allergies, and heritable forms of cancer.

Such human enhancement is private and consensual. No spoliation occurs. No would-be parent is being forced into this. Nobody is being murdered, coercively sterilized, or barred from peaceably traveling from one country to another. To be sure, the embryo that will become the couple's child does not itself consent to the genetic alteration. But it is ontologically impossible that a prenatal entity can offer or withhold consent to anything its parents do that may affect the circumstances into which he is born. I, for instance, never consented to having been born, though I am very grateful that this event occurred.

Some Ashkenazi Jewish couples, in which both partners carry the gene for Tay-Sachs disease, know beforehand that should they choose to have children, there is a 25 percent chance of their child being born with Tay-Sachs disease and dying within the first five years, and a 50 percent chance of their baby being a carrier for the gene.[54] Despite this, no law proscribes them from having children anyway. This is despite the consideration that, should any of their children be born with Tay-Sachs, that child did not consent to having been conceived in the first place. In fact a proposal for any such law forbidding these couples from conceiving children would be properly criticized as an implementation of governist eugenics. Should it be legal for a man and woman to try to have children subsequent to being informed that there is a one-fourth chance of their child having Tay-Sachs disease, when that child could not consent to it? If the answer is yes, it should be legal for a man and woman to enhance their child's genome when he is still in the prenatal stage, when the embryo could not have offered or withheld consent. My point should especially be conceded if one does not consider circumcision a form of spoliation, given that no baby has the contractual competency to offer consent to this particular permanent medical operation. Provided that one understands that it is not spoliation for parents to enhance their future child while he is in embryonic form—when it is ontologically impossible for him to offer or acknowledge consent—it should be clear that private parents arranging consensual agreements with doctors to enhance their future children's genes does not spoliate anyone's Lockean rights.

Princeton University microbiologist Lee M. Silver (b. 1952) dispenses some good advice on the ethics of this issue. Just as parents exercise an ethical prerogative to endow their children with advantages in life by sending them to excellent private schools, Silver reasons, they also possess a right to imbue their children with corresponding advantages at the genetic level. "There is no difference," he advises, "between 'after birth' and 'before birth' if the goal is to give your kids advantages. What difference does it make *when* you give those advantages?"[55] (emphasis his). James Dewey Watson (b. 1928)—one of the Nobel Prize-winning geneticists who co-discovered the double-helix structure of DNA—provides other helpful comments. Like Silver, Watson is frequently faced with the argument that germline genetic engineering would give wealthy patents an "unfair advantage over everyone else." In response Watson comments that there are already people being "born with better abilities than others, and we don't say they shouldn't be born."[56]

Dreadfully, given that private, free-market reprogenetics involves improving humans at the genetic level, hordes of social activists decry it as a revival of eugenics. They predictably liken it to the work of Charles Davenport, Karl Pearson, and the Nazis. Anti-biotech activist and reverend Patrick Mahoney, for one, condemns it as "horrific" and "dangerous." In his opinion, to "think that a parent would do that shows how far our culture and society has fallen. Haven't we learned our lesson from just a generation ago in Nazi Germany? Where this kind of . . . designing your children? I mean, what does that smack of?" Do I want my child "to have blue eyes, blond hair? Do I want my child to have a certain IQ? It's eugenics and we have to say clearly and now, 'We don't engineer human life.'"[57]

Edwin Black introduces that equivocation in his history of eugenics, *War Against the Weak*—the very same book in which he falsely incriminates Herbert Spencer as governist eugenics' intellectual forefather. Remember that Black's book portrays governist eugenics, not as what it really is—governist progressivism taken to its logical conclusion. Nay, Black depicts eugenics as a one-dimensional capitalist conspiracy that wealthy industrialists orchestrated to exterminate the underprivileged. Also bear in mind that Black conceives capitalism to be a heartless system of social stratification. In this book's conclusion, Black infers that in America's near future, germline genetic engineering, which he pejoratively terms "newgenics," will most likely commence in the mostly free marketplace that he considers horrid. He resultantly concludes that both the governist eugenics of the past and the free-market reprogenetics of the future must be equally elitist and therefore malicious. Bio-technicians, he proclaims, "have inherited the spoils" of the governist eugenicists' "war against the weak." He then grieves that germline reprogenetics

> will not come cheap. Only the affluent who can . . . afford personalized elective health care will be able to afford expensive genetic correction. Hence, economic class is destined to be associated with genetic improvement. If the genetically "corrected" and endowed are favored for employment, insurance, credit and other benefits to society, then that will only increase their advantages. But over whom will these advantages be gained? Those who worry about "genelining," "genetic ghettos" and a "genetic underclass" see a sharp societal gulf looming ahead to rival the current inequities of health care and judicial systems. . . . Some of America's leading thinkers on genetic evolution believe that within a few hundred years, the world will indeed be divided into the "genetically endowed"—or "GenRich" as some call them—and those who will serve them, almost like the worker bees Davenport envisioned. . . . This is not the philosophy as much as the *raison d'être* of newgenics.

Here it would seem that Black is simply badmouthing laissez faire. It does not take much time, though, before he once again conflates a market economy with a society wherein a plutocracy practices governism.

> Mass social engineering is still being advocated by eminent voices in the genetics community. Celebrated geneticist James Watson...told a British film crew in 2003, "If you are really stupid, I would call that a disease. The lower 10 per cent who really have difficulty, even in elementary school, what's the cause of it? . . . So I'd like to get rid of that, to help the lower 10 per cent." . . .
>
> Following in the footsteps of Galton, who once amused himself by plotting the geographic distribution of pretty women in England, Watson also told the film crew, "People say it would be terrible if we made all girls pretty. I think it would be great." . . . That is nothing less than a return to the campaign to create a master race—but now aided by computers, digital communications and a globalized commercial infrastructure to accelerate the process.[58]

It is particularly easy to smear Watson as the leader of a cabal of plutocrats conspiring to resurrect governist eugenics. Indeed, the sales of Watson's memoir, *The Double Helix*, from 1968 onto the early 1970s have earned him a net worth of at least one million U.S. dollars.[59]

The Non-Threat of Lockean Reprogenetics

Once again, to uncover the absurdity of Black's and Mahoney's conflation of governist eugenics with private, consensual reprogenetics, one must recall that governist eugenics consisted of governments implementing the following regulations.

- Coercively sterilizing someone, by force of law, who may have had mentally ill ancestors or relatives, or who may carry any other heritable genes the State does not want passed on.
- Exercising government force against immigrant Juan for setting foot on native-born countryman Rick's land at Rick's own invitation. The governist eugenics rationale would be that immigrant Juan, being ethnically different from much of the host country's native-born population, might produce a child with somebody among the native-born population. This interracial mixing would supposedly pollute the native-born population's gene pool with the DNA of his allegedly inferior ethnicity. As we mentioned in chapter 2, that is called *miscegenation*.
- Laws barring Jack and Lisa—both of whom are of differing racial backgrounds—from being on the same plot of real estate as one another at the same time. This ban would be in effect despite the land's owner consenting to both Jack and Lisa being on his land simultaneously. This is Jim Crow segregation and apartheid.
- Laws treating at least one genetic ethnicity differently from others.
- Government-approved murder of one or more persons on the grounds that he or she carries genetic traits that the State does not want delivered to future progeny. This is the governist eugenicist policy of ethnic cleansing and, most famously, the Nazis.

Attempts by human beings to control what genes they endow to future generations of their society, should be illegal only insofar as they spoliate someone. When Black pillories James D. Watson for implementing Francis Galton's plan to utilize "social engineering" to "create a master race," Black irresponsibly neglects to acknowledge a specific contrast. Whereas Galton prevailed upon the State to spoliate individuals on behalf of restructuring England's gene pool in the manner that he preferred, Watson simply asks that the government leave consenting adults free to attempt to maintain the sort of families they want, peaceably. Watson voiced his actual view on the subject of reprogenetics thusly—". . . individuals should direct the evolution of their descendents; don't let the State do it."[60]

Black declined to mention that when Watson admitted that he wanted genetic engineering to raise IQ scores and make girls prettier, Watson was not proposing any law to force anything on anyone. Watson assumed that most parents would voluntarily opt to raise their child's IQ or make her good-looking if the opportunity presented itself. If his advocacy of spoliation is the reason for Francis Galton's treachery, then Black has more in common with Galton than Watson does. It is Black who demands that government spoliate individuals by dictating over what genetic advances they may or may not peaceably bestow upon their own children. Black concludes *War Against the Weak* by sermonizing, "It will take a global consensus to legislate against genetic abuse because no single country's law can by itself anticipate the evolving inter-collaborative nature of global Genomics. Only one precept can prevent the dream of twentieth-century eugenics from finding fulfillment in twenty-first-century genetic engineering: no matter how far or fast the science develops, nothing should be done anywhere by anyone to exclude . . . an individual based on his or her genetic makeup."[61]

Actually it should be legal for someone to avoid associating with others on the basis of their genetic makeup, provided that such shunning does not entail spoliation. Such a basis would be bigoted, but that would not be reason enough to spoliate the person doing the shunning. What the law truly needs to recognize is that every Pure Citizen has the same Lockean moral rights. This is regardless of whether or not someone possesses any genetic enhancements, or whether or not he is the result of reproductive human cloning. It was no doubt monstrous for governist eugenicists to try to alter society's overall genetic makeup by lobbying the State to spoliate peaceful individuals. But it would also be evil for anti-eugenicists to lobby the State to spoliate reprogeneticists who have exercised peaceful methods to modify society's overall genetic makeup. In answer to those who wish to criminalize consensual reprogenetics, Matt Ridley writes that if the State barred the option of human enhancement, that prohibition "would risk increasing the load of suffering in the world . . ." In his perspective, it would be "just as cruel to outlaw" the practice "as it would be to make it compulsory. It is an individual decision, not one that can be best left" to the government. Whereas Lockean reprogenetics "is about giving private individuals private choices on private criteria," governist eugenics "was about nationalising that decision to make people breed not for themselves but for the state. It is a distinction frequently overlooked" by such political collectivists as Edwin Black "in the rush to define what 'we' must allow in the new genetic world. Who is 'we'? We as individuals, or we as the collective interest of the state or the race?"[62]

James Watson correctly evaluates that what the governist eugenicists and anti-biotech governists have in common is their desire to legislate over other people's private decisions on how they may peaceably procreate. In defiance to each of those spoliative positions, the Lockean approach upholds the sanctity of individual freedom against any brand of governism. When it comes to choosing between what the governists call the common good versus the rights of the individual, Watson decides, "My way is always asking, 'What's good for the individual?' . . . It's best to let parents decide" whether they want to utilize reprogenetic techniques or not, and to what extent.[63]

The opinions of Ridley and Watson receive affirmation in that of economics journalist Virginia Postrel, who recognizes the equivocation between Naziism and Lockean reprogenetics to be "both ridiculous and, quite frankly, morally offensive. The Nazis marched people into death camps. They *murdered* people" (emphasis hers). Nazi eugenics "was a state-run program to create one model for the perfect German to be subordinated to the State." On the converse, Lockean reprogenetics entails "letting individual parents make decisions on behalf of the children" for whom they "love and care . . . And to compare that to Nazis is disgusting."[64] As Robert L. Sinsheimer, a biologist and chancellor emeritus at the University of California at Santa Cruz, articulates it, the old government-regulatory "eugenics would have required a continued selection of breeding of the fit, and a culling of the unfit." On the other hand, the new consensualist "eugenics would permit in principle the conversion of all the unfit to the highest genetic level."[65] Thomas R. DeGregori asks the pertinent question. "Who are less like the Nazis—. . . genetic engineers, or those who would straightjacket such sciences?"[66] Far from the individual parents employing germline genetic engineering being akin to Nazis, it is those who would have the State ban this practice who are exercising the force of law to dictate over what other people may or may not peaceably do with their own DNA. Sonia Arrison, the founder of a think tank called Singularity University, points out that it is "present-day opponents" of germline genetic enhancement who "want to use the power of government to force people into certain genetic categories. The authoritarianism that was once associated with racism and sterilization can now be associated with those propping up the status quo."[67] Inventor and former Microsoft software engineer Ramez Naam adds, "It's those who oppose individual and family genetic choice who have, in essence, decided that there's a certain 'correct' genetic heritage for humanity (the one we have today) and that the populace should not be allowed any choice in the matter."[68]

Interestingly, Watson was proudly socialist back when he co-discovered the double helix. But he presently states that the scurrilous equivocation between governist eugenics and Lockean reprogenetics—as well as the PC activists smears' against E. O. Wilson for writing *Sociobiology*—have ultimately "turned me away from the left."[69] Moreover, Watson had publicly described himself as a libertarian,[70] though his continued advocacy of tax funding for scientific research belies that label.[71]

Edwin Black is correct in one aspect. Inasmuch as the free market is permitted to exist in it, a future America may be populated by entrepreneurial fertility doctors who—for a price—will provide couples the service of applying *in-vitro* fertilization techniques and bioengineering to modify their children's DNA at the prenatal stage. The prenatal modifications will render the children stronger, quicker-witted, and healthier than they otherwise would turn out. Regrettably, Black flubs in his assumption that such services will exacerbate social stratification by fostering a wealthy, genetically enhanced over-class that towers above a low-income, genetically inferior underclass. An entrepreneur can increase his profits by widening his customer base, which usually requires that he sell his product or service to varied economic classes, rather than exclusively to the rich. When Henry Ford introduced the Model T in 1908, the average price of an automobile was greater than that of the average house being sold in the United Kingdom—three thousand U.S. dollars. Only the well-heeled could afford motor cars. But within eight years Ford reduced the average price of an auto by more than half.[72] An automobile could be bought in the USA in 1915 for 850 dollars[73] and then for 350 dollars in 1924.[74] Ford became a billionaire by adopting economies of scale to craft a product cheap enough for the middle class to afford. "Does anyone," writes Andrew W. Mellon, "question that Mr. Ford has made more money by reducing the price of his car and increasing his sales than he would have made by maintaining a high price and a greater profit per car," but by selling fewer cars?[75]

This rule applies to cases of modern medicine. At one point in history, no one but the wealthy could afford plastic surgery and LASIK eye surgery. Recently, technological innovations have lowered costs for surgeons to such an extent that these surgeries became affordable to the middle class.[76] The major reason for the drop in price is that health insurance plans seldom pay for these surgeries. That gives cost-conscious patients more of an impetus to shop according to price, and also more incentive for surgeons to slash their prices while maintaining quality service. In 1999, a few years subsequent to the advent of LASIK eye surgery, the procedure was priced at 2,100 dollars per eye. Soon afterward the price lowered to 1,600 dollars each. From 2001 to 2007, the price of the procedure per eye fell 20 percent.[77] Also of note, the inflation- and quality-adjusted price of treatments for heart attacks has dropped at a rate of 1 percent every year from 1983 to 1994.[78] A similar pricing trend has been found with respect to cataract surgery.[79]

Most likely, reprogenetic services will be expensive in the beginning. When economies of scale and technological improvements lower the cost of conferring reprogenetic services, more and more middle-class persons will be able to procure such services. This especially applies if this industry is left alone in a free market, not tax-subsidized by socialized health insurance. Thus, the increased social stratification Black pronounces inevitable would be improbable in a night watchman state. There would not be a single wealthy GenRich Master Race far above a genetically inferior, low-income GenPoor. There would be a wide spectrum of different strata of income levels and genetic enhancement levels, just as contemporary America currently has varying income brackets. Commensurate with the USA's status as home to a burgeoning financial middle class, the future holds both a gigantic financial middle class and a larger genetic middle class.

An attitude toward genetic engineering much healthier than Edwin Black's was displayed by the actor Christopher Reeve (1952–2004) in the final years of his life. Upon suffering from spinal cord injuries that paralyzed him, he acquainted himself with news on the latest efforts of genetic engineers to repair spinal cord tissues and thereby restore locomotive abilities to paraplegics. Although he was a progressive like Black, Reeve came to appreciate the benefits such technologies entailed. At the BIO industry conference in Boston, Massachusetts in March 2000, amidst picketing by anti-globalization, anti-capitalist activists, Reeve announced, "*Biotechnology* is a great big word. For people like me, it's got a simple definition. It's a great big word that means hope."[80]

Ultimately, the free-market reprogenetic revolution will make for an exciting new development in the phenomenon of gene-culture co-evolution. It is based upon the freedom that Spencer and Sumner upheld—not governist eugenics nor some fabled social Darwinism. To leave individuals free to raise and care for their own children peaceably, as they see fit, is the opposite of Naziism and the governist eugenicists' progressivism. Contrary to the social Darwinism myth, Naziism enacted the regulatory-entitlement ideals more than it did Spencerian economics.

I agree with Fereidoun M. Esfandiary, a.k.a., FM-2030. As I do, he takes issue with environmentalists and right-wing bioethicists who chastise genetic engineers for "arrogantly 'playing god.' They warn of dire consequences. These critics are absurd. We humans do not want to be god or to play god. We aspire to much more. God was a crude concept—vengeful, wrathful, destructive. We humans want to evolve beyond god."[81]

Where in the Historical Record—or Fossil Record—Is Social Darwinism?

Some members of the radical political Left grow irate when their critics point out parallels between (1) the policies they advocate and (2) those implemented by the Soviet Union. One of the disturbing ironies, though, is that these same left-wingers can rest comfortably knowing that nobody will detect any similarities between their favorite causes and the Nazis'. Indeed, they have convinced everyone that they, being on the extreme Left, must be the exact opposite of the Nazis on account of the Nazis' place on the extreme Right.

Yet consider the myriad social causes that are presently considered the hallmarks of the contemporary Western Left that were largely ignored by Soviet "left-wingers"[82] and embraced by the National Socialist "right-wingers." Among such causes are environmentalism, an organic produce-eating lifestyle, animal "rights," conserving endangered species, labor regulations, and opposition to synthetic chemicals in agriculture. This demonstrates the irrationality of attempts by such intellectuals as Richard Hofstadter to divide political allegiances between Left and Right. The issue is not whether someone is left-wing or right-wing; the issue is the degree to which any politician or political movement deems it acceptable for government regulations to spoliate individuals and their private property—forms of spoliation that seem to be popular among Democrats, Republicans, and political "centrists" and "moderates" alike. When it comes to political allegiances, observed President Ronald Reagan, there is, in the end, "no such thing as a left or right . There is only an up or down—up to man's age-old dream, the ultimate in individual freedom consistent with law and order—or down to the ant heap of totalitarianism," and those who would forsake "our freedom . . . have embarked on this downward course."[83] Comparable to Reagan, FM-2030 also expresses the right—uh, *correct*—idea on the matter. FM notices that, not merely is the Right "too conservative," but that, in one context, the Left is as well. Sure, left-wingers "still like to view themselves and are viewed by others as progressive." But in "the name of progress . . . the radical Left-winger" has consistently been one to "resist progress." The left-wingers "resist progress because the new breakthroughs" of free innovative enterprise, such as genetic engineering, "do not fit into their highly structured frameworks and confined goals." Any and every "breakthrough" in industrialization "is viewed as a threat" by legions of environmentalists. On such counts, the advocates of "Left radicalism masquerading in the name of progress are putting up the strongest resistances" to the market-based technological changes that could improve our environment. "What effrontery calling themselves progressive . . . . Those who do not believe in progress or in the future" should not arrogate to themselves the privilege "to call themselves progressive." Both FM-2030 and Richard Hofstadter say that left-wing eugenicists were too "conservative," but, as they apply the same locution, they do not level the same accusation. When Hofstadter asserts that left-wing eugenicists were ultimately conservative, he believes that, deep down, they were still too capitalist. By contrast, FM-2030 judges the left-wing eugenicists as too conservative in the literal meaning of the word—too inclined to impose restraints upon entrepreneurial innovation and the unforeseen results it will bring.

For such reasons, FM decides, we should "go far beyond Right and Left . . ." The principles of the liberty to enterprise freely "are . . . *Up*" (italics replace Esfandiary's boldface). FM recommends that we advocates of free enterprise and technological growth think of ourselves as *Up-Wingers*. "Up-Wingers are resigned to nothing." Conversely, conservatism and leftist "radicalism . . . are all Down."[84]

We know which course the genuinely socialist Nazis chose. Paul Roland relates, "Every aspect of the SS man's life from baptism to his funeral was controlled by the state." Thorough dossiers were compiled on all SS members, "detailing every aspect of their public and private lives, including their financial affairs. Even their prospective marriage partners were screened to ensure 'the conditions of race and healthy stock were fulfilled.'"[85] The 1998 television documentary *Nazis: The Occult Conspiracy* similarly mentions that in the Nazi Reich, girls were conscripted into a youth organization known as the League of German Maids, wherein they "were taught self-sacrifice, . . . and that their bodies belong to the nation." Boys faced this conscription as well. "At eighteen," male "cadets surrendered themselves to the Führer . . ." And: "The idea that the few should be sacrificed for the good of the many extended even to Hitler's SA—the storm troopers."[86]

George Mosse discloses that Nazi mass meetings and marches were about denying the individual. "What did it do to these people—these mass meetings?" Should a German "leave his lonely shop and join in the crowd in action, he feels as one" with everyone else. "He has lost his loneliness."[87]

Forasmuch as the Nazis mirrored other anti-capitalist movements in their social collectivism, Isaiah Berlin argues that typical left-wing anti-individualism "is, *mutatis mutandis*, similar to the belief in the primacy of collective patterns taken by those who attribute active properties to race or culture," such as "the ferocious champions of national or racial" struggle, meaning "Gobineau or Houston Stewart Chamberlain or Hitler."[88]

At this juncture Henry Ashby Turner can sum up what should be obvious about Hitler—"the principle of laissez-faire had no place in his thought. As he envisioned and later put into practice in the Third Reich, private ownership of the means of production must always remain contingent on conformity with the purposes of the state." Resultantly, the government must have the "authority to intervene in the economy . . ."[89]

Ian Kershaw, too, stresses that under the Nazis it was always "the state, not the market" that determined "the shape of economic development . . ." Under the Third Reich, commerce becomes "an adjunct of the state."[90]

Alan S. Milward agrees with other scholars that the Nazi "government did not . . . 'preserve the capitalist system,'" but instead tried to thwart it.[91] Niall Ferguson likewise confirms that the Nazi program intended to protect Germans from what Ferguson deems to be "the vagaries of the market."[92]

Stanley G. Payne spells out that Hitler's governing philosophy revolved around "state regulations and controls. Thus it is doubtful that a triumph by Hitler would have 'saved

German capitalism' . . . " Insofar as a relative amount of economic liberty can be called capitalism, ". . . German capitalism enjoyed much more autonomy before and after Hitler. Rather, the reverse of such a notion would be more nearly true: what ultimately saved German capitalism was the defeat of National Socialism in the west" by the much-more-capitalist Allies, "and the incorporation of West Germany into the American sphere of hegemony."[93]

A German academician shares in the aforementioned assessments. Universität Düsseldorf economic historian Karl Hardach judges that the Nazis instituted "a governmentally guided economy where state directives instead of the laws of the market determined production and consumption . . ."[94]

University of Glamorgan psychologist Linda Dubrow-Marshall explicates the reason why anti-capitalist micromanagement of nominally private property, on the part of the State, was important to the Führer and other cult leaders. "If you can control people's money, . . ." and their sexuality, you obtain "personal and private ways of gaining control over" the people themselves.[95]

For decades it has been fashionable to presume that the Nazis spoliated others on account of their being elitists and classists who only cared about sacrificing other people to their own selfish interests. The reality was immensely different. The Nazis shared in the modern American progressives' opinion that those who succeed under capitalism are selfish, classist, elitist social Darwinists that trample over workers, endangered species, and Mother Earth. The narrative continues that the Nazis prescribed that government should therefore enact laws that force the social Darwinian capitalist to subordinate his own interests to those of low-income Aryans, mammals, and the wilderness. As Hitler told Otto Wagener in Spring 1932, "we have seen a constant shifting back and forth between reaction and the true teaching of Christ, between the will of the upper classes to subjugate and the revolt of the masses . . ."[96] And the implementation of Nazi policies instituted the U.S. progressives' aim to maim the freedom of the businessman for the financial betterment of the proletariat and other alleged underclasses. Hitler did precisely that—he curtailed the freedom of Jewish businessmen for the financial betterment of low-income Aryans, endangered species, and other reputed underclasses. In no manner conceiving of their Nazi crusade as the manifestation of elitist capitalist social Darwinism, the Germans of the 1930s attached themselves to the Nazi cause on the basis that they believed it to be the forcible eradication of elitist capitalist social Darwinism. They then accused Jews of practicing said Darwinism.

In short, those who try to incriminate such free-market proponents as Spencer and Sumner as the inspiration for government-imposed eugenics have concocted historical narratives grossly contrary to what transpired. Commensurately, it is unconscionable for them to depict every one of laissez faire's detractors as necessarily the enemies of government-imposed eugenics. To classify Spencer and Sumner with their political opposites—governist eugenicists in the vein of Theodore Roosevelt, Oliver Wendell Holmes, George Bernard Shaw, and Joseph Goebbels—as having common membership in a unified Social Darwinism movement, is a sham. To the extent that the "free-for-all" of capitalist enterprise can be labeled social Darwinism, writes traditional conservative Jonah Goldberg, ". . . Nazism is the opposite of social Darwinism." These national socialists believed "that the state should actively pick winners and losers, and lavish the winners with social benefits, and other forms of government largesse—exactly opposite" of those laissez-faire intellectuals "we call social Darwinists."[97]

What the Social Darwinism Epithet Says About Those Who Apply It

The confusion comes from the inexact manner whereby left-wing governists handle the label of social Darwinism. They employ two very vague definitions of it. The two definitions, in some contexts, clash against one another. The first definition for social Darwinist is "anyone who supports free-market economics, regardless of whether or not he ever invoked metaphors about natural selection to buttress his position." This definition, which is the one utilized by President Obama and Robert M. Rees and the 1982 *New Republic* issue to which I alluded in the first chapter, includes Ronald Reagan but excludes socialist eugenicists, such as Karl Pearson, H. G. Wells, and Joseph Goebbels, from the outset. The other definition of social Darwinist is "anybody who applies natural selection metaphors in order to reinforce political arguments that are considered politically incorrect." This definition includes socialist eugenicists in the vein of Pearson and Wells, while excluding, from the outset, President Reagan and scores of other free-market advocates who never argued that natural selection educes evidence for free enterprise's legitimacy.

Left-wing intellectuals often exploit these two clashing definitions interchangeably, as if there were no line of demarcation between them. The clash between these two definitions accounts for the phenomenon of Hofstadter, Edwin Black, and other governist intellectuals frequently contradicting one another over whether left-wing eugenicists, typified by Louis D. Brandeis and Lester Ward, can be fairly considered social Darwinists. And, as "far as I can tell," admits Thomas C. Leonard, "Hofstadter never applied the epithet 'social Darwinist' to a progressive" while admitting that that progressive was virulently anti-capitalist.

Recall what we noted about Geoffrey Hodgson's research in chapter 2. He embarked on a database search of every English-language reference to *social Darwinism* in over 200,000 articles published from the 1800s to 2004. The search came up with a mere 11 mentions of *social Darwinism* among the articles and reviews appearing prior to 1916. From 1916 to 1943, the expression emerges in 49 articles. Then, from 1944—the year of *SDAT*'s publication—to 2004, the expression makes its way into 4,236 articles. Prior to 1944, the expression was used to describe Spencer in but two articles, one of which was authored by Hofstadter. What, then, asks Thomas C. Leonard,

> are Spencer and Sumner doing in a volume entitled "Social Darwinism in American Thought"? The answer, of course, is that Hofstadter is using "social Darwinist" . . . as an epithet to discredit views he opposed. What is new in the Anglophone literature is Hofstadter's applying the term to free-market economics.
>
> Hofstadter put Spencer and Sumner in the dock less for their putative use of Darwinian ideas than for their defense of economic competition and individualism. As much as Hofstadter rejected biological ideas in social thought, his primary quarrel was with competitive individualism, a position he never abandoned, however much his views of progressivism darkened over time. Like the progressives who had vilified Spencer and Sumner, Hofstadter judged the American Gilded Age economic order a jungle and therefore judged *any* defense of it as "Darwinist," whatever its particulars [emphasis Leonard's].

As Leonard realizes, "the set of Gilded Age and Progressive Era writers who endorse laissez-faire, racism, imperialism, and eugenics is essentially empty, making their putative commonality historically vacuous."[98]

The closest I have come to finding a self-advertised work of social Darwinism from the nineteenth century was *Might Is Right, or Survival of the Fittest*. Published in 1890 by an author employing the pseudonym Ragnar Redbeard, this work proves highly amateurish. The work does not employ the full expression *social Darwinism*, though simple *Darwinism* makes a single appearance. We "literally eat each other with voracity, relish. . . . Thus properly understood, Darwinism is not a very comforting doctrine for fat men." The book praises "Nature" for being "anti-Christ. Darwinism is the mortal form of Hebraism." Redbeard resultantly hates any Christian exhortation to be "altruistic, abandon the world, and love your enemies." He prefers what he judges to be the law of the jungle. "Nature's command is, 'Be egoist, possess the earth and fight it out.'" Redbeard cites Herbert Spencer but once, quoting him saying "a creature not energetic enough to maintain itself must die." Despite the miniscule influence that Spencer's philosophy has exerted on Redbeard's, the editor of a twentieth-century edition of this book provides annotations attempting to cast Redbeard's worldview as a logical extension of Spencer's political economy.[99] But the historical record reveals that no nineteenth-century intellectual has achieved prominence advocating anything near what a laissez-faire social Darwinism doctrine has been imagined to be. That something as paltry as *Might Is Right* amounts to one of the more famous nineteenth-century writings to cite Darwinism in letting the poor die, is evidence that no laissez-faire social Darwinism ever caught on.

Although the political Left originated the social Darwinism bogeyman, the Left is not alone in exploiting it. As mentioned in chapter 2, the Religious Right suggests that anyone who accepts the validity of natural selection is a social Darwinist. Upon hinting as much, the Religious Right implies that the negative consequence of left-wing scientists teaching kids about natural selection is that kids who accept these facts will grow up likelier to enact social Darwinism. Conservatives employ this bogeyman in public debates about physician-assisted suicide—more appropriately known as *aid in dying*—and abortion. These conservatives point out that Planned Parenthood founder and sex-education advocate Margaret Sanger was a racist eugenicist and that the Nazis murdered the elderly under the pretext of euthanasia. On this basis, declare *National Review* contributing editor Jonah Goldberg—who otherwise seemed interested in setting the record straight on social Darwinism—and the conservative *Front Page Magazine*, someone who supports the right

Table 10.1. Comparison between the Explicit Policy Positions of Free-Market Evolutionists (Herbert Spencer and William Graham Sumner) and the Eugenicists

Political Position	Position of Free-Market Evolutionists (Herbert Spencer and William Graham Sumner)	Position of the Eugenicists (including Theodore Roosevelt, George Bernard Shaw, Lester Ward, Louis Brandeis, the Progressive Movement, and Joseph Goebbels)
Government Regulation	Against	For
Antitrust Law	Against	For
Tax-Funded Welfare	Against	For
Government Discriminating on Race	Against	For
Compulsory Surgery on Reproductive Organs of Epileptics and Mental Patients	Against	For
Laws Forbidding Women From Taking Specific Jobs Consensually	Against	For
Immigration Restriction	Spencer For, Sumner Against	For
Whites Militarily Conquering Nonwhite Peoples	Against	For

to abortion and assisted suicide easily risks becoming a social Darwinist and eugenicist.[100]

For the politically left-wing critics of laissez faire, it was convenient that, as Geoffrey Hodgson notices, the eugenics movement was seldom "described as 'Social Darwinism' until after the Second World War."[101] From 1947 to the 1990s, too many academicians conveniently forgot that it was the Progressives and socialists, more than any free-marketer, who championed eugenics legislation. Once they were able to stigmatize free-market economics and eugenics legislation as right-wing, bigoted social Darwinism, it became easy for such writers as Lester Thurow to presume that the social Darwinism of free-marketers inevitably had to metamorphose into the social Darwinism of the eugenicists. In the postwar decades the political Left could implicate free-market advocates as right-wing, bigoted eugenicists while ignoring that the very same progressives who inspired them, and whom they celebrated as the enemies of social Darwinism and eugenics—such as Lester Frank Ward—were actually supportive of the eugenicist measures that would have horrified Spencer.

For such reasons, the man who came up with the best description of the social Darwinism tag is Robert Bannister, who ascertains it to be a "distortion and exaggeration . . ."[102] He notes in *Science and Myth*, "Not only was there no school (or schools) of social Darwinists," but throughout the twentieth century, "the term was a label one pinned on anyone with whom one especially disagreed. The so-called conservative social Darwinists of the 1880s (laissez-faire liberals...and the like) were, *as social Darwinists*, the invention of their opponents to the left. Eventually the label was used, not merely to caricature the 'let-alone-philosophy' (as it was termed), but to denigrate programs of other state activists one happened to oppose, whether New Liberals, fellow socialists, or eugenicists" (emphasis Bannister's). Unlike other "labels that were also initially pejorative," the social Darwinist one "is singular in that virtually no one adopted it as a badge of honor. A social Darwinist . . . was something that nobody wanted to be."[103]

That was true for most of the 1900s, but it might not be true of a particular fringe of society existing in the twenty-first century. Predictably, schooled—but not necessarily educated—members of my generation have heard the contrived myth of a social Darwinism movement that spurred the violence of the Nazis. Strangely, misinterpreting that myth as historical fact, some of the present age's alienated adolescents become fixated on Nazi paraphernalia and proclaim themselves to be modern practitioners of the great social Darwinist tradition. The teenaged perpetrators of the 1999 Columbine High School massacre, Eric David Harris and Dylan Klebold, exalted themselves as the inheritors of the social Darwinism mantle. In his diary Harris penned, "NATURAL SELECTION!!!!!!!!!! [sic; he used that many exclamation points.] . . . Getting rid of all the stupid and weak organisms. I wish the government would just take off every warning label. So then all the dumbasses would either severely hurt themselves or DIE!"[104] Another entry says, "NATURAL SELECTION. Kill all retards, people with bra[i]n fuck ups. . . . Geeeawd! People spend millions of dollars on saving the lives of retards, and why? I don't buy that shit like 'oh, he's my son, though!' so the fuck what, he ain't normal, kill him."[105]

In an April 26, 1998, entry Harris wrote, "It would be great if god removed all vaccines and warning lables [sic] from everything in the world and let natural selection take its course. All the fat ugly retarded crippled stupid fuckheads in the world would die, and oh fucking well if a few of the good guys die to[o]. . . . World war II [sic] is the last time I bet America was proud of it self [sic]. We beat the fuck out of the damn zipperheads"—the Japanese—"and the nazis. We came back, from being bombed and loosing [sic] major battles to nuke that little piece[-]of[-]shit island" nation of Japan "and to take over that cool place called Deutschland."[106] Harris also wrote elsewhere in his journal, "Isn't America supposed to be the land of the free? How come, if I'm free, I can't deprive a stupid fucking dumbshit from his possessions if he leaves them sitting in the front seat of his fucking van out in plain sight and in the middle of fucking nowhere on a Fri-fucking-day night. NATURAL SELECTION. Fucker should be shot."[107] Another day he promised to "kill who ever I deam [sic] unfit for anything at all."[108]

Eventually Harris fantasized about the extinction of every member of the human race, except for those from still-existing hunter-gatherer clans. He wrote in his diary, ". . . just thinking if I want all humans dead or maybe just the quote-unquote 'civilized, developed, and known-of' places on Earth, maybe leave little tribes of natives in the rain forest or something. Hmm, I'll think about that."[109] Harris's nihilism manifested consistently: ". . . the Nazis came up with a 'final solution' to the Jewish problem. Kill them all. Well, in case you haven't figured it out yet, I say 'KILL MANKIND'[;] no one should survive."[110] He sent out an e-mail saying, "I think I would want us [humans] to go extinct."[111] And in another note he said, "I have a goal to destroy as much as possible. . . . I want to burn the world . . ."[112] He had planned to plant a bomb on the campus, killing as large a number of its occupants as possible. Bizarrely, prior to carrying out the attack, he recorded a video of himself addressing some of those same classmates, saying, "Morris, Nate, if you guys live [through the attack] I want you to have whatever you want from my room."[113] On the very day of the massacre, Harris wore a shirt saying *Natural selection*.[114]

Eric Harris had some other traits in common with the Nazis, except that cultural commentators failed to notice that the Nazis so much as possessed these traits. As with the majority of young recruits to the Nazi Party in the early 1930s, Harris expressed resentment toward those who were more affluent than he was. In his diary he railed against "those rich snotty toadies at my school. fuckers think they are higher than me and everyone else with all their $ just because they were born into it?"[115] Citing the *Columbine Documents* of the Jefferson County Sheriff's Office, child psychologist Peter Langman evaluates that Harris "resented people who in his eyes had it easy because they were born with wealth. Keep in mind that Eric's family was far from poor, why was he so bitter and hostile toward rich people? Sensitivity to status is related to paranoia because paranoid people tend to assume that others have achieved their success or status unfairly. Paranoid people," such as Eric Harris and Hitler's earliest political supporters, "think they have gotten a raw deal in life and that other people have manipulated the system to achieve success. To paranoid people, they are once again victims."[116] Joseph Goebbels held capitalism in low regard, and so did Eric Harris. The envy was characteristic of Harris's accomplice, Dylan Klebold, as well. Klebold wrote in his journal, "others' achievements are tormentations [sic]."[117]

Also in common with the Nazis, Harris rejected acknowledgment of objectivity in favor of anti-realist epistemology. In his diary he penned, "there is no such thing as an actual 'real world.'" On account of his rejection of objectivity, Harris concluded that there is no objective morality either; he thought ethics was completely arbitrary and capricious. "Just because your mumsy and dadsy told you blood and violence is bad, you think it's a fucking law of nature? Wrong."[118] He concluded, "There's no such thing as True Good or True evil."[119] And he admitted his own self-hatred. "Everyone is always making fun of me because of how I look, how fucking weak I am. . . . Then again, I have always hated how I looked, I make fun of people who look like me, sometimes even without thinking, sometimes just because I want to rip on myself. That's where a lot of my hate grows from. The fact that I have practically no self-esteem . . ."[120] Eric Harris's diary reveals that, as he has chosen to reject reason, you cannot approach him with reason to persuade him to stand down. When someone of Harris's mentality intends to kill you, he can only be answered with retaliatory force.

The Columbine killers happened to have an inadvertent disciple in Finland by the name of Pekka-Eric Auvinen (b. 1989). Under the username "NaturalSelector" he posted video manifestos on the website YouTube to air his ravings, aggrandizing himself as "a cynical existentialist, anti-human humanist, anti-social social-Darwinist, realistic idealist and god-like atheist."[121] On November 7, 2007, Auvinen went to Jokela High School, which he regularly attended as a student, and shot eight other people before turning his gun on himself. Hours prior to the incident, he posted a video on YouTube[122] in which he stated, "I, as a natural selector, will eliminate all who I see unfit, disgraces of human race and failures of natural selection."[123] I have pointed out many ironies throughout this book, but it is an irony-upon-ironies that these school shooters have proclaimed themselves to be the latest exponents of a nineteenth-century ideological tradition that had not truly existed until the school shooters themselves had decided to entertain it seriously. Predictably the self-fashioned pundits on YouTube immediately buzzed about the Jokela High shooting mere hours after it took place. One of the pundits was Terroja "T.J." Lee Kincaid, better known online as "TheAmazingAtheist." Kincaid had actually argued with Auvinen about his ideology in e-mails and in YouTube videos months preceding the shooting. Not surprisingly, Kincaid ascribes Auvinen's homicidal behavior to "social Darwinism,"[124] which Kincaid thinks "should rightly be called 'social Spencerism.' "[125]

Quite presciently, Spencer estimated in his own lifetime that in about "three cases out of four the alleged opinions of mine condemned by opponents, are not opinions of mine at all, but are opinions wrongly ascribed by them to me; sometimes from carelessness but more frequently from perversity: seeming, not unfrequently, too deliberate."[126] Ever since Hofstadter released *American Thought*, to the day of the present writing, the situation has worsened. As George H. Smith rightly apprehends, "more often than not, slander and misrepresentation have been the stock tools of Spencer critics."[127]

Indeed, in the journal *Victorian Studies*, Richard L. Schoenwald writes that that the best explanation for Spencer's laissez-faire politics is that he resented his toilet training.[128] I wish I were joking, but I am not.

The revisionist historian Harry Elmer Barnes (1889–1968) carps that Spencer's "persistent and ever growing resentment against the extension of governmental activity probably was personally motivated by a subconscious neurotic reaction."[129]

Neo-conservative writer Gertrude Himmelfarb (b. 1922) animadverts upon Spencer for being a "dilettante whose writing was as facile as his thinking"; to her, his "image is comic and pathetic."[130]

Fortunately, not everyone is like Kincaid, Schoenwald, Barnes, or Himmelfarb. Jonah Goldberg is one of the few writers to admit the truth about Spencer. He scribes, "Herbert Spencer...was singled out as the poster boy for all that was wrong in classical" laissez-faire "liberalism. Spencer was indeed a Darwinist . . ., but his interpretation of evolutionary theory reinforced his view that people should be left alone. In almost every sense, Spencer was a good—albeit classical" laissez-faire—"liberal: he championed charity, women's suffrage, and civil liberties. But he was the incarnation of all that was backward, reactionary, and wrong according to the progressive worldview, not because he supported Hitlerian schemes of forced race hygiene but because he adamantly *opposed* them. To this day it is de rigueur among" left-wing governist "intellectuals and historians to take potshots at Spencer as the philosophical wellspring of racism, right-wing 'greed,' and even the Holocaust" (emphasis Goldberg's). Just as I do, Goldberg thus takes Richard Hofstadter and *SDAT* to task for propagating the misconceptions about Spencer and the nineteenth-century's industrialists. Goldberg soundly judges *SDAT* to be a work of "deeply flawed scholarship," resulting in Hofstadter getting "much of his history wrong . . ."[131]

Historian Barry Werth is another figure who notices the irreconcilability of Spencer's consensualism with government-enforced "scientific racism." Though he somewhat annoyingly applies the phrase *social Darwinism* to Spencer's laissez-faireism, Werth establishes Spencer's ideas to be distinct from eugenics doctrine. After 1894, writes Werth, "social theorists turned more aggressively critical" of laissez-faire "social Darwinism"—[sic; actually Spencerian ecosystem economics]. Hence these social theorists embraced "a new, more 'scientific' approach to human betterment by encouraging favored bloodlines through social, legal, and medical means—eugenics. Appealing to the emerging interest in preserving and extending America's 'racial stock' in an era of dawning imperialism, nativist anxiety over rising poverty, disease, and immorality among immigrant groups, and a new zeal for government reforms," the governist "eugenics movement swiftly took root across America." Werth comprehends it as ironic that as "racial science reemerged as a rationale for state social policy . . ., Spencer's theories and social Darwinism were frequently blamed, despite Spencer's lifelong abhorrence of all coercive state control of the individual."[132]

I commend Goldberg and Werth for writing what they have about Spencer, for I cannot stand to witness any more mistreatment of this intellectual giant by Kincaid, Edwin Black, or anyone else. I find John Kenneth Galbraith's case particularly galling.

Recall, from chapter 1, Galbraith pummeling Spencer and Sumner as evil for their wish that the government cease taxpayer-funded welfare spending. Yet, as he derided free-market economics as ruthless Darwinism, Galbraith served as an apologist for the command-economics policies of Mao Tse-Tung. In the year 1972, as Mao ruled China, Galbraith visited the country and then reported in a book published the subsequent year his impression of how the State dictated over economics. Galbraith concludes, "There can be no serious doubt that China is devising a highly effective economic system." On account of Mao's genius, China exhibits "new housing, new industrial plants," and "new building at old plants . . ." Galbraith announces "the impressive figures on the increase in local industrial and agricultural production and employment . . ." Thus he cheers, ". . . the Chinese appear to have developed a plain but remarkably efficient system for the distribution of consumer goods." Yes, "the Chinese economy appears to function very easily and well." To provide the reader an idea of how productive the Red Chinese economy is, Galbraith mentions, "Frank Coe and Sol Adler . . . guess that the rate of expansion in Chinese industrial and agricultural output is now between 10 and 11 percent annually. This does not seem to me implausible though it means a performance . . . rivaling that of Japan." Hence Galbraith cannot fathom the notion of the manufacturing sector being in poorer shape in Red China than in more liberalized nations. "Except that there is no sales, merchandising, marketing or advertising staff . . . not much distinguishes the Chinese from the American or European industrial plant." Just as Michael Moore would argue in 2004 that Cuban socialism provides better health care than would a free-market republic, Galbraith inaccurately suggests that the Maoist State affords greater access to quality health care to citizens than do private providers in the USA. "I am prepared to believe that Greater Shanghai . . . has a better medical service than New York. The average quality of practice is no doubt far higher in New York. But the chance of getting no care is also much higher" in New York as well. Galbraith admits that Josef Stalin was brutal, and denies that the Maoist regime could possibly rival Stalin in rapacity. "Stalin (with Marx, Engels and Lenin) still survives in the Chinese Communist pantheon. His picture…is in most places of public ceremony." Yet Stalin's "policy of planned deprivation associated with his name is strongly rejected" by the Chinese government. Consequently, Galbraith downplays the persecution of political dissidents—a horrifying persecution that would be well-documented in the decades to come. "Dissidents are brought firmly into line in China, but, one suspects, with great politeness. It is a firmly authoritarian society in which those in charge smile and say please." Galbraith goes as far as rationalizing the fact that Chinese citizens largely had to dress alike. "There has been too much snobbish comment about the uniformity of Chinese clothing. General appearance is better" in China "...than on an American campus. . . . In a poor country an arrangement by which every person gets two sets of sound basic garments every year at low prices seems to me an exercise in the greatest good sense." Galbraith therefore contrasts himself against "travelers to the Communist countries" who have "been reluctant to risk hard conclusions. When things went wrong, the skeptics remembered and rejoiced. One should not be craven. The Chinese economy isn't the American or European future. But it is the Chinese future. And let there be no doubt: for the Chinese it works." In the decades subsequent to Mao's demise, we have seen the Chinese people's own verdict on the communist policies that Galbraith praised. Yet, speaking about a visit to a government-run Chinese middle school, Galbraith produces an evaluation that evidently describes his assessment of the trip overall—an "unalloyed success."[133] The supreme irony, then, is the manner in which Galbraith denounced Spencer and Sumner as ruthless as he lionized Mao Tse-Tung as someone who benefited the Chinese people.

I agree entirely with Robert Bannister's assessment that "social Darwinist stereotyping" has "clouded" the "reputations" of honest men.[134] Geoffrey Hodgson himself identifies social Darwinism "as a term of abuse," which has "served not only partisan political ends," but which has, for decades, "foreclosed discussion of the importance of ideas from biology in helping to understand human affairs." For that reason Hodgson concludes, "Overall, the label of 'Social Darwinism' is unhelpful and misleading."[135] There is no fair-minded way to hurl the epithet. In conclusion to our hunt for social Darwinism, the knee-jerk denunciation of anyone who opposes welfare as a social-Darwinist, racist, imperialist fascist who yearns for the annihilation of the financially disadvantaged, is simple name-calling. An exercise in deliberately stigmatizing the advocates of laissez faire, it cannot be classified as honest discourse. I already said this in Book Two, but it is worth repeating: This canard of social Darwinism is so unfit for a climate of rational debate that it ought to act in accordance with the implications of its own name . . . and go extinct.

And what must not go extinct are the insights that Spencer and Sumner have provided. In part III, then, we consider the myriad areas in which Spencer and Sumner have contributed to evolutionary psychology, ecosystem economics, and the science of emergent complexity. We also examine how Spencer and Sumner have been denied rightful credit for such contributions. Subsequent to that examination, we shall explore some final lessons that pertain to the entire trilogy's exploration of liberty.

NOTES

1. McMenamin 1999, 72, accessed online Monday, June 18, 2007.
2. Du Bois, "Germany and Hitler," *Pittsburgh Courier*, December 5, 1936, in Du Bois 1995, 735.
3. Orwell 1941, 73–76. Huber 1994, 61, 72, brought these statements of Orwell's to my attention.
4. M. Parenti 2004, 118.
5. P. du Gay 2000, 49.
6. McMenamin 1999, 73, accessed online Monday, June 18, 2007. The 14 percent unemployment figure for January 1936 comprised 2.5 million people, and the "2.5 million" figure comes from Kershaw 2000, paperback, 582.
7. Stephen Roberts, cited by McMenamin 1999, 73, accessed online Monday, June 18, 2007; and C. Webber and A. Wildavsky 1986, 461–62.
8. Stephen Roberts, cited by McMenamin 1999, 73, accessed online Monday, June 18, 2007; and C. Webber and A. Wildavsky 1986, 461–62.
9. Stephen Roberts, cited by McMenamin 1999, 73, accessed online Monday, June 18, 2007.
10. Kershaw 2000 paperback, 450.

11. Kershaw 2000 paperback, 576. Ian Kershaw further proves the Nazis' economic incompetence in Kershaw 1983.
12. R. Grunberger 1971, 168.
13. Shirer 1990, 262; and C. Webber and A. Wildavsky 1986, 462.
14. Shirer 1990, 262.
15. R. Grunberger 1971, 168–170.
16. Shirer 1990, 262.
17. P. Watson 2010, 361.
18. C. Webber and A. Wildavsky 1986, 462.
19. McMenamin 1999, 74–75, accessed online Monday, June 18, 2007. McMenamin 1999 cites Stephen Roberts. The failures of Hitler's agricultural policy are also highlighted in Kershaw 2000, paperback, 578–579, 581; and Kershaw 1983, 83,306, 315.
20. Kershaw 2000, paperback, 576. A shortened version of this quotation appears in McMenamin 1999, 73–74, accessed online Monday, June 18, 2007. Ian Kershaw further proves the Nazis' economic incompetence in Kershaw 1983.
21. McMenamin 1999, 74–75, accessed online Monday, June 18, 2007. McMenamin 1999 cites Stephen Roberts.
22. Kershaw 2000, paperback, 576. A shortened version of this quotation appears in McMenamin 1999, 73–74, accessed online Monday, June 18, 2007. Ian Kershaw further proves the Nazis' economic incompetence in Kershaw 1983.
23. T. Standage 2009, 211.
24. Paarlberg 2010, 140.
25. G. L. Mosse, ed. 1966, 342–44.
26. C. Webber and A. Wildavsky 1986, 463.
27. G. L. Mosse, ed. 1966, 344.
28. Raymond Fletcher, qtd. by A. Toffler 1981 paperback, 402, from the author's interview with Fletcher.
29. P. A. Samuelson 1980 11th ed., 815, qtd. by M. Krauss 1983, paperback, 43.
30. Shirer 1990, 260.
31. C. Webber and A. Wildavsky 1986, 463.
32. Walther Funk in 1939, qtd. by Shirer 1990, 260.
33. Shirer 1990, 261.
34. Qtd. by Shirer 1990, 261.
35. D. M. Smith 1982, 118. B. Mikkelson and D. P. Mikkelson 2007, accessed online Saturday, October 20, 2012, was the source that first brought this information to my attention.
36. P. Roland 2008, 177.
37. *Schwarze Korps* 1966, 362–63.
38. R. Grundberger 1971, 172.
39. G. L. Mosse, ed. 1966, 345.
40. R. Grunberger 1971, 172.
41. *Frankfurter Zeitung* 1966b , 363–64.
42. R. Grunberger 1971, 172.
43. G. L. Mosse, ed. 1966, 345.
44. *Frankfurter Zeitung* 1966a, 364.
45. Qtd. by R. Grunberger 1971, 170.
46. M. Parenti 2004, 117–18.
47. J. C. Fest 1979 U.S. version, 760.
48. "Reprogenetics" was first used in L. M. Silver 1998 paperback, 8–9.
49. For a more detailed explanation of this process, see L. M. Silver 1998 paperback, 269–273.
50. According to Chorost 2005, Manfred Clynes and Nathan Kline coined the term *cyborg* in 1961.
51. To read the writings wherein Julian Huxley introduced the term "transhumanism," see J. Huxley 2004, accessed online Monday, April 28, 2008, adapted from J. Huxley 1957, 13–17.
52. For information on this, see Glover, prod. 2003
53. G. Dover 2000, 10.
54. "Learning About Tay-Sachs Disease," website of the National Human Genome Research Institute, a Division of the National Institute of Health, March 2007, http://www.genome.gov/10001220, accessed Sunday, May 27, 2007. W. T. Anderson 1996, 104-05; K. Davies 2010, 271; J. Entine 2007, 281; and Zuk 2007, 34; also touch upon this.
55. Lee Silver, interviewed by John Stossel in Colloton et al., prods. 2001.
56. James D. Watson, interview in Glover, prod. 2003.
57. Mahoney interviewed by John Stossel in Colloton et al. prods. 2001.
58. Black 2003, 426-27, 441–43.

59. Narration in Glover prod. 2003. Incidentally, I find it pretty amusing that Edwin Black denounces both Herbert Spencer and James D. Watson as social Darwinist eugenicists, since James Watson himself unfairly derided Spencer for being a Social Darwinist in J. D. Watson ed. 2005, 606.
60. James D. Watson, interviewed in Glover, prod. 2003.
61. Black 2003, 443–44.
62. M. Ridley 1999, 298–99.
63. Watson, interviewed in Glover, prod. 2003.
64. Postrel interviewed by John Stossel in Colloton et al. prods. 2001.
65. Robert L. Sinsheimer, "The Prospect of Designed Genetic Change," in Chadwick, ed. 1992 ed., 145, qtd. by Fukuyama 2002, 87.
66. DeGregori 2001, https://web.archive.org/web/20030802013328/http://www.acsh.org/publications/priorities/1302/coverstory.html, accessed Wednesday, October 1, 2014.
67. S. Arrison 2011, 85.
68. R. Naam 2005, 166–67. S. Arrison 2011, 85, brought my attention to Naam saying this.
69. Dewar 2004, 441.
70. James D. Watson, Question-and-Answer section for "No Restrictions on Genetic Research," *New Perspectives Quarterly*, fall 2005, http://www.digitalnpq.org/archive/2005_fall/13_watson.html; and interview by John H. Richardson, "What I've Learned: James Watson," *Esquire*, October 19, 2007, http://www.esquire.com/features/what-ive-learned/ESQ0107jameswatson, accessed Wednesday, September 24, 2014.
71. According to K. Davies 2010, 20, James D. Watson lobbied Congress in the early 2000s to earmark federal funds for research on mental illness.
72. Herzlinger 2007, 132, points out that in 1908 the average price of an automobile in the USA was that of the average house sold in the United Kingdom, and that by 1916 the price was less than half that. B. W. Folsom 2008, 129, pointed out that the price of an automobile in the USA at the time was 3,000 U.S. dollars.
73. C. Ogden 1999, 60.
74. B. W. Folsom 2008, 129. I have received different figures from M. Wallace, prod. 2006. M. Wallace, prod. 2006 states that from 1912 to 1914, the Ford Motor Company reduced the average price of Model Ts from 600 to 490 dollars. In any case, though, Ford boosted its gross by employing economies of scale to press out larger numbers of units at high volumes and then to sell those units at affordable prices to the masses.
75. Mellon 1924, 13, qtd. by B. W. Folsom 2008, 129.
76. Katie Bushouse, "Plastic Surgery's Affordable to All," *South Florida Sun-Sentinel*, http://tinyurl.com/385ue6, March 17, 2006; Diane Kockler, "Plastic Surgery Moves Beyond Rich and Famous," *Click2Houston*, November 7, 2006, http://tinyurl.com/2ys5s2; Lloyd M. Krieger, "Plastic Surgery Goes Mainstream," *Beauty Fashions*, 2007, http://tinyurl.com/yv35k6; all accessed Saturday, May 26, 2007.
77. Zinser and Hsieh 2007–08, 37, accessed online Wednesday, May 18, 2011.
78. Cannon and Tanner 2007 2d ed., 22, citing Cutler et al. 1998.
79. Cannon and Tanner 2007 2d ed., 22, citing I. Shapiro et al. 1999.
80. Christopher Reeve, keynote speech at the March 2000 BIO conference in Boston, Massachusetts, qtd. by J. F. Crowley 2010, 97.
81. F. M. Esfandiary 1973, 143. I first heard of F. M. Esfandiary from Tuccille 1975, 66.
82. To get a glimpse of the Soviet communists' dismal environmental record, see Feshbach and Friendly 1992.
83. Ronald Reagan, "Rendezvous With Destiny," address on behalf of Barry Goldwater's presidential campaign, October 27, 1964, republished on the website of the Ronald Reagan Presidential Foundation & Library (Simi Valley, California: Ronald Reagan Presidential Library & Museum), http://tinyurl.com/2ukn3, accessed Sunday, June 3, 2007.
84. F. M. Esfandiary 1973, 7–9, 11.
85. P. Roland 2008, 176.
86. Brad Abelle narrating T. Atkinson and J. Baran, prods. 1998.
87. George L. Mosse interviewed in T. Atkinson and J. Baran, prods. 1998.
88. I. Berlin 1970b, 47.
89. H. A. Turner 1985, 78.
90. Kershaw 2000, paperback, 448–49.
91. A. S. Milward 1976, 399.
92. N. Ferguson 2006, 231.
93. S. G. Payne 1980, 91.
94. K. Hardach 1980, 65–66. C. Webber and A. Wildavsky 1986, 463, brought this statement to my attention.
95. Linda Dubrow-Marshall, interviewed in I. Chapman prod. 2012.
96. Hitler, from a conversation with Wagener in Spring 1932, qtd. in Wagener 1985, 316.
97. J. Goldberg 2007, 414 n. 7, 415 n. 7.

98. T. C. Leonard 2009, 43, 41, 45, accessed online Monday, November 7, 2011.
99. Redbeard 2005, 251, 60–61, 5 n.
100. J. Goldberg 2007, 354; and Ponte 2002. Goldberg mentions physician-assisted suicide but Ponte 2002 does not.
101. G. Hodgson 2004.
102. Qtd. by Bannister 1988, 398. This is also quoted in C. Tame 1998, 2.
103. Bannister 1988, xi–xii.
104. Qtd. by D. Cullen 2009, 184.
105. Qtd. by P. Langman 2009, 31–32.
106. E. Harris 1998–1999b, http://acolumbinesite.com/eric/writing/plans.gif, accessed Sunday, January 6, 2008.
107. April 12, 1998 entry in E. Harris 1998–1999a, http://acolumbinesite.com/eric/writing/journal/journal.html, accessed Wednesday, September 24, 2014.
108. April 21, 1998 entry in E. Harris 1998–1999a, accessed online Wednesday, September 24, 2014.
109. June 13, 1998 entry in E. Harris 1998–1999a, accessed online Wednesday, September 24, 2014.
110. June 12, 1998 entry in E. Harris 1998–1999a, accessed online Wednesday, September 24, 2014.
111. Qtd. by P. Langman 2009, 31-32.
112. October 23, 1998 entry in E. Harris 1998–1999a, accessed online Wednesday, September 24, 2014.
113. Qtd. by P. Langman 2009, 34, citing Jefferson County Sheriff's Office, *Columbine Documents*, page 10376.
114. P. Langman 2009, 31–32.
115. April 12, 1998 entry in E. Harris 1998–1999a, accessed online Wednesday, September 24, 2014.
116. P. Langman 2009, 34.
117. P. Langman 2009, 53, citing Jefferson County Sheriff's Office, *Columbine Documents*, page 26397.
118. April 21, 1998 entry in E. Harris 1998–1999a, accessed online Wednesday, September 24, 2014.
119. June 12, 1998 entry in E. Harris 1998–1999a, accessed online Wednesday, September 24, 2014.
120. November 12, 1998 entry in E. Harris 1998–1999b, accessed online Wednesday, September 24, 2014.
121. Phil Black, "Finland in Mourning After Fatal School Shooting," *CNN.Com International/Europe*, November 8, 2007, http://tinyurl.com/yvy9d2 accessed Sunday, January 6, 2008.
122. Alison Godfrey, "Profile of Pekka-Erik Auvinen, The YouTube Killer," *News.Com.Au*, November 8, 2007, http://www.news.com.au/story/0,23599,22723252-2,00.html, accessed Sunday, January 6, 2008.
123. " 'I Will Eliminate the Unfit,'" *IOL: News for South Africa and the World*, November 8, 2007, http://tinyurl.com/2blqp4, accessed Sunday, January 6, 2008.
124. Terrojas "T.J." Lee Kincaid as "TheAmazingAtheist," "I Knew This Would Happen," *YouTube*, November 7, 2007, http://www.youtube.com/watch?v=GIKUWgsIncg, accessed Sunday, January 6, 2008.
125. Terrojas "T.J." Lee Kincaid as "TheAmazingAtheist," "Sample Chapters," *YouTube*, November 7, 2007, http://www. youtube.com/watch?v=8tDh-2AUNUM, accessed Sunday, January 6, 2008.
126. Spencer 1902a, 151. G. H. Smith 1981, 114, brought this statement of Spencer's to my attention.
127. G. H. Smith 1981, 113–14.
128. Schoenwald's paper is Schoenwald 1968. I learned of it from G. H. Smith 1981, 142 n. 6.
129. H. E. Barnes 1948, 129. G. H. Smith 1981, 142 n. 6, brought this statement of Barnes's to my attention.
130. G. Himmelfarb 1959, 183–87. G. H. Smith 1981, 142 n. 6, brought these statements of Himmelfarb's to my attention.
131. J. Goldberg 2007, 257–58, 447 n. 23.
132. B. Werth 2009, 312.
133. Galbraith 1973, 118–120, 125–27, 130–31, 137, 97.
134. Bannister 1988, xxxi.
135. G. Hodgson 2004.

III

The Final Lessons of Liberty

ELEVEN

The Ethologists' Unpaid Debts to Spencer and Sumner

Spencer's Poor Word Choice When Pioneering Biology-Based Economics

No, Herbert Spencer and William Graham Sumner were not eugenicists. No, they did not demand that disease sufferers and the financially downtrodden hasten their journey to the tomb. No, Spencer and Sumner did not proclaim that no one but the wealthiest and healthiest should be able to propagate their genes to successive generations. Some readers may wonder, then, in what manner did these two intellectuals really apply the theories of biological evolution to their studies of society's dynamics. I provided some of the answer in Book Two, but here I can disperse the rest of it. The answer is that both Spencer and Sumner pioneered in evolutionary psychology, evolutionary economics, and the theories of complex adaptive systems and emergent complexity. Heart-wrenchingly, the lead intellectuals in each of these disciplines seem reluctant to admit as much.

I argue that Spencer and Sumner anticipated some aspects of E. O. Wilson's and Charles Lumsden's theory of gene-culture co-evolution. And they further anticipated several modern insights about economies being natural ecosystems. Disregarding such facts, sundry thinkers in the evolutionary-psychology discipline present Spencer's and Sumner's economic insights as their own new discoveries as they simultaneously excoriate Spencer and Sumner as bigoted eugenicists who had no ideas of any value to offer.

The twentieth century boasts such free-market evolutionists as F. A. Hayek, Jane Jacobs, Michael Shermer, and Michael Rothschild. But a hundred years prior to those individuals rising to prominence, Herbert Spencer had already observed that a market economy is a vibrant ecosystem that functions through spontaneous order. Sadly this achievement goes unsung as a consequence of some unfortunate terminology on Spencer's part. When Spencer wanted to explain how spontaneous order was common among peaceful societies and violent shark-infested waters, he groped for a word to describe each of these natural environments and habitats. Because the word *ecosystem* was not known in his time, Spencer instead opted for *organism*.

Of course, if one literally believed that a society or economy was a single organism, then that would imply that every person living and participating in society is but a tiny, expendable cell of a much greater being. Were that the case, it would thus be just for the State to sacrifice the lives and rights of individuals if doing so could overall strengthen the much larger social organism. Spencer did not intend to convey such a message, as he opposed such governism. But because he applied the word *organism* instead of *ecosystem* to characterize the habitat that a night watchman state created, readers ended up assessing Spencer as much more of a utilitarian collectivist than he truly was.[1]

In fact, Spencer prefaced to his readers that although he wished to draw parallels between societies and "organisms," he did not base this theory on politically collectivist assumptions. To some extent, he continued to understand that every human being is a unit unto himself, not an expendable or interchangeable component of a larger animal. To establish that he remained a political individualist committed to Lockeanism, Spencer refrained that

> we have here a tolerably decided contrast between bodies-politic and individual bodies; and it is one which we should keep constantly in view. For it reminds us that while, in individual bodies, the welfare of all other parts is rightly subservient to the nervous system, . . . in bodies-politic the same thing does not hold. . . . It is well that the lives of all parts of an animal should be merged in the life of the whole, because the whole has a corporate consciousness capable of happiness or misery. But it is not so with society; since its living units do not and cannot lose individual consciousness, and since the community as a whole has no corporate consciousness. This is an everlasting reason why the welfares of citizens cannot rightly be sacrificed to some supposed benefit of the State, and why, on the other hand, the State is to be maintained solely for the benefit of the citizens. The corporate life must here be subservient to the lives of the parts, instead of the lives of the parts subservient to the corporate life.[2]

Although he and his contemporaries remained unaware of kin selection, Spencer ascertained that the structures of human societies were the consequence of some sort of natural selection. He therefrom deduced that there was something inherently biological and natural about how customs, mores, economic institutions, and economic organizations have developed in societies. Spencer issued these observations in an 1860 paper with the misleading title of "The Social Organism." In this piece he states that government is like a society's brain. In turn, turnpikes and canals and railroads—whereby goods and services are transported across great distances—are comparable to arteries carrying blood cells to various organs to keep them functional. As a supplement, he discusses similarities between the growing complexities of developing societies with the growing complexities of prehistoric invertebrates as they evolved from single-celled life forms to multi-cellular sea beasts. In this respect, Spencer anticipates the nascent science of Emergent Complexity.

There are some passages of this essay where *ecosystem* appears to be more appropriate a word than *organism* for what Spencer delineates. One instance is where he emphasizes that an economy does not need to be micromanaged by a totalitarian government to run efficiently, as it can largely run itself insofar as the government sticks to upholding consensualist rights. That the most productive market economies are not "put together" predominantly by central totalitarian government control, Spencer wrote in this treatise

> is a truth so manifest, that it seems wonderful [he means "unbelievable"] men should ever have overlooked it. . . . You need but to . . . observe social organization in its leading traits, to see that these are [not] supernatural, . . . but are consequent on general natural causes. The one case of the division of labour suffices to prove this. It has not been by command of any ruler that some men have become manufacturers while others have remained cultivators of the soil. In Lancashire, millions have devoted themselves to the making of cotton-fabrics in Yorkshire, another million lives by producing woollens, and the pottery of Staffordshire, the cutlery of Sheffield, the hardware of Bingham, severally occupy their hundreds of thousands. These are the large facts in the structure of English society; but we can ascribe them neither to [supernatural] miracle, nor to legislation. It is not by "the hero as king," any more than "collective wisdom," that men have been segregated into producers, wholesale distributors, and retail distributors. Our industrial organization, from its main outlines down to its minutest details, has become what it is,

> not simply without legislative guidance, but, to a considerable extent, in spite of legislative hindrances. It has arisen under the pressure of human wants and resulting activities. While each citizen has been pursuing his individual welfare, and none taking thought about the division of labour, or conscious of the need of it, division of labour has yet been ever becoming more complete. . . . Through this combination thus *spontaneously evolved,* every citizen is supplied with daily necessaries; while he yields some product or aid to others [emphasis added].[3]

It is a fact, he concludes, "that under all its aspects and through all its ramifications, society is a [biological] growth and not a manufacture."[4] University of Michigan anthropologist Leslie A. White (1900–1975) voices agreement—"Biological evolution and the evolution of sociocultural systems . . . are characterized by progressive diversification of structure and specialization of function. And both develop structural means of coordinating parts and functions and of regulating (controlling) the behavior of the whole; both move toward higher levels of integration."[5] "It is not so much that society is an organism," Robert L. Carneiro edifies, but that "societies and organisms"—as with the environments that house these organisms, happen to be "*systems*. That is to say, each is a working unit of interdependent parts, organized into structures" that "carry out functions, all leading to the efficient operation of the whole."[6] For this reason, Leslie White judges, "If Spencer had only used the word *'system'* instead of *'organism,'* he would have spared a generation of sociologists and laymen much fruitless and sometimes passionate argument."[7]

1998's *Bionomics* and Its Similarities to Spencer's Social Organism

Compare Spencer's ruminations on economies being spontaneously ordered environments with those of Michael Rothschild in his 1998 book *Bionomics: Economy as Ecosystem*. Writes Rothschild,

> Capitalism, or the market economy, or the free-enterprise system…was not planned. Like life on earth, it did not need to be. . . . Capitalism flourishes wherever it is not suppressed [remember Spencer saying that our market economy "has become what it is, not simply without legislative guidance, but…in spite of legislative hindrances"? –S.H.], because it is a naturally occurring phenomenon. . . .
>
> A capitalist economy can be best comprehended as a living ecosystem. Key phenomena observed in nature—competition, specialization, cooperation, exploitation, learning, growth, and several others—are also central to business life. Moreover, the evolution of the global ecosystem and the emergence of modern industrial society are studded with striking parallels. . . . Each [manmade] organization strives to survive it its niche of the economic ecosystem. . . .
>
> The most difficult concept to accept about the natural world is that it runs itself. No conscious force is needed to keep the ecosystem going. Life is a self-organizing phenomenon. From the interplay of hormones in the human body to the expansions and contractions of the great Arctic caribou herds, nature's intricately linked feedback loops automatically maintain a delicate, yet robust balance. Markets perform the same function in the economy. Without central planning, buyers and sellers constantly adjust to changing prices for commodities, capital, and labor. A flexible economic order *emerges spontaneously* from the chaos of free markets [remember that Spencer said that because the "division of labour" had "spontaneously evolved, every citizen is supplied with his daily necessaries" while "he yields some product or aid to others" –S.H].
>
> Needless to say, this thinking bears little resemblance to conventional economics. Two centuries of economic thought, both capitalist and socialist, are based on the concept of "economy as machine" rather than "economy as ecosystem." Nonetheless, history has demonstrated that no economy behaves like a simple, cyclical machine. Like ecosystems, economies are spectacularly complex and endlessly evolving [emphasis added].[8]

Is that last paragraph not similar to Spencer's assessment that "under all its aspects and through all its ramifications," the market economy "is a [biological] growth and not a manufacture"? Quite logically, Rothschild names his theory *Bionomics*, as it combines "Biology" and "Economics."

Jane Jacobs points out that *bionomics* is a better word than *ecology*, for *eco-* means "house" while *-logy* means "knowledge." Hence *ecology* translates to "house knowledge." On the other hand, *bio-* means "life" and *-nomy* means "management." Therefore *bionomy* refers to "life management" the same way that *economy* alludes to "house management."[9] In concordance with Rothschild's views, Jacobs notes that *ecology* only makes sense when viewed as "the economy" of ontologically given life forms and life systems in "nature."[10]

Rothschild's claim that his Bionomics theory is completely original is undermined by the fact that Spencer's "The Social Organism" provided similar comparisons between society and biology. Spencer writes,

> The orderly progress from simplicity to complexity, displayed by bodies-politic in common with living bodies, is a characteristic which distinguishes living bodies from the inanimate bodies amid which they move. That functional dependence of parts, which is scarcely more manifest in animals than in nations, has no counterpart elsewhere. And in no aggregate except an organic one or a social one, is there a perpetual removal and replacement of parts [such as animals of a particular species in a jungle, or skin cells in an individual creature], joined with a continued integrity of the whole. . . . Societies slowly augment in mass; they progress in complexity of structure; at the same time their parts become more mutually dependent; their living units are removed and replaced without destroying integrity. . . . And further, the analogy between the economical division of labour and the "physiological division of labour," is so striking as long since to have drawn the attention of scientific naturalists....[11]

Milton Friedman and Leonard E. Read observe that this sort of spontaneous order and Emergent Complexity is visible in the usage of language.[12] In the United States there is no one central government authority dictating grammar rules. Where in history there have been governments attempting to legislate over grammar rules, every such effort ultimately failed. Nonetheless, the absence of a central authority has not bred chaos. As the sum consequence of various individual choices, individuals communicating with one another develop their own norms and commonly accepted customs. Hence, the absence of central authority and government force to police and plan verbal expression fully, from the top down, has not caused anarchy or pandemonium. It is by consensual, free-market means that rules and standards emerged on their own from the bottom up. The same principle is visible in open-source software and Internet *wiki*s.

Both Spencer and Rothschild explain how the market economy, despite its artificiality, remains both a voluntarily accepted evolutionary adaptation and self-created habitat. Though Rothschild's explanation is more sophisticated and accurate than the biological metaphors Spencer employed, one must admit that several of Rothschild's key points were touched upon by Spencer beforehand. First is that the market economy foments a spontaneous order—complete with its own set of rules that are commonly agreed-upon by most of its participants—that runs on its own, having no need for micromanagement by a totalitarian State. Second is that the economy undergoes dynamic alterations, and additionally exercises such a degree of unpredictability that it would be more reasonable for social observers to view the economy as being more comparable to a biological process than to a predictable mechanical device. Just as with any institution in a dynamic free economy, grammar rules gradually change over the ages. With that in mind, it would be fair for Rothschild to concede that Spencer preceded him in brainstorming particular aspects of his economic theory. Instead, though, Rothschild furiously disavows any philosophic similarity with "social Darwinists," assuming that all of the defamatory accusations that Hofstadter's disciples have leveled about anti-regulation "social Darwinists" must be accurate.

From Whom and What Does Rothschild Think He's Distancing Himself?

Rothschild is at least implicitly aware that Richard Hofstadter and company have adorned Spencer with the reputation for having been the originator of social Darwinism. Rothschild is evidently not aware, though, that this reputation is undeserved. As a consequence of this false interpretation, Rothschild categorically rejects any ostensibly social Darwinist thought. This rejection most likely encompasses, by implication, Rothschild writing off and dismissing anything Spencer had to say about society being the work of biology. In Rothschild's view, "social Darwinism was a deeply flawed attempt to apply the theory of evolution to human social questions." Rothschild consequently elides the similarities between his own observations about spontaneous order and Spencer's. Rothschild huffs that

> the most familiar form of social Darwinism is an economic philosophy. Politically, a social Darwinist is a denizen of the far Right—a hard-core, laissez-faire capitalist who believes that in the struggle for prosperity, the capable succeed and the incompetent fail. . . .
> . . . this philosophy holds that the rich are generally superior and deserve their wealth. Those crushed underfoot in the scramble for riches are inherently inferior and deserve their hard fate. In the 1830s debate over the reform of England's welfare laws, extreme economic social Darwinists argued that the poor ought to be allowed to starve to death. In their view, society could improve only if the rich reproduced while the poor withered away.

Rothschild's accusation that laissez faire's proponents cited Darwin's theories in the 1830s is a strange one. Darwin did not become famous for his theory until after publishing *On the Origin of Species* in 1859. Nor had Spencer written on evolution by the 1830s. I wonder how laissez-faire partisans could possibly claim to draw conclusions from the renowned Charles Darwin over a decade before he became renowned. As we discovered in Book Two, Rev. T. Robert Malthus himself, whom Darwin considered an influence in his own thinking, was well-known in the 1830s, but he did not argue that the impoverished have a duty to expire without having kids. And as we have acknowledged in *LIME*, Malthus supported tariffs rather than the night watchman state. It is therefore doubtful that Malthus could be among the laissez-faire economists whom Rothschild impugns.

"The onset of the Industrial Revolution," Rothschild persists, "triggered massive social problems. Some businessmen built staggering fortunes, while the great mass of urban workers endured appalling conditions in overcrowded, filthy slums." Then Rothschild continues his inaccurate paraphrasing of the free-market evolutionism attached to the reputations of Spencer and Sumner. "Economic social Darwinists never doubted that the poor and their offspring were condemned to poverty by inborn deficiencies of ability and intellect. . . . Nothing less than the cause of human progress obliged the strong to obliterate the weak. If an inferior group wasn't killed or at least allowed to starve, they would reproduce, leaving more hopeless degenerates to cope with in the future."[13] For the previous passage, Rothschild cites as his information source Allan Chase's *The Legacy of Malthus*,[14] which insinuates that anybody who wants welfare repealed is a social Darwinist who begs the needy to drop dead.[15] Then Rothschild adds that social Darwinists "believed that . . . the laws of heredity demanded ruthlessness if society was to continue its upward course. The alternative was survival of the 'unfittest,' leading to the inevitable decline of civilization." Resultantly,

> white European colonialists seized on Darwin's idea of natural selection to justify their exploitation of dark-skinned natives. They had no trouble convincing themselves that "survival of the fittest" meant it was the white man's duty to subdue the "inferior" races. German philosophers believed nature's brutal struggles for species survival meant an inevitable battle for supremacy against nations. They urged the establishment of a strong central government to build the military machine required for future wars. To a large degree, the German militarism that preceded World War I can be traced to a "scientific" logic that perverted the meaning of Darwin's work. . . . To German social Darwinists, the French, British, and Slavic "races" were inferior forms of humanity.

Rothschild thus concludes that the "perverted logic" of the nineteenth-century's free-market evolutionism ultimately "led to one of the greatest tragedies of human history—the Nazi holocaust." Then Rothschild gets to the point—as an outgrowth "of this horrifying result, biology became a taboo subject for economic thinkers. Many still close their minds to the idea that the insights of modern biology, properly applied, might explain the complexities of the economy."

Rothschild is right that for most of the twentieth-century's latter half, mainstream economists have shied away from drawing prudent comparisons between economics and biology for fear of being labeled social Darwinian apologists. But as with almost everyone else, this engineer-turned-business-consultant errs when he buys into the accusations of Richard Hofstadter's acolytes that Spencer and Sumner's biology-centric economic theories sanctioned or inspired any of the spoliative eugenicist evils that he named. In so doing, Rothschild commits a hypocritical injustice on multiple counts. First he espouses a biology-centric economic theory incredibly similar to the one that Spencer thought up over a century before him. Then, to preempt any attempt by critics to classify him with the same social Darwinist straw man that they have already employed to demean Spencer, Rothschild inconsiderately repeats mainstream academia's misrepresentations about what biology-centric free-market economists like Spencer and Sumner actually said. Rothschild bellows that it was with "Darwin's new theory of evolution then towering over the intellectual landscape" that free-market "political writers"—whom Rothschild brands "quacks"—have sought to "reinforce their arguments by claiming to have proof for their ideas in Darwin's treatise. . . . Sadly, the name of one of the greatest men in the history of science was besmirched by involuntary association with every half-baked ideologue seeking to justify his notion as 'scientifically correct.'"[16]

Rothschild's comments are enveloped by ironies. First is that, by accepting the straw-man depictions of the nineteenth-century's free-market evolutionists as authentic, it is Rothschild who himself besmirches Spencer and Sumner in associating them with every "half-baked ideologue" who espoused a governist eugenicist policy. Rothschild has lots of company, unsurprisingly. Remember from chapter 1 that the libertarian Cato Institute's Brink Lindsey disparaged Sumner as a social Darwinist despite the fact that Lindsey himself compares the untamed market to an ecosystem. The second irony is that, due to *Bionomics*'s own arguments in favor of a mostly capitalist mixed-economy and against a larger encroachment of regulations, Rothschild's governist critics ignore every last one of his disclaimers about opposing old-fashioned nineteenth-century social Darwinism. Rothschild's anti-capitalist critics propound that his message reminds them of social Darwinism anyway. Hence politically progressive journalist Paulina Borsook, who hates the Industrial Revolution still more than Rothschild hates it, sneers at Rothschild's thesis. She refuses to entertain what she dismisses as the "peculiar locus of fake-o biology, technology, and libertarianism represented by Bionomics . . . I can't begin to speculate if it's a harbinger of social Darwinist horrors to come . . . But the next time you hear economics talked about in terms of niches and predators and evolution, know that what you're mostly hearing is the language of the free market, and not very much Mother Nature."[17]

The truth of the matter goes unacknowledged by Rothschild and Borsook. Spencer documented many of the same observations that Rothschild would offer about economies over a century later.

Sumner and Ethnocentrism

Comparable to Rothschild's errors is the manner in which a handful of popular twenty-first-century evolutionary psychologists have congratulated themselves for complete originality when spouting theories that William Graham Sumner had already formulated—albeit in a simpler form—in his 1906 textbook *Folkways*. Yet another gaping irony in this matter is that, though Sumner's outlook on political economy has rendered him one of the most politically incorrect figures in American history, he actually coined a term that the academic "multiculturalist" Left finds indispensable in advancing its PC agenda. That term is *ethnocentrism*. Dictionary.Com states that ethnocentrism is the "belief that one's own culture is superior to all others and is the standard by which all other cultures should be measured."[18] The academic Left denounces, as a right-wing ethnocentrist, practically anyone who pronounces negative judgment upon illiberal practices of non-Western cultures. Yet some militantly Islamic North-African communities insist on mutilating the genitalia of baby girls. Should we believe that the West has no business pronouncing judgment upon any other cultural practice, it follows that pronouncing judgment on forcible clitorectomies is verboten.

A near-consensus of modern anthropologists and sociologists credits Sumner with inventing this neologism, *ethnocentrism*.[19] As *Folkways* introduces the word, "Ethnocentrism is the technical name for this view of things in which one's own group is the center of everything, and all others are scaled and rated with reference to it. . . . Each group . . . boasts itself superior, . . . and looks with contempt on outsiders. Each group thinks its own folkways the only right ones, and if it observes that other groups have other folkways, these excite its scorn."[20] From the late 1960s onward, the countercultural and academic New Left took hold of this term. This decision was supposedly a corrective to the assumption of numerous people that every single institution of Western society was preferable to anything different in any other culture.[21] Such institutions could range from the major ones, such as banking and double-entry bookkeeping, to more innocuous ones, such as eating hamburgers or wearing footwear inside one's own house. Those who assumed that a custom was good solely because it was Western, were accused of ethnocentrism.

To my consternation, the New Left "multiculturalist" intellectuals who pervade academia do not stop there. They go to the other extreme of asserting that no cultural practice can be deemed better than another. Consequently, more than a few "multiculturalists" resort to accusing an American man of bigotry and ethnocentrism if he criticizes as inhumane the treatment of women in Saudi Arabia as second-class citizens (or worse), or the Aztecs' totalitarian government system and imposition of atrocious human sacrifices to avert an imaginary apocalypse.[22]

And lots of multiculturalists go farther, intimating that every institution associated with the West is insipid at best and immoral at worst, particularly insofar as they judge Western culture to be generally individualistic and egoistic. These multiculturalists disparage the respective values of commerce and industrialization and individualism, which they interpret as predominantly Western institutions. Conversely, they uphold, as unconditionally acceptable, every aspect of any non-Western culture that seems to clash with these values—such as social collectivism and animistic religions in various non-industrialized tribes. These particular multiculturalists veer from their original contention that each culture is equally praiseworthy, and end up saying that every culture is equal except for the West, which must rank as inferior. Basically in comparison to other conceivable societies, the MDCs (more-developed countries) are implicitly evaluated as mediocre at best—and, at worst, the epitome of vileness. Such multiculturalists deny that any non-Western custom or tradition that spoliates someone is objectively wrong. Commensurately, they deny that a culture should objectively allow any custom to remain legal if it spoliates no one. These multiculturalists also deny that when life is the standard of human values, scientific rationality is superior to superstition in terms of improving life for everyone in the long haul. From this trilogy's comparisons between relatively liberal and relatively illiberal governmental institutions, we know the multiculturalists are terribly misguided on this point.

When Sumner introduced the concept of ethnocentrism, though, he bore no intention of contributing to anti-Western, PC proselytizing. Sumner pointed out that not just Westerners, but individuals of almost every culture, are wont to hail their own society as greater than any other—an observation that academic multiculturalists commonly overlook as they impute ethnocentrism solely to Westerners.[23] Further, Sumner warned his fellow professors that scientific objectivity required that one guard against his own ethnocentrism while studying alien cultures. *Folkways* inveighed against the notions that (1) all customs observed in the West were inherently rational as a result of the West always being right, and (2) that any difference in customs that other societies have from the West is automatically a consequence of those other cultures being inherently weird and wrongheaded.

Folkways's point is that, to a large extent, the customs that a society presently practices are the culmination of natural selection. Such customs, practices, folkways, mores (Sumner coined the word *mores*, too), traditions, and beliefs either thrived or went extinct according to whether they helped a familial clan secure its long-term genetic legacy. This is explained by the possibility that a familial clan was able to transmit, to the next generation, its genes as a result of practicing some custom that improved its chances for survival. As this custom aided in that clan's survival, that clan would, in turn, pass that same custom down to succeeding generations, inculcating an observance of that beneficial custom into the clan's youngest members.

And most of these customs are not inherently good; they simply contributed to the clan's survival insofar as they helped it adapt to the geographic and environmental circumstances under which it dwelt prior to the Industrial Revolution. This is the same theory that evolutionary psychologists E. O. Wilson and Jared Diamond are inaccurately credited for originating in the late 1900s. Yet these are also the same evolutionary psychologists who, deathly averse to being criticized as social Darwinists, pronounced that their own theories owe no debt to the writings of Sumner or Spencer.

Sumner additionally warned his colleagues about ethnocentrism on the grounds that he did not want them to let their own cultural biases interfere with their objectivity in studying foreign cultures. In *Folkways* he observes, "Special occasion for rules and propriety is offered by eating. . . . The Bakairi are ashamed to see or to be seen eating."[24] Sumner worried that a Western researcher who noticed this oddity about the Bakairi's etiquette rules might permit his own ethnocentrism to get the better of him, chalking the Bakairi's taboo up to this tribe simply being irrational and primitive, and insisting that there was no evolutionary reason why this taboo would persist over multiple generations. *Folkways* articulates that once someone pushes his ethnocentrism aside, and studies other cultures in an unbiased fashion, one may find that the culture might once have benefited from adopting that custom under a specific set of circumstances foisted upon it at one time in history. The Bakairi tribe, for instance, dwelled in a preindustrial environment where rampant food shortages might have motivated individuals in the same village to squabble over their meals. When a man so much as saw a fellow tribesmen eat something, a fight could break out. Under such conditions, Familial Tribe 1, wherein every man continued to eat in front of other men, might have brought itself to extinction. By contrast, Familial Tribe 2, which found that it was appropriate for each tribe member to eat his meals in complete solitude, might have more successfully preserved enough internal peace for the tribe that at least a few of its members lived long enough to endow their progeny with both their genes (through biological processes) and their "eat-your-food-in-private" rules (through conditioning). This should sound familiar, as it is similar to Marvin Harris's theory that the Middle East's anti-pork taboo is a product of gene-culture co-evolution. We went over that theory in Book Two.

Tackling Sociobiology Before There Was *Sociobiology*

Sumner's contributions to evolutionary psychology abound. *Folkways* anticipates the theory of customs being wrought from gene-culture natural-selection. To wit:

> The struggle for existence must be carried on under life conditions and in connection with the competition of life. The life conditions consist in variable elements of the environment, the supply of materials necessary to support life, the difficulty of exploiting them, the state of the arts [technology], and the circumstances of phsyiography, climate, meteorology, etc., which favor life or the contrary. . . . Interests and the relations of action and reaction between the individual and the life conditions, through which relations the evolution of the individual is produced. . . . It would be an error, however, to suppose that all nature is a chaos of warfare and competition. Combination and cooperation are…fundamentally necessary… This combination has well been called antagonistic cooperation. It consists in the combination of two persons or groups to satisfy a great common interest while minor antagonisms of interest which exist between them are suppressed. . . . If a savage puts his hand too near the fire, he suffers pain and draws it back. . . . If he wants to catch an animal for food, he must study its habits and prepare a device adjusted to those habits. If it fails, he must try again, until his observation is "true" and his device is "right." . . . The morality of a group at a time is the sum of the taboos and prescriptions in the folkways by which right conduct is defined. . . . The taboos carry on the accumulated wisdom of generations, which has almost always been purchased by pain, loss, disease, and death. Other taboos contain inhibitions of what will be injurious to the group. The laws about the sexes, about property, about war, and about ghosts, have this character.

Sumner continues that when one examines a Stone Age society's mores, one finds that its mores was "never anything but the consonance" between what was "done and what the mores of the age" needed, and that the same principle applies to the mores of the present era. By contrast, when the Industrial Revolution happened, it brought about a tremendous alteration in "the conditions of life" for the "members of society," which thereupon rendered obsolete lots of the preindustrial aspects of the mores. Among the Stone Age mores that ceased to be useful were (1) the us-versus-them mentality and (2) the conviction, which I disproved in Book Two, that there is a fixed quantity of wealth for humans to consume that is ever-depleting. We recall that the Industrial Revolution altered the environmental circumstances governing the natural selection sequence for those in the West. Remarking upon this, Sumner elucidates that it was necessary for Western customs to "change by adaptation" to the "new conditions . . ." The natural selection sequence whereby the circumstances of a people's biological habit have long determined which of their customs, practices, beliefs, and traditions survive and which die off—especially before industrialization took hold—is an aspect of ontologically given human nature that Sumner dubs "societal selection."[25]

Sumner tackles aspects of sociobiology decades prior to E. O. Wilson penning *Sociobiology*. These evolutionary psychologists should respect Sumner for coming up with so many ideas that gave their discipline a solid foundation with which to work. Two evolutionary psychologists very blatantly borrow from Sumner's theories—such as the theory that learned and conditioned behaviors and customs count as evolutionary adaptations throughout natural selection. They are Ian Tattersall and Niles Eldredge, who state in their book, *The Myths of Human Evolution*, that

> it only seems natural to wonder why the patterns of biological and cultural evolution are so similar . . . A species' distribution is disrupted by shifting patterns of ecological change. . . . Extinction and evolution appear as reflections of a changing physical environment. Classic Darwinian theory sees this response as a direct adaptive reaction: change the environment and the species change to keep pace. But the sort of reaction we have in mind here is *geographic*. Climatic change transforms the distributions of environments, often fragmenting once[-]continuous habitat belts. Evolution and extinction largely mirror changes in physical geography. . . .
>
> Here, then, is a similarity between cultural and biological evolution worth noting: geography seems to play an important role in each. Geography *isolates*. Both cultural and biological evolutionists have for over a century attributed the diversity in the systems they study primarily to the isolation afforded by far-flung geographic distributions [of different organisms]. The histories of particular societies or species are, of course, independent of other societies or species with which they have no contact. . . .
>
> Thus geographic patterns—especially isolation—can foster change and consequent stability in both evolutionary systems [the emphases are from Tattersall and Eldredge].[26]

Tattersall and Eldredge owe Sumner and the mentor of his who ignited his fascination with gene-culture co-evolution, Herbert Spencer,[27] large debts of gratitude. But thanks to their misinformed attitude concerning what Spencer and Sumner advocated, these same sociobiologists subject these two intellectual giants to a litany of insults. In lieu of crediting Spencer, for instance, Tattersall and Eldredge deride him. As we recall from chapter 1, evolutionary psychologist Frans de Waal denigrates Spencer as well, overlooking the precedents that Spencer set for him. Tattersall and Eldredge resent Spancer's advocacy of capitalism, which they assume to be a zero-sum game wherein "there are winners and losers. The winners survive and produce offspring. The losers simply lose." Eldredge and Tattersall proceed to put up this straw-man representation of Spencer's philosophy. "The best survive, and since their offspring resemble them, there is improvement of the population as a whole as time goes by. This is the basis of Adam Smith's laissez-faire, open-competition economics." These two late-twentieth-century scientists then continue that Spencer and other free-marketers in Adam Smith's tradition were "Social Darwinists" and "apologists for the status quo where a privileged few enjoyed the fruits of the Industrial Revolution while the masses labored in frequently miserable conditions . . ."

In the words that Tattersall and Eldredge put in their mouths, Spencer and the nineteenth century's other free-enterprise economists conceived of such drudgery as "the logical, inevitable outcome of laws of nature: survival of the fittest ('natural selection')" that "put the best (including the [laissez-faire] theorists) at the top of the heap in the continuous competitive struggle . . ." The scientific duo then breathes a sigh of relief that Spencer was pilloried by governist opponents who were not "so myopic. Most social reformers also believed in progressive change and improvement; they simply wanted to see more of the populace participate" in communal progress by means of government-imposed wealth redistribution and regulations.[28] And Tattersall and Eldredge omit mention of those reformists' support for governist eugenics.

Shermer Versus Spencer and Sumner

Similarly, science journalist Colin Tudge jeers that "Spencer's social Darwinism was wrong," "misguided," "foul" and definitely the biggest of the "false starts" in the field of studying the extent to which human "behavior has evolved by natural selection . . ." Tudge then goes on to congratulate modern evolutionary psychology for being a complete repudiation of anything Spencer might have dreamt up. "It is amazing," he pouts, ". . . how many modern critics of evolutionary biology apparently conflate it with Spencer's social Darwinism . . ." Thereupon such critics "manage to conclude that a society that acknowledges the genetic roots of behavior is ipso facto red in tooth and claw and can have no respect for human individuality and freedom." More than anything, Tudge opines, modern sociobiology "needs to show that it has shaken off its . . . crude, nineteenth-century forays of social Darwinism." And henceforth "the people who choose to reject the notions of evolutionary psychology out of hand—convinced that Darwinism equals fascism and that's an end of it—must attend to their own education."[29]

Once again we come across an author's flaky accusation that Spencer exhibited "no respect for human individuality and freedom." This demonstrates that Tudge either has not read Spencer's political writings, or, having read them, did not do so for the purpose of comprehension. Contemporary sociobiology does have clear roots in the writings of putative social Darwinists like Spencer and especially Sumner. Additionally, evolutionary psychologists actually concerned about the truth would admit this and correct the public's misconceptions about what Spencer and Sumner meant. It is such evolutionary psychology popularizers as Tudge, accepting and then circulating such misconceptions, who really "must attend to their own education."

Science historian and *Skeptic* magazine publisher Michael Shermer dishes out similar injustices, though in a less severe form. Shermer follows in the tradition founded by Spencer and William Graham Sumner in applying his own era's knowledge about natural selection to gaining insight into the social sciences. Were it the case that anyone who applied evolutionary theories to the social sciences counted as a social Darwinist, that would place Shermer in the social Darwinist camp as well. In fact, Shermer's unfairness is twofold, as he is both an evolutionary psychologist and bionomics theorist at the same time—not a common combination. As Spencer anticipated the theories of bionomics, and both Spencer and Sumner laid the groundwork for evolutionary psychology, it follows that Shermer's has two major intellectual debts to Spencer. But, as with practically everyone else I take to task in this chapter, Shermer spurns recognition of Spencer's innovative spirit.

Shermer prefers to refer to his ideology as *evolutionary economics*, which he defines as "the study of the economy as an evolving complex adaptive system grounded in human nature that evolved functional adaptations to survival as a social primate species . . ." He goes on that "ecologies and economies are *Complex Adaptive Systems* (CAS): systems in which individual particles, parts, or agents interact, process information, learn, and adapt their behavior to changing conditions" (emphasis Shermer's). That has a ring of familiarity. From the beginning of this chapter, we have known that Spencer recognized economies as complex adaptive systems. And Sumner noticed that the same principle applies to societies that adopt customs in response to the geographic and other environmental conditions. Further, Sumner noticed that the natural-selection aspect of the adoption of these customs was particularly visible in forager societies much like those of the Paleolithic period.

But in a book that argues from premises based on both bionomics and evolutionary psychology, *The Mind of the Market*, Shermer regurgitates the same old clichés about Spencer extolling a ruthless social Darwinism. The science historian's work blathers that

Spencer's philosophy was "based on the presumption that animals and humans are inherently selfish . . ." Given that Shermer throws around the terms *selfish* and *altruism* in the invalid manner in which they are conventionally used, one ascertains that Shermer is accusing Spencer of opining that every person cares exclusively about his own welfare and no one else's. This is particularly ironic given that, as we read in Book Two, Herbert Spencer pioneered in the usage of the term *altruism* in the late 1800s, preceded by a few like Auguste Comte and John Stuart Mill. Indeed, the real presumption is the one that Shermer betrays about Spencer. Contradicting Shermer, we have previously observed that Spencer strongly believed in human charity. We recall from Book Two that it was Spencer who started the misconception that *altruism* can be taken as a synonym for benefiting other people and caring about them. That notion perverted the much more specific definition that Comte and Mill provided. Given that Shermer unknowingly applies Spencer's revisionist definition of altruism, Shermer is unknowingly influenced by Spencer in this regard.

Going on, Shermer lambastes Spencer for having "immortalized natural selection in the phrase 'survival of the fittest,' one of the most misleading descriptions in the history of science . . ." Excuse me, Dr. Shermer. There is nothing misleading about that phrase when one understands that it refers to the fact that a species' survival is incumbent upon being able to *fit* into its environment. Then, predictably, Shermer concludes that *survival of the fittest* is the phrase "that has been embraced by social Darwinists ever since, who apply it inappropriately to racial theory, national politics, and economic doctrines." In the paragraphs following that one, Shermer continues that Spencer's theories about natural selection under capitalism are wrong by virtue of the fact that human beings are capable of cooperating and bestowing charity and kindness on one another. Shermer presses this accusation in ignorance of Spencer having stressed that same point.

William Graham Sumner does not receive better treatment from Shermer than Spencer does. Repeating what he did with Spencer, Shermer borrows important ideas from Sumner while disrespecting him, refraining from citing him as an intellectual influence. Anticipating Shermer, and David Berreby, whose derision of summer we quoted in chapter 1, it was Sumner who first posited that the mentalities of *us-versus-them* and *the in-group* were holdovers from the Stone Age. Sumner anticipated Shermer and Berreby in explaining that competition among Stone-Age human beings in the process of natural selection occurred not at the level of one man against another as much as it was one kin group against another. In "primitive society," *Folkways* elaborates, the members of a group "may have some relation to each (kin, neighborhood, alliance, connubium, and commercium) which draws them together and differentiates them from others. Thus a differentiation arises between ourselves, the we-group or in-group, and everybody else, or the others-groups, out-groups. The insiders in a we-group are a relation of peace, order, law, government, and industry, to each other. Their relation to all outsiders, or others-groups, is one of war and plunder, except so far as agreements have modified it."[30]

Now recollect, from Book Two, the strong emphasis that Shermer places on this social-science model in his own writings: "...limited resources led to the selection for within-group cooperation and between-group competition in humans, resulting in within-group amity and between-group enmity. This evolutionary scenario bodes well for our species—if we can continue to expand the circle of whom we consider to be members of our in-group. Recent conflicts are not encouraging, but in the long run there is a trend toward including more people (such as women and minorities) within the in-group deserving of human rights."[31] David Berreby wrote a whole book to advance this same social-science model.

In fact, Oxford University's *Dictionary of Sociology* attributes the very coinages of *in-group* and *out-group* to Sumner himself.[32] To put it simply, Shermer repeatedly uses the term *in-group* and, whether he is cognizant of it or not, his anthropological conception of that term originated with Sumner.[33] Shermer's very point is a repetition of the one from *Folkways* about antagonistic cooperation. What does Sumner receive in exchange? Remember that just a few pages ago, Shermer denounced Herbert Spencer-influenced "social Darwinists" who barfed out some horrendous "economic doctrines."[34] Given that Sumner was influenced by Spencer and promoted Spencer's economic doctrines, the *Skeptic* publisher's derision toward economic social Darwinists is an implicit slap at Sumner. And, as we remember from chapter 1, Berreby refuses to acknowledge Sumner's value.

By repeating the social Darwinism cliché, Shermer evinces in his own argument what is, at best, ignorance about the theories of Spencer and Sumner. This would not trouble me as much as it does except that, to restate a point from Book Two, Shermer received his Ph.D. for being an expert on the history of scientific thought and theories. More precisely, his Ph.D. dissertation was on the philosophy of a man whose theories were heavily influenced by Herbert Spencer, Alfred Russel Wallace.[35] This is galling. Someone cannot truly understand Alfred Russel Wallace's philosophy in the absence of understanding Herbert Spencer's influence on it. Yet Shermer is credited as an authority on Wallace. As the theories of Spencer and Sumner are part of the history of scientific thought and of his own dissertation topic—as are the theories of the eugenicists—Shermer, more than most intellectuals, ought to know and acknowledge that Spencer and Sumner did not want the strong to crush the weak, and that there never was a general movement in favor of a congenitally ruthless social Darwinism.

Excavating Sociobiology's Ancestors

To my sorrow, the practitioners of evolutionary psychology are no better than the evolutionary economists in terms of acknowledging the extent to which their own discipline was pioneered by Herbert Spencer and William Graham Sumner. A case in point is that of the prestigious science journalist Steve Olson (b. 1956). He penned an otherwise interesting sociobiology tome *Mapping Human History*. This work explicates, just as Sumner did decades prior, that natural selection still occurs among human beings and that by means of geographic topography and other environmental factors, natural selection affects which human social customs survive or die out. *Mapping Human History* contains theories about how the natural selection cycle accounts for what are presently regarded as racial differences. The reason why whites, blacks, and East Asians look so different from one another pertains to how each group adapted to its geographic habitat in the Stone Age. Olson gives a particularly plausible explanation for how natural blondes and redheads came to exist.

The first human beings originated in east Africa, and the group of persons living in our era who most closely resemble these first humans are probably the !Kung San (a.k.a., the Kalahari Bushmen). They had about the same height and build, and had brownish bronze (as opposed to "black") skin. As human beings spread throughout Africa and migrated to other continents, the vast majority of our ancestors had this brownish skin color. During the Stone Age, the peoples of northern Europe—the Swedes, Norwegians, Finns, Dutch, Icelanders, and Danes—inhabited a cold northern climate quite similar to that of Eskimos. However, a significant difference between the environments of the far north of the Americas and the far north of Europe accounts for why the Nordics have an appearance distinct from Eskimos. The Eskimos mastered an environment wherein they could eat fish rich in Vitamin D, and Vitamin D protected them from contracting rickets. But such Vitamin D was much harder to come by in northern Europe back then. At some point a mutation occurred that enabled some northern Europeans to survive this Vitamin D deficiency—this mutation being lighter-colored skin that attracted more Vitamin D from the sun than darker skin would. Despite the genes for lighter skin being recessive, lighter skin came to dominate among human beings in northern Europe. This does not suggest that a lighter skin color is inherently better (no skin color is), but it is evident that this genetically induced adaptation provided a greater "fit" for this particular habitat, under this particular set of environmental circumstances, than did genetically induced dark skin.[36] Say, for instance, that both a Light-Skinned Family and a Dark-Skinned Family occupy Stone-Age northern Europe. The Dark-Skinned Family does not get sufficient Vitamin D and it resultantly dies of rickets before most of its members can reproduce. The Light-Skinned Family receives sufficient amounts of Vitamin D from the sun, and hence subsists long enough to replicate its genes. Over time, Dark-Skinned Families dwindle in this environment whereas Light-Skinned Families come to perpetuate their numbers and dominate.

The color of one's eyes and hair correlate with skin color. Someone possessing very little pigment would be an albino with white hair. As an albino's irises contain very little pigment, one can see the blood behind his irises; that is why albinos have red eyes. Someone with much more pigment would have blue eyes, and blonde or red hair. Darker pigments will render one's eyes gray or violet, and still more pigments will render eyes green. That most people have darker pigments explains why most of the Earth's human population has brown eyes and brunette or black hair. The gene for red hair originated in Africa, and had been carried by dark-skinned peoples for millennia before the Nordic races came into existence, and it eventually found its way out of Africa when Africans began migrating into Europe. As red hair is a recessive trait that correlates with light skin, the trait was seldom seen in Africans. A large population of redheads did not emerge except in the environment of northern Europe, where ontologically given ecological circumstances "selected" for the genes that produced light skin.[37]

Between World War Two and the publication of *Sociobiology*, social scientists appeared to be reluctant to discuss that topic so openly, for fear of being stigmatized as social Darwinist eugenicists covertly and cryptically advocating Aryan supremacy and "scientific racism." Of course, every ethnic group on Earth is just as highly evolved—in the biological, genetic context—as any other. The reason why West Africans have specific distinctive physical features is that those were the features that helped West Africans adapt in the Stone Age, just as East Asians have specific distinctive physical features on account of those features being naturally "selected" to persist in the East Asian environment. This happened through the process of kin selection, wherein the presence of particular physical traits in a clan assisted enough members of that clan in surviving long enough to disseminate their long-term genetic legacy.

In terms of "deciding" which familial groups bequeath their DNA to future generations and which do not, such physical "racial" differences are not as big a factor in the modern industrial environment as they were in the Stone Age. Recall from Book Two that modern industrialization has leveled the playing field for every ethnicity on every geographic terrain. Back in the Stone Age these physical differences, such as skin color, could sometimes be one of the biggest factors in determining who was likely or unlikely to transmit his germline. But such superficial racial differences hold miniscule influence in a modern, highly industrialized, commercialistic republic. First the modern industrialization has significantly reduced the statistical death rate and thereupon slowed down the process whereby specific genetically induced physical characteristics in men are eliminated from the gene pool. Moreover, in the modern industrial environment a person primarily maximizes his chances of producing more kin by financially enriching himself. To the extent that that person exists in a free economy, a person enriches himself through maximizing revenue by satisfying consumer demand in a business setting. And this ability to maximize revenue is governed by a rational faculty that is equally keen in every ethnicity and "race." Hence, the physiological features that distinguish one "race" from another are not as big a factor in modern times in deciding which familial groups are "fittest" in perpetuating their genes.[38]

Steve Olson further details how environmental conditions in the Stone Age did more than cause peoples of separate regions to form distinctive "racial" physical features. Additionally, such environmental conditions influence the customs and activities of various cultures. Olson discusses, for instance, the impact that the climate of the Ice Age had from 20,000 B.C. to 16,000 B.C. This climate change encouraged human beings to migrate from northern Europe to southern Europe, where they would find shelter in a cavern and hone their skills at painting on its walls. Olson relates how, when the Ice Age began to end around 13,000 B.C., the new change in climate encouraged men to lose their proficiency in painting while taking up other activities.[39] Predictably, when he applies the science of natural selection as his guide in studying human society, ethnography, and particularly the reasons why particular cultural customs wane or flourish, Olson is merely continuing in the social science tradition pioneered by Spencer and Sumner. Olson's discussion of how a location's environmental circumstances "select" which customs live or die, parallels the drift of Sumner's *Folkways*.

But, as usual, in lieu of bothering to read Sumner and Spencer in context, Olson simply accepts the Hofstadter-inspired falsehoods as indisputable and parrots them. In so doing, he reaps the benefit of the science that Sumner and Spencer co-founded as he downplays Spencer's and Sumner's contributions to social science. In a book published by no less than the National Academy of Sciences, Olson drubs Spencer for being "one of the earliest and most ardent" of the "evolutionary ethicists" he curses for conning humanity. He curses Spencer and other free-marketers for extending evolutionary theory into

> the social and economic realms. They advocated an extreme form of economic *laissez-faire*, in which the strong should prosper and the weak be brushed aside. In its various guises, this philosophy has come to be known as Social Darwinism, though it owes more to Spencer than to Darwin. . . .
>
> At the normative level, Social Darwinism advocates that people acquiesce to or even encourage the workings of evolution, as expressed in its competitive nature. . . . Some people may use it to argue for free market economies in which the government plays little or no role. For instance, as prime minister of Britain, Margaret Thatcher has been reported as saying, "What we need are Darwinian principles." . . .
>
> Social Darwinism can be faulted on a number of grounds, according to [University of Guelph professor of history and philosophy Michael] Ruse. For one thing, in its most competitive forms, it is simply immoral. "Morality does not consist in walking over the weak and the sick, the very young and the very old," Ruse says. "Someone who tells you otherwise is an ethical cretin."

Olson's implication is that Spencer and Sumner must have been "ethical cretins." Olson then tries—as Colin Tudge, E. O. Wilson, and Charles Lumsden have tried—to extricate from evolutionary psychology's reputation any notion that it has any allegiance to Spencer and Sumner. Hence Olson declares, "Biologists have also come to a much broader view of evolution than that espoused by Spencer and other Social Darwinists."[40]

In Place of Gratitude, They Receive Scorn

The most appalling example of ingratitude to Spencer and Sumner on the part of sociobiologists is the case of the man who coined *sociobiology* in the first place and, in the 1970s, revived the discipline of applying knowledge about evolutionary theory to the social sciences—Edward O. Wilson. E. O. Wilson and Charles Lumsden seem to assume that they completely originated the gene-culture co-evolution theory. Specifically, I mean the aspect of it that points out that particular customs emerge through natural selection as a consequence of those customs being adaptations to particular circumstances existing within a human society's biological habitat. But Sumner had already argued that point in 1906. Evolutionary psychologists do not fess up that they built on insights already publicized by Spencer and Sumner. Wilson and Lumsden—as with everybody else—opt to defile the names of their nineteenth-century forebears. Writing together, Lumsden and Wilson accuse Spencer of "Social Darwinism."[41] Writing solo, Wilson contends, "Franz Boas, aided by his famous students Ruth Benedict and Margaret Mead, led a crusade against what they perceived (correctly) to be the eugenics and racism implicit in Social Darwinism. With caution swept aside by moral zeal, they turned opposition into the new ideology of cultural relativism."[42] That statement is misleading. One would glean from it that Drs. Boas, Benedict, and Mead, having been collectively cast by Wilson as enemies of social Darwinism and supporters of cultural relativism, would correspondingly be the enemies of Sumner, whom Wilson christened "the absolute Social Darwinist."[43]

The truth is that Dr. Benedict herself was influenced by Sumner and cited *Folkways* in her monograph *Patterns of Culture*. Moreover, this same work contained a preface by Mead and an introduction by Boas[44] —the same Mead and Boas whom Wilson erroneously characterized as having crusaded against Sumner's teachings. And the "cultural relativism" that Wilson imputes to her was, as we have witnessed previously, a bastardization of Sumner's admonitions against ethnocentrism.

As for Spencer, he also, to some degree, anticipated the gene-culture co-evolution theory propagated by Wilson and Lumsden. To review what we said in Book Two, that is the theory that the human gene pool and cultural customs changed conterminously, affecting one another in a feedback loop. Natural selection led to the emergence of free will, communication, and learning-by-imitating in human beings, thus ensuring the propagation of culture from one generation to another. A familial clan was able to ensure its long-term genetic legacy insofar as its customs were adequate in adapting that clan to its geographic environment. Success in such adaptation likewise determined which genes would be transmitted through the generations, as well as which customs. Alterations in the environment—including technological advancements—produce changes in which customs, behaviors, and other folkways are most conducive to adaptation. Robert L. Carneiro explicates how Spencer anticipated aspects of this model. As Carneiro paraphrases, "Environmental changes, modifying the conditions of existence, lead to a modification of human character. This modified character is no longer congruous with the existing social system. And this incongruity produces a disequilibrium which ultimately results in reequilibration. The reequilibration yields a social system more consistent with the altered nature of its members."[45] Speaking for himself, Spencer phrases it, "the characters of the environment co-operate with the characters of human beings in determining social phenomena . . ."[46] Additionally, "Either the social arrangements are gradually changed until they come into harmony with prevailing ideas and sentiments; or, if surrounding conditions prevent change in the social arrangements, the necessitated habits of life modify the prevailing ideas and sentiments to the requisite extent."[47] This results in "the increasing action and reaction of institutions and character, each slowly modifying the other through successive generations."[48] As such, "the society as a whole has the character of its sustaining system determined by the general character of its environment . . ."[49] Finally, "While spreading over the Earth mankind has found environments of various characters, and in each case the social life fallen into, partly determined by the social life previously led, has been partly determined by the influence of the new environment."[50]

And there are other evolutionary psychologists who refrain from belittling Spencer and Sumner so virulently, but still refuse to thank them for erecting the foundation upon which evolutionary psychology stands. In chapter 2 I mentioned the prominent biology journalist Matt Ridley. He wrote a book entitled *The Origins of Virtue* in order to argue that commerce is a human-invented—and socially conditioned—evolutionary adaptation that has enabled mankind to persist for millennia. In *Origins*'s acknowledgments section, Ridley thanks Colin Tudge, Garrett Hardin, evolutionary psychologist Robert L. Trivers, and "kin selection" coiner W. D. Hamilton. Yet Spencer is almost completely snubbed from the book, being mentioned on but one of its pages, and this mention amounts to this misrepresentation: "Spencer argued that because nature is a pitiless struggle, pitiless struggle must be virtuous." And as we stated in chapter 2, Ridley accused Spencer of founding "social darwinism" and went as far as falsely crediting him for coining the phrase.

There is another aspect in which Ridley builds upon a foundation begun by Spencer and Sumner. It comes with a particularly innovative manner in which Ridley showcases parallels between the realms of biology and cultural change. This is in Ridley's theory that cultural innovation results from "ideas having sex." You will recall that we discussed the ideas-having-sex theory in Book Two. Note that insofar as Spencer and Sumner had the courage to be among the first to apply biological theories in explaining economics and cultural change, they blazed a trail for Ridley to that extent. But, as we noted in Book Two, former *Wired* magazine editor Kevin Kelly arrived at a theory roughly similar to Ridley's around the same time.

Ridley's failures to acknowledge Spencer and Sumner is funny, given that *The Origins of Virtue* purports that evolutionary biology proves this conclusion of the author's—"If we are to recover social harmony and virtue, . . . it is vital that we reduce the power and scope of the state. That does not mean a vicious war of all against all. . . . It means a massive disassembling of the public bureaucracy." The book's final chapter happens to assert that a just society is founded on the free "exchange" of goods and services "between equals. Just as trade between countries is the best recipe for friendship between them, so exchange between enfranchised and empowered individuals is the best recipe for cooperation. We must encourage social and material exchange between equals for that is the raw material of trust, and trust is the foundation of virtue."[51] It would have been nice if Ridley could have conceded that a certain someone argued that same case over a century earlier—the case being that natural selection proves the need for the night watchman state, and that certain someone being the same Spencer whom Ridley blasted for extolling a "pitiless" social Darwinism. Compare the passage of Ridley's just quoted to this summary of Herbert Spencer's thought by E. O. Wilson: "Herbert Spencer, the most ambitious of the nineteenth century evolutionists, argued the necessity of a non-Kantian," logical "approach to ethics. . . . He believed that the human nervous system had been modified through thousands of generations to create certain…faculties of moral intuition, consisting of emotions responding to right and wrong conduct, but that human nature can be molded by 'the rigorous maintenance of the condition of harmonious social cooperation.'"[52] I am saddened by Ridley's failure to grant the fact that his *Origins* was written to prove a thesis already proposed by Spencer and Sumner—that the principles of biological evolution, natural selection, and adaptations to the natural environment can account for much of what occurs in the market economy, and of why it functions as well as it does.

Similarly, Harvard psychologist Steven Pinker makes forays into evolutionary psychology. You will recollect, from Book Two, that he endorsed Michael Shermer's evolutionary psychology book *The Science of Good and Evil*. Writing in *The New Republic*, Pinker pronounces, "Social Darwinism was the misnamed laissez-faire philosophy of Herbert Spencer," which Pinker proclaims to be, at best, "pseudoscientific." Pinker then proclaims that "the term" *social Darwinism* "is often used to smear any application of evolution to the understanding of human beings."[53] Pinker is definitely correct about *social Darwinism* being a smear, but his presumption about Spencer's philosophy being "pseudoscientific" is unfair, given that Spencer had anticipated ideas explored by Pinker and his fellow evolutionary psychologists. Remember, from chapter 3, our observation that Spencer anticipated the thesis of Pinker's book *The Better Angels of Our Nature* which is that as a society advances, it commensurately becomes less prone to violence.

Meanwhile, William Graham Sumner anticipated the main point of a 1997 Pulitzer Prize-winning popular science book that received public applause from E. O. Wilson[54] — *Guns, Germs, and Steel* by governist UCLA geographer Jared Diamond. Conservative commentator Dinesh D'Souza very accurately summarizes Diamond's work by noting that it contends that "cultures are the product of location and natural resources, and whether a culture develops or remains stagnant depends on such factors as the availability of mineral resources, climate, proximity to rivers, and such."[55] In its own words, *Guns, Germs, and Steel*—which I shall abbreviate as *GGS* or *Guns*—states that "the striking differences between the long-term histories of peoples of the different continents have been due not to innate differences in the peoples themselves but to differences in their environments."[56]

Granted, Diamond's book contests the eugenicist tenet of genetics-based psychological determinism, that genetics can render one ethnic group more inherently rational or competent than another. In consideration of that, it is tempting for a PC partisan to propound that Diamond's theory bears no similarity to any of Sumner's ideas. But Diamond's point actually repeats the entire theme of Sumner's *Folkways* in that it professes that a culture's set of customs are a human social group's volitionally chosen adaptations to the geographic environment that this group inhabits. If *Folkways*'s thesis is fundamentally socially Darwinian, then so too is Diamond's Pulitzer-winning tome. *Folkways* and *Guns* advance the same point. The latter work is acclaimed. By contrast, the former work is ignored mostly on account of intellectuals being told, whether they read it or not, that it is bigoted.

There is, however, a huge shortcoming in *GGS* that does not exist in *Folkways*. Diamond does not merely conclude that every ethnic group's customs were largely shaped by the geographic circumstances it lived under prior to the Industrial Revolution. Diamond goes as far as propounding that the ascension of Western commercial society was chiefly the consequence of dumb luck. The luck in this case refers to Diamond's allegation that Western society has come to dominate the globe primarily for the reason that the geography of Europe gave those who settled on the continent a head start. The Europeans' head start, Diamond argues, gave them a privileged position when it came to developing various important technologies such as farming and tool-making.[57] We refuted that claim in Book Two.

We also uncover a much more minor—but still-existing—unpaid debt to Sumner in this incredible statement from *GGS*—"My own impression, from having divided my life between United States cities and New Guinea villages, is that the so-called blessings of civilization are mixed. For example, compared with hunter-gatherers, citizens of modern industrialized states *enjoy better healthcare, lower risk of death by homicide, and a longer lifespan*, but receive much less social support from friendships and extended families"[58] (emphasis added). Diamond, who does not acknowledge any obvious superiority in longer life expectancies and a "lower risk of death by homicide," was over five decades of age when *GGS* hit the bookstores. If he did not have access to the amenities provided by the very same Industrial Revolution he finds ho-hum at best and troublesome at worst, Diamond would not have survived long enough to have *Guns* published. The UCLA geographer's insinuation that industrialization has rendered life more antisocial is exposed as fatuous when one realizes that living in an industrial republic does not preclude a man from associating with extended family members who want to spend time with him.

Industrial civilization does, though, give someone more options to associate with a larger variety of peoples to his liking, rather than just relatives from one's own tribal village.

Incidentally, Diamond is incorrect about there being more social harmony within such tribes than within industrial society. Norman John Grenville Pounds, that historian I cited in Book Two about tribes and the Industrial Revolution, notes that tribal societies have had their shares of intense "internal jealousies and feuds."[59] Although Sumner would not have shared Diamond's PC assessment, Diamond profits from Sumner's work in this context as well. Here in 1997, Diamond inveighed against Western ethnocentrism just as Sumner had 91 years antecedent to Diamond. Not even on this minor point does Sumner receive any credit from the PC professors who borrow so heavily from his scholarship.

Free-market economist Friedrich A. von Hayek likewise came to the same conclusions about gene-culture co-evolution while disavowing intellectual debts to the nineteenth-century individuals who arrived at this same conclusion much earlier. Hayek congratulates himself for recognizing that natural selection determines the life spans of product lines and social customs—every form of "evolution, cultural as well as biological, is a process of continued adaptation . . . to contingent circumstances . . ." And Hayek continues that to "understand our civilisation, one must appreciate that" throughout history, most human social customs arose from most ancient human beings "conforming to certain traditional and largely *moral* practices . . ." (emphasis Hayek's). And the full importance of these customs is something that most men "usually fail to understand . . ." And yet such customs can "spread by means of an evolutionary selection . . . of those groups that happen to follow them. . . . This process is perhaps the least appreciated facet of human evolution."

And then accepting, at face value, everything he was told about nineteenth-century free-market evolutionists like Spencer and Sumner, Hayek commends himself for having rejected what he falsely believes they espoused. "What I have suggested about morals and tradition, about economics and the market, and about evolution, obviously conflicts with . . . the old Social Darwinism," which, thankfully, is a belief that "is no longer widely held . . ." Hayek cautions that he has no respect for such social Darwinists—"Social Darwinism is wrong in many respects . . ." Though he does not name names, he is frustrated with "those 'social scientists' who in the nineteenth century . . . did a lasting disservice to the advance of the theory of cultural evolution, which they indeed brought into discredit." Sounding just like Rothschild and Tudge and E. O. Wilson, Hayek complains that left-wing "students of human affairs often use the inappropriateness" and "plain mistakes" of "Social Darwinism as a pretext for rejecting" any and every "evolutionary approach" to social science.[60]

There are probably more examples of this phenomenon than I can list in a single book. I will name a few more. In their evolutionary psychology book *Why We Get Sick: The Science of Darwinian Medicine*, psychiatrist Randolph M. Nesse and academic ecologist George C. Williams (b. 1926) follow in the footsteps of Spencer and other free-market bionomicists in applying the most up-to-date discoveries about evolution to the study of human society.[61] Akin to sundry other social science academicians, Nesse and Williams take an implicit whack at Spencer and Sumner when they warn their readers that their "enterprise has nothing to do with eugenics or Social Darwinism." From the late 1800s to early 1900s, ". . . Social Darwinist ideology helped to justify withholding medical care from the poor and letting capitalist giants battle irrespective of effects on individuals. These beliefs were intimately linked to those of the eugenicists . . . Such ideology has long ago earned a well-deserved ill repute."[62] Not surprisingly, blurbs of praise for this book arrive from evolutionary psychologists who care not to correct these authors about the Social Darwinism smear: E. O. Wilson, Sarah Blaffer Hrdy, and Richard Dawkins.[63] Two of them—Wilson and Hrdy—join those other renowned evolutionary psychologists, Michael Shermer and Niles Eldredge, in publicly acclaiming another sociobiology text, *Evolution for Everyone*, by David Sloan Wilson.[64] You may recall, from Book Two, D. S. Wilson's putdowns of Spencer and Ayn Rand. And while D. S. Wilson's book *Evolution for Everyone* accuses "Rand's disciples" of dismissing their ideological opponents as foolish, D. S. Wilson himself arbitrarily dismisses Spencer for having governist eugenicist sentiments that the real Spencer had not actually expressed. D. S. Wilson does this with the tacit approval of scientist celebrities who implicitly follow Spencer's teachings. It is high time that Sumner and Spencer be given due recognition for pioneering in many of evolutionary psychology's key concepts, whether the twenty-first century's evolutionary psychologists would prefer to admit this or not.

Some words must also be given concerning Spencer's pioneering work in the theory of Emergent Complexity, also known as the theory of Complex Adaptive Systems and which is peopled by Complexity Theorists. As mentioned in Book Two, the theory concerns how a system that begins as relatively simple can change, stimulated by a few small and relatively simple steps, into a system of elaborate complexity. As Spencer enunciated this progression, "The advance from the simple to the complex, through a process of successive differentiations, is seen alike in . . . every single organism on its surface; it is seen in the evolution of Humanity, whether contemplated in the civilized individual, or in the aggregation of races; it is seen in the evolution of Society in respect alike of its political, its religious, and its economical organisation; and it is seen in the evolution of all" the "endless concrete and abstract products of human activity."[65]

Here is an example of how much Complexity Theory takes after ideas already promulgated by Spencer. One partisan of Complexity Theory—Paul Ormerod, a professor and board member of *The Economist* magazine—writes that "conventional economics is mistaken when it views the economy and society as a machine, whose behavior . . . is ultimately predictable and controllable. On the contrary, human society is much more like a living organism—a living creature, whose behaviour can only be understood by looking at the complex interactions of its individual parts."[66] Where have we heard this before? Ormerod gives credit to F. A. Hayek and Joseph A. Schumpeter for arguing, decades prior to himself, that the economy is an ecosystem,[67] but no acknowledgment is paid to Spencer or Sumner.

Sadly, unaware of what Spencer and Sumner truly stood for, Complexity Theorists do their best to distance themselves from those pioneering intellectuals, a weakness that Complexity Theorists share with evolutionary psychologists and twentieth-century free-market economists. Physicist and *Science* magazine correspondent M. Mitchell Waldrop writes that Complexity Theorists wish to avoid "the stigma from the time of social Darwinism in the nineteenth century, when people were defending both war and gross inequity on the grounds of 'the survival of the fittest.'"[68] Such free-market evolutionists as Spencer are thereupon denounced by Complexity Theorists—again, by practitioners in a field pioneered by Spencer himself. Take 1969 physics Nobel laureate Murray Gell-Mann (b. 1929), who co-founded the very think tank most famous for studying emergent complexity, the Santa Fe Institute.[69] Gell-Mann clarifies that he does not want the theory of emergent complexity to go down the same road as its primitive forebear, free-market evolutionism. Thus he pontificates, "The Nazi racial theories are, of course, a horrible example of misapplying metaphors from science. Nineteenth-century ideas of social Darwinism are another example."[70] More ignorance is betrayed by another advocate of Complexity Theory—Ph.D. physicist and chemist Philip Ball. Ball implicitly denounces Spencer when he inveighs about the nineteenth-century's laissez-faire "liberals" who "espoused pseudo-Darwinian arguments for eugenics . . ." Ironically, despite his own obvious hatred for free-market economics, this same Philip Ball admits that Friedrich A. von Hayek was correct in arguing for the existence of spontaneous order—"Hayek is right at least to believe that there *are* spontaneous forces of society"[71] (emphasis Ball's). Philip Ball gives no credit to Spencer for providing this argument decades prior to Hayek.

Defenders of human enhancement also sometimes misrepresent Spencer. In the previous chapter, we defended the right of parents to employ germline bioengineering to improve their children's lives, despite such critics as Edwin Black deriding this option as social Darwinism. One proponent of the right to human enhancement very sensitive to the social Darwinism accusation is Meridian Institute fellow and *Mother Jones* magazine contributor Walter Truett Anderson. In his 1996 book *Evolution Isn't What It Used to Be*, Anderson lays out an impressive argument for enhancement. This same book also presents an argument similar to that of gene-culture co-evolution, wherein Anderson points out, as I did throughout Book Two, that customs counts as evolutionary adaptations, and therefore such technologies as germline genetic engineering are wholly natural exercises of man's nature.[72] In that aspect, *Evolution* unknowingly articulates the case that William Graham Sumner presented in *Folkways*. Anderson does not credit Sumner, and yet, anticipating that critics will accuse him falsely of peddling social Darwinism, W. T. Anderson attempts in *Evolution* to distance himself from Spencer. We recall, from chapter 1, W. T. Anderson quoting Spencer out-of-context to have it appear that Spencer discouraged smallpox vaccination on account of his desire that smallpox wipe out the underprivileged of England. In W. T. Anderson's phrasing, Herbert Spencer opposed vaccination because he believed "it would enable people who might otherwise have perished to survive instead and reproduce." On the matter of compulsory sterilization and scientific racism, Anderson also wails that government-imposed eugenics "was championed with equal enthusiasm by right-wing social Darwinists and left-wing socialists such as George Bernard Shaw."[73] Anderson's misrepresentation of Spencer vexes me. Anderson's support for voluntary human enhancement happens to be a consistent application of the classical-liberal principles Spencer espoused.

Rare Shows of Appreciation for Spencer

Yes, the overwhelming trend over the twentieth century—and, as of this writing, the beginning of the twenty-first century—has been undigested hatred toward Spencer and Sumner. Still, I would be remiss in failing to mention some historical figures still widely revered in my own age who adored Spencer. He was much more respected in his own era than he is in ours. In chapter 1 I quoted Darwin's praise for Spencer. Similar accolades arrive from John Stuart Mill, who admires Spencer as "a thinker" of great "force and depth," one whose "value...is such that I can hardly overstate."[74] Although electrical engineers Thomas Edison and Nikola Tesla were rivals in business, they both shared an appreciation for Spencer. Tesla was an avid reader of Spencer's writings on civil engineering. Tesla also shared some opinions of Spencer's concerning free enterprise. The Serbian inventor criticized the New Deal for being parasitic of other "people's capital." He thought it "destructive of established industries . . ." And, as an entrepreneur, Tesla did not appreciate "the distribution of wealth by excessive taxation . . ."[75] Edison, on the other hand, appreciated Spencer's political philosophy—the very same political philosophy for which Spencer became reviled from the New Deal hence. "Herbert Spencer . . ." Edison wrote in 1914, ". . . had the right scientific idea of investigating economics. He hired thirty clerks to run down those laws and see what their results were."[76] Upon examination of what Spencer's clerks found, Edison rejected socialism as undesirable.[77] Crediting Spencer for his outlook, the Wizard of Menlo Park decided, "I do not believe the Government should do anything" but allow every citizen "a free swing" in his peaceful endeavors "and see that every man is protected in that which he produces." Moreover, the inventor particularly desired that "the courts" pay heed to "the man who makes inventions" and "protect him" against infringement.[78] Unlike Hofstadter, Edison gave Spencer an accurate reading.

Also appreciative of Spencer is the acclaimed New York University historian Will Durant (1885–1981), co-author with his wife Ariel (1898–1981) of the book series *The Story of Civilization*, for which the couple won a Pulitzer Prize. From 1926 to 1961, Mr. Durant maintained that Spencer "summed up his age as no man had ever summed up any age since Dante; and he accomplished so masterly a coordination of so vast an area of knowledge that criticism" of him "is almost shamed into silence by his achievement. We are standing now on heights which his struggles and labors won for us; we seem to be above him because he raised us on his shoulders." Then Mr. Durant, twentieth-century man that he was, predicted, "Some day, when the sting of opposition is forgotten, we shall do him better justice."[79] That prediction was not to manifest itself in Durant's own lifetime; I hope that it finally comes true in the twenty-first century.

Among the few prominent citizens who continued voicing support for Spencer throughout the mid-twentieth century was Henry Hazlitt, whom we quoted in *The Freedom of Peaceful Action* about the peaceful sector versus the violent sector. As I write this, hardly anyone knows of him, but Hazlitt was famous in the early 1900s for his economics columns in the *New York Times*. President Ronald Reagan held his work in high esteem.[80] Hazlitt is too familiar with the left-wing cliché that, in Hazlitt's paraphrasing, the Industrial Revolution and Inventive Period were respectively "pictured as periods when no one 'cared,' . . . when everybody who did not succeed in the cutthroat competition . . . was allowed to starve." As for Spencer's "political views," they "are regarded by most present day writers, who bother to mention him at all, as 'extreme *laissez faire*' and hence 'discredited.'" Nonetheless, an active-minded, unprejudiced "person who takes the trouble . . . to read or reread *The Man Versus the State* will probably be startled by two things." The first is the remarkable foresight whereby Spencer predicted "what the future encroachments of the State were likely to be on individual liberty, above all in the economic realm. The second is the extent to which these encroachments had already occurred in 1884, the year in which he was writing."[81] Hazlitt was grateful for Sumner as well.[82]

A bit of justice from ethologists arrives by way of Robert L. Carneiro of the American Museum of Natural History. He discloses, "The fact is that Herbert Spencer was a towering figure who could rightly be called the father, or co-father, of *both* sociology and anthropology"[83] (emphasis added). Although Auguste Comte coined the expression *sociology* in 1839,[84] it was Spencer who tirelessly worked for decades to bring scientific rigor to sociology. Following the first steps taken by Comte and John Stuart Mill, Spencer ventured past them in establishing the field that is social science—of scientific, comparative studies of societies. "In an incisive and readable style, Spencer marshaled the arguments for a social science in front of the literate English-speaking public, contending that human society was part of nature and could be studied and explained scientifically." Moreover, the "reading of Spencer's work by such men" as William Graham Sumner, Lester F. Ward, and progressive eugenicist Charles Cooley "between 1870 and 1890 led directly to the founding of academic sociology in the United States." For such reasons, Carneiro states unreservedly, "The present writer...freely proclaims his debt to Spencer, and has made Spencer's writings the focus of much of his own."[85] The academic mathematician Charles Dodgson, more famously known as *Alice in Wonderland* author Lewis Carroll, also bore a qualified admiration for Spencerian thought. Despite his devout religious views, Carroll respected Charles Darwin's work and owned copies of five of Spencer's books.[86]

Amidst the few theorists of Emergent Complexity to admit Spencer's role in his own discipline is physicist and stock speculator J. Doyne[87] Farmer; the middle name is pronounced *DOH-ann*. Farmer concedes that Complexity Theory "isn't new. It was articulated in the nineteenth century by Herbert Spencer . . . Spencer argued in a very articulate way for the commonality of these processes of self-organization, and used his ideas to make a theory of sociology."[88] Farmer praises the evolutionary biologist Stephen Jay Gould by saying of him that "he's perhaps the Herbert Spencer of our day."[89] This is ironic, given that Gould was a member of Science of the People—the group that denounced E. O. Wilson, explicitly for being too socially Darwinian and implicitly for being too much like Spencer. Another irony is that both Murray Gell-Mann and Richard Dawkins[90] have publicly praised Farmer. I find this ironic, as they both denounced and tried to distance themselves from social Darwinism—the same social Darwinism they probably believe was started by the Herbert Spencer whom Farmer admires.

Shamefully, Farmer still uncritically accepts the worst lies about the great thinker—I mean Spencer, not Gould or Gell-Mann. "One of the factors," Farmer continues, "that caused Spencer's ideas to lose popularity was social Darwinism—the idea that those who were wealthy and powerful had become that way because they were naturally 'fit,' while the downtrodden were unfit—which was a poor extension from biological to social evolution, based on a simpleminded understanding of how biological evolution works."[91] For Farmer to have gotten this part of Spencer's philosophy so wrong, while noticing Spencer's correctness about emergent complexity, Farmer must have paid more attention to Spencer's natural science writings than those that were on political economy. Writing about complexity theory and the Santa Fe Institute, science journalist and philosophy Ph.D. Ed Regis also credits Herbert Spencer for pioneering in ideas about Emergent Complexity. Sadly, Regis, too, misidentifies Spencer as an apologist for "dog-eat-dog" cruelty.[92]

Another Complexity Theorist to recognize Spencer's pioneering work is *Wired* magazine cofounder Kevin Kelly, whom we quoted at length in the previous two books of our trilogy. Despite Kelly's previously mentioned sophisms about the nature of invention, I am grateful that he refrains from the social Darwinism smear. He correctly notes that Herbert Spencer was one who "understood" Emergent Complexity and this particular tenet of the sociobiologists' gene-culture co-evolution theory—"that the physical environment," through natural selection, ultimately "shapes" the "creatures" within it and that, in turn, those same creatures shape that same "environment . . ." [93] My great hope is that more social scientists benefitting from Spencer's work shall follow Kelly's lead in this instance.

A Complexity Theorist's Case for Free Will

Incidentally, the principles of emergent complexity hold implications for an issue we examined in Book Two—that of free will. Human volition is itself an emergent property. The enemies of free will often present their argument for determinism through what University of California-Santa Barbara neuroscientist Michael S. Gazzaniga identifies as the "causal chain claim." Gazzaniga summarizes it: "(1) The brain enables the mind and the brain is a physical entity; (2) The physical world is determined" by immutable natural principles that are predictable, and therefore "our brains must also be determined" in the same predictable manner; "(3) If our brains are determined, and if the brain is the necessary and sufficient organ that enables the mind, then we are left with the belief that the thoughts that arise from our mind also are determined; (4) Thus, free will is an illusion, and we must revise our concepts of what it means to be personally responsible for our actions."

As Gazzaniga observes, claim 2 is actually untrue. Over billions of years, natural processes build up chain reactions that cause unprecedented events to occur. As an example, there was once a time when Earth did not exist. Now Earth does exist. There was once a time when there was no life on Earth. Now there is life on Earth. Finally, there was once a time when no mammals were on Earth. Now there are. That natural processes can occasionally bring forth unprecedented events does not belie the Law of Identity. Nor does it contradict my argument that metaphysically given facts could not have turned out any differently than they did turn out. That a set of natural processes, *L*, can bring forth an unprecedented natural event, *M*, does not mean that either *L* or *M* contradicts itself—that either *L* or *M* is both itself and not-itself at the same time and in the same respect. No, it merely means that *L*, behaving as itself and acting according to its nature, at one point eventually brings forth an unprecedented event, *M*, which also has its own set of discernible attributes and acts according to its own nature. That there was once no mammals, and that now there are, in no manner means that mammals clash against the Law of Identity—mammals have their own nature and behave accordingly. Moreover, that metaphysically given facts, not being able to have turned out any differently than they did, does not preclude such metaphysically given facts from being *emergent* results that at one time did not exist.

To repeat a point from Book Two, each of the following is an example of emergence. (A). The time when the chemical reactions of inorganic matter on Earth formed the first self-replicating strands of proteins that would evolve into the earliest primitive organisms. (B). The evolution of fish into amphibians. (C). The bottom-up emergence of normalized customs that govern the communicative intercourses carried out every day in Standard American English. (D). The similarly normalized customs of commercial transactions resulting, bottom up, as the sum of various compromises made between consumers and vendors. "Emergence," defines Gazzaniga, "is when micro-level complex systems that are far from equilibrium (thus allowing for the amplification of random events) self-organize (creative, self-generated, adaptability-seeking behavior) into new structures, with new properties that previously did not exist, to form a new level of organization on the macro level."[94]

Compare Gazzaniga's definition of emergence with Herbert Spencer's implicit proclamation that previously unprecedented phenomena are emergent within nature: "The advance from the simple to the complex, through a process of successive differentiations, is seen alike . . . in the . . . evolution . . . of every single organism . . .; it is seen in the evolution of Humanity, whether contemplated in the civilized individual, or in the aggregation of races; it is seen in the evolution of Society in respect alike of its political, its religious, and its economical organisation; and it is seen in the evolution of all those endless concrete and abstract products of human activity which constitute the environment of our daily life. From the remotest past which Science can fathom, up to the novelties of yesterday, that in which progress essentially consists, is the transformation of the homogenous into the heterogeneous."[95] Observe that Spencer applies the term *evolution* to just about any emergent "transformation from the homogeneous to the heterogeneous," not merely to the transmutation of species. It was actually readers of Spencer's, such as Charles Darwin, who took the term *evolution* upon seeing it in Spencer's works and then applied it specifically in the context of emergent new species. Anyhow, with respect to Spencer's views on how complex systems emerge from simpler systems, it should be evident that J. Doyne Farmer is right to credit Spencer as a pioneer in this discipline.

With respect to proving the existence of emergent complexity, Gazzaniga points out that the rules applicable to Newtonian mechanics—the physical properties of the realm of objects readily perceivable to humans—are not always applicable to the realm of quantum mechanics, which involves the physical properties of objects on the atomic level. The laws of Newtonian mechanics are contextually absolute, meaning that they are merely absolute in their own context and do not inexorably apply "to all levels of organization; it depends which level of organization you are describing, and new rules apply when higher levels emerge. Quantum mechanics are the rules for atoms, Newton's laws are the rules for objects, and one couldn't completely predict the other." Likewise, that neurons and neurotransmitters, studied in isolation, may behave in a manner predictable to neuroscientists does not somehow prove that there is no spontaneity when it comes to the human brain acting as a whole. "I think," Gazzaniga concludes, "that conscious thought is an emergent property."[96] Think of the various parts of a human brain as being analogous to the connected components of a computer—only more advanced than the computer. Just as the linkages of many otherwise simple-binary switches can produce results that are increasingly complex and decreasingly predictable, the same principle applies to how the linkages of otherwise-simple brain cells can eventually produce an unprecedented, emergent property. That emergent property is the faculty of volition, which allows for sapient beings to devise and enact plans that were never devised or enacted before.

It is therefore ironic that, as will be recalled from Book Two, Spencer disbelieved in free will. Although Spencer refused to acknowledge free will, it is the study of emergent complexity—a science that Spencer himself pioneered in—that ultimately provides us insight on how free will and natural change can coexist with a universe governed by such principles as the Law of Identity. Given that I have defended Spencer against the invalid criticisms that anti-capitalists have leveled against him, it is in the interest of fairness that I shall survey what I ascertain to be Spencer's actual faults. Spencer's dismissal of free will is but one of the various errors of Spencer's I must address.

When Sumner Is Challenged By Facts

As we have reckoned, Herbert Spencer and William Graham Sumner are seldom criticized on account of the shortcomings for which they actually deserve criticism. Now is an opportune moment to expound upon the actual errors in their writings. This is in addition to the shortcomings of Spencer's I have already discussed in this trilogy's two previous installments:

1. His concessions to Kantian epistemology (Book One).
2. His altruism-based ethics (Book Two).
3. His denial of free will (Book Two).
4. His proclamation that achievement happens on the collective, not individual, level (Book Two).

First, very poor word choices on the part of clergyman turned Yale University sociologist William Graham Sumner give the appearance that he completely disagrees with me about the veracity of the Rule of Peace. For this reason, Objectivist philosopher Leonard Peikoff writes that Sumner "denied natural rights."[97] One could easily gain the same impression as Peikoff upon reading an essay by Sumner titled "The Challenge of Facts." In it the Yale sociologist decries the "doctrine that men come into the world endowed with 'natural rights,' or as joint inheritors of the 'rights of man,' which have been 'declared' times without number during the last century"[98] by Thomas Jefferson, Jean-Baptiste Say, and other Enlightenment free-marketers.

Actually Sumner did largely embrace the Rule of Peace. Still, in lieu of describing spoliation as a violation of rights—as Peikoff, Spencer, and I do—Sumner instead deplored various forms of spoliation as "violations of civil liberty."[99] That is, he employed the phrase *civil liberty* in the same manner that Peikoff, Spencer, and most other laissez-faireists use *rights*. That can be gleaned from this passage of Sumner's. "What we mean by liberty is civil liberty, or liberty under law; and this means the guarantees of law that a man shall not be interfered with while using his own powers for his own welfare. It is, therefore, a civil and political status; and that nation has the freest institutions in which the guarantees of peace for the laborer and security for the capitalist are the highest." You may remember those words from Book One.

One reason why Sumner dismissed natural rights is that he was dissatisfied with the original definition provided by John Locke and Locke's Enlightenment-Era fellow travelers. And Sumner was convinced that the new "definition" of *rights* being introduced by anti-capitalists in the late 1800s was becoming the "correct" definition. It was around this time that U.S. progressives had already begun to steal the word *rights* (just as they were stealing *liberal*) and misusing it in the same manner that President Franklin D. Roosevelt and his New Dealers would just a few decades later. Sumner naïvely surrendered the word *rights* to these people. Thus, when Sumner disputed the existence of natural rights, he did not mean that Lockeans were misguided to disapprove of government infringements on private ownership. Such spoliation is precisely what Sumner deplored as a gross abuse of civil liberty. Instead Sumner ardently protested what would, by the late twentieth century, be dubbed a "right to housing" or a "right to socialized healthcare." Like any laissez-faireist, Sumner rejected the New Dealers' platitude that a poor man's "right to private property" meant that if that poor man did not have a lot of money, it followed that the government was obligated to confiscate money from wealthier people and then give it to him.

Should one keep this in mind while reading the following Sumner-authored denunciation of natural rights, one can judge that his position favors Lockean privatization against the forms of spoliation that FDR extols on behalf of the needy. "If it is true, then, that a man is born with rights, he comes into the world with claims on somebody besides his parents. Against whom does he hold such rights? . . . if men have rights by birth, these rights must hold against their fellow-men and must mean that somebody else is to spend his energy to sustain the existence of the persons so born. What then becomes of the natural rights of the one whose energies are to be diverted from his own interests?"[100] Had Sumner fought against the socialists' contortion of the term *rights*—to remind everyone that natural rights, as originally defined by Locke and the free-market Enlightenment *philosophes*, was synonymous with what he called civil liberty—he would not have fostered the misconception that he opposed the same night watchman state that Locke defended.

Sumner is, dispiritingly, guilty of something far more egregious than faulty semantics. And as we shall learn shortly, Sumner's confusion over individual rights is actually the direct result of this graver misjudgment. Tragically, Sumner creates a self-contradiction in his writings on political philosophy as a result of his tacit acceptance of the cases made by Plato, David Hume, and Immanuel Kant. I am referring to the claim of these philosophers that the realm of ethics (which Sumner writes off as pedantic "ideology") is completely unrelated to the realm of facts. Sumner does so because he assimilates the assumptions of Kant and Hume that morality cannot be objective unless moral principles are context-less and categorical. The Yale sociologist and ex-clergyman correctly ascertains that no principle can be valid if it comes with no context. Yet he erroneously denies the existence of any objective moral principles as such. Sumner failed to appreciate or acknowledge the ethical interpretation that Ayn Rand would provide in the decades trailing Sumner's death: that objective moral absolutes do exist, but that, for a principle to be absolute, its absoluteness exists exclusively within its own proper context.

When he expounds upon politics, Sumner dresses up his evaluations as mere statements of fact, free of the arbitrary subjectivism that he mistakenly assumes goes hand-in-hand with ideology and ethics. All the while he irrationally classifies ethics and ideology as topics that are beneath the dignity of his paradigm. Eluding that the same civil liberty and Rule of Peace that he treasures are themselves ideals,[101] Sumner ridiculously asserts that "ideals," *qua* absolute principles, have done "great harm."[102]

"Ideals," the onetime clergyman proclaims, "are necessarily phantasms. They have no basis in fact."[103] And he goes on that every ideal "is formed by giving up one's hold on reality and taking a flight into the realm of fiction."[104] Were we to employ the term *ideals* strictly as Plato does, I would have to agree. When Plato referred to an idea as an ideal, he alluded to the idea to be understood on the noumenal level and not to be taken as any sort of concrete in phenomenal existence. Yet there is nothing wrong in having "ideals" as Ayn Rand spoke of them. Ideals, in Rand's meaning, are rational goals and values that a moral person consistently preserves and works toward. Elsewhere Sumner writes, "Dogmatic ideals like perfect liberty, justice, or equality, especially if economic and not political liberty, justice, and equality are meant, can never furnish rational or scientific motives of action or starting points for rational effort. They can never enter into scientific thinking since they admit no analysis and can be tested by no canons of truth. They have no footing in reality. . . . All those who talk about the 'power of ideas' are more or less under this fallacy." Note that by political liberty he means the liberty of the night watchman state and that, by economic equality, he refers to government redistributionism. Terribly, Sumner considers the ideals of political liberty to be capricious in the end.

The former priest then plays into the hands of every anti-capitalist who grumbles that free-market advocates cast morality and idealism aside in favor of cynical, amoral materialism. Sumner thereupon pleads *no contest* to one of the anti-capitalists' favorite charges. It is the charge that free-marketers support capitalism for no reason other than that free-marketers elevate so-called "practicality" above any concern about scruples or noble aspirations. In lieu of conceding this false assumption, Sumner really should be pointing out that free-marketers understand that capitalism is the most palpable sociopolitical-economic system precisely on account of its being the most humane and visionary sort of society. Hence the Yale sociologist offers, "Serious study of human society shows us that we can never do anything but use and develop the opportunities which are offered to us by the conditions and conjures of the moment."[105] In contrast to conducting oneself by any firm moral principles, "we need . . . to know, with the greatest exactitude, what is, and then plan to deal with the case as it is by the most approved means."[106]

"Means" that are "most approved" by whom? Sumner does not extend a plausible explanation. Following Hume's lead, this Yale professor first refuses to acknowledge the existence of consistent ethical principles, and then concludes from this errant premise that every concern about personal honor is tenuous and largely illusory. Sumner interprets rules about good and evil as not being a legitimate field of philosophic study from which men may conceivably capture objective truths. For him, ethics can be nothing but "motives of action" that "are derived" from "dogmatic precepts . . ."[107] Accordingly, "Ethics belong to the folkways of time and place; they can be kept sound and vigorous only by the constant reaction between traditional rule and the individual judgment."[108]

But do not forget that such dogged epistemological Skeptics as Hume and Sumner are unable to operate in the absence of applying any normative rules in deciding upon their actions. Hence Sumner somewhat relents that men need their own set of scruples. Despite that, Sumner refrains that he develops his own implicit ethical judgments for one simple reason. It is that no one can help but live by a moral code, despite any and every moral code being inescapably capricious. Much as we observed David Hume doing in Book One, Sumner thereupon advises his readers to conform unthinkingly to whatever ethical code is popular in their current era and location. ". . . [T]hese current rules of action which are traditional and accepted in our society should be adopted and obeyed. This is true, however, only because it is impossible for nearly all of us to investigate for ourselves and win personal convictions, and it is impossible for any of us to do so except in a few special matters."[109] In fact this Hume-influenced Skepticism sheds some light on Sumner's misbegotten excoriation of rights. To some extent Sumner did understand that individual rights, as defined by the U.S. Founding Fathers and Locke's other philosophic descendents, were the same as what he considered civil liberty. In an 1887 article entitled "Some Natural Rights," he grumbles, "The eighteenth-century notion of 'natural rights,' or of the 'rights of man,' was a revolt against the notion that a man had nothing and was entitled to nothing until some other men had given him some rights here. The rights of man meant that a man, as a man, entered human society, not under servitude and constraint to other men, or to social traditions, but under a presumption of non-servitude and non-obligation to other men, or to social organization. Natural rights . . . meant that the fundamental presumption must be changed, and that every man must, in the view of social order and obligation, be regarded as free and independent, until some necessity had been established for restraining him, instead of being held to be in complete subjection to social bonds, until he could provide that some established authority had emancipated him. When so regarded, it is evident that the notion of natural rights is one of great value and importance."[110]

How can one reconcile that admission with Sumner's general hostility to the principles of individual rights? In that same article Sumner continues following the Hume-ish Skeptic partly line by asserting that Lockean rights are groundless, and that Westerners conceived of them for the sole reason that they serve some fleeting momentary expediency, just as worshiping totems might have once served tribal societies. The Yale scholar therefore contests a "notion" that would later be championed by Robert Nozick and other Kant-influenced libertarians who presume morality to be based on context-free, unconditional, inherently good strictures that do not pertain to serving human life—"that rights have independent authority in absolute right, so that they are not relative or contingent, but absolute."[111] As stated throughout this trilogy that you are reading, Sumner is correct that individual rights need not be of paranormal origin, and that they are not the contextless Categorical Imperatives that Kant presumes every moral principle to be. Sumner sagely contends that if rights are the result of Kantian commandments that must be followed unconditionally, absent of any regard for how they affect anybody's life or interests, rights become "arbitrary dogmatism, and empty declamation." It is at that point, imparts Sumner, that governists are granted the opportunity to redefine individual rights to include the New Dealers' "right" to receive goods and services that the government obtained by forcibly expropriating the rich. "It has become one of the favorite methods of modern schemers, when they find it difficult to provide means by which men may get what they need in order to enjoy earthly comfort, to put all those necessary things among 'natural rights.' . . . It is the duty of the state to secure natural rights. Therefore, if there is anything which a man wants, he is entitled to have it so long as there is any of it."[112]

Expectedly, Sumner errs just as much as Nozick when he leaps to the false conclusion that if individual Lockean rights are not context-less, intrinsic, platonic absolutes, they must be inherently whimsical delusions of a still-primitive Western society. In keeping with the Hume-ish Skeptics' notion that any moral precept—including rights—is malleable if not illusory, Sumner opines that one should "think of rights as rules of the game of social competition which are current now and here. They are not absolute. . . . They must be enjoyed under existing circumstances, that is, subject to limitations of tradition, custom, and fact."[113] As established in *The Freedom of Peaceful Action*, the existence of nightwatchman-state rights *are* contingent upon the actions of human beings, but that does not preclude them from being objective, contextual absolutes. What renders them objective, context-laden absolutes is that they are the logical corollaries of objectively discernible aspects of the human species' ontological nature. As much as Sumner goes along with Hume in denying that *ought* comes from *is*, the science of political philosophy remains inextricable from ethics. This is why, subsequent to denying the applicability of *ought* to his own study of politics and economics, Sumner inadvertently smuggles *ought* and ethics back into his writings.

The consequence of this is that Sumner delivers endless proclamations about what the government ought to do, or on why activists and businesspeople should or should not engage in some specific activities. When Sumner does this, he does it with the utmost certainty in his moral convictions—the same moral certainty that he criticizes when he spots it in other writers. This explains the morally righteous tone found in much of Sumner's works, such as in his diatribes against Western imperialism. Hence the former priest rationalizes this hypocritical moralizing by claiming that, though his defense of civil liberty and the Rule of Peace are morally right, his moral rightness, too, stems merely from the malleable social mores. "Civil liberty is really . . . not a matter of resolutions, or 'declarations'"— Sumner means the Declaration of Independence, as it stresses absolute rights—as the U.S. Founding Fathers and other Enlightenment *philosophes* "seemed to think" in the 1700s.[114] Sumner resultantly deduces, "Civil liberty" must "be an affair of positive law, of institutions, and of history. It varies from time to time, for the notion of rights is constantly in flux."[115] In a tone of sheer absolutism, Sumner praises his own worldview for being "hostile to absolutism . . ."[116] What Sumner overlooks is that the same argument he advanced against the validity of natural rights can be thrown back at him in rebuttal to his own moral argument for "civil liberty."

Consider that Sumner implicitly capitulates that the laissez-faire, anti-spoliation "civil liberty" he defends has no objective foundation, but is an empty abstraction formulated from the majority of citizens in Sumner's own society. Should one follow that logic, then Sumner has to relinquish his own concept of "civil liberty" when America's most prominent community leaders chose to redefine that phrase arbitrarily. The same governists who co-opted the term *rights* in Sumner's lifetime have also co-opted *civil liberties*. Just as they had redefined "property rights" as a derelict's "right" to have the government provide him cash it extorted from somebody else, so too have they laced *civil liberties* with a similar meaning. This explains why, as we discussed in chapter 8, the ACLU gets away with arguing that a low-income artist has a "right" to demand that the government mulct money from wealthier citizens and then bankroll or purchase his artwork. It also explains why, when some Lockean-sympathizing politician seeks to reduce taxpayer financing of such artists, the ACLU squeals that this is a violation of the artist's civil liberties. To countless lawyers associated with this organization, civil liberties include an artist's entitlement to capital that was extracted from other human beings at gunpoint. Sumner would have protested the late twentieth-century's government-funded arts programs if he had been around to hear of them, but he would not have had a credible grievance against them. He already conceded that true morality was subjectively and capriciously determined by the majority opinion of his society, and majority opinion has often come on the side of the ACLU whenever such a controversy erupts. It is not credible for Sumner to maintain that individuals should sanction the free market on the basis that the free market is what the majority of Westerners wanted in his own day. Indeed, the free market is not as widely revered by the majority in my own era, as I type these words. Sumner's position has no more credibility than does a Robert Nozick-influenced libertarian when he argues, quite contrary to Sumner, that individuals should accept that capitalism's greatness is intrinsic and requires no inductive, objective explanation.

The Yale sociologist could have avoided this intellectual implosion had he admitted that individual rights and civil liberty are the same. Another aspect of avoiding that error—a still-more-important aspect—would involve the admission that rights and civil liberty are contextually absolute moral principles that formed not from societal consensus but from scientifically observable facts of man's nature as a rational, volitional being. For these reasons, I cannot quibble with Peikoff's ultimately bleak assessment that Sumner "often acted unwittingly to undermine" the free enterprise system that "he admired . . . Thus he . . . proclaimed that laissez-faire is not 'a rule of science,' but a matter of mere expediency; and ended as a skeptic, holding that there are no moral objective standards, and that 'an absolute philosophy of truth and right . . . is a delusion.'"[117]

Sumner, quite expectedly, is far from alone in taking large philosophic missteps. His mentor, Herbert Spencer, has also subscribed to some debilitating fallacies.

The Unfit Aspects of Spencer's Theories

Given that Spencer, unlike Sumner, recognized laissez-faire rights to private property as objective moral absolutes, it may initially appear paradoxical that he was greatly admired by the following famous figures, even after they had firmly pledged allegiance to socialism:[118] Jack London,[119] Beatrice Webb,[120] Alfred Russel Wallace,[121] and the anti-eugenics biologist named for him, Herbert Spencer Jennings. Spencer additionally received accolades from celebrities who did not appreciate being called socialists but who nevertheless agitated for progressive governist measures that Spencer never would have supported—attorney Clarence Darrow[122] and multimillionaire industrialist Andrew Carnegie.

The reason for this stems from Spencer's unwise concessions to his socialist adversaries. The most egregious of these was his concession that the moral ideal for society would be for the individual to subordinate his own interests entirely to the well-being of the social collective. That is a concession to the moral doctrines of altruism when one speaks the word as Spencer's contemporary—French philosopher-sociologist Auguste Comte—defined it. And, although—as we discovered in Book Two—Spencer played a large role in trying to redefine *altruism* as merely any person's beneficence toward others, the subordination of self to non-self in priority remains present in Spencer's homilies about altruism.

Spencer believes that we modern human beings tend to pursue our own financial gain, and that this is perfectly fine and moral when one does so in the absence of spoliation. That is right on, as far as it goes. Unfortunately Spencer veers off-course by then declaring that though such an ethical stance is fine for the present age, it is still morally primitive. The more human civilization advances and evolves, the more morally sophisticated our civilization will become. To Spencer, this more highly evolved morality will not entail the greater amount of individualism and individual freedom that Ayn Rand would associate with moral progress, but a greater preponderance of the social collectivism for which nineteenth-century socialists clamored. Much as Noam Chomsky and other far-Left anti-corporation activists would, Spencer envisions the perfectly virtuous society as one that is both anarchistic and collectivist.

Spencer constructs an absurdly self-contradictory argument by claiming that in his future altruist-collectivist anarchist arcadia, the sacrifice of a man's own self-interests for the social collective will create so much happiness for him that to sacrifice himself will be in his own self-interest. This would especially apply as everyone will already be so rich, and have such high living standards, that one will seldom find an the opportunity to reduce his own living standard for anyone else's benefit. Hence, "there will disappear that apparently permanent opposition between egoism and altruism . . . Subjectively looked at, the conciliation will be such that the individual will not have to balance between self-regarding impulses and other-regarding impulses; but, instead, those satisfactions of other-regarding impulses which involve self-sacrifice, becoming rare and much prized, will be so unhesitatingly preferred that the competition of self-regarding impulses with them will scarcely be felt. And the subjective conciliation will also be such that though altruistic pleasure will be attained, yet the motive of action will not consciously be the attainment of altruistic pleasure; but the idea present will be the securing of others' pleasures."[123] As this contention makes very little sense, it is not surprising that Spencer's description of how such a society would run itself is incredibly jumbled and perplexing. He cannot deliver specific, concrete examples of how this perfect society is supposed to function. This failure stems from the fact that that system cannot work. Spencer's argument is terrible, and it undermines the very case he rendered for laissez-faire freedom. It is for the reason that, as Leonard Peikoff phrases it, Spencer's vision makes concessions to governist premises along the lines of "men are helpless in the face of their own motivation" and "laissez-faire appeals to the evil in men, but men are stuck with it" for the present and for the next few centuries, until we evolve (sic) beyond that.[124]

Spencer prepared such an outlandish word salad in order to disavow the contention widespread among governists that there exists a "permanent antagonism between" the "claims of self and the claims of others."[125] Remember that Spencer's governist opponents wanted to undermine the entrepreneur's "claims of self" and redirect the entrepreneur's wealth to "the claims of others." Had Spencer simply wanted to call attention to the fact that one man peaceably promoting his own interest does not physically harm anyone else, but in fact benefits others, then he would have been far better off explaining, as Rand did, that the market by its very nature does not entail that one man's gain must come at the forcible material expense of another. It would be clearer to say that, rather than one man's gain being another man's loss, the material gain that an entrepreneur realizes through peaceful methods spoliates no one. And then Spencer could add, as Adam Smith did, that the entrepreneur's self-interested actions in fact benefit parties other than the entrepreneur. Such a world where one man's gain begets the gain of everyone else, cannot be won through a (d)evolution toward social collectivism, but through unshackling the market's peaceful properties. Once de-controlled, the market ecosystem can shower financial rewards upon entrepreneurs who peaceably sell life-enhancing goods and services to those willing and able to pay for them.

Scores of nineteenth-century collectivists believed that their socialism had taken the implicit premises of Spencer's philosophy to their logical conclusion. This idea becomes less startling when one learns that Spencer agreed with Andrew Carnegie, Arthur Brooks, and practically every other collectivist, socialist, and governist that a rich man deserves to live no more than to the extent to which he provides utility for persons other than himself. Spencer's socialist fans nodded along as he stated that an anarchist, collectivist society was the ethical ideal, whereas they clashed with Spencer when it came to deciding what was the most effective means of attaining that goal. When the early twentieth-century's governists voiced their resentment toward Spencer's opposition to spoliative welfare laws, Spencer lacked the courage to point out that the governists' moral premises were unfounded. He was not brave enough to argue that the poor man's need for material sustenance can never justify coercively depriving a rich man of the most basic survival need (the absence of spoliation).

Spencer proposed that the perfect collectivist world order cannot reach fruition unless contemporary society bans spoliation and lets self-interested entrepreneurs go about their business. Once spoliation is eliminated, Spencer explained, human society will evolve over the course of several thousand years until a quasi-collectivist workers' paradise comes into being. Spencer's socialist admirers held much more internally consistent philosophies. They reasoned that if social collectivism is the moral ideal, whereas individual rights and entrepreneurial autonomy are morally lesser concepts, then it is moral to violate the entrepreneur's individual rights to bring about the perfect collectivist society. And a violent revolution would bring about a perfect society within the immediate future; socialists need not adhere to Spencer's recommendation that they wait millennia for this brave new world to come about. Spencer tried to win his socialist detractors over to his side by countering that though he could never approve of welfare laws, the poor would still receive charity in his ideal sort of political system. He said that rich men who engaged in no philanthropy would be subjected to much-deserved social ostracism. There "meets us the objection that there if there is no compulsory raising of funds to relieve distress, and everything is left to the promptings of sympathy, people who have little or no sympathy, forming a large part of the community, will contribute nothing... It is forgotten that in the absence of a coercive law there often exists...public opinion. . . . Hence it may reasonably be concluded that if private voluntary relief of the poor replaced public compulsory relief, the diffused sentiment which enforces the one would go a long way toward maintaining the other. The general feeling would become such that few, even of the unsympathetic, would dare to face the scorn which would result did they shirk all share of the common responsibility; and while there would probably be thus insured something like contributions from the indifferent or the callous, there would, in some of them, be initiated, by the formal practice of beneficence, a feeling which in course of time would render the beneficence genuine and pleasurable."[126]

By this logic a man would deserve to be hated if he chose to become a hermit (possibly living in a part of wilderness not claimed as the private property of anyone but himself) violating nobody's rights while providing material benefit to no one other than himself. Spencer is explicit about this. "Is each person under obligation to carry on social intercourse? May he...lead a solitary life, or a life limited to the family circle? Or does positive beneficence dictate the cultivating of friendships and acquaintanceships to the extent of giving and receiving hospitalities? . . . We may indeed say that, peremptory," i.e., mandatory, "claims permitting, some amount of social intercourse is obligatory; since, without it, general happiness would fall short." That is really just Spencer's assumption. Then he continues, "If a community of solitaries, or families leading reclusive lives, would be relatively dull—if gatherings for the interchange of ideas and mutual excitation of emotions add, in considerable measure, to the gratifications of each and all; then there seems to be imposed on each the duty of furthering such gatherings. Of course, this duty is less peremptory than most other duties; and when it can be fulfilled must be fulfilled in subordination to them."[127] Western civilization would have to wait for decades before someone such as Rand finally spoke the truth that neither Spencer nor his free-market allies dared to utter: that it is morally permissible for a peaceful individual to live primarily—or exclusively—for his own sake in a peaceable manner. Spencer will not countenance your right to be a bird which flies solo. Sadly, in his arguments for the night watchman state in present day (that is, what was "present day" for Spencer, not those of us reading my book), Spencer largely fell back on collectivist precepts.

Spencer justly apprehended that Manchesterism produced a sociopolitical ecosystem that was far more conducive than any other in facilitating the proliferation of new technologies, products, and services that improved current living standards, and would provide the foundation for yet more improvements for the lives of those in future generations. Additionally, as with Adam Smith, Spencer wisely articulated that such social improvements were the result of entrepreneurship, even when the socially beneficent entrepreneur was more concerned about his own well-being than he was about the side benefits that his actions would inevitably confer upon his clients and employees. As Adam Smith put it in his *Theory of Moral Sentiments*, entrepreneurs "divide with the poor" and their employees the productive results of "their improvements" in technology and services. Smith provided special mention of how this happens when, in "their natural selfishness," these entrepreneurs are interested in nothing more than "the gratification of their own vain and insatiable desires. . . . They are led by an invisible hand to make nearly the same distribution of the necessaries of life, which would have been made, had the earth been divided into equal portions among all its inhabitants and thus without intending it, without knowing it, advance the interests of the society, and afford means to the multiplication of the species."[128]

Unfortunately to his detriment, Spencer parroted Smith's ostensive assumption that the entrepreneur's self-interested profit-seeking was no more than a means to the greater end of collective societal enrichment. Doubly, Spencer went along with Smith in taking it for granted that the end of benefitting others was the sole justification for profiting oneself. As Rand has shown, this assumption is hogwash. It cannot be justly denied that an entrepreneur's peaceful profit-motivation is of benefit to persons other than that entrepreneur. But such peaceful self-interest does not have to be justified as a means to some other end. The fact is that a person does not need to benefit anybody else for his own peaceful pursuit of his own self-interest to be scrupulous. To my disappointment, this truth has been missed not by Spencer and Sumner alone but also by evolutionary psychologists—the same evolutionary psychologists who are unacknowledged successors to Spencer and Sumner. By this point I should have fully wiped out the misconception that Spencer's laissez-faire politics in any manner inspired governist eugenics. There is but one respect wherein Spencer cleared the path for the further onset of governism. In failing to contest the validity of altruist ethics—and, besides, contorting the word's definition in effort to promote its usage—Spencer ceded the grounds whereby he could have otherwise refuted governism's main contention. That would be the contention that the State could rightly prioritize the social collective above the individual. Insofar as one concedes that altruist-collectivist premise, one has little reason to object to forcible sacrifices of individuals on the social collective's behalf.

It is unfortunate that most successful businesspeople mouth the sort of platitudes that are aligned with Spencer's. Although Book Two lauds former Xerox CEO Joseph Wilson for valiantly assisting Chester F. Carlson with protecting his patents on xerographic photocopying, I take issue with Wilson's homilies about altruism. "The happiest development," Wilson proclaimed, ". . . about the relationship between business and society, is that gain is no longer thought to be the sole motivating force of businessmen." Then, conflating the Scottish philosopher's utilitarian argument with a defense of self-interest, Wilson propounds, "The kind of escape from responsibility prescribed by Adam Smith is now considered by many businessmen as unrealistic, a concept of the utmost sterility."[129]

Despite how, in Book Two, I provided quotations of Ludwig von Mises voicing partial agreement with the Objectivist ethics, Mises unfortunately proves inconsistent in this area. He ultimately falls back on the same Adam Smithian utilitarianism as Spencer. As with Spencer, Mises correctly observes that life is a positive-sum game and then incorrectly concludes that there is no ultimate conflict between egoism and altruism. "The idea of a dualism of motivation assumed by most ethical theorists, when they distinguish between egoistic and altruistic motives of action, cannot therefore be maintained. This attempt to contrast egoistic and altruistic action springs from a misconception of the social interdependence of individuals. The power to choose whether my actions and conduct shall serve myself or my fellow beings is not given to me—which perhaps may be regarded as fortunate. . . . Thus the famous scientific dispute as to the possibility of deriving the altruistic from the egoistic motives of action may be regarded as definitely disposed of." Additionally akin to Spencer, Mises presumes that there is something inherently wrong—something inherently psychologically disturbed or ethically troubling—in someone who wishes to be a hermit or recluse.[130]

More heartbreaking still, Spencer would have looked askance at the very theme of the trilogy you presently read. My trilogy's point is that political change is directed by the culture's dominating philosophy. A culture dominated by an Aristotelian approach and Objectivist tenets—reason, eudaemonic self-interest, the Rule of Peace, and the individualism that unites them—make for a cultural milieu that yields prosperity and progress. By contrast, a culture whose animating philosophy is Platonism—driven by fideism, altruism, governism, and social collectivism—shall result in stagnation and want. The overall attitude of the culture is the sum of various individual choices. That applies particularly to the degree that the culture is one of freethinking, as opposed to one where the government censors dissent and financially subsidizes propaganda. Ignoring these considerations, Spencer proclaimed, "Ideas do not govern and overthrow the world."[131] He attributed impersonal factors—completely external to conscious human decision—to be the driving forces behind cultural evolution.[132] He particularly pronounced that the main motivation of human action is emotion, not reason, and that philosophy was therefore not a major factor—as if people were not emotional about their philosophies.[133] It is ironic that Spencer again indirectly disproved one of his own conclusions. Contrary to Spencer, individual human choices do matter, such as the choice of various scholars to impugn Spencer's political support for free enterprise and Lockean rights. To the degree that important Western opinion leaders have chosen to reject Spencer's political conclusions, we have suffered terribly for it.

As Spencer recognized, governism is a tremendous problem. As noted in Book One, some misguided libertarians therefore conclude that to preserve liberty against governism, we must opt for what they misjudge to be tyranny's opposite: anarchy. I discussed the inadequacy of anarchy in *FOPA*. Next I shall address the other failures of the libertarian anarchists' proposals.

NOTES

1. Spencer does not seem to be the first intellectual who used the term *organism* to describe society, however. Spencer's precursor and ideological opponent, socialist philosopher Auguste Comte, also described society as an organism, as can be seen in Comte 1983b, 385; and Comte 1983c, 420, 427. I should note that unlike Spencer, Comte really did intend to convey a pro-collectivism message when describing all of society as a single unified organism. It is also telling that Comte used the title *Social Statics*, in Comte 1983c, before Herbert Spencer did. In this context, the term *statics* refers to facts and figures; the title *social statics* roughly means "facts and figures on society."
2. Spencer, "The Social Organism," *The Westminster Review*, January 1860, republished as Spencer 1981h, 397, para. 17, accessed online Saturday, May 26, 2007. He made a list of "the leading differences between societies and individual organisms" in *ibid.*, 394–96, para. 13–17, http://tinyurl.com/2xruy6, accessed Saturday, May 26, 2007.
3. Spencer 1981h, 384-85, para. 3, http://tinyurl.com/2xruy6, accessed Saturday, May 26, 2007.
4. Spencer 1981h, 388, para. 5, http://tinyurl.com/2xruy6, accessed Saturday, May 26, 2007.
5. L. A. White 1968, 15. R. L. Carneiro 1981, 186, brought this statement of White's to my attention.
6. R. L. Carneiro 1981, 186.
7. L. A. White 1975, 123.
8. M. Rothschild 1990, xi–xiv.
9. J. Jacobs 2000, 10.
10. J. Jacobs 2000, 10.
11. Spencer 1981h, 393–416, para. 18- 35, http://tinyurl.com/2xruy6, accessed Saturday, May 26, 2007.
12. M. Friedman and Rose Friedman 1990, trade paperback, 25.
13. M. Rothschild 1990, 343–47.
14. For the citation, see M. Rothschild 1990, 345, and then *ibid.*, 407 n. 3.
15. To see what I mean, read Allan Chase 1980, paperback.
16. M. Rothschild 1990, xii, 344–45. In *ibid.*, xiii, he also tries to distance himself from E. O. Wilson's sociobiology, which he accuses of being a modern revival of the sort of politically incorrect eugenicism with which he does not want to be associated.
17. Borsook 2000, 72.
18. "Ethnocentrism," *The American Heritage Dictionary of the English Language* 4th ed. 2006, qtd. by "Ethnocentrism," *Dictionary.Com*, http://dictionary.reference.com/browse/ethnocentrism, accessed Wednesday, October 1, 2014.
19. Shone 2004, http://tinyurl.com/3dh22t and http://tinyurl.com/363mwx, accessed Tuesday, May 29, 2007. University of Michigan psychologist Richard E. Nisbett is one of the few scholars who disagrees. According to Richard E. Nisbett 2003, 4, "The word 'ethnocentric' is of Greek origin. The term resulted

from the Greeks' recognition that their belief that their way of life was superior to that of the Persians might be based on mere prejudice. They decided it was not." Even if Nisbett is right, it remains true that it was still Sumner who popularized the concept of "ethnocentrism" as a pejorative throughout academia in the twentieth century.

20. Sumner 1940, 13. "Folkways"—yet another term Sumner coined—refers to a society's set of customs, practices, beliefs, and traditions.

21. R. H. Gabriel 1956, 239, makes note of how *Folkways* admonishes social scientists not to allow their own concepts of moral social custom to interfere with their unbiased studies of the customs of foreign cultures.

22. I should make clear what I mean by "multiculturalism" when I criticize it. I live in Hawaii, where people practice customs from many different societies. The Western influence is very strong, but people eat lot of East Asian and Hawaiian food, and many island residents have adopted the Japanese practice of removing one's footwear before entering a house. Some people think of this as "multiculturalism." I do not have any problem with people adopting other societies' customs when they find such customs congenial and harmless. I do not think of this as "multiculturalism," though; I think of it as "cosmopolitanism." Indeed, one can adopt many Japanese customs without assuming the inherent evilness of private property rights in particular or Western civilization in general.When academicians on the continental United States preach "multiculturalism," they are usually trying to disparage Western civilization in general, and they specifically take issue with the following institutions they see as predominantly Western: private ownership, individualism, the profit motive, technological progress, inductive reasoning, rational self-interest, commerce, free enterprise, and big business.

23. D'Souza 2002 provides numerous historical examples of non-Western cultures proclaiming themselves superior to every other culture.

24. Sumner 1940, 459. M. Harris 1985, 241, talks of a similar phenomenon wherein low-income mothers in some parts of India eat out and try not to have their children see them eating. If they were seen eating, they would be asked to share the food. That they have as much as they can to themselves is actually important for the whole family, though, as it is necessary for the mother to eat as much as she can in order for her to maintain enough strength to engage in the grueling labor with which she supports her family.

25. Sumner 1940, 27–31, 173–260.

26. Eldredge and Tattersall 1982, 176–181.

27. Sumner indeed acknowledges Spencer as an important influence upon his scholarship. Sumner 1940, 8, cites Herbert Spencer's *Principles of Sociology*.

28. Eldredge and Tattersall 1982, 33, 172.

29. Tudge 2000, 171, 180-81, 224, 229.

30. Shermer 2009, paperback, 3, 12, 20–22.

31. Shermer 2003b, http://tinyurl.com/2aok9f, accessed Tuesday, May 29, 2007.

32. See the entry "Sumner, William Graham" in John Scott and Marshall ed. 2007. This is also stated in Robert C. Bannister's introduction to Sumner 1992, http://tinyurl.com/3dvc43, accessed Tuesday, May 29, 2007.

33. For another example of Michael Shermer using the W. G. Sumner-coined word *in-group*—and using it the same anthropological way that W. G. Sumner did—see Shermer 2009, paperback, 13. For an example of Michael Shermer saying *out-group*—another W. G. Sumner-originated term—and describing the tribal "us against them" mentality previously described by W. G. Sumner, see Hudgins 2007, 32, http://www. objectivistcenter.org/ct-1852-M_Shermer.aspx, accessed Sunday, July 5, 2009. For an example of W. G. Sumner using it in 1906—perhaps for the first time in print—see Sumner 1940, 12–13.

34. Shermer 2009 paperback, 20–21.

35. "About: Meet Michael Shermer," *Skeptics Society* website, 2009, http://www.skeptic.com/about_us/meet_michael_shermer.html, accessed Friday, July 3, 2009.

36. The observation about sunlight, rickets, and Vitamin D playing a role in the emergence of light skin, blue eyes, and blonde hair in human beings is made in S. Olson 2002, 40-41. It can also be found in Jared Diamond, 1993 paperback, 115; R. Dunn 2011, 204; and M. Harris 1985, 145.

37. The theories relating to red hair are in M. Roach 2005.

38. I emphatically disagree with the eugenicist argument laid out in *The Bell Curve. The Bell Curve* proclaims: (1) IQ number plays the biggest role in determining how economically successful a person is, and even plays the biggest role in determining the person's ethical conduct, such as one's tendency to commit violent crime. Generally, the higher the IQ, the less likely someone is to initiate violence. (2) IQ is genetically inheritable, and IQ is unchanging. (3) Some races, on average, have higher IQs than others. Hence *The Bell Curve* tries to argue that race, playing a pivotal role in determining one's IQ number, still plays an essential role in determining how "fit" someone is being able to best perpetuate his genetic legacy in the modern industrial environment. But Steve Johnson's book *Everything Bad Is Good for You* disproves that. The fact of the matter is that over the past 50 years, the average IQ numbers of blacks and whites have both been increasing. Were one of the main premises of *The Bell Curve* true—that IQ number is genetically-determined and that it cannot change immediately over a single generation—then there

would be no accounting for this finding. Hence the argument of *The Bell Curve* is built on flawed assumptions. See S. Johnson 2005.

39. S. Olson 2002, 162–63.
40. S. Olson 1989, 75–77.
41. Lumsden and Wilson 1983, 171.
42. E. Wilson 1998, 184.
43. Edward O. Wilson calls William Graham Sumner "the absolute Social Darwinist" in E. Wilson 1978, 208.
44. R. Benedict 1959, xii–x, xv–xvii, 253.
45. R. L. Carneiro 1981, 176.
46. Spencer 1910a vol. 1, 35–36. R. L. Carneiro 1981, 180, brought this statement of Spencer's to my attention.
47. Spencer 1978 vol. 1, 136. R. L. Carneiro 1981, 176, brought this statement of Spencer's to my attention.
48. Spencer 1886, 337. R. L. Carneiro 1981, 176, brought this statement of Spencer's to my attention.
49. Spencer 1910a, 615. R. L. Carneiro 176, brought this statement of Spencer's to my attention.
50. Spencer 1897, 331. R. L. Carneiro 1981, 176, brought this statement of Spencer's to my attention.
51. M. Ridley 1997, 155, 264–65.
52. E. Wilson 1978, 224 n. 5.
53. Steven Pinker, "Science Is Not Your Enemy," *The New Republic*, August 6, 2013, http://www.newrepublic.com/node/114127/print, accessed Monday, June 9, 2014.
54. For E. O. Wilson's praise of this book, see the front cover flap of Jared Diamond 1997.
55. D'Souza 2002, 45.
56. Jared Diamond 1997, 25.
57. D'Souza 2002, 45–46.
58. Jared Diamond 1997, 18.
59. Pounds 1989, 4–5.
60. Hayek 1988, 25, 6, 52, 27.
61. Zuk 2007, 14.
62. R. Nesse and G. C. Williams 1995, 11.
63. See the back cover of R. Nesse and G. C. Williams 1995.
64. See back cover of D. S. Wilson 2007.
65. Spencer, "Progress: Its Law and Causes," *Westminster Review* vol. 11, April 1857, p. 465, qtd. by R. L. Carneiro 1981, 157.
66. P. Ormerod 2000, trade paperback, vii.
67. P. Ormerod 2005, 223.
68. M. M. Waldrop 1993, trade paperback, 214.
69. J. Brockman 1995b, 318, points out that Gell-Mann helped start the Santa Fe Institute.
70. J. Brockman 1995b, 324.
71. P. Ball 2004, 462, 464.
72. W. T. Anderson 1996, 50–57.
73. W. T. Anderson 1996, 105, 109.
74. J. S. Mill 1887 8th ed., 201–02. G. H. Smith 1981, 142–43 n. 6, brought this statement of Mill's to my attention. In G. H. Smith 1981, 143 n. 6, George H. Smith provides three other quotations of Mill's in praise of Spencer.
75. Nikola Tesla, letter, to John "Jack" Pierpont Morgan, Jr., November 29, 1934, qtd. by M. Cheney 1983, paperback, 248; and M. J. Seifer 1998, 435.
76. Edison 1968b, 162.
77. Edison 1914, 25; and Edison 1968b, 165–66.
78. Edison 1914, 26; and Edison 1968b, 161.
79. W. Durant 1961, 300.
80. On March 20, 1981, Ronald Reagan said in his address to the Conservative Political Action Conference (CPAC), "There are so many people and institutions who come to mind for their role in the success we celebrate tonight. Intellectual leaders like Russell Kirk, Friedrich Hayek, Henry Hazlitt, Milton Friedman, James Burnham, Ludwig von Mises—they shaped so much of our thoughts." For that, see R. Reagan 2004a, 96.
81. H. Hazlitt 1993b, 174.
82. Henry Hazlitt interviewed in Zupan interviewer 1984, 40.
83. R. L. Carneiro 1981, 153.
84. R. L. Carneiro 1981, 154; and Spencer 1910a, v.
85. R. L. Carneiro 1981, 164–65, 200.
86. Morton N. Cohen 1996, paperback, 350.
87. T. A. Bass 2000 paperback, 55, says *Doyne* is pronounced "do-an." *Ibid.*, 75, quotes Doyne Farmer describing himself as a "Zen Buddhist businessman." *Ibid.*, 130, also states that the actor Jeff Goldblum

interviewed Farmer as research for his character in the 1993 movie *Jurassic Park*, as his character, Ian Malcolm, was an expert in the Chaos Theory that Farmer co-founded.

88. J. Doyne Farmer interviewed in J. Brockman 1995c, 367–68. I first learned of Farmer's acknowledgment of Spencer from M. M. Waldrop 1993 trade paperback, 286. With respect to Farmer's ideas about self-organizing systems, the source says, "In fact, he [Farmer] says, you could even find the same issues latent in the work of Herbert Spencer, the English philosopher who helped popularize Darwin's theory back in the 1860s by coining such phrases as 'survival of the fittest,' and who saw Darwinian evolution as just a special case of a broader force driving the spontaneous origin of structure in the universe."

89. J. Brockman 1995a, 67.

90. J. Brockman 1995c, 373–74.

91. J. Doyne Farmer interviewed in J. Brockman 1995c, 369, in J. Brockman, ed. 1995d, 369.

92. E. Regis 2003, 154–55.

93. K. Kelly 1994, 81.

94. Gazzaniga 2011, 129, 124.

95. Spencer 1907f, 20–21. R. L. Carneiro 1981, 157, brought this statement of Spencer's to my attention.

96. Gazzaniga 2011, 129–130.

97. Peikoff 1993a, 122. Sumner's opposition to natural rights is also noted in R. H. Gabriel 1956, 233.

98. William Graham Sumner, "The Challenge of Facts," written in the 1880s, published as Sumner 1969-A, 103.

99. He uses the exact phrase "violations of civil liberty" to describe State acts of spoliation in "The Influence of Commercial Crises on Opinions About Economic Doctrines," an address to the Free Trade Club in New York City, May 15, 1879, published as Sumner 1969-E, 62.

100. William Graham Sumner, "The Challenge of Facts," written in the 1880s, published as Sumner 1969-A, 93–96, 103–104.

101. Not only did Sumner support laissez faire in principle, but he also favored using the term to describe his own views, as can be seen in Sumner, "Laissez-Faire," fragment of essay written in the 1880s, published as Sumner 1969-F, 468–477.

102. Sumner, "The Scientific Attitude of Mind," address to the initiates of the Sigma Xi Society at Yale University, March 4, 1905, published as Sumner 1969-L, 51.

103. Sumner, "Purposes and Consequences," written sometime between 1900 and 1906, published as Sumner 1969-J, 17–18.

104. Sumner 1969-L, 51–52.

105. Sumner 1969-J, 17–18.

106. Sumner 1969-L, 52.

107. Sumner 1969-J, 18.

108. Sumner 1969-L, 52.

109. Sumner 1969-J, 18–19.

110. Sumner, "Some Natural Rights," *The Independent*, July 28, 1887, republished as Sumner 1969-M, 364.

111. Sumner, "Rights," written between 1900 and 1906, published as Sumner 1969-K, 358.

112. Sumner 1969-M, 364.

113. Sumner 1969-K, 362.

114. Sumner, "Liberty and Responsibility, Part 1 of 9," *The Independent* on November 21, 1889, republished as Sumner 1969-G, 314.

115. Sumner, "Liberty and Responsibility, Part 3 of 9: Liberty and Discipline," *The Independent*, January 16, 1890, republished as Sumner 1969-I, 320.

116. Sumner 1969-G, 314.

117. Peikoff 1993a, 122.

118. Socialists finding inspiration in Herbert Spencer's philosophy is mentioned in Mowry 1958, 47.

119. Hofstadter 1959, 34.

120. Beatrice Webb's early admiration for Spencer is mentioned in G. Allen 1894, section labeled PRS.32, http://praxeology.net/GA-PRS.htm, accessed Thursday, November 22, 2007; W. Petersen 1979, 227; and B. Webb 1980 paperback, 26–27.

121. In honor of Spencer, Alfred Russel Wallace even named his son Herbert Spencer Wallace, according to Desmond and Moore 1991, 534; and Shermer 2004, 269.

122. Hofstadter 1959, 34.

123. Spencer 1978 vol. 1, pt. 1, ch. 14, sec. 97, 283–84, http://tinyurl.com/29bt4r, accessed Sunday, June 17, 2007.

124. Peikoff 1993a, 122.

125. Spencer 1978 vol. 1, pt. 1, ch. 14, sec. 92, 271, http://tinyurl.com/29bt4r, accessed Sunday, June 17, 2007.

126. Spencer 1978 vol. 2, pt. 6, ch. 7, sec. 456, 403–404, http://tinyurl.com/ypfn99, accessed Sunday, June 17, 2007.

127. Spencer 1978 vol. 2, pt. 6, ch. 8, sec. 459, 411, http://tinyurl.com/34p5pk, accessed Sunday, June 17, 2007.

128. A. Smith 1790 6th ed., pt. 4, ch. 1, para. 10, http://tinyurl.com/3n2myt, accessed Tuesday, April 22, 2008.

129. Joseph Wilson, qtd. by C. D. Ellis 2006, 373.

130. Mises 1981, pt. 4, ch. 27, sec. 2, Para. 3–5, accessed online Saturday, September 14, 2013. I learned about Mises's views on this topic from P.-O. Samuelsson 2012, http://perolofsamuelsson1.wordpress.com/2012/04/29/ludwig-von-mises-on-egoism-vs-altruism/, accessed Saturday, September 14, 2013.

131. Spencer 1891c, 69. R. L. Carneiro 1981, 171, brought this statement of Spencer's to my attention.

132. R. L. Carneiro 1981, 170–71.

133. G. H. Smith 1981, 135.

TWELVE

Overthrowing the Anarchists

One of the Most Specious Ideas in the History of Political Philosophy

In Book One we briefly touched upon the ideology touted by libertarian anarchists, which is sometimes called *anarcho-capitalism, market anarchy, anarcho-pluralism,* and *agorism*. Its most well-known adherents include the economists Murray N. Rothbard and David D. Friedman (Milton Friedman's son). It is mostly on account of this ideology that I do not attach the *libertarian* label to this trilogy's argument; I do not intend to drag the libertarians' anarchist baggage along with me. And to affix *capitalism* or *market* to the name of this ideology is to provide a misnomer; capitalism can persist and avoid self-destruction; anarchy cannot. For the most part, I shall try to refer to the libertarians' model simply as *anarchy* and its proponents as *anarchists* or, once again, *Rothbardians*. This chapter shall provide further elaboration on why anarchy is not a viable alternative to the night watchman state.

The anarchist ideology regards the institution of government as inherently evil. Thus, the anarchist proclaims that any and every government should be abolished. A night watchman state remains unacceptable. In its stead, people should be able to subscribe to private, competing, revenue-seeking, protection rackets—uh; make that "protection-service-providing companies." Once you sign up for protection under such a company, you contractually agree to follow its rules, which are the protection companies' equivalent of legislative statutes. The anarchist announces that if you are dissatisfied with the competence or laws of Defense Agency *A*, then you should be able to unsubscribe from it and then sign up your family with Defense Agency *B*. Rothbardian anarchists, remember, frequently rail about the U.S. Constitution replacing the Articles of Confederation, as the U.S. Constitution supposedly placed greater central authority over the federal government than did the pre-1789 confederation. The implication is that decentralization of authority is inherently preferable. Yet, as Fareed Zakaria reminds us, feudalism was itself a highly oppressive government system in which the monarch's authority was decentralized. Each landlord remained largely independent of the monarch. "From the Middle Ages until the seventeenth century, European sovereigns were distant creatures who ruled their kingdoms mostly in name. The king of France, for example, was considered only a duke in Brittany and had limited authority in that region for hundreds of years. In practice if monarchs wanted to do anything—start a war, build a fort—they had to borrow and bargain for money and troops from local chieftains, who became earls, viscounts, and dukes in the process."[1] Were it valid to regard the quantity of liberty as inversely proportional to the centralization of executive authority, then one should regard the Dark

Age's feudalism as providing greater liberty than the U.S. republic has any time post-1789.

The crucial demarcation between feudalism and Rothbardian anarchy is that, under the latter, you are supposedly allowed to unsubscribe from one fief and then take refuge with another, whereas peasants under feudalism could not vacate their fief. A system similar to Rothbard's proposal actually existed among the hunter-horticulturalist culture of ancient Hawaii—you could leave the jurisdiction of one chief and swear loyalty to another.[2] That did not encourage freedom; there was still uniformity in the existence of repressive, theocratic laws.[3] University of Hawaii historians Ralph S. Kuykendall and A. Grove Day provide a description of the ancient social structure that anarcho-pluralism would mirror in its best possible form—"The common people were still completely subjected to the nobles, without any rights" being respected, "except that of moving to the land of another . . . chief."[4] Under anarcho-pluralism, though, the system allegedly would not be confined to geography. You can dwell in the same house for three decades and go through twelve different protection services in that duration. In that same period, your next-door neighbor might have been signed up with six defense agencies you never tried.

You can ask the anarchist why he would not be satisfied with a night watchman state. These are his two main objections to it.

(1) The Taxation Argument—The anarchist and I agree that the institution of compulsory taxation is barbaric and would not exist in the best plausibly functional society. But a significant disagreement divides us. The anarchist asserts that the institution of compulsory taxation is inherent to the very existence of any government, including a night watchman state. He insists that the one avenue for eliminating compulsory taxation would be to establish Rothbardian anarchy. I refuted that presumption in Book One. The purpose of this chapter is to address the anarchist's other favorite argument against the night watchman state, coming up next.

(2) The "Monopolistic Hegemony" Argument—The anarchist shouts that it is disgusting for a night watchman state to maintain a monopoly, over a specific geographic region, on the use of retaliatory force. Recall, from Book One, that a free society punishes the initiation of the use of physical force, but not the exercise of retaliatory force. Suppose an entrepreneur started Revenge, Inc. You pledge a subscription fee to Revenge, Inc. Imagine some crook—we will call him Morty—beats you up. Then, instead of going to the police, you go to Revenge, Inc., and hire the business to rough up Morty in return. According to the Rothbardian, this business deal with Revenge, Inc., is morally acceptable. Moreover, if the night watchman state tries to prosecute Revenge, Inc., for this vigilante action, the anarchist considers the night watchman state to be acting abhorrently, initiating the exercise of force against Revenge, Inc. Thus, proclaims the anarchist, the night watchman state has not truly banned the initiation of the use of force. It, in fact, initiates the use of force against "competing" retribution-service businesses in order to maintain its monopoly on the industry that is retaliatory violence-dispensation. The night watchman state therefore acts as a monopolistic hegemony or empire. This is the argument I challenge in this chapter.

David D. Friedman stresses an additional selling point in his attempt to market the ideology of anarchy. When businesses compete against each other for consumer dollars, one of their main methods of competition is product differentiation. In the beverage market, it is not as if every beverage is cola. Both Coke and Pepsi are cola-flavored soft drinks. Should you dislike the taste of cola, you can purchase orange juice instead. Friedman says that such differentiation also applies to the different *flavors* of protection service. *Flavor* is my term for it, to be sure, not Friedman's. In any case, in a society such as the one Friedman advocates, there "might be many courts and even many legal systems."[5] As there would be a variety of disparate tastes among clients, one protection agency might have age-of-consent laws whereas another does not. Friedman trumpets that you should be able to choose between a Protection Service that forbids abortion versus one that permits the procedure. Insofar as there are a variety of models from which to select, that indicates the industry is competitive. By contrast, the more disparate firms in an industry grow similar from one another, offering products or services that are hardly distinguishable from each other, the less competitive they seem. For a firm to distinguish itself from its competitors by offering a legal system with rules vastly discrepant from its competitors' rules, is for that firm to be very competitive. When every firm acts alike, that is a form of collusion, and is no less a decrease in competitiveness than is collusion on price. The more homogenous the rule systems of the separate agencies become, the more they behave as a cartel, and they will, in practice, seem to be a single regional monopoly.

Note that this entire attempt to sell the system of competing protection services is based on a false idea we explored in *The Freedom of Peaceful Action*. It is the presumption that a political administration's moral legitimacy is contingent solely upon whether it forged a Social Contract with its citizens—in this case, subscribers—prior to performing any action in presiding over them.

The Inevitability of Monopolistic Hegemony

The proponents of anarchy assert that if the private defense agency to which you are currently subscribed is treating you crummy, you can unsubscribe from it and purchase a subscription from its competitor. Fortunately, Robert James Bidinotto observes the problem with there being "competing justice systems" in the same geographic region. Such a system "posits 'competition' in the use of force, but more: 'competition' in *defining* the rightful uses of force. To whom must these competing 'protection agencies' ultimately answer? To what *standard* are their own actions and verdicts to be held? . . . Should sex between adults and young children be allowed? Should kids be freed from the control of their parents, and should they be permitted to walk out at whim? If yes in each case, should parents who interfere be stopped—by deadly force, if necessary?"[6] (emphases Bidinotto's). Anarchists just take it for granted that every protection service will allow you to unsubscribe from it. They do not anticipate the possibility that a protection service would threaten to initiate the exercise of force against you if you try to unsubscribe from it. Rothbardians have yet to deliver a truly compelling answer to this question: "What happens when I try to unsubscribe from my defense agency, and my defense agency resultantly tells me that if I try to unsubscribe from it, it will break my legs?" Were a private defense agency to issue that threat, I will not be able to call the police for help. This is anarchy, remember—there are no police other than the private defense service businesses themselves.

Massachusetts-based scholars Daron Acemoglu and James A. Robinson describe a real-life situation that parallels that hypothetical scenario: in one part of the globe, "political power" is "widely distributed—almost pluralistic. Indeed there is no real authority that can control or sanction what anyone does. Society is divided into . . . clans that cannot dominate one another. This distribution of power leads not to inclusive institutions but to chaos, and at the root of it is the . . . lack of any kind of political centralization, or state centralization, and its inability to enforce even the minimal amount of law and order to support economic activity, trade, or even the basic security of its citizens."[7] Acemoglu and Robinson were describing the anarchy of Somalia, and that is what an anarcho-pluralistic society would resemble, at least prior to a single protection agency gaining a hegemony and becoming the new monopolistic central government (more about that later).

Regarding my question of how I could address Protection Service *X* threatening me for trying to unsubscribe from it, the closest Rothbardians have come to giving a coherent reply is this. Should Protection Service *X* threaten me, then I can still plea for Protection Service *Y*'s help. Maybe Protection Service *Y* will threaten to use retaliatory force against Service *X* if this continues. Protection Service *Y* can go as far as forming a coalition with other protection services. This coalition will exercise retributive force against any protection service that refuses its customers the right to unsubscribe. Behind Service *X*'s back, I can hire other protection agencies to gang up on the evil agency and threaten to destroy the evil agency if it does not let its customers unsubscribe.

But that retort is fallacious. It assumes that every agency will be roughly equal in power and size. The anarchists overlook that, under competition, some competitors become bigger and richer than the others. Both a lone petty purse-snatcher and the Mafia count as private agencies, but the Mafia has more money and manpower than the purse-snatcher does. What typically goes unmentioned by anarchists is that, historically, the most powerful empire on a continent could gain the power to set rules for other governments on that continent . . . and sometimes whole other continents as well. It need not set the rules by direct conquest. Once other nations become aware of the empire's military might, the empire can simply exercise "soft rule"—it can intimidate other countries into exacting the policies that it wants, much in the way that a well-established Mafioso reaches a point where he can get what he wants without lifting a finger. When the Ottoman Empire was at its peak, it was able to set standards for the rest of the Middle East. During the nineteenth century the British Empire was able to set the rules for India, Ireland, Hong Kong, Australia, New Zealand, and many parts of Africa, such as Zimbabwe (which was then Rhodesia, as it was conquered by Englishman Cecil Rhodes). The same principle would apply in anarchy. The richest, most powerful defense agency could declare war on the smaller ones and physically beat them into submission. And should the dominating defense agency be a highly illiberal one, that would destroy freedom for everyone. That is the great irony of self-described libertarians—were they ever to institute the societal structure of their dreams, their precious anarchy would simply be a transitional phase that would likely revert to a new imperial order. This is exactly the sort of monopolistic hegemony that anarcho-pluralism is intended to stop. As noticed by Harry Binswanger, life under anarchy would be comparable to a baseball game where the team that had the most points after every two innings could alter the rules for both teams for the next two innings. It could apply one set of rules for itself and another set of rules for the other team. Should every private defense agency look out for its own financial survival—and this is exactly what the anarchists count on—then why should the wining team not rig the rest of the game to perpetuate its lead?[8]

This is another gaping contradiction in the anarchists' rejoinder: the anarchy model is supposed to preclude any protection service from dictating the rules or business model of another. Recall the anarchist's objection to the idea that a single party can forcibly dictate the extent to which other parties can deal in the issuance of physical force. But that is exactly what will happen if Protection Service *Y* or its coalition can unleash force to pressure Service *X* to amend its policies. Consider the implications. Under this circumstance, either Service *Y* alone, or Service *Y* and its coalition, are behaving as a monopolistic hegemony or imperial cartel.

Some forms of violence produce consequences that cannot be personalized, and therefore cannot be privatized. This issue refutes more than market "anarchy"; it refutes a common contention by libertarians that civil law should replace criminal law. Some libertarians object to the notion of criminal law, which stresses that if you initiate the use of force on me by beating me, I am not your sole victim. According to the philosophy of criminal law, everyone in our community has been victimized by this spoliation. Therefore, when a prosecutor files criminal charges against you, the prosecutor acts not merely on behalf of me but on behalf of everyone in the community. No, proclaim a number of libertarians led by Murray Rothbard. This libertarian argument proclaims that Stuart Hayashi was the sole victim of your beating, and therefore the State should not claim to act at the behest of the entire community in prosecuting you. In lieu of that, proclaim such libertarians, there should be no criminal prosecution of you. Rather, Stuart should initiate a tort lawsuit against you, probably bringing his private defense agency to act on his behalf and to reach some settlement with the private defense agency that you have hired.[9] Inspired by Rothbard, libertarian engineer and entrepreneur W. Alan Burris asserts, "A fundamental mistake of the present system is the ridiculous idea that a crime is committed against that imaginary being, 'society,' or 'the people.'" Should someone be bludgeoned by a criminal, no one but the specific victim has had his rights "violated," and no one but the specific "victim has a claim against the criminal."[10] Allegedly, should you be a recluse who has no friends or family, you can still rely on tort law to seek justice on your behalf if you are murdered. You purchase homicide insurance. Your insurance policy would stipulate that, in the event of your murder, the insurance company would hire a specific law firm to seek out the culprit and initiate a wrongful death lawsuit against the culprit. Should the law firm prevail, the damages would be awarded to the private party specified in your will. Surely, once you are dead, you would not know how competently the homicide insurer has acted on your behalf. Still, as there would probably be more than one recluse in this community, such firms can remain competitive. Should one firm prove incompetent in seeking justice for one murdered recluse, other recluses in the community can observe this happening and therefore opt for other homicide-insurance firms instead.

Here is where the Rothbardian libertarians are wrong: the violent victimization of one person does indeed victimize everyone in the community. The reason why the police, military, and courts cannot and should not be completely privatized is that violence itself cannot be completely privatized.

In Books One and Two, we explored the concept of the Coase Theorem. The idea is that if one party, such as a business, initiates the use of force upon other people inadvertently, such as through polluting the air of landholders on adjoining real estate, this involves that business cutting its own costs by imposing those same costs upon other parties. That is called an *externality*. A consistent system of laissez faire rectifies that situation by recognizing that individuals own their own private property. In this situation, the victims whose air was polluted can file a lawsuit against the polluter, transferring those externalized costs back to the polluter. That is called *internalizing* the externalities. This can be done to rectify many initiations of the use of force. That is what civil suits are for—corrective justice. As a private individual was harmed, the spoliative party doing the harm corrects the action, as much as possible, to make the private individual victim whole again. There are some cases, though, where torts are not sufficient to rectify an initiation of the use of force. When it comes to such acts as rape and murder, such an act necessarily carries externalized costs. No matter the extent to which the violent agencies have supposedly been privatized, the externalized costs of violence cannot be fully internalized. Any time someone exercises considerable violence—and this includes violence in self-defense—he demonstrates himself to be a potential violent threat to everyone else. That is where criminal law comes in. There are some cases where the spoliator can be subject to both a criminal trial and a civil suit. The notorious football player O. J. Simpson, for instance, underwent both a criminal trial and a civil suit for the death of his wife Nicole Brown Simpson. Criminal law is justified because some spoliations are so severe that they cannot be privatized or sufficiently internalized. In these cases the original violence committed is, by its nature, a public matter, as is its rectification.

If Mr. X tries to kill me and I violently subdue him, this is not "just between Mr. X and me." It is everyone's rightful business to know about this altercation; the other citizens in the neighborhood possess a right to know whether I will be violent toward them as well. Someone who initiates the use of force on one person may do this to anyone else, and thereby threatens the entire municipality. In that respect, *violence initiated on merely one person produces socialized losses for all*. That is why it is just for the police and courts, acting as a representative of the neighborhood, to question and investigate me, and to go through Due Process to determine whether I was in fact acting in self-defense. If I initiated the use of force against Mr. X, I prove myself to be a threat to other people, not just Mr. X. That is why initiators of the use of force deserve to be imprisoned or punished. If, however, the police and courts logically ascertain, through Due Process, that I was acting in self-defense, then they leave me be, having rationally determined that there is not sufficient reason to believe I pose a threat to other people in the neighborhood.

When two people get into a violent dispute, it is just that the government establishes this as a matter of public record. Third parties do have a right to access such information to learn whether I will pose a physical threat to them. Every act of violence is a "monopolistic imposition" of one's own will. Acts in self-defense prove no exception. If Mr. X attacks me, and I violently defend myself, I am "monopolistically imposing" my own liberal ethos on Mr. X.[11] Should anarchists take their anti-"monopolistic" pluralism to its logical conclusion, they should consider self-defense to be morally verboten.

Vigilantes versus Due Process

There is another great flaw in the anarchist argument. Many libertarians agree with me that every individual has the right to defend himself violently against spoliation, as well as the right to defend other innocent parties violently against spoliation. Unfortunately, such libertarians conclude that it logically follows from this that a perfectly consensualist society should allow for private organizations to exercise retaliatory violence whenever it catches the private organizations' fancy. This means that vigilantism—the practice of vigilante behavior—is always justified as long as the vigilante is acting strictly on behalf of the innocent and against the guilty. And, therefore, when the anarchist utters the expression *private defense agency*, it serves, in effect, as a euphemism for *vigilante*.

There are three different types of parties in a society that dispense violence: (1) governments, (2) violent criminals, and (3) vigilantes. These are "violence-dispensing agencies," or just "violence agencies." The distinctions among these three categories are not always clear-cut; there can be some overlap among them. I will employ a binary categorization for the wielders of violence: *spoliator* and *avenger*. A spoliator is a party who initiates the use of physical force. Note that a spoliator can be a violent criminal, a governmental law enforcer, or a vigilante. By contrast, an avenger is a party who exercises retaliatory force against the spoliative party. An avenger cannot be a violent criminal, for, in the context of our discussion, one necessarily has to be a spoliator to be classified as a violent criminal. When acting properly, a policeman can be an avenger. If a violent criminal mugs me, and then the policeman rains blows upon the mugger, the policeman is a government agent who has acted in this scenario as an avenger.

Vigilantes are a tricky category. A vigilante is a private party who still wields violence when there is no immediate emergency. Predictably, the vigilante believes himself to be an avenger and not a spoliator. Nevertheless, as this chapter shall elucidate, the vigilante is more likely to be a spoliator than an avenger anyway. If an assailant is chasing after Mary with his axe, and she kills him, that is self-defense, and Mary is an avenger. But in the context of our discussion Mary is not a vigilante. Party *1* is a vigilante when Party *1* issues violence—violence Party *1* considers retaliatory—against Party *2* a considerable duration (perhaps a day or more) subsequent to Party *1* or a valued member of Party *1*'s social circle being spoliated.

Ayn Rand implores us to consider "what would happen if a man missed his wallet, concluded that he had been robbed," and "broke into every house in the neighborhood to search it . . ."[12] Harry Binswanger sagely subjoins, "The attempt to invoke individual rights to justify 'competing' with the government collapses at the first attempt to concretize what it would mean in reality. Picture a band of strangers marching down Main Street, submachine guns at the ready. When confronted by the police, the leader of the band announces: 'Me and the boys are only here to see that justice is done, so you have no right to interfere with us.' According to the 'libertarian' anarchists, in such a confrontation the police are morally bound to withdraw, on pain of betraying the rights of self-defense and free trade."[13]

Picture a scenario wherein I do not get along with a man named Riley, who lives a block from my house. We have longstanding animosities. Visualize that one day, my house is burglarized. I immediately jump to the conclusion that my neighbor must have done it. In the absence of any objective investigation, I decide to become a private defense agency. I am my own defense agency "competing" against the night watchman state. I, as a vigilante, spray-paint obscenities on the walls of Riley's house. Then you come and tell me that I have initiated the use of force against my neighbor. I say, "Nope. I was employing retaliatory force against him for the usage of force that he initiated! He burglarized my house. Therefore, I exacted retribution." Take a gander at how I took this action in the absence of solid proof that Riley burglarized me. Is this justified retaliatory force? Retaliatory violence, on the part of private citizens, is justified solely in cases of emergency where immediate violent action is required in order to save someone's life from extermination. When a woman is assaulted, for example, it makes sense that, rather than simply wait for the police to arrive, she utilizes violence in her own defense, possibly slaying the assailant. However, retributive aggression by civilians is *not* justified in cases outside of emergencies; the reason is that whenever individuals choose to undertake vigilante actions, there is an extremely high likelihood that the vigilantes will spoliate innocent parties whom the vigilantes wrongly assume to be guilty.

When it comes to the exercise of retaliatory force, a fact of especial importance is that a suspect—someone who is suspected of spoliating another party—might actually be innocent. We might suspect that Riley has burglarized homes when, in fact, he has not. What is relevant is that there are many cases where people might exercise violence against Riley—believing their violence to be retaliatory in nature—when Riley is actually innocent, which would render Riley himself the victim of spoliation. Any violent punishment of an innocent man, under the hasty presumption that he is guilty of spoliation, would result in the spoliation of the innocent man.

Return to the anarchists' proclamation that a truly free society would not prevent a vigilante from issuing supposedly retaliatory force against persons presumed to have wronged him. Their assumption is that on any occasion on which a vigilante issues what he believes to be justified retaliation, the vigilante perforce acts in accordance with the pertinent facts. In effect, the anarchist position assumes that whenever someone seeks retribution for spoliation, the identity of the culprit is obvious. The fact is that, in the case of most crimes, the culprit's identity is not obvious. Imagine that a husband and wife are alone when the husband kills the wife and disposes of her body. That is how most forms of spoliation are committed—in secret. That a suspect's guilt or innocence is not obvious is what necessitates Due Process of Law. There must be due process in order to prevent innocent people from being punished for crimes they did not actually commit. As I will explain, insofar as a society is a night watchman state, its Due Process of Law prohibits vigilante activity. There is a principle more important than the violent punishment of spoliators. The principle is that one refrains from exercising heavy-handed force against people who turn out to be innocent. That is why, should a vigilante strongly suspect Mr. Y of having wronged him, the vigilante himself is the spoliator if he acts outside the bounds of Due Process. And the vigilante cannot act in accordance with Due Process unless there remains a hegemonic party that enforces the Due Process rules on everyone in its dominion.

Yes, when someone is accused of spoliation, whether that suspect actually committed spoliation is seldom straightforward. And because force must be exercised against the true spoliators, we ought to be sure that, before extreme deadly force is actually directed against a suspected spoliator, we know for a fact that the suspected spoliator is actually guilty of the spoliation. Because an accused person can be innocent, every party that thinks of itself as an avenger (whether vigilante or government agent) is really a probable spoliator. Any time Mr. X claims to be issuing "retaliatory" force, when he did not comply with Due Process procedures, Mr. X should be considered someone who actually initiates the use of force. In the case of my engaging in vandalism against Riley, I should be considered a spoliator even if it later turns out that Riley did burglarize me. When violence-dispensing agencies, who think of themselves as mere avengers against spoliation, accuse Riley of burglary, and then proceed to exercise violence against him, how can we ensure that the violence agency is not spoliating an innocent man? This is why we need Due Process of Law. As any would-be avenger could easily be a spoliator when attacking Riley the Accused, the would-be avenger should not be allowed to issue the full brunt of violent punishment until after the would-be avenger has proven, beyond any reasonable doubt, that Riley is guilty of the spoliation for which he has been accused.

Again, I accuse Riley of breaking into my residence. Observe that when Due Process requires that I prove Riley's guilt—guilt in terms of his having spoliated me—before I be allowed to ask for any severely violent punishment for Riley, the law holds objective facts as paramount in importance. *Feeling* that someone is guilty is not adequate; the public has to be informed of the factuality of Riley having initiated the use of force to such a degree as to warrant the exercise of comparable violence against him. This is the principle that Objectivists identify as *objective law*. What separates a night watchman state from either a violent criminal or a tyrannical government is that the night watchman state's acts of force are necessarily guided by inductive reason and objective investigation. It is only following an objective search for truth—and following the acquisition of facts about Riley initiating the use of force—that the night watchman state will tolerate the unleashing of any comparable violence against Riley. To decipher whether imprisonment of Riley is retaliation against the guilty, not spoliation of the innocent, we need to launch a fact-finding mission. We need an objective body to conduct an objective investigation of whether my neighbor actually burgled me. Moreover, we want to minimize any abuses against a suspect who can turn out to be innocent of the charge for which he was accused. For that reason, we need an objective body to ensure that the accused is treated humanely. That is where Due Process comes in. When a night watchman state implements Due Process, it is the case of objective reason placing proper restraints on the use of violence by any party, whether private or governmental. Recall the Due Process procedures we went over in Book One. A night watchman state properly ensures such Due Process rules. That requires that the night watchman state function as a "hegemony" that imposes those Due Process rules on private agencies, such as me, that fall within its regional jurisdiction.

When an avenging party uncovers proof that a suspect committed the spoliation for which he was fingered, it is essential that this proof be made available to the general public. Imagine that I collect substantial evidence that Riley burgled me, but I keep every bit of it to myself. I avail none of the evidence to the general public for scrutiny. Then I brutally beat up Riley or raze his domicile. In such a scenario it is not unlikely that some members of the public will conclude that I fabricated accusations against Riley as a pretext for spoliating him. Then those members of the public can themselves become vigilantes who wield violence upon me, believing their own action to be avenging the spoliation that I committed against Riley. And the violence would go on.

And there is additional significance to every court case being public information; this is a safeguard against wrongful prosecution. When a public tribunal proves Riley's guilt, that is intended to satisfy the onlookers that Riley's consequent imprisonment is just. That discourages would-be vigilantes from attempting to rectify misperceived injustice by springing guilty men from prison. Should Riley be falsely convicted, based on errors of the jury or prosecutorial procedure, any vigilant attorney can examine the records of the case and file an appeal.

I love movies and comic books about heroic vigilantes. Unfortunately, I am guessing that people watch too many of these films and read too many of these comics, given that they seem to assume that vigilantes are generally heroic. When we examine history, vigilantes usually turn out to be dirty rotten cutthroats who hurt innocent people. Al Capone was a real-life vigilante. A real-life vigilante organization is the Ku Klux Klan. It promises to "protect" whites from people of other races. Every angry mob of private citizens that falsely accused a black man of rape—and immediately lynched him—was comprised of vigilantes. Such lynch mobs kill innocent people as a direct result of the lynch mob's refusal to abide by the rules of due process that no party but a constitutionally republican government can administer. This illustrates that our own society would be an oppressive one if the government did not require everyone—policemen and private citizens alike—to abide by the Fifth and Fourth Amendments during their investigations.

My opinion on civil disobedience is this. The degree to which someone is morally right to engage in vigilante actions is inversely proportional to the extent to which he dwells in a night watchman state. To the extent that he suffers under corruption or governism, such as in a Third-World kleptocracy, a person cannot rely on the State to punish spoliation. For that reason, it is incumbent upon private citizens to try to mete out whatever private justice is possible to them. Yet, the degree to which a society is already a night watchman state is the degree to which the government is already doing a good job in enforcing justice and quashing spoliation. For someone to rebel against spoliative laws—such as Apartheid—by breaking them, is to strike a blow for freedom. Conversely, for a Rothbardian anarchist to rebel against a night watchman state by setting up his own vigilante organization, in the absence of the night watchman state's oversight, is to make a mockery of freedom.[14]

In the Absence of Due Process, There Is No Long-Term Freedom

As I argued in Book One, it is true that a society *can* have due process in the long run in the absence of freedom. But, by that same token, a society cannot have freedom in the long run in the absence of due process. Conjure in your imagination a male prosecutor who would throw a woman in prison—and exercise the force of law against her without granting her any due process. Imagine the prosecutor does this in a context outside of states of national emergency or wartime. That prosecutor is behaving in a spoliative manner. The same principle applies to private citizens who are violent. Any group of private citizens that would beat you up, in the absence of granting you any night-watchman-state-mandated due process, is an oppressive group.

To reject a night watchman state in favor of the anarchy of competing vigilante groups is to say that police services should be ceded to competing oppressive mobs. Supposing that the anarchist has a right to become a vigilante who need not answer to the night watchman state's due-process laws, that anarchist would essentially arrogate to himself the authority to try and execute people, devoid of permitting them any chance to prove that they are innocent of spoliating others. Does the anarchist deserve a moral right to behave as a self-appointed grand inquisitor, throwing people into kangaroo courts? Should the accused be stripped of their right to demonstrate their innocence of the crimes for which they are accused? Granting that the answer to those queries is no, freedom requires that a night watchman hegemony exists, and that it monopolistically holds any would-be avenger within its borders to the standards of due process. A Noncitizen, too, must receive due process if a Pure Citizen presses charges against him.

Let us return to the scenario of my suspecting Riley of trespassing onto my premises and stealing my belongings. In this new scenario, though, I do not hastily enact violent vengeance. Maybe I, acting as a private citizen, actually do conform to the night watchman state's Due Process rules. Staying within the bounds of law and civil liberties, I gather evidence of Riley's guilt. If I am following those due-process rules, then I am not actually competing against the night watchman state. Should a private organization act in strict accordance with the rules of the night watchman government, then it, in effect, becomes an extension of the night watchman state. Private citizens may be deputized, for instance. It is logical that once a private citizen actually demonstrates himself to be responsible enough to become an issuer of avenging violence, he necessarily proves himself to be a sort of adjunct or appendage of the constitutional-republican state. Harry Binswanger explains that although "a proper government does not prohibit a man from using force to defend himself in an emergency, when recourse to the government is not available," it still "does, properly, require him to prove objectively, at a trial, that he was acting in emergency self-defense. Similarly, the government does not ban private guards; but it does, properly, bring private guards under its supervision, and does not grant them any special rights or immunities: they remain subject to the government's authority and legal procedures."[15]

The administering of justice does not necessarily require that there be a single party issuing night watchman rules—there can be an entire coalition of parties that enforce night-watchman rules. Yet such a coalition would be acting as a cartel—a cartel whose members collude in enforcing uniformly night-watchman republic rules. Were they to do that, they might develop a clearinghouse where laws are standardized among various protection service businesses. That single body of law, though, would result in the private protector agencies behaving as a single hegemonic government.

Insofar as the Night Watchman State Is Absent, Freedom Is Absent

Think about David D. Friedman's pitch for anarchy. One client might have high demand for a protection agency that strongly adheres to due-process (DP) rules. Conversely, I might be a "get-tough-on-crime" conservative who thinks that Miranda rights and other DP rules have greatly impeded the U.S. constitutional republic's ability to punish guilty criminals. Hence, my fellow get-tough conservatives and I might have strong demand for a protection service agency that that puts very little emphasis on due process, considering accused parties guilty until proven innocent. We might not only want that agency to enforce such anti-DP procedures upon its clients, but also upon the clients of pro-DP agencies whom we believe to have wronged us. First, if there are some private agencies that enforce due-process rules, while other agencies do not, then that creates a situation where liberty is not truly maximized, because the pro-DP, more-liberalized agencies tacitly tolerate the oppression of people (including non-consenting minors) under anti-DP agencies.

An anarchist might counter that in any dispute between clients who are respectively under pro-DP and anti-DP agencies, their differences can be resolved by both agencies compromising when selecting a third-party arbitrator. In the process, both sides would each acquiesce a little in its demands. Take me as someone aligned with the anti-DP agency. I want it to prosecute someone who is subscribed to a pro-DP agency. Here, my anti-DP agency would have to acquiesce and allow for my enemy to be treated with more Due Process than I prefer. If market forces pressure the anti-DP agency to become more like the pro-DP agencies, then that actually means that market forces push the maverick agencies toward the direction of conforming to a standardization of industry practices. But such a market-driven standardization of industry practices actually undermines David D. Friedman's attempt to exploit, as a selling point, the idea that anarchy is attractive for the reason that it would allow for a variety of different agencies to engage in radically different practices. The truth is that the arbitrator's verdict is meaningless if either side can reject it. This evinces that the verdict cannot be binding unless (1) a central government enforces it, or (2) one of the defense agencies is capable of driving the other into submission. In the case of 2, the dominant defense agency becomes a monopolistic-hegemony-in-practice in this case, and such a *de-facto* monopolistic hegemony is exactly what the anarchist wants abolished.

I wish for people to see through the fallacy that Arbitration Solves Everything. Someone can point out that a system of competing, private defense agencies will stir up skirmishes. There will be cases where a legal dispute arises between the clients of two competing violence agencies. The anarchist will reply, "No, everything can be settled by private Dispute Resolution Organizations—DROs." Too many people believe that government is supposed to be the solution for everything they interpret as a problem—they want the government to be God, or *GOD*vernment. Believing themselves to reject the irrationality of this view, anarchists propound that private arbitration must necessarily be the solution for everything and private arbitration must be God (Allah-narchy?). Private arbitration is not any more of a cure-all than is governism. There are cases where private arbitration fails and the disputing parties end up going to court anyway. An attorney told me of a case where two brothers fought over ownership over the two-story house they inherited. They went through multiple levels of private arbitration, only to end up summoning their lawyers to appeal the previous arbitrator's ruling. Eventually, the last arbitrator decreed that one brother gain full ownership over the first story and the other brother gain full ownership over the second story. Naturally, that facilitated additional vagaries; every time the brother on the second story walked down to the first story, he was technically trespassing. The brothers had to go to a "monopoly government" court anyway.

There is a contradiction in a libertarian who proclaims the inherent greatness of both alternative dispute resolution (ADR)—private mediation, private arbitration, and DROs—and enacting tort reform through implementation of a *loser-pays* tort system in the USA. As we recall from *Life in the Market Ecosystem*, I favor the loser-pays system. Under the current U.S. tort system, corporations tend to favor ADR over going to court. They figure that no matter what happens, the plaintiff's action automatically imposes a cost on them, and, statistically speaking, paying their lawyers to settle the case is more often cheaper than winning the case in court. However, if the loser-pays system is implemented that dynamic will change. The corporation will notice that if it works out a settlement with the plaintiff, it will necessarily lose money, whereas, if the corporation lets the case go to court, it might not lose any money. For that reason, when the loser-pays system goes into effect, the corporation will have more of an incentive to let the case go to a governmental court—where it can win and thereby recoup its legal costs—and *less* of an incentive to resort to alternative dispute resolution. That will reduce demand for ADR. Many anarchists cite the proliferation of ADR as proof of ADR's inherent superiority over the court system. In reality a large part of companies' current preference for ADR over government courts has to do with the current governist setup of the tort system. Were the tort system to be further liberalized by the incorporation of *loser pays*, much of the companies' preference for ADR over government courts would shrink. It would drive up businesses' usage of the night watchman state's courts and reduce their demand for ADR. When anarchists cite the popularity of ADR as evidence of the supremacy of anarchy, they overlook that a large part of ADR's popularity was stimulated by the governism they wish were nonexistent.

It can be argued that a single, powerful, pro-DP private agency—or an alliance of such agencies—can pressure the anti-DP agencies into becoming more-DP oriented. Such pressure would be exerted through the threat of violence. Even a threat of a mere economic boycott would be exercised through violence if the pro-DP agency threatened violent punishment (contract-based or otherwise) upon any client of its own that disobeyed the economic sanction rules. But if such a pro-DP agency—or an alliance of such agencies—exerted such pressure on other agencies to amend their policies, that pro-DP party would be behaving in a manner comparable to that of a monopolistic hegemony. If the pro-DP party did not exert pressure upon other agencies to allow due process to non-consenting minors, then it would be like a national-government hegemony that tolerates the oppression of minors in other jurisdictions, and that would mean anarchy has not eliminated the inequitable conditions of the geopolitical status quo. Contrariwise, if the pro-DP party *does* exert pressure on other agencies to allow Due Process to their minors, then it succumbs to acting as a regional monopoly government.

Some anarchists proclaim that an eight-year-old boy should be recognized as contractually competent enough to decide for himself which protection agency he wants, regardless of what his parents expect. It can be said that if a boy's parents want him subscribed to Agency *G*, while the boy wants to be under Agency *H*, then the boy should have the right to be under Agency *H* against his parents' wishes. The boy can get his way either through violence or nonviolent market forces, and neither option appears particularly convenient for the anarchist's case. Visualize a single agency—or coalition of agencies—wanting every eight-year-old boy to be recognized as contractually competent enough to choose his own protection agency. Suppose this position is called Child Emancipation (CE). Should the Child Emancipating agencies be unable to tolerate so many other agencies forcibly holding eight-year-old boys as "clients," the pro-CE party can threaten violence to exert pressure on the protection agencies that allegedly oppose CE. They would do so in order to intimidate the anti-CE agencies into conforming to the pro-CE standards. Again, that turns the pro-CE party into a kind of monopolistic hegemony. The other alternative is for all or most of the private agencies to agree voluntarily with one another that an eight-year-old boy should be allowed to subscribe or unsubscribe from whatever agency he wants, regardless of whether it contradicts his parents' wills. Yet such industry standardization conflicts with David Friedman's point about there being a lot of variety in having various social systems from which to choose. The more a market anarchist region moves toward adopting a mostly peaceful equilibrium, the more it will seem to be a move toward regional monopolization by one firm or a cartel of firms. By contrast, for various firms to compete by distinguishing themselves from one another by setting up dissimilar social systems, is to invite more chaos and disequilibrium that renders inter-agency skirmishes more likely.

To be certain that due process becomes the standard operating procedure of every entity that advertises itself as a defense agency, a night watchman government must necessarily have the final word on adjudication. This is why a society that minimizes spoliation must necessarily have a dominating night watchman order.

When anarchists insist that it is a violation of your rights for a night watchman state to crack down on vigilantes, they overlook the following points.

1. Due process of law is essential for fighting against spoliation and for preventing would-be crime-fighters—be they government employees or private citizens—from spoliating innocent people they wrongly suspect of spoliation.
2. Nothing but a central night-watchman-government can exercise the power to require every single person—including the night watchman government's own employees—to abide by due-process procedures.
3. Any vigilante unbound by due-process rules has free reign to spoliate innocent persons.
4. The facts inescapably point to the conclusion that a night watchman government—bound by due process—minimizes the infliction of spoliation in a society, whereas a society that lets vigilantes run rampant will *maximize* spoliation.

Yes, Rothbardians howl that they can achieve a laissez-faire, spoliation-minimizing society in the absence of universally government-mandated due process. They gloss over the fact that a laissez-faire, spoliation-minimizing society cannot exist in the absence of that universally government-mandated due process. Despite the intrinsicism of his faulty epistemology and faulty ethics, Robert Nozick is correct about anarchy—insofar as human beings achieve livable social conditions, the result will not be competing protection firms under anarchy but their having delegated the authorized exercise of retaliatory force to a single, hegemonic body which is government.[16] Someone can give a society the label of *libertarian* until he turns blue in the face. The bottom line is that if that society has no due process of law, then it is not a free society, but a spoliative one. As we realized in Book One, there is no proper substitute for the night watchman state.

What the Label *Minarchy* Tries to Minimize

By now I should have made myself clear on why I have avoided the employment of a particular neologism thrown around by libertarian activists—*minarchism*. According to anarchist libertarian Samuel Edward Konkin III (d. 2004), laissez-faire advocates fall into two camps—anarchists and minarchists—with Konkin proudly falling into the former. Konkin and other libertarians employ the *minarchist* tag because of their misplaced emphasis on what can be termed the "size" of government. Government's "size," in this context, refers to the scope of its powers—powers pertaining both (1) to its night-watchman-state duties and (2) to the governist initiatives that reach beyond the night-watchman-state functions. Governmental size also here pertains (3) to the government's periodic budgets and (4) to the number of its employees. Those who stress a minarchism-anarchism spectrum wish to emphasize that the "size" of a society's government is always inversely proportional to the degree of liberty which that society enjoys. According to this standard, maximum liberty entails a government of absolute minimum size. Henry Hazlitt, to his discredit, emphatically proclaimed himself to be an advocate of "minimal government."[17] Robert Nozick falls for this fallacy as well, referring to the perfect night watchman government as an "ultraminimal state."[18]

I knew a rather pushy libertarian anarchist who kept trying to pressure me to accept the designation *minarchist*. At that moment, I did not know why. At this writing, I do. The anarchist tries to get his sparring partners to describes themselves as supporters of minimal government. At that juncture, the anarchist pats himself on the back for having cornered these people. He points out that to shrink government to an absolute minimum in size is to have *zero* government. Thus, in Konkin's view, purely consistent advocates of liberty are invariably anarchistic. Konkin and other anarchists then condescendingly convey that most self-proclaimed free-marketers are too timid and psychologically insecure to concede that perfect liberty requires anarchy. In this interpretation, non-anarchist free-marketers—supporters of the night watchman state—are in a wishy-washy position, preferring (a) that there generally be "less government" than that which currently exists while (b) they are not brave enough to take the pro-liberty position to its supposedly logical (anarchist) conclusion. Allegedly, instead of being bold purists and embracing anarchy, non-anarchist free-marketers cowardly hide themselves under the security blanket of "minimal government" or "small government."

Therefore, to Konkin, the "minarchist" position is a wimpy milquetoast compromise between the extremes of totalitarianism (100-percent government and therefore pure evil) versus anarchism (0 percent government and therefore pure goodness).[19]

A comparable attitude can be found in various left-wing atheists with respect to people who are gradually losing their faith. Throughout the Enlightenment period, various people tried to adopt a middle ground between religion and atheism. That middle ground was deism. A deist is someone who repudiates every organized religion but still clings to a belief in some God. A variety of atheists, myself included, believe that the deist position is a transitional stage for someone brave enough to leave behind his church but still not brave enough to leave behind any belief in a supernatural entity. Likewise, a libertarian anarchist can regard, with some condescension, a putative minarchist as someone in that same transitional stage. He can proclaim, "You have made great progress, but, as long as you insist on the need for police and a military, you remain a caterpillar. The day on which you accept anarchism is the day on which you will have matured into a beautiful butterfly."

The pushy libertarian anarchist of my acquaintance asks other people to quantify the amount of government involvement they want in their lives. He probes, for example, whether you would prefer to have "Zero-percent government involvement? Twenty-five percent? Seventy-five percent? One-hundred percent?" Should you answer that you want "25-percent government involvement" in your life, he immediately concludes that you are for "75-percent liberty, as you cling to 25-percent tyranny." The inappropriateness of this line of questioning can be revealed in another example. Imagine that you go to the hospital for an appendectomy. The surgeon in charge does not know specifically where to apply the scalpel, but simply inquires, "How much of your body do you want sliced up? Do you want me to slice up 15 percent? Thirty percent? Ninety percent? Would it be better if I applied no incision anywhere? Should your entire body be minced and ground into little pieces?" Such a doctor would be evading the matter at hand. The issue would not be what percentage of your body must be cut up; the issue would be the specific part of your body to which the surgeon would direct the operation. Likewise, the issue is not what percentage of your life requires the application of government force; the issue is what specific activities government force must be applied. Should there be hundreds of gangsters in your community spoliating the innocent, a large amount of government involvement would have to be directed to stopping them. Conversely, should there be nothing more than a petty theft once every ten years, and no more crime than that, a legitimate government would correspondingly do little.

I therefore submit that the connotation behind the "minarchism" expression is a straw man. Insofar as the "size" of government refers to (1) its budget, (2) its number of employees, and (3) the effectiveness and firepower of its night-watchman-state agencies, a government's size does not directly correlate with tyranny. Should a night watchman state have a large, well-funded military, that could maximize liberty by guarding against foreign spoliators. Suppose Country *J*, already a night watchman state, were to "shrink" its government—"minimizing" its government or "minarchizing" itself—by gutting its own military's budget and drafting Rules of Engagement that prioritize enemy nations' civilians and their collateral at the cost of Country *J* being able to achieve the swiftest military victories. That reduction in the size of Country *J*'s government would result in foreign dictatorships conquering Country *J* and thereby maximizing oppression within Country *J*. For Country *J* to minimize its government in this context would likewise minimize liberty. Nay, the size of a country's government is not inversely proportional to the degrees of Domestic Peace and liberty that that country experiences. Rather, the degree of liberty accumulated is directly proportional to two other measures.

1. The degree to which the government prevents and punishes spoliation.
2. The degree to which the government abstains from spoliating.

Position *1* requires a night watchman state's presence and thereby precludes anarchy. As we have read, libertarian anarchists counter that Position 2 is impossible. They virulently insist that every government—including a night watchman state—is, by definition, a commission of spoliation. The irony is astonishing. The night watchman state is not the contradiction of liberty, but its prerequisite. I am emphatically not a "minarchist." That entire phrase, and the ideological implications behind its coinage, amount to a red herring that leads free-marketers astray. My position is that of support for the night watchman state. I want the government "big" and possibly maximized to the extent that this bigness assists in the government's night-watchman duties. I want the government small, limited, and minimized insofar as that minimization deters the government from spoliative measures. Contrary to my libertarian anarchist friend, the degree of liberty you find in life cannot be properly quantified according to some numerical measure of how much "government involvement" there is in your existence. I want a high level of government involvement in my affairs inasmuch as I want the government to enforce each and every contract I choose to enter. I want no government involvement insofar as such involvement refers to the degree to which the State interferes in dictating contractual terms that should be left up to the free negotiation of consenting adults. No, my friend is mixed up when he presumes that wanting "25 percent government involvement" in your doings amounts to your supporting "75 percent liberty and 25 percent tyranny," especially with no distinction being made in that "25 percent involvement" between night-watchman functions and spoliative governist functions.

It has been difficult for me to imagine the anarchists being more wrong on anything than they are on this issue. Nevertheless, they also misunderstand libel law and the essential need for truthfulness in consensual human intercourse. Such matters must be examined as well. Upon completion of that examination, we will wrap up this trilogy.

NOTES

1. Zakaria 2003, 37.
2. S. Banner 2007, 132, citing M. Kelly 1956, 1–49; D. P. McGregor 1996, 6–8; and J. Linnekin 1983, 169–171. This is also observed by Valeri 1985, 155, 372 n. 63, citing W. Shaler 1935, 95, and E. Townsend 1921, 17. Also see Kuykendall and Day 1961, 54. The similarities between the ancient Hawaiian system and European feudalism are pointed out in S. D. Porteus 1945, 40.
3. University of Chicago anthropologist Valerio Valeri observes in this in Valeri 1985, 130-171. The same applies to Kuykendall and Day 1961, 27–28, 54.
4. Kuykendall and Day 1961, 54.
5. D. Friedman 1989 2d ed., paperback, 117.
6. Bidinotto 1994, accessed online Tuesday, August 6, 2013.
7. Acemoglu and J. A. Robinson 2012, 80.
8. Binswanger 1981c, http://hblist.com/anarchy.htm, accessed Friday, January 18, 2008,
9. A. Burris 1983 2d ed., 334–35; and Rothbard 1973b, 289–290.
10. A. Burris 1983, 2d ed., 334.
11. Binswanger 2011b, accessed online Tuesday, January 3, 2012.

12. Rand 1963d, 45, republished in Rand 1970 paperback, 127.

13. Binswanger 1981c, 12; and Binswanger 2011c, http://hblist.com/anarchy.htm, accessed Tuesday, January 3, 2012.

14. R. D. McGrath 1984 argues that vigilantes did create a peaceful social order in the Old West. However, to the extent that that may be true, it is actually because those vigilantes increasingly came to act like a monopolistic night-watchman-state.

15. Binswanger 1981c, 12; and Binswanger 2011c, accessed online Tuesday, January 3, 2012.

16. R. Nozick 1974, 118.

17. H. Hazlitt 1993a.

18. R. Nozick 1974, 27.

19. B. Doherty 2007, 248, 688 n. 17, reports about how Samuel Konkin III was the libertarian anarchist who coined *minarchism*. In this interpretation, anarchism is the perfect ideology, and minarchism is an imperfect but necessary step to becoming fully mature and intellectually perfect by embracing anarchism. In this regard, one can think of a "minarchist" as a boy riding his bicycle with his training wheels on. The boy finally proclaiming himself an anarchist is the moment that he is able to ride his bicycle in the absence of training wheels.

THIRTEEN

Natural Liberty Requires Adherence to Truth

Libertarian Defamation of Libel Law

Laws against libel and slander, which fall under the category of defamation, are justified for the reason that libel and slander constitute fraud. With some embarrassment, though, I admit that for several years the anarchist libertarians had convinced me that there should be no such laws. The anarchist libertarian's argument is as follows. Every lie, no matter how malicious and damaging, counts as free speech.[1] Suppose that Roy concocts the allegation that Wally is a child rapist, and convinces everyone in Wally's town that Wally had committed this crime but had escaped prosecution for it. This leads to the community shunning Wally and refusing to provide him any job. The anarchist libertarian claims that Roy practiced free speech, and did not spoliate Wally or his private property. The libertarian anarchist says that if community members let this lie affect their treatment of Wally, the fault lies not in Roy for having carried out this deception, but in the townspeople for having uncritically swallowed it in lieu of checking up on it. Though acknowledging that Roy defrauded gullible community members who otherwise may have hired or married Wally, the anarchist libertarian proclaims Roy did not defraud Wally himself. Hence, Wally should not be able to sue Roy on the basis that Roy exercised fraud to harm him.

Just as he does in other matters, Murray N. Rothbard here gives the standard anarchist libertarian view. "What the law of libel and slander does, in short, is to argue a 'property right' of someone in his own reputation. Yet someone's 'reputation' is not and cannot be 'owned' by him, since it is purely a function of the subjective feelings and attitudes held by other people." As "no one can ever truly 'own' the mind and attitude of another, this means that no one can literally have a property right in his 'reputation.'"[2] W. Alan Burris echoes, "Slander and libel are unethical dirty tricks. But . . . there is no force involved, and no one has a property right in the opinions of others. Therefore, it is not moral to use force to stop slander and libel, or to collect damages."[3] And academic economist Walter Block (b. 1941) agrees that "a person's reputation does not 'belong' to him . . . A person's reputation is what other people think of him; it consists of the thoughts which other people have. A man does not own his reputation any more than he owns the thoughts of others," and a man's reputation is no more than those very thoughts about him. "Whether his reputation was 'taken from him' by fair means or foul, by truth or falsehood, he did not own it in the first place and hence, should have no recourse to the law for damages."[4]

That whole argument is a straw man. A man does not need to hold statutory ownership over his reputation in order for libel law to be valid. The issue is not that a man's reputation is his private belonging, but that libel is a species of fraud that does serious damage to people's lives—and a person's life is his own exclusive property. Moreover, as we shall later examine, uncorrected defamation obliterates the contractual competency that is a prerequisite for contract law to function properly.

Of course, anarchist libertarians produce other points. They propound that when the government takes action against slander and libel, the government becomes a sort of Truth Police. Given the record of governments lying to their citizens, does one want a government court to legislate over what private citizens consider truthful or untruthful? Upon further reflection, I reject the anarchist libertarians' argument. First, I will address the aforementioned argument about Truth Policing. It is another straw man. In reality, courts render decisions every day that require the courts to decide under the law, officially, whether a specific accusation can be substantiated as truthful or not. When a court deems that Gustave is guilty of murdering his wife, it publicly and officially decides that the accusation "Gustave murdered his wife" is true. Once the criminal justice system has presented damning evidence of Gustave's guilt to the public, the legal system must act upon Gustave accordingly by imprisoning him.

Obviously, court judgments are not always the same as that which later prove ontologically true. The ontological truth is that which is true in fact. Were Gustave innocent of murder in factual reality, but convicted of this crime by an unintelligent jury, then he would be guilty in "legal truth" but innocent in ontological truth. Naturally, ontological truth is more important, and the due-process system was set up to ensure that "legal truth" comes as close as humanly possible to matching ontological truth. Contrary to criticisms from anarchist libertarians, the ability of courts to err in their decisions does not render the court system, per se, to be morally or practicably illegitimate. That courts can deliver erroneous verdicts is what gives rise to the fact that the due process system grants defendants and respondents the ability to appeal rulings made against them, thereby enabling the judges and juries of appeals courts to double-check previous decisions handed down by other judges and juries.[5]

At any rate, any court case that involves anyone accused of spoliation, happens to be a case in which the court must offer an official declaration of what it judges to be ontologically true, and its decision is "legally true." A court that dispenses legally true verdicts is not the same as a governmental Truth Police trying to force anyone to believe anything to be true or false. Any member of the public is free to believe that a court falsely concluded an innocent man to be guilty, or vice versa. By that same token, the outcome of a libel or slander suit does not force anyone to believe anything. Hence the thought-police argument does not stick. For that very reason, when a court, upon objective examination of the evidence, rules that Roy is guilty of libeling and slandering Wally, the court's assessment is a legal truth that does not force anyone to agree with the court ruling, with Roy, or with anyone or anything else.

Libel law avoids being Orwellian thought control on yet another count. An Orwellian government that tries to police thought is one that tries to police people's opinions—that is, their emotional evaluations of the facts, such as "Wally is much ruder and brusquer than I would prefer." Libel law legislates exclusively over serious accusations that pertain to empirically demonstrable facts, such as "It is verifiably true that Wally stole my cat and killed her." When a false accusation about matters pertaining to facts circulates throughout the public, it can have a devastating impact on Wally's ability to live and act peaceably by his own judgment. Such a devastating impact is not comparable to people forming negative opinions about Wally based upon their own interactions with him. Furthermore, it is a straw-man argument when anarchist libertarians say that the existence of libel law provides plaintiffs with the opportunity to sue people over ridiculously hyperbolic and unbelievable statements, such as "Wally is the Devil," or with accusations pertaining to inconsequential matters, such as "Wally chews with his mouth open." Such a suit will not stand up in court. Likewise, trivial white lies, such as Roy telling a woman she is stunningly beautiful when he thinks her physical appearance is mediocre at best, are not subject to libel law. Little white lies are not severe enough to influence the woman's life in the most adverse fashion. Wally can win a defamation lawsuit against Roy solely on the basis that Roy's false allegations were plausible, unproven, and had a harrowing effect on Wally's life that would not have occurred had Roy not publicly disseminated his fabrications.

More importantly, serious forms of libel and slander are initiations of the use of force for the same reason that the law legitimately classifies harassment and stalking as forms of spoliation. To harass and stalk a woman systematically does not cause obvious physical damage to her health or her property, but a chronic manifestation of this phenomenon can seriously disrupt her daily routine, undermine her earning ability, and pose an enormous strain on years of her life. Grave forms of libel and slander can produce the same effect. To those who believe that the consequences of libel are not severe enough to legislate against, I would inquire as to whether they think that stalking and harassment are no big deal either.[6]

Slander and libel are fraud for the same reason that identity theft—which, in legalese, translates to *unauthorized appropriation* under tort law[7]—is rightfully considered fraud. Imagine that I memorize your Social Security Number and craft a fake ID with which I can impersonate you. Then I go to your bank, pretend to be you, and withdraw everything from your savings and checking accounts. Some anarchist libertarians may concede that in this scenario, I defrauded the bank by directly fooling its tellers. On that basis, the libertarians might concede that the bank may rightfully take action against me. But anarchist libertarians may still have the gall to add that because I did not directly fool you or directly breach any contract I made with you, you yourself should have no legal authority to file charges against me. That would be a ludicrous conclusion, given that my defrauding the bank has obviously caused demonstrable personal harm to you. The same logic applies to a libeler.

Consider that Julianne would hire—or marry—Wally had she known his true character. But Julianne purchases a newsmagazine published by Roy, which scurrilously impugns Wally as a pedophile. And Roy publishes this on the World Wide Web as well, presenting the false claim to millions of Web surfers. Julianne does not know whether she fully believes this charge, but it stirs enough doubt in her mind to refrain from hiring or marrying Wally. For this same reason, no one else will hire or marry Wally. Should it be that most people who have heard or read the lie about Wally are not fully convinced of its authenticity, it can still wreak havoc on Wally's life. Again, the anarchist libertarian may concede that since the subscribers of Roy's magazine subscribed on the implicit contractual understanding that they were paying for verifiable truth instead of arbitrary lies, the subscribers were defrauded and may take legal action against Roy. Indeed, considering that Julianne would have been much likelier to hire or marry Wally had she known the truth, Roy had defrauded *her*. But the same anarchist libertarian will self-righteously profess that, given Wally's status as a third party who did not remunerate Roy for truthful information, Wally should have no legal recourse against Roy. The anarchist libertarian would insist as much, despite Roy's lies damaging Wally's life more than they did the lives of Julianne and his other readers. Such a claim is ridiculous for the same reason that it would be ridiculous to say that my draining your bank account, while masquerading as you, victimizes your bank but not you.

Contractual Competency Drowns in an Ocean of Lies

While anarchist libertarians pat themselves on the back for their putative defense of private-enterprise economics, they neglect to note that the rule of law—which underpins free-market activity—cannot sustain itself except under the fulfillment of specific conditions. One of these conditions is that the legal environment recognizes civil suits that financially penalize individuals who disseminate life-destroying lies. The example of trademark infringement highlights this point. The entire basis of competitive capitalism functioning properly is that you can identify poor service with a poor brand, and good service with a good brand. When a company does well, you reward it with repeat business. When a company serves you poorly, you shun it. Suppose the Zyber Pharmaceuticals brand earns a great reputation among HIV patients for providing them drugs that effectively preserve their lives. Then I, deciding to profit off of Zyber's reputation parasitically, sell sugar pills in packages featuring Zyber's brand name and logo. I do this to foster the impression that I am selling genuine HIV pills approved by Zyber's own inspection team. This, too, is a form of identity theft—rather than for me to give the false impression that I am some other bank depositor, I here instead convey the false impression that my products are Zyber's. Doing that, I can unscrupulously profit from that company's decades-long record of providing solid service. Suppose libertarians opposed to libel law ran the government, and refused to crack down on these and other knockoffs. This would destroy the very foundation of the free market. Consumers would no longer be able to reward good brands financially and boycott bad brands. The inferior brands can simply steal the logos and names of the good brands.

This would destroy Zyber, its reputation, and the income that it otherwise would have rightfully accrued. Were HIV patients to recognize that Zyber did not authorize the faulty pills, the patients' inability to distinguish real Zyber pills from knockoffs would still teach them to avoid anything bearing the Zyber label. This can send shockwaves throughout the entire supply chain, as distributors, wholesalers, and retailers are just as capable of purchasing counterfeits unknowingly. Trademark-infringing parties could destroy every reputable brand if they faced no prosecution. Should Zyber spend millions of dollars to create new logos and brand names for itself, and to redesign its products' packaging accordingly, counterfeiters will swipe those new symbols and brand names as well. Inasmuch as the sale of knockoffs goes unpunished, quality brand names become meaningless. This, in turn, erases the benefits of competing businesses under capitalism—if you cannot recognize a business by its brand name, then you cannot recognize any large-scale transnational business as such, eliminating your ability to remain loyal to quality companies and to avoid their inept rivals.

An anarchist libertarian may concede that the HIV patients may have legal recourse against the trademark-stealer. But, to a consistent free-marketer such as me, that does not go far enough. In this example, Zyber is obviously a victim. When a counterfeiter lies about Zyber in order to defraud Zyber's potential customers, the victims include not merely the customers who are directly deceived, but also the company that faces financial harm from the deception. The same principle applies to Roy circulating devastating lies about Wally. Libertarians put so much of their confidence in contract law. But the veracity of contract law, as such, bases itself upon the assurance that good faith goes into the signing of most contracts. As you may recall, Book One explains the importance of contractual good faith. *Good faith,* in the context of contract law, is not the epistemological faith criticized in Book One. People can reliably set up contracts with one another on the basis that they have the contractual competency discussed in this chapter. Contractual competency is contingent upon the fact that, though some people may tell relatively harmless white lies to each other, a contract-maker can be reasonably assured by law enforcement that nobody is spreading seriously life- or career-eroding untruths about him or anyone else with whom he might contract. The existence of libel law buttresses the contractual competency of legally sane adults for the reason that it encourages among contract-makers a valid presumption of innocence not unlike the judicial presumption of innocence. Julianne can judge a potential business partner to be reasonably innocent of spoliation, and she can rest assured as much until authorities catch her potential business partners for spoliation.

Were there no libel law, and any party could promulgate career-eroding untruths about other parties undeterred, the environment would become one in which lies circulate about everyone. A significant portion of the population would be unable to do business on the basis that it could not contest the devastating and scurrilous allegations aired about it. The wide dissemination of mendacious accusations also evinces that it is significantly more difficult for a businessperson to separate true claims from false ones. When an anarchist libertarian states that it is incumbent upon Julianne to do her own homework about Wally and the rest of the thousands of persons with whom she might conduct business, devoid of any libel law to help her separate crooks from innocent persons, that anarchist libertarian is, in effect, insisting to Julianne that she waste years of her life. He expects her to embark on a grueling task that she would not have had to undergo had defamation laws been in place. As usual, the libertarian anarchist's expectations prove unrealistic. Our system must structure the laws according to practicality. Rather than letting this cumbersome process eat away her time, Julianne ought to be living her life freely.

When no one can trust anyone else, as too many false accusations run rampant, it is no longer rational to expect that individuals form contracts among one another. There is no good faith, placing every legally sane, consenting adult into a state of contractual incompetence. The chaotic conditions cancel the ability to offer contractual assent. This argument of mine is in no manner a concession to the governist's assumption that there need to be government-imposed occupational licensing requirements. A normal government says that no one should be able to do business as a doctor, plumber, or barber until one has gone through the hurdles of obtaining a license from the State to certify that one is capable in this field. Such occupational licensing laws are a holdover from the guilds of the high Middle Ages and Renaissance,[8] and their true purpose is to restrict business competition in these industries. Licensing ensures that doctors, lawyers, and plumbers may form something of a government-regulation-protected cartel. What would do much more to assure patients that a doctor is competent than government certification is industry competition. It should be legal for a surgeon to advertise his record and rate of success in past surgical procedures.

A legally competent adult already possesses the capacity to choose her own doctor, plumber, or barber under laissez-faire competition. Such a patient or client is capable of assessing for herself whether she is satisfied with the service she has been provided. That people can formulate contracts in good faith is a vital condition for contractual competency. And this good faith cannot exist if libel and slander laws go unenforced. In other words, when libel, slander, trademark infringement, and other forms of life-threatening mendacity are legal, the very institution of free commerce ceases to exist. In its place arises a Hobbesian "war of all against all" whereby everyone may fall prey to identity thieves, product counterfeiters, libelers, and other such charlatans. That becomes an environment heavy on spoliation. As with any other form of fraud, libel is an initiation of the use of force against intangible and genuinely existing individual rights, and laws against libel are merely retaliatory force in defense of rights.

In Book One we introduced spoliation as direct physical brutality. We then mentioned John McCaskey's explanation of how contract breach and fraud constitute indirect initiations of the use of force, wherein the contract-breaker or fraudster manipulates the victim into applying physical force to himself in a harmful manner. Two people—Tad and Lewis—develop an unwritten, yet binding, contractual agreement. Tad promises that if Lewis hands him a specific item, Tad will give him 10,000 dollars a month later in a lump sum. Lewis hands Tad the item but, a month later, Tad does not deliver the 10,000 dollars. He never does. It is a breach of their contractual arrangement. As McCaskey articulates, Tad's default on his promise involves harm resulting indirectly from the initiation of the use of force—the same *force* which physicists speak of in their own scientific discipline. The physical force harming Lewis is the physical act of Lewis handing his item over to Tad when, had he known that Tad would fail to deliver on the promised 10,000 dollars, Lewis never would have engaged in such a physical action. Therefore, in breaching the contract, Ted has manipulated Lewis into applying force in a manner harmful to Lewis's own life.

We can extend that logic to Wally, Julianne, and Roy. Again, Wally and Julianne are engaged to be married. Roy goes to Julianne and tells her the completely fabricated story that Wally molested children and has escaped criminal prosecution. The very possibility that this might be true disturbs Julianne. She breaks off her engagement to Wally. Should Wally approach her to try to explain himself, Julianne shall interpret that as stalking and harassment, and she shall summon the police to detain Wally.

Part of Wally's freedom to act peaceably on his own judgment requires that other people deal with him in good faith—not on false pretenses that might lead to violent interaction. In the absence of Roy's duplicity concerning Wally, both Julianne and Wally would be free to act peaceably on their own judgments. When Roy feeds devastating lies about Wally to Julianne, Julianne acts on that false information. It is not as if beliefs are held in a vacuum. Contrary to David Hume, people must necessarily act on what they believe to be true. Rejecting the proclamations of Rothbard and Walter Block, the use of physical force is initiated in this scenario. The physical force is the physical act of Julianne shunning Wally, physically avoiding his company, and calling the police on Wally if he approaches her peaceably. Thus, Roy has manipulated Julianne into becoming an agent that applies force that harms Wally.

When Tad breaches contract with Lewis, manipulating Lewis into relinquishing an item to Tad under false pretenses, Tad transforms Lewis into the agent applying force that harms Lewis. Likewise, if Roy feeds devastating lies to Julianne about Wally, Roy transforms Julianne into an agent applying force that harms Wally. That is why libertarians are incorrect and why defamation counts as spoliation that a night watchman state can rightfully deem actionable under civil law.[9]

Never Count on a Broken Clock—Not Twice a Day

As we are on the subject of truth and libertarian dismissals of its importance, it is opportune for us to return to the topic that Book One's opening chapter focused upon. Should two men each proclaim a need for a laissez-faire government limited to protecting people from force, should we judge them to have the same goals when their epistemologies differ? Yes, insist such libertarians as Murray N. Rothbard. According to Rothbard, if it is the case that someone has reached the correct conclusions on politics through an invalid methodology, that is fine, as nothing matters except that the person's conclusions are correct. One can cite an old saying that goes, "Even a broken clock is right twice a day." There is a philosophic implication in that statement that is wholly misleading. Suppose my clock breaks and is stuck at 10:43. That clock is never right, given that it has lost the causal means whereby it once provided accurate timekeeping. My broken clock proves invariably unreliable for telling time. That principle continues to apply at 10:43 a.m. and 10:43 p.m.

Likewise, consider the following. I will ask heads-or-tails questions to a dime and then flip it to get the answer. I say, "Was Thomas Jefferson the third president of the United States? Heads means yes and tails means no." The dime lands on heads. Does this mean the coin told me the truth in this case? Does it mean that if I flipped the coin in one million instances, the coin would tell me the right answer in about half of those instances? It would divulge the right answer none of the time, because it never gave an answer; flipping a coin is not a reliable means of obtaining the truth on matters that are not coin-related.[10] Here is the principle.

> The intellectual contents of a conclusion are comprised of more than just the words that express that conclusion. Your conclusion's intellectual contents are the consequence of the epistemological method whereby you formed that conclusion. Therefore, a person's conclusion consists not merely of the words expressing the conclusion, but also the intellectual basis that the person acquired in arriving at that conclusion. This principle applies regardless of whether one holds the conclusion tentatively or intransigently.

The Objectivist theory of concepts, discussed in Book One, helped us derive that principle. Suppose two teachers approach a little girl. Both teachers tell the child, "A caterpillar will spin a chrysalis and later emerge from that structure as a winged creature called a butterfly." Nonetheless, it was very dissimilar lines of thought that led these two teachers to holding this sentence as true. Teacher *1* finds sensory evidence irrelevant. Teacher *1* tells the girl,

> I have a good reason why you should believe that this caterpillar will turn into a butterfly. I have never seen this happen, and I don't believe anyone who claims to have seen this happen. But I know it is factual because the children's book *Butterflies* says so. This book is the source of all truth, and therefore its factuality stands in the absence of any corroborating data that you obtain independently of reading the book. Should any data outside the book prove inconsistent with the book, then we know your interpretation of the data has to be wrong. Whenever your interpretation is at variance with the contents of the book, revise your interpretation to conform always to that of the book's contents.

Teacher *2* takes a contrasting approach. Teacher *2* says to the little girl, "Let's go outside and catch a caterpillar." The two do so, and they put the caterpillar in a terrarium. Teacher *2* shows the little girl a picture of a butterfly from that *Butterflies* book. Teacher *2* says, "This is a butterfly." Over time, they observe the caterpillar enclosing itself in an organic mass—a chrysalis. Then on one beautiful day, they observe the butterfly breaking free of the structure. Teacher *2* says to the girl, "Aha! A caterpillar will spin a chrysalis and later emerge from it as a winged creature called a butterfly."

Clearly, Teacher *1* is not a reliable source of information. Still, can we say that, by a lucky guess, Teacher *1* was right in his conclusion? Can we say that Teacher *1* is wrong in most conclusions, but was correct in the conclusion about the butterfly? No.

Note that two different words can have the same spelling, but can actually be considered different words insofar as they refer to unrelated concepts. On the one hand, one can say *measure* when referring to a quantitative magnitude of something. Alternatively, one can say *measure* when referring to an action that someone takes, or some rule of conduct the State imposes. *Leaves* can be a plural noun that refers to flat green objects found growing on a tree. *Leaves* can also be a verb and refer to the removal of someone or some object from some location. We would not conclude that the first *leaves* is the same as the second merely based on the spelling and pronunciation; the intellectual contents of every concept are wider than that. That same logic applies to entire sentences. Both Teacher *1* and Teacher *2* can affirm to the little girl, in complete seriousness, "A caterpillar will spin a chrysalis and later emerge from the chrysalis as a winged creature called a butterfly." Their concepts of what that sentence means, though, are discrepant. When Teacher *1* utters that sentence, there is this deeper, implicit meaning behind it: "I am saying this sentence, and I stubbornly hold the sentence as true. But this stubbornness is not based on any serious commitment to adducing empirical observations and inducing open-ended-yet-consistent principles therefrom."

When Teacher 2 utters that sentence, there is this deeper, implicit meaning behind it:

> I have confidence in this sentence based on inductive reason. It is the consequence of observing reality and inducing conclusions therefrom. A conclusion being based on empirical observations renders it a contextual certainty. The conclusion being a contextual certainty does not entail it being a "closed" conclusion. Concepts—including concepts deemed certain by repeated empirical observation—are open-ended. Further observations will be incorporated into the concept, expanding it. Nuances and exceptions will be observed, and such qualifications and revisions will be integrated into the increasingly complex concept. Such conceptual revisions do not undermine the observed data that have already been priorly validated. When I do not engage in observations myself, first-hand, I rely upon other sources. Nevertheless, my reliance on those other sources is contingent upon there being evidence that those other sources have a solid record behind them—a record of adducing reliable observations and then inducing coherent explanations from them, as I have done.

Teacher *2*'s approach is the one I find valid. Of course, when I was a very small boy, my parents told me that caterpillars pupate and metamorphose into butterflies. I immediately believed them, prior to having seen this myself. That is fine, as they had already established a record of reliability in telling me truths that I did have the chance to test first-hand. At some point I did observe caterpillars in terrariums myself. It is perfectly valid that one trusts the judgments of other people, in the absence of first-hand observations, under these conditions.

1. One keeps in mind the limits of one's knowledge about the other sources' reliability.
2. When it comes to convictions based on others' observations—convictions strongly held—those other information sources have shown a record of adequate reliability in adhering to the epistemological method of observational rationality.
3. One allows the concepts to be open-ended, as new data prove it possible for the concepts to be expanded and revised. Those revisions do not destroy the certainty or absoluteness of the already inductively validated data; the revisions simply reveal additional facets to the data.

The concept that a person holds—the conclusion to which that person arrives—implicitly subsumes all of this. Thus, when Teacher *1* says, "A caterpillar will spin a chrysalis and later emerge from the chrysalis as a winged creature called a butterfly," the conceptual contents of Teacher *1* saying that are dissimilar from the intellectual contents of what Teacher *2* means when uttering that sentence. Naturally, if Teacher *1* tells me that caterpillars metamorphose into butterflies, that consideration should not immediately lead me to conclude, in blind rebellion, that caterpillars never change into butterflies. The rational approach is to realize that the concepts that Teacher *1* explicates are cognitively meaningless. Independent of Teacher *1*'s assertions, I am to search for the truth about butterflies from sources that do respect the inductive method of knowledge acquisition.

As one might dismiss the scenario of the two teachers as far-fetched fantasy, I shall cite a scenario from history. Democritus and other ancient Greek philosophers arbitrarily speculated that every bit of matter was comprised of microscopic, indivisible units that they called *atoms*. For this reason, some readers might be under the misapprehension that Democritus and these other philosophers had conjured up modern atomic theory in the absence of evidence. That is a false notion that conflates semantics with evidentiary understanding. What happened was as follows. As physicist and science historian David Harriman narrates, the chemists of the Enlightenment had to demarcate the "clear distinction between pure substances and mixtures. Unlike mixtures, pure substances have well-defined and invariant properties. Under the same conditions, every sample of a pure substance will melt (or boil) at precisely the same temperature. Every such sample has the same hardness and the same mass density, and the same amount of heat will always cause the same rise in temperature (for a unit mass). Furthermore, when a portion of a substance undergoes a chemical reaction, the properties of the remaining part do not change." Conversely, the "concept 'mixture' could then be defined as a material composed of two or more substances."

The "next key step," Harriman continues, "was the division of substances into elements and compounds. This was made possible by the discovery that mass is conserved in chemical reactions, i.e. the weights of the reactants are always equal to the weights of the products."[11] Antoine Lavoisier (1743–1794)[12] proved this. He constructed a glass container with weighing scales on the bottom. He placed a chunk of metal inside the container and then measured the weight of both the metal and the air around it. When he stimulated chemical reactions on the metal so that it would rust, the rusting effect caused the metal to gain in weight. The weight that the metal gained was exactly the weight lost to the air around the metal. At both the beginning and end of the experiment, the total weight of both the metal piece and the air around it remained unchanged. This proved (1) that matter changes form while (2) the total quantity of matter in the universe does not change.[13] After this, Harriman adds, it "was found that some substances can be decomposed into two or more other substances with the same total weight; on the other hand, other substances resisted all such attempts at chemical decomposition. Those that cannot be decomposed are elements . . ." By contrast, those that can be decomposed into two or more separate elements are compounds. Then Joseph Louis Proust (1754–1826) demonstrated that copper carbonate invariably consists of copper, oxygen, and carbon in the same mass ratio (5:4:1). He demonstrated that this principle applies however the sample was prepared and no matter how it was isolated from the wilderness environment. From this empirical observation, Proust was able to devise by 1777 the Law of Constant Composition, which is that separate samples of a chemical compound invariably consist of the same chemical elements in the same proportion, mass-wise. This discovery proved essential to the eventual realization that every chemical element is comprised of microscopic particles known as atoms, each of these atoms being of the same element. It was likewise essential to the discovery that a chemical compound like copper carbonate is comprised of microscopic particles known as molecules, with each molecule being comprised of at least two smaller particles, each particle being an atom.

In 1803, John Dalton contributed to further understanding of the atomic theory. As David Harriman narrates, that was the year that Dalton

> analyzed three gases composed of nitrogen and oxygen. The first is a colorless gas that has a pleasant odor and the capacity to cause hysterical laughter when inhaled; the second is colorless, nearly odorless, and has a significantly lower mass density; the third has the highest mass density of the three and a deep brown color at high temperatures. Quantitative analysis showed that the three gases are also distinguished by the relative weights of the two combining elements. If we consider samples of each gas that contain 1.75 grams of nitrogen, then we find that the laughing gas contains one gram of oxygen, the second gas contains two grams, and the third gas contains four grams. Dalton found a similar result when he analyzed two different gases that are both composed of carbon and hydrogen; for samples containing the same weight of carbon, the weight of hydrogen contained in one of the gases is precisely twice that of the other. On the basis of such data, Dalton arrived at a new law: *When two elements combine to form more than one compound, the*

weights of one element that combine with identical weights of the other are in simple multiple proportions.

Dalton's discovery was crucial to science's ability to distinguish carbon monoxide from carbon dioxide. Each of these chemical compounds are made up of no more than two elements—carbon and oxygen. Yet, in samples of both gases containing carbon in the same weight, one finds that the weight of oxygen in CO_2 is twice that of the oxygen in CO_1. Enlightenment-era scientists recognized that a theoretical model that would account for their observations would be one envisioning each form of matter to be comprised of microscopic particles. These scientists remembered the word *atom* from Democritus and the Greek philosophers and, in tribute to them, appropriated that term for themselves. Once they had such terms as *molecule* and *atom*, the chemists were better able to produce names and theoretical models to explain their findings; such theoretical models withstood the test and retest of various experiments. As Democritus and those other Greeks possessed no evidentiary basis for their claims about atoms, their concept of *atom* was less than miniscule in size; it was vacuous. "If one tries to think about the implications" of an idea as arbitrary as Democritus's, notes Harriman, "one draws a blank; implications depend upon connections to the rest of one's knowledge." Consider what this means in respect to Ayn Rand's analogy that a mind is like a file cabinet. A concept is like a file folder. Collecting more sensory data about a concept is like filling the file folder with more important documents. Sometimes when you fill the file folder with a book's worth of documents of various sub-categories, you might have to divide that single file, putting more file folders into the cabinet and placing them close to each other. In the case of Democritus's idea, "the word 'atom' was just a label on an empty file folder."[14] By contrast, despite their appropriation of the old word, the chemists devised an entirely new concept filled with genuine intellectual contents derived from sensory data. Although individual atoms and molecules are too small for the unaided senses to observe, atomic theory does account for scientific observations. Should you visit a laboratory and encounter samples of carbon dioxide and carbon monoxide, with the carbon part being of identical weights in both samples, you will find that the weight of the oxygen in the CO_2 sample is exactly double in proportion to the weight of the oxygen in the CO_1 sample. You will notice that this is always the case, whether the carbon in each sample weighs one gram or five kilograms. For this reason, Democritus's conception of "atom" is invalid whereas John Dalton's conception of "atom" *is* valid. Additionally, this demonstrates that Dalton's concept of "atom" is not the same as Democritus's.

The same principle applies to other areas of life. Suppose I hold strong convictions in private property rights and free markets. I have a philosophic basis in saying this. What is most important is not how verbally sophisticated one's arguments are; what is most important is that one has, at least implicitly, a logical and evidentiary support for the conclusions. Suppose Politician *X* says that he, too, upholds the absoluteness of private property rights and free markets "because that's what I deduced from the Eighth Commandment. The Bible is the ultimate authority on truth." Suppose Politician *X* always deduces, unlike me, that free-market principles prove that an artist does not deserve the right to ask the State to enforce the copyrights on his original artwork. Just because Politician *X* and I use the same sentences, such as "I oppose the initiation of the use of physical force against person and property," it does not prove that our conceptual meanings behind those sentences are the same. The concept that is one's conclusion subsumes the epistemological method whereby one arrived at that conclusion. It is also because of our very concepts being distinct that we inevitably formulate wholly opposite applications of the sentence "I oppose the initiation of the use of physical force against person and property." Hence my support for copyrights and Politician *X*'s opposition to them.

Two men, both employing rational epistemology, can come to honest disagreements in their conclusions. They have different life experiences that lead to different observations, thus resulting in different inductions. However, insofar as Person *A* exercises an epistemology far more rational than Person *B*, the intellectual contents of their conclusions will differ: even when, in complete seriousness, they say the same sentences. Rand proved that individual rights are justified by man's rational nature on these grounds.

1. Man derives values according to the standard of what benefits or enhances his eudaemonia, the state of survival being a prerequisite to the maintenance of eudaemonia.
2. To live and service this eudaemonia, man needs to appropriate sustenance for his own survival.
3. A person's proper appropriation and utilization of this sustenance requires one hold the ultimate rightful control over the sustenance. This entails that no one else violently wrestles such control away from this person. This is where the concept of rights begins.
4. The institution of a police department is the most effective method of ensuring the security of everyone's rights.

Locke also observed Points *2*, *3*, and *4* to some degree, but not so much Point *1*. Although Rand built upon Locke's discoveries and used his terminology—"rights to life, liberty, and estate"—Rand took the concepts much farther and constructed a new, superior theory, just as Dalton's explanation about atoms is complete in a manner that Democritus never matched. Particularly relevant to this discussion is that someone's arbitrary proclamation that "Rights to life, liberty, and property come from God—and that's it," despite any semantic similarities that statement might hold with Rand's writings, does not fall into the same category as the Objectivist understanding of rights.

This is why, although my political rhetoric may share some superficial similarities with those of other self-proclaimed free-market advocates, we do not perforce agree on the application of laissez-faire principles. Nor are we necessarily working toward the same ends. For that reason, one should not assume that I am fighting for the same sort of society as a libertarian who proclaims that Christianity and the Eighth Commandment serve as the bases of his libertarianism. One should not presume that I agree with a libertarian who airs arbitrary suppositions about God giving us rights that justify laissez faire. Finally, I am not in agreement with such libertarians as Robert Nozick, who stake their case for laissez faire on arbitrarily chosen deontological premises from which they deduce, in the tradition of Kant, their entire political philosophies.

We cannot decipher what is morally right or wrong unless we exercise the proper methodology: objective epistemology. Deviation from objective epistemology is irrationalism—going by whims and arbitrary pronouncements. Once people abandon rationality as their guide, they have no constraints in developing rationalizations for illiberalism and spoliation. Nothing but inductive reason can refute such rationalizations. I hope that part II's discussion of the Third Reich made forthright the causal connection between irrational epistemology and generally politically illiberal results. As Voltaire explains, "Those who can make you believe absurdities can make you commit atrocities."[15] The Nazis' violent illiberalism was the consequence of the Nazis practicing their own irrational romanticist epistemology.

Here we can employ another analogy involving teachers. When children take math class in middle school, their textbook usually includes, in its back section, the answers to the math homework problems that the teacher assigns from the book. Most math teachers expect their students to show their work in solving each assigned math exercise. This is how the teacher discerns that the students understand the principles involved in the solution to each homework exercise. Suppose you have two teachers. One is a libertarian and the other is me, an Objectivist. The problem we assign is, "What is the best social system for governing society?" The libertarian teacher says he does not care what epistemological method anyone employs in finding solutions. This libertarian will be satisfied if other people repeat, back to him, what he believes to be the correct solution: "We should reject overbearing government power. I oppose the initiation of the use of force against person or property, especially when that use of force is initiated by the State." When the libertarian exhibits his indifference to how people arrive at that conclusion, that is akin to a math teacher being okay with students not showing their work on their home assignments. That libertarian teacher remains content if students peek at the back section of the textbook for the answers, not bothering to apprehend the principles necessary for solving the math problems. Yet it would be better if the students could learn the principles involved. In that case, when they are confronted with math problems of a similar type in their everyday experiences, they can deductively apply the principles they were supposed to have learned in math class. In contrast to the libertarian example, as an Objectivist I am the math teacher who expects students to show their work. When someone provides the "correct answer," absent of employing the rational method for finding and validating it, that "answer" is meaningless. Person *L* "saying" the words which express truth, while not knowing how to validate that truth, exemplifies no more understanding on Person *L*'s part than does a parrot's squawking. One understands liberty and rights insofar as one understands their secular, empirical, inductive validation.

This principle provides further evidence of that point stressed in Book One—that reason and liberty are coterminous, each relying upon the other. It is through inductive reasoning that we justify the necessity of liberty. Likewise, we require the absolute freedom of peaceful action in order to reason fully—to think our own unique thoughts, to express them, and to maturate them fully by translating those same thoughts into peaceable actions. This should be the ultimate rebuke to that assumption we examined in part II and in Books One and Two—that secular Enlightenment reason, taken to its final extreme, leads to illiberality and totalitarianism. The Religious Right and Postmodernist Left would have us believe that to exercise cold, atheistic reason consistently is to support a government that outlaws every religion. In fact, the reverse is true—the one sort of system that allows for consistently rational atheism is one that allows for people to think any thought, wise or unwise, peaceably. That invariably includes the freedom of religion—to hold any religious belief, or no religious belief, peaceably. A nation that outlaws every religion is a nation that denies you the liberty to come to atheism honestly.

This is how I figure that. To say that, upon reflection and meditation, you have decided upon atheism requires that you possessed and possess the freedom to mull over this topic and examine every pertinent option. For you to have enacted the freedom to reject Hinduism, for instance, you would have needed the legal option to *embrace* Hinduism. It would have meant that you could hold any beliefs in your head—including religious beliefs—absent of the threat of violent reprisal. Were you deprived of the option to denounce atheism, you would be forcibly deprived of the capacity to decide upon and proclaim your atheism by your own free will. Therefore, a communist nation that coerces you to profess atheism diminishes atheism by attempting to proscribe you from arriving at an atheistic conclusion through honest means. It is irrational to proclaim that people becoming "extremists for rationality" will result in their enacting illiberal legislation to force everyone to be atheistic and rational, for that legislation would be the pinnacle of unreason. It would be unreason on two grounds—it would be self-contradictory on the grounds mentioned, and it would threaten violence against peaceful people, thereby denying them the ability to act on their own thinking. A religious person must be able to think and act freely and peaceably in order to think through the matter if he is to accept atheism honestly one day.

To attempt to coerce—by force of governmental law—anyone to be wiser or more rational, is a contradiction in terms. As wisdom is accumulated through inductive learning from experience, which includes experiencing the consequences of unwise choices, there is no freedom to hold or exercise wisdom unless there is the complementary freedom to be unwise and to exercise foolish-but-peaceful choices. As we have recognized throughout this chapter, of course, your freedom to behave peaceably but unwisely is contingent upon those unwise actions being based on beliefs you genuinely hold and beliefs others genuinely held when exposing you to them. When someone deliberately provides you false information in order to harm you or someone else, that obstructs your liberty to act peaceably on your own judgment, and therefore does not count as an example of someone exercising the freedom to be unwise-but-peaceful. Still, liberty and inductive reason are conjoined.

Journalist Brooke Allen correctly observes that the laissez-faire liberal founding of the USA arose not from religious dogma or other irrationalities, but arose from a specific philosopher's inductive reasoning. "The most important political thinker of the Enlightenment and certainly the one who had the greatest impact in America"—indeed, the "man who had the greatest single influence on the United States Constitution"—"was not Jesus but John Locke . . ."[16] And Locke "did as much as any theorist has ever done to loosen politics from the bonds of organized religion." Allen also adds that the virtues of George Washington "owed more to Aristotle than to Christian writings..."[17] But we need not take Allen's word for it; we can consult the writings of Thomas Jefferson himself. When asked about the source of the night-watchman-state principles within the Declaration of Independence, Jefferson attributed the history of rational, natural-law philosophy, specifically naming Aristotle, Marcus Tullius Cicero, John Locke, and Locke's laissez-faireist contemporary Algernon Sidney.[18] And as we have observed throughout my trilogy, both Cicero and Locke found inspiration in Aristotelian inductive reason and formed pro-private-property-rights conclusions.[19] Moreover, Algernon Sidney was of the early Enlightenment tradition that emphasized inductive reason.[20] John Adams, as well, considers Aristotle—alongside Sidney and Locke and classical liberals—to be an architect of the philosophy of freedom. The principles of liberty, imparts the second U.S. President, amount to "revolution principles. They are the principles of Aristotle . . ., of Livy and Cicero, and Sydney, . . . and Locke. The principles of nature and eternal reason. The principles on which" any legitimately free republic ". . . stands."[21]

That Jefferson and Adams would cite this set of philosophers as the early adopters laying the foundation for the Declaration's Enlightenment-Era laissez-faire thesis, suggests that this duo of Founders agrees with me that laissez-faire liberal philosophy derives from Aristotelian inductive reason. It may be on a similar basis that, despite Aristotle's recommendations for specific and regrettable governist precepts, H. L. Mencken considers Aristotle to be a hero of pro-liberty philosophy in the final analysis. Hence, Mencken places the Athenian sage in the same category as figures more famously associated with laissez faire—"The ideal government of all reflective men, from Aristotle to Herbert Spencer, is one which lets the individual alone . . ."[22] And in agreement with Jefferson, Mencken, and myself is Ayn Rand. Aristotle's philosophy, Rand writes, was the original "intellectual Declaration of Independence. Aristotle, the father of logic, should be given the title of the world's first *intellectual,* in the purest and noblest sense of that word. . . . If we consider the fact that to this day everything that makes us civilized beings, every rational value that we possess—including the birth of science, the industrial revolution, the creation of the United States, even the structure of our language—is the result of Aristotle's influence, of the degree to which, explicitly or implicitly, men accepted his epistemological principles, we would have to say: never have so many owed so much to one man"[23] (emphases Rand's). For that reason, Rand concludes, "If there is a philosophical Atlas who carries the whole of Western civilization on his shoulders, it is Aristotle. . . . Whatever intellectual progress men have achieved rests on his achievements."[24]

Conversely, we discern that there is a chain of influence going in the other direction as well. By Jean-Jacques Rousseau's own admission, Rousseau was influenced by Plato.[25] Plato also influenced Immanuel Kant. Rousseau, in turn, influenced Kant,[26] Karl Marx,[27] Rev. T. Robert Malthus,[28] and the French Revolution's Maximilian Robespierre. Rousseau and Kant influenced the German romanticist school of philosophy. German romanticism influenced the authoritarianism of Germany in the 1800s and 1900s and also the early-twentieth-century U.S. Progressive movement. It was not Herbert Spencer but Kant who influenced Houston Chamberlain, who went on to promote racism in Germany in the 1920s. Kant, Hegel, the romanticists, and the nineteenth-century socialists William Godwin and Charles Fourier influenced Joseph-Arthur de Gobineau and Chamberlain. Gobineau and Chamberlain influenced the Third Reich's eugenicists. Eugenicists and romantics influenced Jacques Derrida and the postmodernists of the twentieth and twenty-first centuries. Robespierre and the other Jacobins served as an inspiration to the nineteenth-century socialist movement. In kind, both the nineteenth-century socialist movement and the romanticists influenced the political Left of the twentieth and twenty-first centuries. T. Robert Malthus, Marx, and socialists influenced the modern population control and anti-globalization movements. Whatever G. E. Moore's arguments against them are, Kant and David Hume implicitly influenced G. E. Moore, particularly regarding his claims about a Naturalistic Fallacy. G. E. Moore[29] and Malthus[30] influenced John Maynard Keynes, and Keynes provided an obtuse rationalization for the tax-and-spend regulatory-entitlement state policies that have burdened governments throughout the twentieth century and continue to mislead them at the date of this writing. Thus, through their tenets, such philosophers as Plato and Rousseau and Kant and Marx have indirectly contributed to the governism that has menaced Western civilization in the twentieth century and continues to menace it as of this writing.

There have indeed been philosophers who advocated laissez faire but failed to provide adequate inductive justifications for laissez faire. David Hume and Frédéric Bastiat come to mind. For laissez faire to be on a sound footing in philosophy, it needs an adequate, inductively rational basis, such as what Ayn Rand provided in her writings. Conversely, when societies embrace epistemological methods other than strict observational rationality, they leave themselves direly vulnerable to oppression. To review the point from Book One, it is dangerous to presume that a proposition can be theoretically possible in the absence of observational support. Should we concede that a proposition can be theoretically possible in the absence of observational support, a would-be dictator can declare that his arbitrary assertions—including his assertion that his will should be imposed by law—can be validly accepted at face value. And once we believe that a proposition can be accepted in the absence of evidence, we tacitly convey that our logical arguments against the would-be dictator cannot wholly nullify the would-be dictator's declarations. Again, recalling Book One, consider David Hume and Immanuel Kant. These two philosophers argued in favor of liberal republicanism and free international commerce. Yet, on account of their successors believing Hume and Kant to have invalidated observational reason as a means of obtaining any objective knowledge—including knowledge about ethics—Hume and Kant indirectly undermined individual liberty. As we recall from chapter 9, Kant's successors—Johann Gottfried Herder and Johann Gottlieb Fichte—believed that Hume and Kant had successfully invalidated any acceptance of inductive reason. Then Herder, Fichte, and other successors pointed out that it is through observational reason that any person ascertains that he or she is an individual whose volitional consciousness is separate from that of any other person's. Herder and Fichte figured that if observational reason is not the sole method of ascertaining truth, then any proposition one can imagine can be accepted as theoretically possible. They then proposed that there is no individual, and instead that there is but a single collective consciousness among people. Once they and Georg Hegel got that idea accepted in Germany, it became easy for Germans to rationalize that the individual could be justly sacrificed for a mystical State. That is why, despite their professed sympathies for laissez-faire liberal republicanism, Hume and Kant ultimately cleared the path for mystical governism in nineteenth-century Germany. To undermine cultural acceptance in observational reason is ultimately to undermine the logical arguments for liberty. And as we also recollect from chapter 9, it was the philosophers of nineteenth-century Germany who inspired the likes of Richard T. Ely and Simon Patten and John Dewey to establish the regulatory-entitlement state in the USA. And, on a far more disastrous scale, the mystical governism of nineteenth-century Germany metamorphosed into the mystical governism of the Nazi Reich.

In sum, nothing can justify a philosophy or civilization resulting in Peaceful Choice, save for an epistemology of inductive reason. Volitional reason—*Homo sapiens sapiens*'s most prominent evolutionary adaptation—is what provides us knowledge about the nature of liberty.

NOTES

1. W. Block 1976, 59-60, argues such.
2. Rothbard 1973b, 106.
3. A. Burris 1983 2d ed., 57
4. W. Block 1976, 60.
5. That someone can be falsely convicted of murder is also the reason why I support having a moratorium on the death penalty. Should a man wrongly convicted of homicide later be exonerated, he can be set free; a state execution of an innocent man is irreversible.
6. Under tort law, stalking and harassment come under the names *intrusion upon solitude or seclusion* (noted in B. Fisher 2004 8th ed., 265) and *intentional infliction of mental or emotional distress* (*ibid.*, 272). Note that someone can be successfully sued for inflicting mental distress only under the condition that such an infliction is of an extremely outrageous and devastating nature, such as harassment; one cannot sue someone for having his feelings hurt as a result of some normal, peaceful everyday action.
7. B. Fisher 2004 8th ed., 265. Since both identity theft and harassment infringe upon property rights that are not necessarily tangible, it is fitting that they are both classified as *invasions of privacy* (see B. Fisher and Phillips 2004 8th ed., 265–66), privacy being an intangible property right.
8. For descriptions of the guild system from the Middle Ages and European Renaissance, see Acemoglu and J. A. Robinson 2012, 187-88; and C. Webber and A. Wildavsky 1986, 243.
9. C. Biddle 2011, 26, also properly judges defamation to be an initiation of the use of force.
10. I first read this example and observation in G. H. Smith 1989, trade paperback, 120–22.
11. D. Harriman 2010, 153–54.
12. In terms of Antoine Lavoisier's contributions to liberty, his legacy is something of a mixed one. Interestingly, he was an investor in a business known as a tax farm. A tax farm is a private business, and the monarch outsources to the tax farm the authority to collect taxes forcibly. In exchange to the tax farm's services, the king allows it to take a pre-set percentage of the loot for itself. Because of Lavoisier's co-ownership of a tax farm—along with Lavoisier's rejection of the scientific theories of the violent French Revolution instigator Marat—Marat had Lavoisier guillotined in the French Revolution. This is according to Bodanis 2000, 29–30. Also interesting, Lavoisier was a mentor to the entrepreneur and chemical engineer Eleuthère Irénée Du Pont de Nemours, founder of the DuPont Company and son of the free-market theorist Pierre Samuel Du Pont de Nemours.
13. Bodanis 2000, 29–30.
14. D. Harriman 2010, 154–56, 152.
15. That quotation, in the original French, is "Certainement qui est en droit de vous rendre absurde est en droit de vous rendre injuste." That is in Voltaire 1792, 249–250. I used the English translation at http://en.wikiquote.org/wiki/Voltaire#Questions_sur_l.27Encyclop.C3.A9die_.281770-1774.29, accessed Saturday, June 15, 2013.
16. B. Allen 2006, xv.
17. B. Allen 2006, 165, xv, 37.

18. For that, see Thomas Jefferson's letter to Henry Lee, May 8, 1825, republished by the University of Virginia Library 2012, http://etext.virginia.edu/etcbin/toccer-new2?id=JefLett.sgm&images=images/modeng&data=/texts/ english/modeng/parsed&tag=public&part=280&division=div1, accessed Wednesday, April 18, 2012.

19. For Aristotle's influence over Cicero, see Book One. Aristotle's influence on Cicero is mentioned in Blakesley 1839, 109-140; D. J. Furley 1989, 203; Isle 2006, 87; and J. S. Reid ed. 1874, xxiv. To quote Book 2 of the Cicero-authored *Academica,* Cicero lauded Aristotle's writings as "a golden flood of words." For that, see Cicero 2006a, 70. For Cicero's defenses of private property, see Books One and Two. See Cicero defend private property in Cicero 1913, Bk. 2, sec. 21, p. 73, accessed online Wednesday, August 3, 2011. J. Powell 2000, 8, brought this statement of Cicero's to my attention. Also see Cicero defend private property against welfare-state redistributionism in Cicero 1913, Bk. 1, Sec. 14, accessed online Friday, November 16, 2007.

20. We saw Algernon Sidney defend private property rights and the night watchman state in Book One. Sidney's writings in defense of laissez faire are A. Sidney 1750 vol. 2, 126; and A. Sidney 1996, ch. 3, sec. 6, para. 2, accessed online Wednesday, April 18, 2012. A. Burris 1983 2d ed., 67, brought this statement of Sidney's to my attention.

21. John Adams 1819a, 12.

22. Mencken 1922d, 292.

23. Rand 1961c, 22–23.

24. Rand 1963e, 18; and Rand 1990c, 6.

25. P. Gay 1966 vol. 1, x.

26. W. Durant 1961, 195–96; and P. Gay 1969 vol. 2, 130, 217.

27. Marx 1844, accessed online Wednesday, July 29, 2009.

28. T. G. Buchholz 1995, 225; and R. Porter 2004 U.S. version, 429.

29. I first heard of G. E. Moore and learned of G. E. Moore's influence on Keynes in W. C. Biven 1989, 7–8, 11. This is also noted in J. B. Davis 1994, 9–49, 53–103, 148–157, 180–85, 190–94.

30. John Maynard Keynes acknowledges Malthus's influence on him in Keynes 1920, 10, 21; Keynes 1921, 83 fn. 1; and Keynes 1964, 32–33, 362–64.

Conclusion of Trilogy

Rue That Rousseau

We started this trilogy with a discussion of how political liberty cannot be justified in the absence of a rational epistemology. Upon justifying that rational epistemology, we applied it throughout our exploration of the ramifications of free industry and its foil, governism. Given that we have established that true agreement with this book's political conclusions can proceed from no source other than agreement with its epistemology, we have come full circle. With our journey nearing its end, it is time for us to reflect.

No more than a tiny minority of entrepreneurs would feel motivated to perform their socially beneficent activities were there a complete absence of security for their private property. Acknowledgement of such a fact casts no negative light on businesspeople—it is to their credit, morally, that they peaceably pursue their own self-interest. And by this stage in our journey I hope I have exposed the reasons why it is erroneous for Jean-Jacques Rousseau and other advocates of social collectivism to denounce the very institution of private property as base. Recall from Book One's opening chapter that Rousseau depicted life for cavemen as happy, for they were under anarchy and had no inequality. Contrary to Rousseau, though, men were never actually subjected to long-term anarchy, and life prior to the Industrial Revolution was fleeting and miserable. Moreover, social hierarchy did and does exist among hunter-gatherers.

Rousseau's *Discourse on Inequality* rued the occasion on which history's earliest capitalist "enclosed a piece of ground, bethought himself of saying *This is mine*, and found people simple enough to believe him . . ." This allegedly "bound new fetters on the poor, and gave new powers to the rich." It "eternally fixed" economic "inequality . . . and, for the advantage of a few ambitious individuals, subjected all mankind to perpetual labour, slavery, and wretchedness."[1] The Huguenot philosopher's statements contain a false assumption. It is that when some individuals began to claim private ownership over specific parcels of land, they had impoverished other men, as those other men no longer held the same level of access to that same land as they did earlier. As with every Malthusian, Rousseau begs the question that wealth is a zero-sum game, wherein one person having a higher standard of living inexorably deprives everyone else of that same high quality of life. I guess it is appropriate enough that Rousseau would maintain that fallacy. Rousseau was friends with Malthus's father,[2] and held influence over Malthus when he was still a boy. To review the point of Book Two, capitalism is no such zero-sum game. Remember that wealth is not fixed according to a supposedly static quantity of natural resources available on earth. Recall that capitalism itself mitigates resource depletion. In a free market, a peaceful entrepreneur increases his profits by extracting greater value for consumers out of ever-smaller and ever-fewer inputs of natural resources, human labor, and manmade equipment. And that manmade equipment was itself the product of prior mixtures of natural resources and human labor and entrepreneurship.

Consider how natural resources have been increasingly conserved by improvements in the thermal energy efficiency of engines, and of how the profit motive has successfully encouraged entrepreneurs to implement such improvements. After Thomas Newcomen and Thomas Savery invented their steam engine in the early 1700s, the issue of thermal efficiency cropped up. Thermal efficiency refers to the amount of thermal energy the entrepreneur's operations put into motion, versus the stored energy that remains untapped, as the fuel is burned. The more energy goes untapped, the more inputs go to waste as the fuel is burned. By reducing the waste byproducts—putting more energy to work per every unit of inputs—the entrepreneur commensurately reduces his costs while maintaining or increasing his productive output and revenue. When Newcomen died in 1729, the thermal efficiency of his steam engine was no more than half of one percent. By 1821, James Watt's steam engine operated at the thermal efficiency of 4 percent.[3] In terms of percentage points, that sounds small. But this means that from 1729 to 1821, the thermal efficiency of steam engines increased eightfold. Thanks to Watt improving the thermal efficiency of steam engines, it followed that industrialists could produce greater and greater value for customers while using up ever-smaller and ever-fewer inputs of natural resources for every unit of product. And Rudolf Diesel and subsequent engineers further improved the thermal efficiency of engines. By January of 1897, Rudolf Diesel's internal combustion engine operated at a thermal efficiency of 26.2 percent.[4] Thermal efficiency in 1897 was therefore over six-fold what it was in 1821 and over fifty-fold what it was in 1729. Insofar as freedom remains, the trend over time is for more units of utility—value—to be produced per the same quantity of material inputted.

Yet, despite the economic realities, Rousseau's misconceptions persist as I type these words, both in radical activist groups and, to a lesser but still frightening extent, mainstream politicians in power. First, let us observe the misconceptions alive with radical activist groups. An anti-globalization outfit calling itself the Green Anarchy Collective declares, "Prior to civilization there generally existed ample leisure time, considerable gender autonomy and equality, a non-destructive approach to the natural world, the absence of organized violence,... and strong health and robusticity. Civilization inaugurated warfare, the subjugation of women, population growth, drudge work, concepts of property, entrenched hierarchies, and virtually every known disease, to name a few of its devastating derivatives. Civilization begins with and relies on an enforced renunciation of instinctual freedom."[5]

The declamation of Rousseau's that I quoted takes no inventory of the facts. The same applies to the Green Anarchy manifesto I quoted. Were there never an implicit form of personal property, nobody would be able to nourish himself with the fruits of the Earth. In this particular recounting of history, Rousseau doubly fails to consider that early tribal societies had in some cases an implicit conception of private ownership. This is particularly odd, considering that the Huguenot philosopher would half-heartedly concede the rightness of homesteading years later, only then to reassert that economic and political collectivism must trump individual property rights. Moreover, we spot the error in Rousseau and Marx and Osama bin Laden blaming private property rights for economic "inequality" and for helping the rich become richer and the poor poorer. The amount of freedom in the economy directly correlates with the ability of the poor to exercise their own private property rights so that they may raise their own living standards.

Rousseau surmised that John Locke was basically correct that private property rights could not exist in the absence of a Social Contract. Rousseau judged from this that the Social Contract itself was the source of rights, and that a proper Social Contract could nullify these rights. He also inferred that those who first established the legal institutions of land title had authored this original Social Contract of which Locke had written. But whereas Locke considered a pro-property Social Contract to be just, Rousseau did not. Hence the Huguenot philosopher called for the original Social Contract that Locke embraced to be shredded and replaced with a new Social Contract that forced social collectivism and censorship on everyone.[6] Heed the words of Victor Klemperer (1881–1960), a Jewish scholar and Enlightenment historian who resided in Germany during the Nazis' ascension to dictatorial supremacy, and who studied Rousseau's *Discourse on Political Economy*. On Thursday, July 30, 1936, Klemperer wrote in his diary that "whole passages" of Rousseau's work "could be from Hitler's speeches."[7]

As I stated in this trilogy's chapter of inception, Rousseau spoke of the supposedly pro-privatization Social Contract as if it were Original Sin,[8] despite Rousseau's own denial of the Original Sin touted in Christianity since Augustine's day. Rousseau evidently thought a new socialist Social Contract would be mankind's redemption. A gaggle of environmentalists appear to share in the assessment that the advent of private property, agribusiness, and industrialization was the Original Sin, and that humanity will be smitten for this by means of an ecological Judgment Day. As an example, I offer David M. Graber of the National Parks Service. In an adulatory review of environmentalist Bill McKibben's book *The End of Nature*, Graber argues that the ascent of capitalism was man's great initial betrayal of everything sacred. He proclaims that anthropogenic climate disruption has been the most significant "side effect of the Industrial Revolution practically upon its birth." Then Graber delivers one of the strangest invocations of the Social Contract theory I have heard. Not surprising for a politically correct Green, he employs the word *nature* as a synonym for *wilderness*, whereas I refer to *nature* as the constant principles of scientifically validated reality. Graber congratulates himself for being among those who "value wilderness for its own sake, not for what value it confers upon mankind." We radical environmentalists "are not interested in the utility of a particular species, or free-flowing river, or ecosystem, to mankind. They have intrinsic value, more value—to me—than another human body, or a billion of them. Human happiness, and certainly human fecundity, are not as important as a wild and healthy planet. I know social scientists who remind me that people are part of nature, but it isn't true. Somewhere along the line—at about a billion years ago, maybe half that"—we humans "quit the contract and became a cancer." Pardon me, but quit what contract? This is a contract with whom? Nonhuman animals that are not contractually competent? This is evidently Graber's own variant of Social Contract theory.

Graber continues, "We have become a plague upon ourselves and upon the Earth. It is cosmically unlikely that the developed world will choose to end its orgy of fossil-fuel consumption, and the Third World is suicidal consumption of landscape. Until such time as Homo sapiens should decide to rejoin nature, some of us can only hope for the right virus to come along."[9] Note that these words were not published in some underground hippie 'zine, but in the mainstream *Los Angeles Times Book Review*. According to Graber's analysis, the first act we must perform to be saved from ecological cataclysm is to repent for our self-interested utilization of Earth's resources. Then we must pledge eternal allegiance to a higher power. That higher power is a collectivist governmental authority, legislating over which technologies and economic institutions are permissible and which are not.

Nor is Graber alone in such sentiments. Despite his own sympathies for environmentalism, U.C. Santa Barbara ecologist Daniel Botkin concedes that much of the environmental movement's presumptions about the inherent sinfulness of reshaping the wilderness is a retread of Original Sin. Those who say you should experience remorse over industrial production and commercial consumption, notes Botkin, are "casting up from the sea of Judeo-Christian tradition the belief that man is a sinner, bound to sin; that most recently we have sinned against nature, and we are being punished for it (as we should be) by Mother Nature… Therefore we must do penance, suffer for our sins, which in this case means living minimally, using only enough energy to provide the bare necessities of life and disallowing us enough energy to be creative, to develop more science and technology . . .; nor should we, by implication (perhaps unintentionally) have enough energy to be otherwise creative . . ."[10] Given the outlook of Graber and the advocates of deep ecology, it comes as little surprise that the author who inspired Graber's words, Bill McKibben, reviles a certain philosopher of liberty and technological entrepreneurial innovation. It happens, McKibben sniffs, that "every time I contemplate Ayn Rand I find myself wishing I'd been born to a different species."[11]

It is also appropriate that enemies of capitalism, industrialization, and unregulated technological innovation would begrudge man's initial steps toward land privatization and farming. It must be remembered that farming on private land was an early example of humanity technologically restructuring its surroundings for its own profit. Hence Rousseau bemoans the inception of agriculture, as that was when "property was introduced . . ., and vast forests became smiling fields, which man had to water with the sweat of his brow," and where "misery" was to "germinate and grow up with the crops. Metallurgy and agriculture were the two arts which produced this great revolution. . . . it was iron and corn, which first civilised men, and ruined humanity."[12] It will be noticed that the process whereby entrepreneurs harnessed natural phenomena for industrial ends would not have occurred were it not for their self-interested exercise of rationality. It would not have happened were it not for science. On these grounds, Rousseau determines it urgent to denounce rationality as such. He faults ancient Greece for being "the inventor of the sciences." Then he flatly asserts, providing not a whiff of evidentiary support, that science was born of the "moral philosophy of human pride." By *pride*, he does not mean the healthy self-esteem that comes with efficaciously and peaceably pursuing one's one happiness—the sort of pride that gained a positive reputation in the late twentieth century thanks in some part to Ayn Rand. On the contrary, Rousseau is alluding to the sort of arrogance that insecure men have always tried to pass off as pride, and which Christian conservatives have repeatedly used as their straw man in attempt to discredit any legitimate concept of healthy self-esteem.

The Huguenot philosopher ultimately concludes, "Thus the arts and sciences owe their birth to our vices . . ."[13]

Cambridge University biochemist Terence H. Kealey aptly paraphrases Rousseau's anti-mind, anti-science, anti-individualist, anti-technology, anti-commerce perspective. "The enslavement of Man occurred through capitalism, which originated in private property . . . the ownership of property encouraged its owners to improve it, which caused them to foster science and technology . . ." This is bad, as "under the distortions induced by reason (which is emotionally sterile), by capitalism (which inflames greed) and by private property (which isolates and destroys sympathy) a man's personality degrades, and he becomes psychologically alienated...from other people. His selfishness grows into a devouring monster of vanity and egotism, *amour-propre*." Rousseau's influence has been pernicious. Kealey notes that prior to Rousseau, Renaissance poets such as John Milton celebrated Galileo and other scientists. Subsequent to Rousseau's ascension to fame, Romanticist philosophy emerged and it became fashionable for litterateurs to "despise scientists" and entrepreneurs as "emotional cripples, in hock to materialism, vainglory and greed."[14]

Rousseau evidently did not have the courage, however, to admit that he fully opposed the institution of private property rights. He contradicted himself by stating that he believed private property rights to be extremely important, but that the government could still justly nullify such rights when the common good required it. "It is certain that the right of property is the most sacred of all the rights . . . On the other hand, it is no less certain that the maintenance of the State and the government involves costs and outgoings; and as every one who agrees to the end must acquiesce in the means, it follows that the members of a society ought to contribute from their property to its support. Besides, it is difficult to secure the property of individuals on one side, without attacking it on another; and it is impossible that all the regulations which govern the order of succession, will, contracts, &c. should not lay individuals under some constraint as to the disposition of their goods, and should not consequently restrict the right of property."[15] The concession that property rights are "sacred," then, is insincerity on Rousseau's part. Were it ethical for the State to assail private property rights on behalf of the collectivist credo, such rights would lose their meaning. Thus the irony of Rousseau's appeal that through some free contractual agreement, a man can assent to having the State deny his freedoms for the rest of his life. The irony does not go unnoticed by Seton Hall University communications professor Agnes Hooper Gottlieb and her coauthors. They chide Rousseau for having "called for the subjugation of the individual to the will of the state. That sounds like fascism to us."[16] We have indeed learned of how fascism and Naziism were not manifestations of laissez-faire Social Darwinism, but were instead Romanticist philosophy and governism combined and then taken to their inevitable conclusion.

"Rousseau's doctrines," Kealey admits, "are immoral."

Ode to Liberty

Auditing the information we have absorbed, I am saddened that Rousseau's agenda of instituting this new Social Contract has been at least partially achieved in modern America and Europe. It has been achieved in that the regulatory-entitlement state—complete with environmental regulations—remains powerful and entrenched in Western society. The farther that Rousseau's philosophy progresses in establishing total political collectivism in the West, the worse our lives will become. We must reverse the course that Rousseau has set us upon, for it is definitely the wrong path. On every count, philosophic criticisms against the free market are unfounded. Laissez faire does not plant the seeds of its own demise. Industry will continue to be a boon for us to the extent that we keep it free. In lieu of falling for Rousseau's homilies, we can heed the advice of a much wiser contemporary of his—that American firebrand whose very words opened this trilogy, Patrick Henry. Henry recognized the truth very early on—"If a man is in chains, he droops and bows to the earth, for his spirits are broken. . . . but let him twist the fetters from his legs, and he will stand erect. . . . Fetter not commerce, let her be as free as the air."[17]

As we reach the dock at the end of our intellectual odyssey, we recognize that, contrary to the dismissals of David Hume and utilitarian collectivist economists, invocations of individual rights to one's own life and private property are not capricious. Your individual rights to your own life, your body, your mind, and other forms of private property are rooted in the principles of nature and sociobiological evolution. Such rights induce the need for the sort of night watchman state I have described—one focused on recognizing personal self-ownership. This night watchman state is to protect its Pure Citizens' rights against physical assault, poisoning, property damage, vandalism, theft, extortion, contractual breach, fraud, defamation, patent infringement, and every other form of spoliation. And the night watchman state must also tolerate any childless adult's right to abnegate Pure Citizenship for himself. There is no ethical excuse for compulsory taxation.

There are no real rights other than consensualist, free-enterprise rights. The alternative "rights" that President Franklin D. Roosevelt's ilk have touted, such as a "right" to force others to provide for one's healthcare, are found hopelessly wanting. Laissez-faire liberal rights are what are rational and palpable. They are indispensable for the continued survival of our civilization, whether or not Malthusian environmentalists and anti-globalization activists persist in denying these considerations. Freedom and peaceful entrepreneurship are what best nurture the soils of the marketplace ecosystem, providing it the nutrients with which it can flourish. The sooner that our policymakers come to grips with this, the more healthfully will our civilization—whether we consider it a monopolistic empire or not—will be able to endure. Recognizing this reality is not ruthless social Darwinism, but it is a form of cultural evolution.

Examining our nation's history, we discern that not even Patrick Henry comprehended the full significance of the famous phrase of his that I quoted at the launch of this trilogy. This is a fuller quotation of the public address in which it was spoken: "Is life so dear, or peace so sweet, as to be purchased at the price of chains and slavery? Forbid it, Almighty God! I know not what course others may take; but as for me, give me liberty or give me death!"[18] Though the words are eloquent, they facilitate a misleading conclusion. It appears that Henry inferred that one could practicably live under draconian governism, though such a condition would be so horrid that it would be preferable to risk one's life to cast it off. Yet, as we have discovered, to dwell under tyranny—rather than under free industry—drastically shortens one's life expectancy. For the same reason, tyranny increases one's chances of perishing from homicide at the hands of the State. Therefore, there is ultimately a far more pronounced dichotomy between liberty and violent death than Patrick Henry realized. Every gain in political freedom reduces the probability that a person will prematurely meet his doom.

Moreover, human life—taken to its logical maximization—is Aristotelian eudemonia, and achieving Aristotelian eudemonia requires the exercise of personal autonomy. Therefore, insofar as there is adherence to the Rule of Peace, a society's members maximize their opportunities to attain eudemonia. And to attain eudemonia is to live life as properly as is possible to one's nature as a human. For that reason, the Rule of Peace can be thought of as the Public Policy of Life. By contrast, the more governism pervades a society, the more it obstructs that society's members from living life to its full potential, as government regulation itself threatens the physical destruction of innocent parties. Therefore, governism can be justly branded the Public Policy of Death. To favor human life, then, is to favor human liberty. A just social system is one that codifies the absolute freedom of peaceful action.

NOTES

1. Rousseau, *A Discourse On a Subject Proposed By the Academy of Dijon: What Is the Origin of Inequality Among Men, and Is It Authorised By Natural Law?*, 1754, republished in Rousseau 1923, pt. 2, para.1, 14, http://tinyurl.com/2loygu, accessed Wednesday, October 3, 2007.
2. R. Porter 2004 U.S. version, 429.
3. M. Grosser 1978, 16.
4. J. F. Moon 1974, 47.
5. Green Anarchy Collective, "What Is Green Anarchy?: An Introduction to Anti-Civilization Anarchist Thought and Practice," *Green Anarchy*, Summer 2004, republished in *Sacred Lands*, http://www.sacredlands.org/green_anarchy.htm, accessed Thursday, November 7, 2013. I learned about this manifesto from K. Kelly 2010, 209-210.
6. Younkins 2002, 165–68. For a first-hand example of Rousseau's collectivism, see Rousseau, *The Social Contract, or Principles of Political Right*, Bk. 1, Ch. 7, http://tinyurl.com/5t5fem, accessed Thursday, July 31, 2008, in Rousseau 1923.
7. Victor Klemperer, diary entry, Thursday, July 30, 1936, in Klemperer 1998, 180. In *ibid.*, Klemperer elaborated on what Rousseau's writings had in common with Hitler's speeches—"the prostitution of reason in the service of subjective feeling . . ."
8. I have heard an "anarcho-capitalist" libertarian spin on Original Sin as well, in which the formation of the first Social Contract counts as the Original Sin. In this version of the tale, human beings frolicked in blissful anarchy and traded with one another under "anarcho-capitalism." Unfortunately, some human beings sold out to villainy by agreeing with one another to form a government. This was a Faustian bargain in which these men sold everyone's freedom down the river in an attempt to gain more security. Since then, the very institution of government has been the greatest source of mankind's woes. And humanity can only redeem itself—and restore the freedom that was once man's natural state—by casting

off government and going back to "market anarchy." This story is false, since there has never been an era of true long-term anarchy for mankind.

9. Graber 1989, 1, 9, http://articles.latimes.com/print/1989-10-22/books/bk-726_1_bill-mckibben, accessed Tuesday, January 18, 2011. This article came to my attention through J. L. Simon 1998, trade paperback, 1998, 547; and P. Schwartz 1999, 221.

10. D. B. Botkin 2012, xvii–xviii.

11. Bill McKibben, "Let Them Build Seawalls," *The Dish: Biased & Balanced*, edited by Andrew Sullivan, August 28, 2014, http://dish.andrewsullivan.com/2014/08/28/let-them-build-seawalls/, accessed September 9, 2014.

12. Rousseau, *A Discourse On a Subject Proposed By the Academy of Dijon: What Is the Origin of Inequality Among Men, and Is It Authorised By Natural Law?*, 1754, republished in Rousseau 1923, Pt. 2, Para. 21-22, accessed online Wednesday, August 24, 2011. T. Kealey 1996 U.S. paperback, 312, brought this assessment of Rousseau's to my attention.

13. Rousseau, *A Discourse on the Moral Effects of the Arts and Sciences*, 1761, republished in Rousseau 1923, Pt. 2, Para. 1, accessed online Wednesday, August 24, 2011. T. Kealey 1996 U.S. paperback, 312, brought this statement of Rousseau's to my attention.

14. T. Kealey 1996, U.S. paperback, 312-15.

15. Jean-Jacques Rousseau, "A Discourse on Political Economy," para. 43, 1755, http://tinyurl.com/5jfdld, accessed Wednesday, August 20, 2008, in Rousseau 1923.

16. A. H. Gottlieb et al. 1998, 7.

17. Patrick Henry, qtd. by W. W. Henry 1891 vol. 2, 192. I first heard of this quotation from J. Appleby 1996 3d ed., 184.

18. Patrick Henry, address delivered in St. John's Church in Richmond, Virginia, on March 23, 1775, republished on the website of the University of Oklahoma College of Law at http://www.law.ou.edu/ushistory/henry.shtml, accessed Friday, June 15, 2007. Sadly, not even Patrick Henry was fully consistent in his philosophy. He supported spoliation inasmuch as he actually advocated that people be taxed, and that their tax money be used in the dissemination of Christian teachings. Fortunately, Thomas Jefferson and James Madison opposed him on this. For information on this matter, see J. Meacham 2006, 85.

Bibliography

Sources that have a single * asterisk are recommended reading or viewing.

Sources with two ** asterisks are so fascinating that reading them counted among the best experiences of my life.

Acemoglu, Daron, and James A. Robinson. 2012. *Why Nations Fail: The Origins of Power, Prosperity, and Poverty*. New York, New York: Crown Business.

Achbar, Mark, et al., prods. 2003. *The Corporation: The Pathological Pursuit of Profit and Power*. Mark Achbar and Jennifer Abbott, dir. DVD. Big Picture Media Corporation. Transcript available online as a PDF in two installments, at http://hellocoolworld.com/files/TheCorporation/ Transcript_finalpt1 copy.pdf and http://hellocoolworld.com/files/TheCorporation/Transcript_ finalpt2 copy.pdf.

Ackerman, Carl W. 1930. *George Eastman: Founder of Kodak and the Photography Business*. New York, New York: Houghton Mifflin.

Adams, James Truslow. 1931. *The Epic of America*. Boston, Massachusetts: Little, Brown and Company.

Adams, John. 1819a. "Addressed to the Inhabitants of the Colony of Massachusetts Bay, January 23, 1775." In John Adams 1819c: 9-14.

———. 1819c. *Novanglus, and Massachusettensis; or Political Essays, Published in the Years 1774 and 1775, on the Principal Points of Controversy, Between Great Britain and Her Colonies*. John Adams, ed. Boston, Massachusetts: Hews & Goss.

Allen, Arthur. 2008. *Vaccine: The Controversial Story of Medicine's Greatest Lifesaver*. New York, New York: W. W. Norton & Company.

Allen, Brooke. 2006. *Moral Minority: Our Skeptical Founding Fathers*. Chicago, Illinois: Ivan R. Dee.

Allen, Frederick (Lewis), (III). 1994. *Secret Formula: How Brilliant Marketing and Relentless Salesmanship Made Coca-Cola the Best Known Product in the World*. New York, New York: Harper Business.

Allen, Grant. 1894. *Personal Reminiscences of Herbert Spencer*. Republished on the website of the Molinari Institute. Roderick T. Long, ed. Auburn, Alabama: Molinari Institute. http:// praxeology.net/GA-PRS.htm.

Aly, Götz. 2008. [2005.] *Hitler's Beneficiaries: Plunder, Racial War, and the Nazi Welfare State*. New York, New York: Henry Holt & Company.

Aly, Götz, et al. 1994. *Cleansing the Fatherland: Nazi Medicine and Racial Hygiene*. Baltimore, Maryland: Johns Hopkins University Press.

American Cancer Society (ACS). 2006. "Common Questions About Diet and Cancer." Website of the American Cancer Society. September 28. Atlanta, Georgia: American Cancer Society. http://tinyurl.com/5tmsw.

American Council on Science & Health (ACSH). 1998. "Health Panel Calls Attack on Milk from rBST-Treated Cows an Unwarranted Distortion of Science." Press Release. March 1. Republished on the website of the American Council on Science & Health. New York, New York: American Council on Science & Health. http://acsh.org/1998/03/health-panel-calls-attack-on-milk-from-rbst-treated-cows-an-unwarranted-distortion-of-science/.

———. 2002. "Warning: Starbucks Protestors Spread False Fears About Safe Foods." Press Release. February 25. Republished on the website of the American Council on Science & Health. New York, New York: American Council on Science & Health. http://acsh.org/2002/02/warning-starbucks-protestors-spread-false-fears-about-safe-foods/.

The American Heritage Dictionary of the English Language. 4th ed. 2006. Boston, Massachusetts: Houghton Mifflin Company.

Ammann, Daniel. 2009. *The King of Oil: The Secret Lives of Marc Rich*. New York, New York: St. Martin's Press.

*Ames, Bruce N. 1995. "Pesticides, Cancer, and Misconceptions." In J. L. Simon ed. 1995, trade paperback: 588-594.

*Ames, B. N., et al. 1987. "Ranking Possible Carcinogenic Hazards." *Science* vol. 236 (no. 4799, April 17): 271-280.

Anderson, Walter Truett. 1996. *Evolution Isn't What It Used to Be: The Augmented Animal and the Whole Wired World*. New York, New York: W. H. Freeman and Company.

Anti-Defamation League (ADL). 2003. "PETA's Appeal for Jewish Community Support 'the Height of Chutzpah.'" Website of the Anti-Defamation League. Press Release. October 14. http://archive.adl.org/nr/exeres/654aaeac-709a-4f4a-aadb-4c70bf65705c,0b1623ca-d5a4-465d-a369-df6e8679cd9e,frameless.html.

———. 2005. "Animal Rights Group Should Take the Lead of PETA and Stop Using Holocaust Imagery." Website of the Anti-Defamation League. Press Release. August 2. http://63.146.172.78/NR/exeres/DD0CA7E4-9644-45FA-B564-718D7F85BC36,0B1623CA-D5A4-465D-A369-DF6E8679CD9E,frameless.htm.

Appleby, Joyce. 1996 3d ed. *Liberalism and Republicanism in the Historical Imagination*. Cambridge, Massachusetts: Harvard University Press.

Arluke, A., and Boria Sax. 1992. "Understanding Nazi Animal Protection and the Holocaust." *Anthrozoös* vol. 5 (no. 1): 6–31.

Arrison, Sonia. 2011. *100 Plus: How the Coming Age of Longevity Will Change Everything, From Careers and Relationships to Family and Faith*. New York, New York: Basic Books.

Asimov, Isaac. 1964 paperback. [1964 hardcover.] *A Short History of Biology*. Garden City, New York: The Natural History Press.

Ask.Com Communications. 2014. *Dictionary.Com*. Oakland, California: Ask.Com Communications. http://dictionary.reference.com.

Atkinson, Tracy, and Joan Baran, prods. 1998. *Nazis: The Occult Conspiracy*. Tracy Atkinson and Joan Baran, dirs. Written by Tracy Atkinson, Joan Baran, and Leslie Shane. Brad Abelle, narrator. Cable television broadcast: Discovery Channel.

Baeumler, Alfred. 1937. *Studien zur Deutschen Geistesgeschichte* (*Studies in German History*). Berlin, Germany: Junker und Dünnhaupt Verlag (Junker & Dünnhaupt Publishing House).

———. 1966. [1937.] "Nietzsche and National Socialism." In G. L. Mosse ed. 1966: 97–101.

*Bailey, Ronald. 1994 paperback. [1993.] *Eco-Scam: The False Prophets of Ecological Apocalypse*. New York, New York: St. Martin's Press.

———. 2000b. "Strands of Life." *Reason* vol. 32 (no. 4, August/September): 58–61. Republished on *Reason Online*. Washington, D.C.: Reason Foundation. http://www.reason.com/news/show/ 27776.html.

———. 2002. "Organic Alchemy: Organic Farming Could Kill Billions of People." Reason Foundation op-ed. *Reason Online*. June 5. Washington, D.C.: Reason Foundation. http://www.reason.com/news/show/34820.html.

Baker, Ray Stannard. 1964 rpt. [1908.] *Following the Color Line: American Negro Citizenship in the Progressive Era*. New York, New York: Harper & Row.

Baldwin, Neil. 2001. *Henry Ford and the Jews: The Mass Production of Hate*. New York, New York: PublicAffairs.

Ball, Philip. 2004. *Critical Mass: How One Thing Leads to Another*. New York, New York: Farrar, Strauss and Giroux.

Bander, Edward, ed. 1991. [1966.] *Justice Holmes Ex Cathedra*. Unknown city: William S. Hein Publishing.

Banner, Stuart. 2007. *Possessing the Pacific: Land, Settlers, and Indigenous People from Australia to Alaska*. Cambridge, Massachusetts: Harvard University Press.

Banning, Lance, ed. 2004. *Liberty and Order: The First American Party Struggle*. Indianapolis, Indiana: Liberty Fund. Republished in the *Online Library of Liberty*. Indianapolis, Indiana: Liberty Fund. http://oll.libertyfund.org/index.php?option=com_staticxt&staticfile=show.php%3Ftitle=875&Itemid=99999999.

Bannister, Robert C. 1970. "'The Survival of the Fittest Is Our Doctrine': History or Histrionics?" *Journal of the History of Ideas* vol. 31 (no. 3, July–September): 377–398.

———. 1973. "William Graham Sumner's 'Social Darwinism': A Reconsideration." *History of Political Economy* vol. 5 (no. 1): 89–109.

*———. 1988. [1979.] *Social Darwinism: Science and Myth in Anglo-American Social Thought*. Philadelphia, Pennsylvania: Temple University Press.

Barbash, Jack. 1989. "John R. Commons: Pioneer of Labor Economics." *Monthly Labor Review* vol. 112 (no. 5, May): 44–49. Republished on the website of the Bureau of Labor Statistics. Washington, D.C.: Department of Labor. http://stats.bls.gov/mlr/1989/05/art4full.pdf.

Barkan, Elazar. 1992. *The Retreat of Scientific Racism: Changing Concepts of Race in Britain and the United States Between the World Wars*. Cambridge, United Kingdom: Cambridge University Press.

Barnard, Harry. 2002. [1958.] *Independent Man: The Life of Senator James Couzens*. Detroit, Michigan: Wayne State University Press.

Barnes, Harry Elmer. 1948. *An Introduction to the History of Sociology*. Chicago, Illinois: The University of Chicago Press.

Barnes, James Strachey. 1928. *The Universal Aspects of Fascism*. London, United Kingdom: Williams & Norgate.

Bass, Thomas A. 2000, paperback. [1999.] *The Predictors*. New York, New York: Henry Holt and Company.

Bast, Joseph L., et al. 1994. *Eco-Sanity: A Common-Sense Guide to Environmentalism*. Lanham, Maryland: Madison Books.

Battles, Matthew. 2004, paperback. [2003.] *Library: An Unquiet History*. New York, New York: W. W. Norton & Company.

Bean, Jonathan, ed. 2009. *Race and Liberty in America: The Essential Reader*. Lexington, Kentucky: The University Press of Kentucky.

Bell, Bernard W., et al., eds. 1996. *W.E.B. Du Bois on Race and Culture*. New York, New York: Routledge.

Bellamy, Edward. 1951. [1888.] *Looking Backward, 2000–1887*. New York, New York: The Modern Library.

———. 2000. [1888.] *Looking Backward, 2000–1887*. Edited and placed online by Dennis G. Jerz. Greensburg, Pennsylvania: Seton Hall University. http://jerz.setonhill.edu/resources/texts/bellamy_e_lb/index.html.

Bellomy, Donald C. 1984. "'Social Darwinism' Revisited." *Perspectives in American History* (N.S.) 1: 1-129.

Bemis, Edward W. 1888. "Restriction of Immigration." *Andover Review* (March): 251–264.

Benedict, Ruth. 1959. [1934.] *Patterns of Culture*. Boston, Massachusetts: Houghton Mifflin Company.

Berlin, Isaiah. 1970a. [1969.] *Four Essays on Liberty*. New York, New York: Oxford University Press.

———. 1970b. [1953.] "Historical Inevitability." In I. Berlin 1970a: 41–117.

———. 1980a. *Against the Current: Essays in the History of Ideas*. Henry Hardy, ed. New York, New York: Viking.

———. 1980b. "Hume and the Sources of German Anti-Rationalism." In I. Berlin 1980a: 162–187.

———. 1998a. "Kant As an Unfamiliar Source of Nationalism." In I. Berlin 1998b paperback: 232–248.

———. 1998b paperback. [1996.] *The Sense of Reality: Studies in Ideas and Their History*. Henry Hardy, ed. New York, New York: Farrar, Straus and Giroux.

Bernstein, David E. 2011. *Rehabilitating Lochner: Defending Individual Rights Against Progressive Reform*. Chicago, Illinois: University of Chicago Press.

Bernstein, Richard. 2001. "*IBM and the Holocaust:* Assessing the Culpability." *New York Times*. March 7. Republished on the New York Times website. New York, New York: The New York Times Company. http://tinyurl.com/368j5j.

Bernstein, William J. 2008. *A Splendid Exchange: How Trade Shaped the World*. New York, New York: Atlantic Monthly Press.

Berreby, David. 2005. *Us and Them: Understanding Your Tribal Mind*. New York, New York: Little, Brown and Company.

Berry, R. 2004. *Hitler: Neither Vegetarian Nor Animal Lover*, Brooklyn, New York: Pythagorean Books.

Bessel, Richard. 1992. "The Formation and Dissolution of a German National Electorate: From Kaiserreich to Third Reich." In L. E. Jones and Retallack eds. 1992: 399–418.

Biddiss, Michael. 1970. *Father of Racist Ideology: The Social and Political Thought of Count Gobineau*. New York, New York: Weybright and Talley.

**Biddle, Craig. 2011. "Ayn Rand's Theory of Rights: The Moral Foundation of a Free Society." *The Objective Standard* vol. 6 (no. 3, Fall): 13–36. Republished online on the *Objective Standard* website. Craig Biddle, ed. Glen Allen, Virginia: The Objective Standard.

Biddle, Wayne. 2009. *Dark Side of the Moon: Wernher von Braun, the Third Reich, and the Space Race*. New York, New York: W. W. Norton & Company.

**Bidinotto, Robert James. 1994. "The Contradiction in Anarchism." Republished on the website of the Objectivism WWW Service. Jeff Schachter and Raymie Stata, eds. Cambridge, Massachusetts: The Objectivism WWW Service. http://mol.redbarn.org/objectivism/writing/RobertBidinotto/ContradictionInAnarchism.html.

*———. 2003. "Contra Anarchism, Part 3." *Bidinotto Blog*. Washington, D.C.: Robert James Bidinotto. http://bidinotto.journalspace.com/?entryid=57. Note: Due to a malfunction on the *Journalspace* domain, every entry from this weblog was deleted from the World Wide Web. Robert James Bidinotto retains his own personal copy of it, however.

*———. 2006. "Corporate Cash for Collectivist Causes." *The New Individualist* vol. 9 (no. 11, November): 40–42. Republished on the website of the Atlas Society. Washington, D.C.: Atlas Society. http://www.atlassociety.org/ct-1814-cash_left.aspx.

Bin Laden, Osama. 2002. "Full Text: Bin Laden's 'Letter to America.'" *The Guardian*. Sunday, November 24. http://tinyurl.com/kva1.

**Binswanger, Harry. 1981c. "Q & A Department." *The Objectivist Forum* vol. 2 (no. 4, August): 11–14. Republished on the website of *The Harry Binswanger List*. New York, New York: TOF Publications, Inc.

**———. 2011c. "Anarchism Vs. Objectivism." [A revision of Binswanger 1981 with additional material added on July 20, 2011.] July 20. *HBL: The Harry Binswanger List*. New York, New York: The Harry Binswanger List. http://www.hblist.com/anarchy.htm.

Biven, W. Carl. 1989. *Who Killed John Maynard Keynes: Conflicts of the Evolution of Economic Policy*. Homewood, Illinois: Dow Jones-Irwin.

Black, Edwin. 2003. *War Against the Weak: Eugenics and America's Campaign to Create a Master Race*. New York, New York: Four Walls Eight Windows.

Blakesley, Joseph Williams. 1839. *A Life of Aristotle; Including a Critical Discussion of Some Questions of Literary History Connected With His Works*. London, United Kingdom: John W. Parker.

Blakey, Clayton B., and Moorfield Storey. 1918. [1916.] "Argument for Plaintiff in Error." In E. Knaebel ed. 1918a vol. 245: 60–64.

Block, Walter. 1976. *Defending the Undefendable: The Pimp, Prostitute, Scab, Slumlord, Libeler, Moneylender, and Other Scapegoats in the Rogue's Gallery of American Society*. New York, New York: Fleet Press Corporation.

Blumenthal, Sidney. 1986. *The Rise of the Counter-Establishment: From Conservative Ideology to Political Power*. New York, New York: Times Books.

Boaz, David, ed. 1997. *The Libertarian Reader: Classic and Contemporary Writings from Lao-Tzu to Milton Friedman*. New York, New York: The Free Press.

Bodanis, David. 2000. *E = mc2: A Biography of the World's Most Famous Equation*. New York, New York: Berkeley Publishing Group.

Bork, Robert. 2003a. "The Case for Censorship." In Bork ed. 2003b: 140–153.

Bork, Robert, ed. 2003b. *Slouching Toward Gomorrah*. New York, New York: HarperCollins.

Bormann, Martin. 1953. *Hitler's Secret Conversations, 1941–1944*. New York, New York: Farrar, Straus, and Young.

Borsook, Paulina. 2000. *Cyberselfish: A Critical Romp Through the Terribly Libertarian Culture of High Tech*. New York, New York: PublicAffairs.

Botkin, Daniel B. 2012. *The Moon in the Nautilus Shell: Discordant Harmonies Reconsidered*. New York, New York: Oxford University Press.

Bova, Ben. 1998. *Immortality: How Science Is Extending Your Lifespan—And Changing the World*. New York, New York: Avon Books.

Bovard, James. 1994. *Lost Rights: The Destruction of American Liberty*. New York, New York: St. Martin's Press.

*Bowden, Thomas A. 2009. "Justice Holmes and the Empty Constitution." *The Objective Standard* vol. 4 (no. 2, Summer): 21–46. Republished online on the *Objective Standard* website. Craig Biddle, ed. Glen Allen, Virginia: The Objective Standard. http://www.theobjectivestandard.com/issues/2009-summer/justice-holmes-empty-constitution.asp.

Bowen, Harold G., and Charles F. Kettering. 1958. *A Short History of Technology*. New York, New York: Thomas Alva Edison Foundation.

Bowers, David F. 1945. "Review of *Social Darwinism in American Thought* by Richard Hofstadter." *The Pacific Historical Review* 4: 103–04.

Bowler, Peter J. 1996. *Charles Darwin: The Man and His Influence*. New York, New York: Cambridge University Press.

Bradbury, K. E., et al. 2014. "Organic Food Consumption and the Incidence of Cancer in a Large Prospective Study of Women in the United Kingdom." *British Journal of Cancer* vol. 110 (March 27): 2321–2326.

Brady, Ivan A., and Barry L. Isaac, eds. 1975. *A Reader in Cultural Change* (2 vols.). New York, New York: Schenkman Publishing.

Brady, Kathleen. 1989. *Ida Tarbell: Portrait of a Muckraker*. Pittsburgh, Pennsylvania: University of Pittsburgh Press.

Brandeis, Louis D. 2013. [1908.] *The Brandeis Brief—In Its Entirety*. [Brandeis's brief in *Muller v. Oregon* 208 U.S. 14.] Louisville, Kentucky: Louis D. Brandeis School of Law at the University of Louisville. http://www.law.louisville.edu/library/collections/brandeis/node/235.

Branden, Nathaniel. 2009. *The Vision of Ayn Rand: The Basic Principles of Objectivism*. Gilbert, Arizona: Cobden Press.

Brands, H. W. 1997. *T.R.: The Last Romantic*. New York, New York: Basic Books.

———. 1998. *What America Owes the World: The Struggle for Foreign Policy*. New York, New York: Cambridge University Press.

Brantlinger, Patrick. 1977. *The Spirit of Reform: British Literature and Politics, 1832–1867*. Cambridge, Massachusetts: Harvard University Press.

Bressler, Marvin. 1981. "Biological Determinism and Ideological Indeterminacy." In E. White ed. 1981: 181–192.

Brewer, Holly. 2003. "Power and Authority in the Colonial South: The English Legacy and Its Contradictions." In J. P. Ward ed. 2003: 27–52.

Briggs, Vernon M., Jr. 2001. "American Unionism and U.S. Immigration Policy." Cornell University Law School: 1–12. Republished as a PDF on the website of the Center for Immigration Studies. Washington, D.C.: Center for Immigration Studies. http://www.cis.org/sites/cis.org/files/ articles/2001/back1001.pdf.

**Britting, Jeff. 2004. *Ayn Rand*. New York, New York: Overlook Duckworth.

Brockman, John. 1995a. "The Pattern of Life's History." Interview with Stephen Jay Gould. In J. Brockman ed. 1995d: 51–73.

———. 1995b. "Plectics." Interview with Murray Gell-Mann. In J. Brockman ed. 1995d: 316-332.

———. 1995c. "The Second Law of Organization." Interview with J. Doyne Farmer. In J. Brockman ed. 1995d: 359–378.

Brockman, John, ed. 1995d. *The Third Culture*. New York, New York: Simon & Schuster.

**Brook, Yaron. 2007. "The Morality of Moneylending: A Short History." *The Objective Standard* vol. 2 (no. 3, Fall): 9–45. Craig Biddle, ed. Glenn Allen, Virginia: The Objective Standard.

Brooks, Arthur C. 2006. *Who Really Cares: The Surprising Truth About Compassionate Conservatism*. New York, New York: Basic Books.

Brown, Travis. 1994. *The First United States Patent for Many Everyday Things*. Lanham, Maryland: The Scarecrow Press.

Bruce, Thomas R., ed. 2013. Legal Information Institute website. Ithaca, New York: Cornell University Law School. http://www.law.cornell.edu/.

Buchanan, Patrick. 2009. "Making a Monkey Out of Darwin." *Real Clear Politics*. Tuesday, June 30. Syndicated from Creators Syndicate. New York, New York: Creators Syndicate. http://www.realclearpolitics.com/articles/2009/06/30/making_a_monkey_out_of_darwin_97230.html.

Buchholz, Todd G. 1999 Revised 2d ed. [1990.] *New Ideas from Dead Economists: An Introduction to Modern Economic Thought*. New York, New York: Plume.

Bullock, Alan. 1991. [1952.] *Hitler: A Study in Tyranny*. New York, New York: HarperCollins.

Burns, Kate, ed. 2006. *Is the American Dream a Myth?* Farmington Hills, Michigan: Greenhaven Press.

Burris, Alan. 1983 2d ed. [1981.] *A Liberty Primer*. Rochester, New York: Society for Individual Liberty.

Buruma, Ian, and Avishai Margalit. 2005, paperback. [2004.] *Occidentalism: The West in the Eyes of Its Enemies*. New York, New York: Penguin Books.

Buss, David M. 2000. *The Dangerous Passion: Why Jealousy Is as Necessary as Love and Sex*. New York, New York: The Free Press.

Caine, Stanley P. 1974. "The Origins of Progressivism." In Gould ed. 1974: 11–34.

Cairncross, Frances. 1992 U.S. version. [1991.] *Costing the Earth: The Challenge for Governments, The Opportunities for Business*. Cambridge, Massachusetts: Harvard Business School Press.

Calic, Edouard, ed. 1971. *Unmasked: Two Confidential Interviews With Hitler in 1931*. London, United Kingdom: Chatto & Windus.

Campbell, Robert L. 2006. "Altruism in Auguste Comte and Ayn Rand." *The Journal of Ayn Rand Studies* vol. 7 (no. 2, Spring): 357–369. Republished on the website of Robert L. Campbell. Clemson, South Carolina: Clemson University. PDF at http://hubcap.clemson.edu/~campber/ altruismrandcomte.pdf.

Cannadine, David. 2006. *Mellon: An American Life*. New York, New York: Alfred A. Knopf.

Cannon, Michael F., and Michael D. Tanner. 2007 2d ed. *Healthy Competition: What's Holding Back Health Care and How to Fix It*. Washington, D.C.: Cato Institute.

Carlin, George, et al., prods. 2006. *Life Is Worth Losing*. Rocco Urbiski, dir. George Carlin, writer and performer. Released January 10, 2006. Audio album. Eardrum/Atlantic.

Carnegie, Andrew. 1886. *Triumphant Democracy, or Fifty Years' March of the Republic*. New York, New York: Charles Scribner's Sons.

———. 1901. *The Gospel of Wealth, And Other Timely Essays*. New York, New York: The Century Co. Republished online. Mountain View, California: Google Books. http://tinyurl.com/3adwcc.

———. 1986. [1920.] *The Autobiography of Andrew Carnegie*. Boston, Massachusetts: Northeastern University Press.

Carneiro, Robert L. 1981. "Herbert Spencer as an Anthropologist." *The Journal of Libertarian Studies* vol. 5 (no. 2, Spring): 153–210. Murray N. Rothbard, ed. New York, New York: The Center for Libertarian Studies.

Caton, Hiram. 1981. "Domesticating Nature: Thoughts on the Ethnology of Modern Politics." In E. White ed. 1981: 99–134.

Chadwick, Ruth F., ed. 1992 rev. ed. *Ethics, Reproduction, and Genetic Control*. New York, New York: Routledge.

Chamberlain, Houston Stewart. 1911. *Foundations of the Nineteenth Century* (2 vols.). New York, New York: John Lane Company.

———. 1912. *Foundations of the Nineteenth Century* (2 vols.). New York, New York: John Lane Company.

———. 2005. [1915.] *Political Ideals*. Alexander Jacob, trans. Lanham, Maryland: University Press of America.

Chanda, Nayan. 2007. *Bound Together: How Traders, Preachers, Adventurers, and Warriors Shaped Globalization*. New Haven, Connecticut: Yale University Press.

Chapman, Izzy, prod. 2012. *The Nazi Gospels*. Paul Copeland, writer and dir. Eric Meyers, narrator. December 30. Cable television broadcast: History Channel.

Charles, Daniel. 2005. *Master Mind: The Rise of Fall of Fritz Haber, The Nobel Laureate Who Launched the Age of Chemical Warfare*. New York, New York: HarperCollins.

Chase, Allan. 1980, paperback. [1976.] *The Legacy of Malthus: The Social Costs of the New Scientific Racism*. Chicago, Illinois. University of Illinois Press.

———. 1982. *Magic Shots: A Human and Scientific Account of the Long and Continuing Struggle to Eradicate Infectious Diseases By Vaccination*. New York, New York: William Morrow and Company.

Cheney, Margaret. 1983, paperback. [1981.] *Tesla: Man Out of Time*. New York, New York: Dell Publishing.

Chernow, Ron. 1998. *Titan: The Life of John D. Rockefeller, Sr.* New York, New York: Random House.

———. 2001, trade paperback. [1990.] *The House of Morgan: An American Banking Dynasty and the Rise of Modern Finance*. New York, New York: Grove Press.

Chew, William L., III, ed. 1994. *National Stereotypes in Perspective: Americans in France, Frenchmen in America*. Rodopi: Amsterdam, The Netherlands.

Choate, Pat. 2005. *Hot Property: The Stealing of Ideas in an Age of Globalization*. New York, New York: Alfred A. Knopf.

Chorost, Michael. 2005. *Rebuilt: How Becoming Part Computer Made Me More Human*. Boston, Massachusetts: Houghton Mifflin Company.

Cicero, Marcus Tullius. 1913. *De Officiis* (*On Duties*). Walter Miller, trans. Cambridge, Massachusetts: Harvard University Press. Republished online on the Constitution Society website. Austin, Texas: The Constitution Society. http://www.constitution.org/rom/de_officiis.htm.

———. 2006a. "Lucullus (*Academica* Book 2)." In Cicero 2006b: 3–86.

———. 2006b. *On Academic Scepticism*. Charles Brittain, trans. and ed. Indianapolis, Indiana: Hackett Publishing Company.

Cohen, Morton N. 1996 paperback. [1995.] *Lewis Carroll: A Biography*. New York, New York: Vintage.

Cohen, Patricia. 2007. "A Split Emerges as Conservatives Discuss Darwin." *The New York Times*. Saturday, May 5. Published on the *New York Times*'s website. New York, New York: The New York Times Company. http://tinyurl.com/ysmm4l.

*Colloton, Deborah, et al., prods. 2001. *Tampering With Nature*. John Stossel, reporter. Roger Goodman, dir. Friday, June 29. Television broadcast. American Broadcasting Company (ABC).

Colonial Williamsburg Foundation. 2010. *Colonial Williamsburg Official History Site*. Williamsburg, Virginia: Colonial Williamsburg Foundation. http://www.history.org.

Commager, Henry Steele. 1950. *The American Mind: An Interpretation of American Thought and Character Since the 1880's*. New Haven, Connecticut: Yale University Press.

———. 1967. *Lester Ward and the Welfare State*. Indianapolis, Indiana: Bobbs-Merrill.

Commager, Henry Steele, ed. 1964 2d ed. [1951.] *Living Ideas in America*. New York, New York: Harper & Row.

Commons, John R. 1897. "Natural Selection, Social Selection, and Heredity." *Arena* no. 18: 90–97.

———. 1907. *Races and Immigrants in America*. New York, New York: Macmillan.

Compton, James. 1968. *The Swastika and the Eagle: Hitler, the United States, and the Origins of the Second World War*. London, United Kingdom: Bodley.

Comte, Auguste. 1983a rpt. [1975.] *Auguste Comte and Positivism: The Essential Writings*. Gertrud Lenzer, ed. Chicago, Illinois: University of Chicago Press.

———. 1983b. [1851–54.] *Système de Politique Positive* vol. 1 (out of 4). In Comte 1983a rpt.: 309–389.

———. 1984c. [1851–54.] *Système de Politique Positive* vol. 2 (out of 4). In Comte 1983a rpt.: 390–441.

Cooke, Alistair. 1973. *Alistair Cooke's America*. New York, New York: Alfred A. Knopf.

Cooley, Charles Horton. 1909. *Social Organization: A Study of the Larger Mind*. New York, New York: Charles Scribner's Sons.

———. 1920. "Reflections Upon the Sociology of Herbert Spencer." *American Journal of Sociology*, vol. 26 (no. 2, September): 129–145.

Coolidge, Calvin. 1919e. *Have Faith in Massachusetts: A Collection of Speeches and Messages*. Boston, Massachusetts: Houghton Mifflin Company.

———. 1919h. [1918.] "Written for the *Sunday Advertiser and American*—September 1, 1918." In C. Coolidge 1919e: 132–37.

Copenhaver, Brian P., and Rebecca Copenhaver, trans. and eds. 2012. *From Kant to Croce: Modern Philosophy in Italy, 1800–1950*. Toronto, Canada: University of Toronto Press.

Corballis, Michael C. 1991. *The Lopsided Ape: Evolution of the Generative Mind*. New York, New York: Oxford University Press.

Cort, John C. 1988. *Christian Socialism: An Informal History*. Maryknoll, New York: Orbis Books.

Coser, Lewis A. 1977 2d ed. [1971.] *Masters of Sociological Thought*. New York, New York: Harcourt Brace Jovanovich.

Cousins, Norman, ed. 1958. *In God We Trust: The Religious Beliefs and Ideas of the American Founding Fathers*. New York, New York: Harper & Brothers.

Cowen, Tyler. 2000 paperback. [1998.] *In Praise of Commercial Culture*. Cambridge, Massachusetts: Harvard University Press.

Croly, Herbert. 1914. *The Promise of American Life*. New York, New York: The Macmillan Company.

Crowley, John F. 2010. *Chasing Miracles: The Crowley Family Journey Through Strength, Hope, and Joy*. New York, New York: Newmarket Press.

Cullen, Dave. 2009. *Columbine*. New York, New York: Twelve.

Cutler, David M., et al. 1998. "Are Medical Prices Declining?: Evidence from Heart Attack Treatments." *Quarterly Journal of Economics*, vol. 113, no. 3 (November): 991–1024.

Darwin, Charles. 1871. *The Descent of Man, and Selection in Relation to Sex* (2 vols.). New York, New York: D. Appleton and Company.

———. 1872. 6th ed. [1st edition published in 1859.] *The Origin of Species*. London, United Kingdom: John Murray. Republished by *Project Gutenberg*. Oxford, Massachusetts: The Project Gutenberg Literary Archive Foundation. http://tinyurl.com/8zb7b.

———. 1896. *The Life and Letters of Charles Darwin* (2 vols.). Francis Darwin, ed. New York, New York: Appleton and Company.

———. 1958 rpt. of 6th edition. [1st edition published 1859; 6th edition published 1872.] *The Origin of Species: By Means of Natural Selection or the Preservation of Favoured Races in the Struggle for Life*. New York, New York: New American Library.

Davidowicz, Lucy S. 1986. *The War Against the Jews, 1933–1945*. New York, New York: Bantam Books.

Davie, Maurice R., ed. 1940. *Sumner Today: Selected Essays By William Graham Sumner, With Comments By American Leaders*. New Haven, Connecticut: Yale University Press.

Davies, Kevin. 2010. *The $1,000 Genome: The Revolution in DNA Sequencing and the New Era of Personalized Medicine*. New York, New York: The Free Press.

Davis, John B. 1994. *Keynes's Philosophical Development*. New York, New York: Cambridge University Press.

Dawkins, Richard. 2008. "Lying for Jesus?" *The Richard Dawkins Foundation for Science and Reason*. Saturday, March 22. Arlington, Virginia: Richard Dawkins Foundation for Science and Reason. http://old.richarddawkins.net/articles/2394-lying-for-jesus.

Degler, Carl N. 1991. *In Search of Human Nature: The Decline and Revival of Darwinism in American Social Thought*. New York, New York: Oxford University Press.

*DeGregori, Thomas R. 1999. "Nazis Versus Cancer: The Flip Side of Fascism?" *Priorities*, vol. 11 (no. 4, December 1). Republished on the website of the American Council on Science & Health. New York, New York: American Council on Science & Health. http://acsh.org/1999/12/nazis-versus-cancer-the-flip-side-of-fascism/.

*———. 2001. "Environmentalism, Animal Rights Activism, and Eco-Nazism." *Priorities*, vol. 13 (no. 2, April 1). Republished on the website of the American Council on Science & Health. New York, New York: American Council on Science & Health. https://web.archive.org/web/20030802013328/http://www.acsh.org/publications/priorities/1302/coverstory.html.

*de Kruif, Paul. 1932. *Men Against Death*. New York, New York: Harcourt, Brace & World, Inc.

Denfield, Rene. 1996. "Old Messages: Ecofeminism and the Alienation of Young People from Environmental Activism." In P. R. Gross et al. eds. 1996: 246–255.

Derber, Charles. 1998. *Corporation Nation: How Corporations Are Taking Over Our Lives and What We Can Do About It*. New York, New York: St. Martin's Press.

Desmond, Adrian, and James Moore. 1991. *Darwin*. New York, New York: Warner Books.

Dewar, Elaine. 2004. *The Second Tree: Stem Cells, Chimeras, and Quests for Immortality*. New York, New York: Carroll & Graf Publishers.

Dewey, John. 1987a. *John Dewey* vol. 11: *The Later Works, 1925–1953—Essays, Reviews, Trotsky Inquiry, Miscellany, and* Liberalism and Social Action, *1935–1937*. Jo An Boydston, ed. Carbondale, Illinois: Southern Illinois University Press.

———. 1987b. *Liberalism and Social Action*. In Dewey 1987a: 3–65.

Diamandis, Peter, and Steven Kotler. 2012. *Abundance: Why the Future Is Better Than You Think*. New York, New York: The Free Press.

Diamond, Jared. 1993, paperback. [1992.] *The Third Chimpanzee: The Evolution and Future of the Human Animal*. New York, New York: Harper Perennial.

———. 1997. *Guns, Germans, and Steel: The Fates of Human Societies*. New York, New York: W. W. Norton & Company.

*Diamond, Jeff, prod. 1994. *Are We Scaring Ourselves to Death?* John Stossel, reporter. George Paul, dir. April 21. Television broadcast. American Broadcasting Company (ABC). Videocassette.

Diederichs, Eugen. 1936. *Leben und Werks*. Lulu von Strauß (Strauss) und Torney, ed. Germany: Jena Diederichs.

Diggins, John P. 1972. *Mussolini and Fascism: The View from America*. Princeton, New Jersey: Princeton University Press.

Diner, Steve J. 1998. *A Very Different Age: Americans of the Progressive Era*. New York, New York: Hill and Wang.

Doak, Robin S. 2003. *Calvin Coolidge*. Mankato, Minnesota: Compass Point Books.

Doherty, Brian. 2007. *Radicals for Capitalism: A Freewheeling History of the Modern American Libertarian Movement*. New York, New York: PublicAffairs.

Doll, Richard, and Richard Peto. 1981. "Causes of Cancer: Quantitative Estimates of Avoidable Risks of Cancer in the United States Today." *Journal of the National Cancer Institute*, vol. 66 (no. 6, June): 1191–1308.

Dover, Gabriel. 2000. *Dear Mr. Darwin: Letters on the Evolution of Life and Human Nature*. Berkeley, California: University of California Press.

Dray, Philip. 2005. *Stealing God's Thunder: Benjamin Franklin's Lightning Rod and the Invention of America*. New York, New York: Random House.

Dreher, Robert Edward. 1970. "Arthur de Gobineau: An Intellectual Portrait." Unpublished dissertation. Madison, Wisconsin: University of Wisconsin.
Drumont, Édouard. 1968. *Édouard Drumont ou L'Anticapitalisme Nationale*, Beau de Loménie, ed. Paris, France: J. J. Pauvert.
D'Souza, Dinesh. 2002. *What's So Great About America*. Washington, D.C.: Regnery Publishing.
Du Bois, W. E. B. 1920. *Darkwater: Voices from Within the Veil*. New York, New York: Harcourt Brace.
———. 1923. "Back to Africa." *Century Magazine* (no. 105, February): 539–548.
———. 1935. "A Negro Nation Within a Nation." *Current History* (no. 42, June): 264–270.
———. 1995. *W. E. B. Du Bois: A Reader*. David Levering Lewis, ed. New York, New York: Henry Holt and Company.
———. 2008. [1897.] *The Conservation of the Races*. Hong Kong: Forgotten Books.
du Gay, Paul. 2000. *In Praise of Bureaucracy: Weber, Organization, Ethics*. Thousand Oaks, California: SAGE Publications.
Dunn, Rob. 2011. *The Wild Life of Our Bodies: Predators, Parasites, and Partners That Shape Who We Are Today*. New York, New York: HarperCollins.
Durant, Will. 1961. [1926.] *The Story of Philosophy: The Lives and Opinions of the Great Philosophers of the Western World*. New York, New York: Simon & Schuster.
Durrett, Deanne. 2002. *Jonas Salk*. Detroit, Michigan: Thomson Gale.
D.W.B. 1946. "Review of *Social Darwinism in American Thought, 1860–1915* by R. Hofstadter." *The English Historical Review* 61: 123–24.
Dykes, Nicholas. 1996. "A Tangled Web of Guesses: A Critical Assessment of the Philosophy of Karl Popper." *Philosophical Notes* no. 37: 1–38. London, United Kingdom: Libertarian Alliance. Online as a PDF at http://www.libertarian.co.uk/lapubs/philn/philn037.pdf.
Edison, Thomas Alva. 1914. "Today and Tomorrow." *The Independent*, vol. 77 (no. 3396, Monday, January 5): 24–27.
———. 1968a. [1948.] *The Diary and Sundry Observations of Thomas Alva Edison*. Dagobert David Runes, ed. New York, New York: Greenwood Press.
———. 1968b. [1914.] "Man and Machine." In T. A. Edison 1968a: 161–66.
Edmunds, L. Francis. 1982 2d ed. [1979.] *Rudolf Steiner Education: The Waldorf Schools*. London, United Kingdom: Rudolf Steiner Press.
Ehrlich, Paul R. 1968. *The Population Bomb*. New York, New York: Ballantine Books.
Einstein, Albert. 1993b. [1956.] *Out of My Later Years*. New York, New York: Wing Books.
———. 1993d. [1938.] "Why Do They Hate the Jews?" In Einstein 1993b: 245–53.
———. 2005a. *The New Quotable Einstein*. Alice Calaprice, ed. Princeton, New Jersey: Princeton University Press.
———. 2005d. [1940.] "On Woodrow Wilson (1856–1924)." In Einstein 2005a: 102.
Eisenach, Eldon J., ed. 2006. *The Social and Political Thought of American Progressivism*. Indianapolis, Indiana: Hacket Publishing.
Ekirch, Arthur A., Jr. 1974. *Progressivism in America: A Study of the Era from Theodore Roosevelt to Woodrow Wilson*. New York, New York: New Viewpoints.
Eldredge, Niles, and Ian Tattersall. 1982. *The Myths of Human Evolution*. New York, New York: Columbia University Press.
Ellis, Charles D. 2006. *Joe Wilson and the Creation of Xerox*. Hoboken, New Jersey: John Wiley & Sons.
Ellis, Richard J. 2005. *To the Flag: The Unlikely History of the Pledge of Allegiance*. Lawrence, Kansas: University Press of Kansas.
Ely, Richard T. 1885. "Recent American Socialism." *Johns Hopkins University Studies in Historical and Political Science* third series (no. 4, April): 5–74.
———. 1889. *An Introduction to Political Economy*. New York, New York: Chautauqua Press.
———. 1894a. "Thoughts on Immigration, No. I." *The Congregationalist*. June 28.
———. 1894b. "Thoughts on Immigration, No. II." *The Congregationalist*. June 28.
———. 1898. "Fraternalism vs. Paternalism in Government." *Century* vol. 55 (no. 5): 780–84.
———. 1901. "Social Progress." *Cosmopolitan* vol. 31 (no. 1): 61–64.
———. 1903. *Studies in the Evolution of Industrial Society*. New York, New York: Chautauqua Press.
———. 1914. *Property and Contract in Their Relation to the Distribution of Wealth* vol. 2. New York, New York: Macmillan.
———. 1918. *The World War and Leadership in a Democracy*. New York, New York: Macmillan.
———. 1964. [1886.] "The Doctrine of *Laissez-Faire* Is Unsafe in Politics and Unsound in Morals." In Commager ed. 1964 2d ed.: 335–37.
———. 2006. [1903.] *Studies in the Evolution of Industrial Society* excerpts. In E. J. Eisenach ed. 2006: 50–58.
English, Daylanne. 2000. "W. E. B. Du Bois's Family Crisis." *American Literature* vol. 72, (no. 2, June): 291–319.
Entine, Jon. 2007. *Abraham's Children: Race, Identity, and the DNA of the Chosen People*. New York, New York: Grand Central Publishing.

Esfandiary, F. M. 1970. *Optimism One: The Emerging Radicalism*. New York, New York: W. W. Norton & Company.
———. 1973. *Up-Wingers*. New York, New York: The John Day Company.
Fallon, James. 2013. *The Psychopath Inside: A Neuroscientist's Personal Journey Into the Dark Side of the Brain*. New York, New York: Viking.
Faux, Jeff. 2006. *The Global Class War: How America's Bipartisan Elite Lost Our Future—And What It Will Take to Win It Back*. Hoboken, New Jersey: John Wiley & Sons, Inc.
Faye, Emmanuel. 2009 U.S. version. [2005.] *Heidegger: The Introduction of Nazism Into Philosophy in Light of the Unpublished Seminars of 1933–1935*. Michael B. Smith, trans. New Haven, Connecticut: Yale University Press.
Federal Reserve Bank of Minneapolis. 2014. *Consumer Price Index (Estimate), 1800–*. Minneapolis, Minnesota: The Federal Reserve Bank of Minneapolis. http://www.minneapolisfed.org/community_education/teacher/calc/hist1800.cfm.
Feldman, Anthony, and Peter Ford. 1979. *Scientists & Inventors*. New York, New York: Facts on File.
Ferguson, Niall. 2004, paperback. [2002.] *Empire: The Rise and Demise of the British World Order and the Lessons for Global Power*. New York, New York: Basic Books.
———. 2006. *The War of the World: Twentieth-Century Conflict and the Descent of the West*. New York, New York: The Penguin Press.
———. 2008. *The Ascent of Money: A Financial History of the World*. New York, New York: The Penguin Press.
Feshbach, Murray, and Alfred Friendly, Jr. 1992. *Ecocide in the USSR: Health and Nature Under Siege*. New York, New York: Basic Books.
Fest, Joachim C. 1974 U.S. version. [1973.] *Hitler*. Richard and Clara Winston, trans. San Diego, California: Harcourt Brace Jovanovich.
Fichman, Martin. 2002. *Evolutionary Theory and Victorian Culture*. New York, New York: Humanity Books.
Fine, Sidney. 1956. *Laissez-Faire Thought and the General Welfare State: A Study of Conflict in American Thought, 1865–1901*. Ann Arbor, Michigan: University of Michigan Press.
Fisher, Bruce D., and Michael J. Phillips. 2004 8th ed. *The Legal, Ethical, and Regulatory Environment of Business*. Mason, Ohio: West Legal Studies in Business.
Fisher, Irving. 1909. "Report on National Vitality, Its Wastes and Conservation." *Bulletin of the Committee of One Hundred on National Health*. Washington, D.C.: Government Printing Office.
Fiske, John. 1940. *The Letters of John Fiske*. Ethel F. Fisk, ed. New York, New York: Macmillan.
*Folsom, Burton W., Jr. 1996 3d ed. [Originally published in 1987 as *Entrepreneurs vs. the State*.] *The Myth of the Robber Barons*. Herndon, Virginia: Young America's Foundation.
*———. 1998. *Empire Builders: How Michigan Entrepreneurs Helped Make America Great*. Traverse City, Michigan: Rhodes & Easton.
*———. 2008. *New Deal or Raw Deal?: How FDR's Economic Legacy Has Damaged America*. New York, New York: Threshold Editions.
*Folsom, Burton W., Jr., ed. 1996. *The Industrial Revolution and Free Trade*. Irvington-on-Hudson, New York: Foundation for Economic Education.
Fosdick, Raymond B. 1956. *John D. Rockefeller, Jr.: A Portrait*. New York, New York: Harper.
Frank, Robert H. 2011. *The Darwin Economy: Liberty, Competition, and the Common Good*. Princeton, New Jersey: Princeton University Press.
Frankfurter Zeitung. 1966a. [October 15, 1939.] "A Butcher Resists." In G. L. Mosse ed. 1966: 364.
———. 1966b. [January 11, 1939.] "The Price Police." In G. L. Mosse ed. 1966: 363–64.
Fraser, Steve. 2005. *Every Man a Speculator: A History of Wall Street in American Life*. New York, New York: HarperCollins.
Freund. 1991. [1956.] "Mr. Holmes Had Cure for Most of Our Ills." In E. Bander ed. 1991: 228.
Friedman, David. 1989 2d ed., paperback. [1973.] *The Machinery of Freedom: Guide to a Radical Capitalism*. La Salle, Illinois: Open Court Publishing Company.
Friedman, Milton, and Rose Friedman. 1990. [1980.] *Free to Choose: A Personal Statement*. San Diego, California: Harcourt Brace Jovanovich.
Fritzsche, Peter. 1992. "Weimar Populism and National Socialism in Local Perspective." In L. E. Jones and Retallack eds. 1992: 287–306.
Fukuyama, Francis. 1992. *The End of History and the Last Man*. New York, New York: The Free Press.
———. 1999. *The Great Disruption: Human Nature and the Reconstitution of Social Order*. New York, New York: The Free Press.
———. 2002. *Our Posthuman Future: Consequences of the Biotechnology Revolution*. New York, New York: Farrar, Straus and Giroux.
———. 2011. *The Origins of Political Order: From Prehuman Times to the French Revolution*. New York, New York: Farrar, Straus and Giroux.
Fumento, Michael. 1996. "Cancer Charge Against Milk Udderly Ridiculous." Website of Michael Fumento. Arlington, Virginia: Michael Fumento. http://www.fumento.com/milk.html.

———. 1999. "Samuel Epstein: Science Meets the X-Files." *National Post*. May 8. Republished on the website of Michael Fumento. Arlington, Virginia: Michael Fumento. http://www.fumento.com/epames.html.

———. 2004. "The 'Cancer Epidemic' That Never Was." *TCS Daily*. June 23. Nick Schulz, editor-in-chief. Washington, D.C.: Tech Central Station. http://www.fumento.com/disease/cancer.html.

Furley, David J. 1989. "Aristotelian Material in Cicero's *De Natura Deorum*." In Fortenbaugh and Steinmetz eds. 1989: 201–219.

Furner, Mary. 1975. *Advocacy and Objectivity: A Crisis in the Professionalization of American Social Science, 1865–1905*. Lexington, Kentucky: University Press of Kentucky.

Gabriel, R. H. 1956. [1940.] *The Course of American Democratic Thought*. New York, New York: The Ronald Press Company.

Galbraith, John Kenneth. 1973. *A China Passage*. Boston, Massachusetts: Houghton Mifflin Company.

———. 1987. *Economics in Perspective: A Critical History*. Boston, Massachusetts: Houghton Mifflin Company.

———. 2002. "Stop the Madness." *Globe and Mail*. Interview conducted by Rupert Cornwell. July 6. Toronto, Canada: Globe and Mail, Inc.

Galton, Francis. 1873. "Hereditary Improvement." *Every Saturday: A Journal of Choice Reading*, vol. 3 (no. 8, Saturday, February 22): 207–213.

———. 1883. *Inquiries Into Human Faculty and Its Development*. New York, New York: Macmillan.

———. 1908. *Memoirs of My Life*. London, United Kingdom: Methuen.

Gandhi, Mohandas K. 1960. *Trusteeship*. Ravindra Kelekar, ed. Ahmedabad, India: Navajivan Mudranalaya. Republished on the *Trusteeship* website. Ahmedabad, India: Bombay Sarvodaya Mandal/Gandhi Book Centre. http://www.mkgandhi.org/trusteeship/trusteeship.htm.

Gann, Marjorie, and Janet Willen. 2011. *Five Thousand Years of Slavery* . Toronto, Canada: Tundra Books.

Gardner, Chris, et al. 2006, paperback. [2006.] *The Pursuit of Happyness*. New York, New York: HarperCollins.

Gardner, Howard. 1995. *Leading Minds: An Anatomy of Leadership*. New York, New York: Basic Books.

Garraty, John A., ed. 1968. *The Transformation of American Society, 1870–1890*. Columbia, South Carolina: University of South Carolina Press.

Gartner, John D. 2005. *The Hypomanic Edge: The Link Between (a Little) Craziness and (a Lot of) Success in America*. New York, New York: Simon & Schuster.

Gay, Peter. 1969. *The Enlightenment: An Interpretation* vol. 2: *The Science of Freedom*. New York, New York: Alfred A. Knopf.

Gay, Peter, and R. K. Webb. 1973. *Modern Europe*. New York, New York: Harper & Row.

Gazzaniga, Michael S. 2011. *Who's in Charge?: Free Will and the Science of the Brain*. New York, New York: HarperCollins.

Geisst, Charles R. 2004. [1997.] *Wall Street: A History—From Its Humble Beginnings to the Fall of Enron*. New York, New York: Oxford University Press.

Gellately, Robert. 2001. *Backing Hitler: Consent and Coercion in Nazi Germany*. New York, New York: Oxford University Press.

———. 2007. *Lenin, Stalin, and Hitler: The Age of Social Catastrophe*. New York, New York: Alfred A. Knopf.

George, Henry. 2005. [1879.] *Progress and Poverty: An Inquiry in the Cause of Industrial Depressions and of Increase of Want with Increase of Wealth—The Remedy*. New York, New York: Cosimo.

Ghent, William J. 1902. *Our Benevolent Feudalism*. New York, New York: The Macmillan Co.

Giblin, James Cross. 2002. *The Life and Death of Adolf Hitler*. New York, New York: Clarion Books.

Gillespie, Nick, ed. and debate moderator. "Who's Afraid of Human Enhancement?: A *Reason* Debate on the Promise, Perils, and Ethics of Human Biotechnology." *Reason*, vol. 37 (no. 8, January). Republished on *Reason Online*. Washington, D.C.: Reason Foundation. http://www.reason.com/news/show/33064.html.

Glover, David, prod. 2003. "Pandora's Box." *DNA*, Episode 5 of 5. Jeff Goldblum, narrator. David Glover, dir. Sunday, February 1. Television broadcast. Public Broadcasting Service (PBS).

Glynn, Kathleen, et al., prod. 1997. *The Big One*. Michael Moore, dir. Michael Moore, writer. Michael Moore, host. Motion picture. Dog Eat Dog Productions.

Godwin, William. 1820. *Of Population*. London, United Kingdom: Longman, Hurst, Rees, Orme, & Brown.

Goebbels, Joseph. 1929. *Michael: Ein Deutsches Schicksal* (*Michael: A German Fate*) in *Tagebuchblätten* (*Diaries*). Munich, Germany: Zentralverlag der NSDAP (Central Publishing of the National Socialist German Workers Party), Frz. Eher Nachf.

———. 1966. [1929.] "Michael: A German Fate." In G. L. Mosse ed. 1966: 104–112.

———. 1987. [1929.] *Michael: A Novel*. Joachim Neugroschel, trans. New York, New York: Amok Press.

Goldberg, Jonah. 2007. *Liberal Fascism: The Secret History of the American Left from Mussolini to the Politics of Meaning*. New York, New York: Doubleday.

Goldman, Eric F. 2001. [1956.] *Rendezvous With Destiny: A History of Modern American Reform*. New York, New York: Ivan R. Dee.

Goodall, Jane. 1971. *In the Shadow of Man*. New York, New York: Dell Publishing Company.

———. 1986. *Chimpanzees of the Gombe: Patterns of Behavior*. Cambridge, Massachusetts: The Belknap Press of Harvard University Press.

Gooding-Williams, Robert. 1996. "Outlaw, Appiah, and Du Bois's 'The Conservation of Races.'" In B. W. Bell et al. eds. 1996: 39–56.

———. 2010. *In the Shadow of Du Bois: Afro-American Political Thought in America*. Cambridge, Massachusetts: Harvard University Press.

Goodrick-Clarke, Nicholas. 1992, paperback. [1985.] *The Occult Roots of Nazism: Secret Aryan Cults and Their Influence on Nazi Ideology—The Ariosophists of Austria and Germany, 1890–1935*. Washington Square, New York: New York University Press.

Goran, Morris. 1967. *The Story of Fritz Haber*. Norman, Oklahoma: University of Oklahoma Press.

Gordon, John Steele. 2004. *An Empire of Wealth: The Epic History of American Economic Power*. New York, New York: HarperCollins.

———. 1999. *The Great Game: The Emergence of Wall Street as a World Power, 1653–2000*. New York, New York: Scribner.

Gottfredson, Linda, et al. 1994. "Mainstream Science on Intelligence: An Editorial With 52 Signatories, History, and Bibliography." *Wall Street Journal*. December 13. Republished online as a PDF on Dr. Linda Gottfredson's University of Delaware website at http://www.udel.edu/educ/gottfredson/reprints/1997mainstream.pdf.

Gottlieb, Agnes Hooper, et al. 1998. *1,000 Years, 1,000 People: Ranking the Men and Women Who Shaped the Millennium*. New York, New York: Kodansha America.

Gould, Lewis L., ed. 1974. *The Progressive Era*. Syracuse, New York: Syracuse University Press.

Graber, David M. 1989. "Mother Nature as a Hothouse Flower." *Los Angeles Times Book Review* (October 22): 1, 9. Republished on the website of the *Los Angeles Times*. Los Angeles, California: The Los Angeles Times Media Group. http://articles.latimes.com/print/1989-10-22/books/bk-726_1_bill-mckibben.

Grant, Madison. 1918 revised ed. [1916.] *The Passing of the Great Race; or The Racial Basis of European History*. New York, New York: Charles Scribner's Sons.

Gray, John. 1997. *Enlightenment's Wake: Politics and Culture at the Close of the Modern Age*. London, United Kingdom: Routledge.

———. 2013. "Communism, Fascism and Liberals Now." *The Times Literary Supplement* (no. 5727, Friday, January 4): 7. Published online Wednesday, January 2, 2013 on the *Times Literary Supplement* website. London, United Kingdom: The Times Literary Supplement. http://www.the-tls.co.uk/tls/public/article1186584.ece.

Greenberg, Jonathan. 1990. *Staking a Claim: Jake Simmons and the Making of an African-American Oil Dynasty*. New York, New York: Atheneum.

Gregg, Alan. 1955. "A Medical Aspect of the Population Problem." *Science* vol. 121 (no. 3150, May 13): 681-82. First page republished on official website of *Science*. Washington, D.C.: American Association for the Advancement of Science (AAAS). http://www.sciencemag.org/content/121/3150/681.extract.

Gregor, A. James. 1979. *Young Mussolini and the Intellectual Origins of Fascism*. Berkeley, California: University of California Press.

Gross, Michael. 2003. *Genuine Authentic: The Real Life of Ralph Lauren*. New York, New York: HarperCollins.

Gross, Paul R., et al., eds. 1996, trade paperback. *The Flight from Science and Reason*. Baltimore, Maryland: The Johns Hopkins University Press.

Grosser, Morton. 1978. *Diesel: The Man and His Engine*. New York, New York: Atheneum.

"Group Blasts PETA 'Holocaust' Project." 2003. *CNN.Com*. Friday, February 28. http://www.cnn.com/2003/US/Northeast/02/28/peta.holocaust/.

Gruber, Howard E. 1974. *Darwin on Man: A Psychological Study of Scientific Creativity, Together With Darwin's Early and Unpublished Notebooks*. Paul H. Barrett, transcribed and annotated. New York, New York: E. P. Dutton & Co., Inc.

Grunberger, Richard. 1971. *The 12-Year Reich: A Social History of Nazi Germany, 1933–1945*. New York, New York: Holt, Rinehart and Wilson.

Günther, Hans K. 1966. [1929.] "The Nordic Race as Ideal Type." In G. L. Mosse ed. 1966: 61-65.

Gurlitt, Ludwig. 1906. Der Deutsche und sein Vaterland (The German and His Fatherland). Berlin, Germany: Wiegandt & Grieben.

Haeckel, Ernst. 1900. *The Riddle of the Universe, at the Close of the Nineteenth Century*. Joseph McCabe, trans. New York, New York: Harper & Brothers.

———. 1904. *The Wonders of Life: A Popular Study of Biological Philosophy*. Joseph McCabe, trans. New York, New York: Harper & Brothers.

Hager, Thomas. 2008. *The Alchemy of Air: A Jewish Genius, a Doomed Tycoon, and the Scientific Discovery That Fed the World But Fueled the Rise of Hitler*. New York, New York: Harcourt Books.

Halpin, John, and Conor P. Williams. 2010. *The Progressive Tradition Series* Pt. 1: *The Progressive Intellectual Tradition in America*. April. Washington, D.C.: Center for American Progress. http://www.americanprogress.org/wp-content/uploads/issues/2010/04/pdf/progressiveintellectualism_execsumm.pdf.

Hardach, Karl. 1980. *The Political Economy of Germany in the Twentieth Century*. Berkeley, California: University of California Press.

Hardin, Garrett. 1951. [1949.] *Biology: Its Human Implications*. San Francisco, California: W. H. Freeman and Company.

———. 1963. "A Second Sermon on the Mount." *Perspectives in Biology and Medicine*, vol. 6 (no. 3, Spring): 366–371.

Hardt, Michael, and Antonio Negri. *Empire*. 2000. Cambridge, Massachusetts: Harvard University Press.

Harford, Tim. 2008. *The Logic of Life: The Rational Economics of an Irrational World*. New York, New York: Random House.

Harrell, David Edwin, Jr., et al. 2005. *Unto a Good Land: A History of the American People* (2 vols.). Grand Rapids, Michigan: Wm. B. Eerdmans Publishing Co.

**Harriman, David. 2008b. "Proof of the Atomic Theory." *The Objective Standard*, vol. 3 (no. 2, Summer): 67–97. Republished online on the *Objective Standard* website. Craig Biddle, ed. Glen Allen, Virginia: The Objective Standard. http://tinyurl.com/5zhert.

**———. 2010. *The Logical Leap: Induction in Physics*. New York, New York: The New American Library.

Harris, Abraham L. 1942. "Sombart and German (National) Socialism." *The Journal of Political Economy* vol. 50 (no. 6, December): 805–835.

Harris, Eric. 1998–1999a. Journal. In C. Shepard ed. 1999–2008: http://acolumbinesite.com/eric/writing/journal/journal.html.

———. 1998–1999b. "Plan." In C. Shepard ed. 1999–2008: http://acolumbinesite.com/eric/writing/plans.gif.

———. 1998–1999c. "REB's Wisdom." In C. Shepard ed. 1999–2008: http://acolumbinesite.com/wisdom2.html.

Harris, Marvin. 1985. *Good to Eat: Riddles of Food and Culture*. New York, New York: Simon & Schuster.

Hart, Harold H., ed. 1973. *Population Control: For and Against*. New York, New York: Hart Publishing Company, Inc.

Hartshorn, Peter. 2011. *I Have Seen the Future: A Life of Lincoln Steffens*. Berkeley, California: Counterpoint Press.

**———. 2014. *The Freedom of Peaceful Action: On the Origin of Individual Rights*. Lanham, Maryland: Lexington Books.

Hayek, F. A. 1988. *The Collected Works of F. A. Hayek* vol. 1, *The Fatal Conceit: The Errors of Socialism*. W. W. Bartley III, ed. Chicago, Illinois: The University of Chicago Press.

Hazlitt, Henry. 1993a. "The Case for the Minimal State." In H. Hazlitt 1993d: 98–108.

———. 1993b. [1979.] "From Spencer's 1884 to Orwell's 1984." In H. Hazlitt 1993d: 174–182.

———. 1993d. *The Wisdom of Henry Hazlitt*. Hans F. Sennholz, ed. Irvington-on-Hudson, New York: Foundation for Economic Education.

Heiber, Helmut. 1972. *Goebbels*. John K. Dickinson, trans. New York, New York: Hawthorne Books.

Heilbroner, Robert L. 1986 6th ed. [1953.] *The Worldly Philosophers: The Lives, Times, and Ideas of the Great Economic Thinkers*. New York, New York: Simon & Schuster.

———. 1999 7th ed. paperback. [1953.] *The Worldly Philosophers: The Lives, Times, and Ideas of the Great Economic Thinkers*. New York, New York: Touchstone.

Heins, Marjorie. 1993. *Sex, Sin, and Blasphemy: A Guide to America's Censorship Wars*. New York, New York: The New Press.

Henderson, Charles Richmond. 1896. "Business Men and Social Theorists." *American Journal of Sociology*, vol. 1 (no. 4, January): 385–397.

———. 1900. "Science in Philanthropy." *Atlantic Monthly* vol. 85 (no. 508): 249–254.

———. 1909. "Are Modern Industry and City Life Unfavorable to the Family?" *American Economic Association Quarterly* 10: 217–232.

Henderson, William O. 1983. *Friedrich List—Economist and Visionary, 1789–1846*. London, United Kingdom: Frank Cass.

Henry, William Wirt. 1891. *Patrick Henry: Life, Correspondence, and Speeches* (3 vols.). New York, New York: Charles Scribner's Sons.

Herman, Arthur. 1997. *The Idea of Decline in Western History*. New York, New York: The Free Press.

Hern, Warren M. 1993. "Is Human Culture Carcinogenic for Uncontrolled Population Growth and Ecological Destruction?" *BioScience* vol. 43 (no. 11, December): 768–773. Republished as a PDF on Warren M. Hern's website. Boulder, Colorado: Warren M. Hern. http://www.drhern.com/pdfs/culturecarc93.pdf.

Herrnstein, Richard J., and Charles Murray. 1996, trade paperback. [1994.] *The Bell Curve: Intelligence and Class Structure in American Life*. New York, New York: The Free Press.

Herschbach, Dudley R. 1996. "Imaginary Gardens With Real Toads." In P. R. Gross et al. eds. 1996: 11–30.

Herzlinger, Regina. 2007. *Who Killed Health Care?: America's $2 Trillion Medical Problem—And the Consumer-Driven Cure*. New York, New York: McGraw-Hill.

Herzog, Hal. 2010. *Some We Love, Some We Hate, Some We Eat: Why It's So Hard to Think Straight About Animals*. New York, New York: Harper.

Hession, Charles H. 1984. *John Maynard Keynes: A Personal Biography of the Man Who Revolutionized Capitalism and the Way We Live*. New York, New York: Macmillan.

Higham, John. 1967. *Strangers in the Land: Patterns of American Nativism, 1860–1925*. New York, New York: Atheneum.

*Hill, James J. 1910. *Highways of Progress*. New York, New York, Doubleday.

Himmelfarb, Gertrude. 1959. *Darwin and the Darwinian Revolution*. London, United Kingdom: Chatto and Windus.

Hinks, Peter, et al., eds. 2007. *Encyclopedia of Antislavery and Abolition* vol. 2: *J-Z*. Westport, Connecticut: Greenwood Press.

Hirsen, James. 2003. *Tales from the Left Coast: True Stories of Hollywood Stars and Their Outrageous Politics*. New York, New York: Crown Forum.

Hitler, Adolf. 1942. *The Speeches of Adolf Hitler, 1922–1939* vol. 1 of 2. N. H. Baynes, ed. New York, New York: Oxford University Press.

———. 1966. [1923.] "Education, Instinct, and Will." In G. L. Mosse ed. 1966: 10.

———. 2000a. *Hitler's Table Talk*. Hugh R. Trevor-Roper, ed. New York, New York: Enigma Books.

———. 2000b. "1942—11 April, 1942, at Dinner." In Hitler 2000a: 318–321.

———. 2001. [1924.] *Mein Kampf*. Ralph Manheim, trans. Boston, Massachusetts: Mariner Books.

———. 2003. *Hitler's Second Book: The Unpublished Sequel to "Mein Kampf."* Gerald L. Weinberg, ed. Krista Smith, trans. New York, New York: Enigma Books.

———. 2012. [1924.] *Mein Kampf*. James Murphy, trans. New York, New York: Hurst & Blackett.

Hobson, John M. 2004. *The Eastern Origins of Western Civilization*. New York, New York: Cambridge University Press.

Hochschild, Adam. 2005. *Bury the Chains: Prophets and Rebels in the Fight to Free an Empire's Slaves*. Boston, Massachusetts: Houghton Mifflin Company.

Hodgson, Adam. 1823 2d ed. *A Letter to M. Jean-Baptiste Say on the Comparative Expense of Free and Slave Labor*. Liverpool, United Kingdom: Hatchard & Son.

Hodgson, Geoffrey M. 2004. "Social Darwinism in Anglophone Academic Journals: A Contribution to the History of the Term." *Journal of Historical Sociology* vol. 17 (no. 4, December): 428–463.

Hoffman, Bruce. 2006, paperback. [1998.] *Inside Terrorism*. New York, New York: Columbia University Press.

Hoffman, Frederick L. 1915. *Mortality From Cancer Throughout the World*. Newark, New Jersey: Prudential Press.

Hofstadter, Richard. 1955. *The Age of Reform: From Bryan to FDR* . New York, New York: Vintage.

———. 1959. [1944.] *Social Darwinism in American Thought*. New York, New York: George Braziller.

———. 1992. [1944.] *Social Darwinism in American Thought*. Ypsilanti, Michigan: Beacon Press.

Hofstadter, Richard, et al. 1972 3d ed. [1957.] *The United States*. Englewood Cliffs, New Jersey: Prentice-Hall.

*Hogberg, David, and Sarah Haney. 2006. "Funding Liberalism With Blue-Chip Profits: *Fortune* 100 Foundations Back Leftist Causes." *Foundation Watch* (August): 1–8. Matthew Vadum, editor. Republished on the website of the Capital Research Center. Washington, D.C.: Capital Research Center. PDF at http://www.capitalresearch.org/pubs/pdf/v1185995021.pdf.

Holbrook, Stewart Hall. 1957. *Dreamers of the American Dream*. Garden City, New York: Doubleday and Company.

Hollander, Paul. 1981. *Political Pilgrims: Western Intellectuals in Search of the Good Society*. New Brunswick, New Jersey: Transaction Publishers.

———. 2003. [1993.] *Anti-Americanism: Irrational and Rational*. New Brunswick, New Jersey: Transaction Publishers.

Holmes, Oliver Wendell, Jr. 1873. "The Gas-Stokers' Strike." *American Law Review* 7: 582–84.

———. 1988. [1873.] " 'Masters and Men': The Gas-Stokers' Strike." In M. Lerner ed. 1988: 48–51.

Horan, James D. 1967. *The Pinkertons: The Detective Dynasty That Made History*. New York, New York: Crown Publishers.

Howe, Mark De Wolfe, ed. 1942. *Holmes-Pollock Letters: The Correspondence of Mr. Justice Holmes and Sir Frederick Pollock 1874–1932* (2 vols.). Cambridge, Massachusetts: Harvard University Press.

Hrdy, Sarah Blaffer. 1981. *The Woman That Never Evolved*. Cambridge, Massachusetts: Harvard University Press.

Huber, Peter W. 1994. *Orwell's Revenge: The* 1984 *Palimpsest*. New York, New York: The Free Press. Republished on the website of Peter Huber. New York, New York: Peter W. Huber. http://web.archive.org/web/20050404032459/http://www.phuber.com/huber/orwell/orw_rev.pdf.

**Hull, Gary, ed. 2005. *The Abolition of Antitrust*. New Brunswick, New Jersey: Transaction Publishers.

Humboldt, Wilhelm von. 1854. [1792.] *The Sphere and Duties of Government*. Joseph Coulthard, ed. London, United Kingdom: John Chapman. Republished in the *Online Library of Liberty*. Indianapolis, Indiana: Liberty Fund. http://oll.libertyfund.org/index.php?option=com_staticxt&staticfile=show.php%3Ftitle=589&Itemid=99999999.

———. 1969. [1792.] *The Limits of State Action*. J. W. Burrow, ed. New York, New York: Cambridge University Press.

Huxley, Aldous. 1958. *Brave New World Revisited*. New York, New York: Harper & Row.

———. 1962 paperback. [1936.] *Brave New World*. New York, New York: Bantam Books.

Huxley, Julian. 1957. *New Bottles for New Wine*. London, United Kingdom: Chatto & Windus.

———. 2004. "Transhumanism." Published on the website of the World Transhumanist Association. Posted by J. Hughes. August 27. Willington, Connecticut: World Transhumanist Association. http://www.transhumanism.org/index.php/WTA/more/huxley/.

Inaugural Addresses of the Presidents of the United States: From George Washington to Barack Obama. 2008. [1989.] Washington, D.C.: U.S. Government Printing Office. Republished on *Bartleby.Com*. New York, New York: Bartleby.Com, Inc. http://www.bartleby.com/br/124. html.

Irvine, William. 1959. *Apes, Angels, and Victorians: Darwin, Huxley, and Evolution*. Cleveland, Ohio: Meridian Books.

Isle, Mick. 2006. *Aristotle: Pioneering Philosopher and Founder of the Lyceum*. New York, New York: The Rosen Publishing Group.

Jackman, Simon. 1996. "Liberalism, Public Opinion, and Their Critics: Some Lessons for Defending Science." In P. Gross et al. eds. 1996, trade paperback: 346–368.

Jackson, Robert B. 1968. *The Gasoline Buggy of the Duryea Brothers*. New York, New York: Henry Z. Walck.

Jacobs, Jane. 2000. *The Nature of Economies*. New York, New York: The Modern Library.

Jacoby, Susan. 2004. *Freethinkers: A History of American Secularism*. New York, New York: Metropolitan Books.

———. 2008. *The Age of American Unreason*. New York, New York: Pantheon Books.

James, D., ed. 1972. *Outside, Looking In*. New York, New York: Harper & Row.

Jareski, Eugene. 2010. *The American Way of War: Guided Missiles, Misguided Men, and a Republic in Peril*. New York, New York: Simon & Schuster.

Jarrett, H., ed. 1958. *Perspectives on Conservation*. Baltimore, Maryland: Johns Hopkins Press.

Jefferson, Thomas. 1829a. [1825.] "Letter 188: To William B. Giles—Monticello, December 26, 1825." In T. Jefferson 1829b vol. 4: 421–23.

———. 1829b. *Memoir, Correspondence, and Miscellanies from the Papers of Thomas Jefferson* (4 vols.). Thomas Jefferson Randolph, ed. Charlottesville, Virginia: F. Carr and Co.

Johnson, George. 2013. *The Cancer Chronicles: Unlocking Medicine's Deepest Mystery*. New York, New York: Vintage.

Jones, James H. 1997. *Alfred C. Kinsey: A Public/Private Life*. New York, New York: W. W. Norton & Company.

Jones, Larry Eugene, and James Retallack, eds. 1992. *Elections, Mass Politics, and Social Change in Modern Germany*. New York, New York: Press Syndicate of the University of Cambridge.

———. 1993. *Between Reform, Reaction, and Resistance: Studies in the History of German Conservatism from 1789 to 1945*. Providence, Rhode Island: Berg.

Kaczynski, Theodore. 1995. *Industrial Society and Its Future*. Republished on the website of *The Washington Post*. Washington, D.C.: The Washington Post Company. http://www. washingtonpost.com/wp-srv/national/longterm/unabomber/manifesto.text.htm.

———. 2010a. "Letters to David Skrbina." In Kaczynski 2010b: 254–359.

———. 2010b. [2008.] *Technological Slavery: The Collected Writings of Theodore J. Kaczynski, a.k.a., "The Unabomber."* David Skrbina. Port Townsend, Washington: Feral House.

———. 2010c. "The Truth About Primitive Life: A Critique of Anarcho-Primitivism." In Kaczynski 2010a: 126–189.

Kaku, Michio. 1997. *Visions: How Science Will Revolutionize the 21st Century*. New York, New York: Doubleday.

Karier, Clarence J., ed. 1975. *Shaping the American Educational State: 1900 to the Present*. New York, New York: The Free Press.

Kasun, Jacqueline. 1988. *The War Against Population: The Economics and Ideology of World Population Control*. San Francisco: Ignatius Press.

Kater, Michael H. 2004. *Hitler Youth*. Cambridge, Massachusetts: Harvard University Press.

Kaye, Howard. 1986. *The Social Meaning of Modern Biology*. New Haven, Connecticut: Yale University Press.

*Kealey, Terence. 1996, U.S. paperback. [1996 United Kingdom version.] *The Economics of Scientific Research*. New York, New York: St. Martin's Press.

*Keeley, Lawrence H. 1997, paperback. [1996.] *War Before Civilization*. New York, New York: Oxford University Press.

*Keller, Albert Galloway, and Maurice R. Davie, eds. 1969. *Essays of William Graham Sumner* (2 vols.). Hamden, Connecticut: Archon Books.

Kelly, Alfred. 1981. *The Descent of Darwin: The Popularization of Darwinism in Germany, 1860–1914*. Chapel Hill, North Carolina: The University of North Carolina Press.

Kelly, Kevin. 1994. *Out of Control: The Rise of Neo-Biological Civilization*. Reading, Massachusetts: Addison-Wesley Publishing Company.

———. 2010. *What Technology Wants*. New York, New York: Viking.

Kelly, Marion. 1956. "Changes in Land Tenure in Hawaii, 1778–1850." Master of Arts thesis at the University of Hawai'i at Manoa: 1–49.

Kennedy, James G. 1978. *Herbert Spencer*. Boston, Massachusetts: Twayne Publishers.

Kennedy, Robert F., Jr. 2005, paperback. [2004.] *Crimes Against Nature: How George W. Bush and His Corporate Pals Are Plundering the Country and Hijacking Our Democracy*. New York, New York: Harper Perennial.

Kern, Louis J. 1994. "'Slavery Recedes But the Prejudice to Which It Has Given Birth Is Immovable': Beaumont and Tocqueville Confront Slavery and Racism in Ante-Bellum America and Orléanist France." In W. L. Chew ed. 1994: 143–186.

*Kershaw, Ian. 1983. *Popular Opinion and Political Dissent in the Third Reich: Bavaria, 1933–1945*. New York, New York: Clarendon Press/Oxford University Press.

*———. 2000 paperback. [1998.] *Hitler, 1889–1936: Hubris*. New York, New York: W. W. Norton and Company.

Kevles, Daniel J. 1985. *In the Name of Eugenics: Genetics and the Use of Human Heredity*. New York, New York: Alfred A. Knopf.

Keynes, John Maynard. 1920. *The Economic Consequences of the Peace*. New York, New York: Harcourt, Brace and Howe.

———. 1921. *A Treatise on Probability*. New York, New York: Macmillan.

———. 1964. [1953.] *General Theory of Employment, Interest, and Money*. Orlando, Florida: Harcourt.

———. 2009a. [1926.] *The End of Laissez-Faire*. In Keynes 2009b: 13-46.

———. 2009b. *The End of Laissez-Faire: The Economic Consequences of the Peace*. Amherst, New York: Prometheus Books.

———. 2013. [1953.] *General Theory of Employment, Interest, and Money*. Edison Martin Imprint.

Kidd, Benjamin. 1894. *Social Evolution*. London, United Kingdom: Macmillan & Co.

King-Hele, Desmond. 1999. *Erasmus Darwin: A Life of Unequalled Achievement*. London, United Kingdom: Giles de la Mare.

Klemperer, Victor. 1998. [1995.] *I Will Bear Witness: A Diary of the Nazi Years, 1933–1941*. Martin Chalmers, trans. and ed. New York, New York: Random House.

Knaebel, Ernest, ed. 1918a. "*Buchanan v. Warley*: Error to the Court of Appeals of the State of Kentucky. 245 U.S. 60." In E. Knaebel ed. 1918b vol. 245: 60–82.

———. 1918b. *United States Reports* vol. 245: *Cases Adjudged in the Supreme Court at October Term, 1917, From October 1, 1917, to March 4, 1918*. New York, New York: The Banks Law Publishing Co.

Knapp, Alistair, prod. 2011. "H. G. Wells." *Prophets of Science Fiction*, episode 3. November 23. Declan Whitebloom, dir. Jonathan Adams, narrator. Cable television broadcast: The Science Channel.

Kolko, Gabriel. 1963. *The Triumph of Conservatism: A Reinterpretation of American History, 1900–1916*. New York, New York: The Free Press of Glencoe.

Konkin, Samuel Edward, III. 2008. *An Agorist Primer: Counter-Economics, Total Freedom, and You*. Huntington Beach, California: KoPubCo.

Kopel, David, and Richard Griffiths. 2003. "Hitler's Control: The Lessons of Nazi History." *National Review Online*. May 22. Rich Lowry, editor-in-chief. New York, New York: National Review. http://www.nationalreview.com/kopel/kopel052203.asp.

Kotzin, Daniel P. 2007. "Tocqueville, Alexis de (1805–1859)." In P. Hinks et al. ed. 2007 vol. 2: 680–81.

Krass, Peter. 2002. *Carnegie*. Hoboken, New Jersey: John Wiley & Sons.

*Krauss, Melvyn B. 1983 paperback. [1981.] *Development Without Aid: Growth, Poverty, and Government*. Lanham, Maryland: University Press of America.

Kühl, Stefan. 2002. *The Nazi Connection: Eugenics, American Racism, and German National Socialism*. New York, New York: Oxford University Press.

Kurland, Philip B., and Ralph Lerner, eds. 2000. *The Founders' Constitution* (10 vols.). Chicago, Illinois: The University of Chicago Press. Republished online at http://press-pubs.uchicago.edu/founders/.

Lacey, Robert. 1986. *Ford: The Men and the Machine*. Boston, Massachusetts: Little Brown.

Lambert, Tim, prod. 2009. *What Darwin Didn't Know*. Armand Leroi, host. Tim Lambert, dir. Television broadcast. British Broadcasting Corporation.

La Monte, Robert Rivers, and H. L. Mencken, eds. 1910. *Men Versus the Man: A Correspondence Between Robert Rivers La Monte, Socialist, and H. L. Mencken, Individualist*. New York, New York: Henry Holt and Company.

Langman, Peter. 2009. *Why Kids Kill: Inside the Minds of School Shooters*. New York, New York: Palgrave Macmillan.

Laquer, Walter, ed. 1976. *Fascism: A Reader's Guide*. Berkeley, California: University of California Press.

Larkin, Jack. 1988. *The Reshaping of Everyday Life, 1790–1840*. New York, New York: Harper & Row.

Lears, Jackson. 1981. *No Place of Grace: Antimodernism and the Transformation of American Culture, 1880–1920*. New York, New York: Pantheon Books.

LeDrew, Stephen. 2013. "Scientism, Humanism, and Religion: The New Atheism and the Rise of the Secular Movement." Ph.D. dissertation. Toronto, Canada: York University. Republished as a PDF on the York University website. Toronto, Canada: York University. http://yorkspace.library.yorku.ca/xmlui/bitstream/handle/10315/27589/LeDrew_Stephen_H_2013_PhD.pdf.

Lee, Albert. 1980. *Henry Ford and the Jews*. New York, New York: Stein and Day.

Lee, J. J. 1978. "Labour in German Industrialization." In Mathias and Postan eds. 1978: 442–491.

Legal Information Institute. 2010. *U.S. Code*. Ithaca, New York: Cornell University Law School. http://www.law.cornell.edu/uscode/.

Lenard, Philip. 1936. *Deutsche Physik* vol. 1: *Einleitung und Mechanik* (*German Physics* vol. 1: *Introduction and Mechanics*). Munich, Germany: Lehmanns Verlag (Lehmann's Publishing).

———. 1966. [1936.] "The Limits of Science." In G. L. Mosse ed. 1966: 201–05.

*Leonard, Thomas C. 2003. "'More Merciful and Not Less Effective': Eugenics and American Economics in the Progressive Era." *History of Political Economy*, vol. 35 (no. 4, Winter): 709–734. Republished as a PDF on Thomas C. Leonard's university website. Princeton, New Jersey: Princeton University. http://www.princeton.edu/~tleonard/papers/Eugenics.pdf.

*———. 2005a. "Mistaking Eugenics for Social Darwinism: Why Eugenics Is Missing from the History of American Economics." *History of Political Economy*, vol. 37 (supplement): 200-233. Republished as a PDF on Thomas C. Leonard's website. Princeton, New Jersey: Princeton University. http://www.princeton.edu/~tleonard/papers/mistaking.pdf.

*———. 2005b. "Retrospectives: Eugenics and Economics in the Progressive Era." *Journal of Economic Perspectives*, vol. 19 (no. 4, Fall): 207–224. Republished as a PDF on Thomas C. Leonard's website. Princeton, New Jersey: Princeton University. http://www.princeton.edu/~tleonard/ papers/retrospectives.pdf.

*———. 2009. "Origins of the Myth of Social Darwinism: The Ambiguous Legacy of Richard Hofstadter's *Social Darwinism in American Thought*." *Journal of Economic Behavior & Organization*, no. 71: 37–51. Republished as a PDF on Thomas C. Leonard's university website. Princeton, New Jersey: Princeton University. http://www.princeton.edu/~tleonard/papers/myth.pdf.

Lerner, Max, ed. 1988. [1943.] *The Mind and Faith of Justice Holmes: His Speeches, Essays, Letters, and Judicial Opinions*. New Brunswick, New Jersey: Transaction Publishers.

Leuchtenburg, William E. 1963. *Franklin D. Roosevelt and the New Deal, 1932–1940*. New York, New York: Harper & Row.

Levine, I. E. 1959. *The Discoverer of Insulin: Dr. Frederick G. Banting*. New York, New York: Julian Messner.

Lewis, David Levering. 1994, paperback. [1993.] *W. E. B. Du Bois: Biography of a Race, 1868–1919*. New York, New York: Henry Holt and Company.

Lewis, Martin W. 1992. *Green Delusions: An Environmentalist Critique of Radical Environmentalism*. Durham, North Carolina: Duke University Press.

Ley, Robert, ed. 2007. *The Avalon Project at Yale Law School*. New Haven, Connecticut: Yale University. Published online at http://www.yale.edu/lawweb/avalon/avalon.htm.

Lietz, Hermann. 1934. [1919.] *Des Vaterlandes Not und Hoffnung* (*The Fatherland's Need and Hope*). Haubinda, Germany: Hermann Lietz School.

Lind, Michael. 2005. *What Lincoln Believed: The Values and Convictions of America's Greatest President*. New York, New York: Doubleday.

Lindsey, Brink. 2002. *Against the Dead Hand: The Uncertain Struggle for Global Capitalism*. New York, New York: John Wiley & Sons, Inc.

*———. 2004. "Job Losses and Trade: A Reality Check." *Trade Briefing Paper* (no. 19, March): 1-16. Republished as a PDF on the website of the Cato Institute. Washington, D.C.: Cato Institute. http://www.freetrade.org/pubs/briefs/tbp-019.pdf.

———. 2007. *The Age of Abundance: How Prosperity Transformed America's Politics and Culture*. New York, New York: Collins.

Linnekin, Jocelyn. 1983. "The *Hui* Lands of Keanae: Hawaiian Land Tenure and the Great Mahele." *Journal of the Polynesian Society*, no. 92: 169–171.

Lloyd, Henry Demarest. 1894. *Wealth Against Commonwealth*. New York, New York: Harper Brothers.

Lochmüller, Benedikt. 1935. *Hans Schemm* vol. 1. Bayreuth, Germany: Deutscher Volksverlag (German People Publishing House).

**Locke, Edwin A. 1995. "Animal 'Rights' and the New Man-Haters." Ayn Rand Institute's website. Marina del Rey, California: Ayn Rand Institute. https://web.archive.org/web/19961018083137/http://www.aynrand.org/animals.htm.

*Locke, John. 1764 rpt. [1689.] *Two Treatises of Government*. Thomas Hollis, ed. London, United Kingdom: A. Millar et al. Republished in the *Online Library of Liberty*. Indianapolis, Indiana: Liberty Fund. http://tinyurl.com/yuou9r.

Löhr, Hans. 1935. *Über die Stellung und Bedeutung der Heilkunde im Nationalsozialistischen Staate* (*About the Position and Role of Medicine in the National Socialist State*). Berlin, Germany: Nornen-Verlag (Nornen Publishing House).

———. 1966. [1935.] "The Physician Must Come to Terms With the Irrational." In G. L. Mosse ed. 1966: 227–234.

Lombard, Herbert L., and Carl R. Doering. 1928. "Cancer Studies in Massachusetts 2: Habits, Characteristics, and Environment of Individuals With and Without Cancer." *New England Journal of Medicine* 198: 481–487. Republished as a PDF on the website of Alfredo Morabia. Geneva, Switzerland: Epidemiology.CH. http://www.epidemiology.ch/history/PDF bg/Lombard HL and Doering CR 1928 cancer studies in mass II habits, characteristics.pdf.

Lombardo, Paul A. 2008. *Three Generations, No Imbeciles: Eugenics, the Supreme Court, and "Buck v. Bell."* Baltimore, Maryland: The Johns Hopkins University Press.

*Lomborg, Bjørn. 2001 paperback. [1998.] *The Skeptical Environmentalist: Measuring the Real State of the World*. New York, New York: Cambridge University Press.

Lukács, Georg. 1959. *The Destruction of Reason* (*La Déstruction de la Raison*). Paris, France: La Nef de Paris.

Lumsden, Charles J., and Edward O. Wilson. 1983. *Promethean Fire: Reflections on the Origin of Mind*. Cambridge, Massachusetts: Harvard University Press.

Lush, Minnie, and David Sirota. 2003 5th ed. [1993.] *California Real Estate Finance*. Chicago, Illinois: Dearborn.

Luther, Martin. 1848. *The Table Talk, or Familiar Discourse*. William Hazlitt, trans. London, United Kingdom: David Bogue.

———. 1879a. *Dr. Martin Luther's Sämtliche Werke* vols. 19–20 [*The Complete Works* vols. 19–20]. Frankfurt, Germany: Bender & Zimmer.

———. 1879b. [1537.] "Predigt über das Evangelium am Sonntag, Weihnachten, 30 December 1537 [Preaching the Gospel on the Sunday After Christmas, 30 December 1537]." In Luther 1879a vols. 19–20: 471–482.

McCraw, Thomas K. 1974. "The Progressive Legacy." In L. Gould ed. 1974: 181–201.

———. 1984. *Prophets of Regulation: Charles Francis Adams, Louis D. Brandeis, James M. Landis, Alfred E. Kahn*. Cambridge, Massachusetts: The Belknap Press of Harvard University Press.

McGerr, Michael. 2003. *A Fierce Discontent: The Rise and Fall of the Progressive Movement in America, 1870–1920*. New York, New York: The Free Press.

McGrath, Roger D. 1984. *Gunfighters, Highwaymen, & Vigilantes: Violence on the Frontier*. Berkeley, California: University of California Press.

McGregor, Davianna Pomaika'i. 1996. "An Introduction to the *Hoa'aina* and Their Rights." *Hawaiian Journal of History* no. 30: 6–8.

McInerney, Thomas J., and Fred L. Israel, eds. 2013 2d ed. [1999.] *Presidential Documents: Words That Shaped a Nation from Washington to Obama*. New York, New York: Routledge.

McMenamin, Michael. 1999. "Nazi Economics." *Reason* vol. 31 (no. 4, August/September): 72–75. Republished on *Reason Online*. Washington, D.C.: Reason Foundation. http://www.reason.com/news/show/31084.html.

MacDougall, Kent A. 1996. "Humans As Cancer." *Wild Earth* vol. 6 (no. 3, Fall): 81–88. Republished as a PDF on Professor Bron Taylor's environmental ethics website. Gainesville, Florida: Bron Taylor. http://www.brontaylor.com/courses/pdf/MacDougall--HumansCancer.pdf.

MacFarquhar, Larissa. 2004. "The Populist: Michael Moore Can Make You Cry." *The New Yorker*. February 16. David Remnick, editor-in-chief. New York, New York: Condé Nast Publications. http://tinyurl.com/2nz66j.

Machan, Tibor R., ed. 1974. *The Libertarian Alternative: Essays in Social and Political Philosophy*. Chicago, Illinois: Nelson-Hall Company.

———, ed. 1982b. *The Libertarian Reader*. Totowa, New Jersey: Rowman & Littlefield.

Maier, Pauline, et al. 2003. *Inventing America: A History of the United States*. New York, New York: W. W. Norton & Company.

Manes, Christopher. 1990. *Green Rage: Radical Environmentalism and the Unmaking of Civilization*. Boston, Massachusetts: Little, Brown and Company.

Maney, Kevin. 2003. *The Maverick and His Machine: Thomas Watson, Sr. and the Making of IBM*. New York, New York: John Wiley & Sons.

Marks, Jonathan. 2002. *What It Means to Be 98% Chimpanzee: Apes, People, and Their Genes*. Berkeley, California: University of California Press.

Marx, Karl. 1844. [Written in Autumn 1843.] "On *The Jewish Question*." *Deutsch-Französische Jahrbücher* (February). Republished on *Marxists.Org*. Andy Blunden, et al., eds. http://www.marxists.org/archive/marx/works/1844/jewish-question/.

Meacham, Jon. 2006. *American Gospel: God, the Founding Fathers, and the Making of a Nation*. New York, New York: Random House.

Mellon, Andrew. 1924. *Taxation: The People's Business*. New York, New York: Macmillan.

Mencken, H. L. 1910a. "Mencken's Reply to La Monte's Second Letter." In R. R. La Monte and Mencken eds. 1910: 61–80.

———. 1910b. "Mencken's Reply to La Monte's Fifth Letter." In R. R. La Monte and Mencken eds. 1910: 185–206.

———. 1922c. "Education." In Mencken 1922f: 238–265.
———. 1922d. "Matters of State." In Mencken 1922f: 289–298.
———. 1922f. *Prejudices: Third Series*. New York, New York: Alfred A. Knopf.
———. 1926. *Notes on Democracy*. New York, New York: Alfred A. Knopf.
———. 1990-A. [1937.] "An American Dictator." In Mencken 1990-Q: 367.
———. 1990-E. [1910.] "Competition." In Mencken 1990-Q: 246.
———. 1990-I. [1905.] "Darwin's Influence." In Mencken 1990-Q: 287.
———. 1990-J. [1926.] "Demagogue as Statesman." In Mencken 1990-Q: 372.
———. 1990-N. [1933.] "FDR's Inaugural Address." In Mencken 1990-Q: 408–09.
———. 1990-Q. *The Gist of Mencken: Quotations from America's Critic*. Mayo DuBasky, ed. Metuchen, New Jersey: The Scarecrow Press.
———. 1990-EE. [1917.] "Ludwig van Beethoven." In Mencken 1990-Q: 744.
———. 1990-LL. [1919.] "Progeny of Great Men." In Mencken 1990-Q: 276–77.
———. 1990-OO. [1925.] "Repellant Men." In Mencken 1990-Q: 408.
———. 1990-VV. [1905.] "The War Against Orthodoxy." In Mencken 1990-Q: 281–82.
———. 2006-H. [1918.] "Heredity." In Mencken 2006-N: 128–29.
———. 2006-I. [1919.] "Jack London." In Mencken 2006-N: 276–77.
———. 2006-N. [1995.] *A Second Mencken Chrestomathy*. H. L. Mencken and Terry Teachout, eds. Baltimore, Maryland: Johns Hopkins University Press.
Meyers, Morton A. 2007. *Happy Accidents: Serendipity in Modern Medical Breakthroughs*. New York, New York: Arcade Publishing.
Michaud, Eric. 2004. *The Cult of Art in Nazi Germany*. Janet Lloyd, trans. Stanford, California: Stanford University Press.
Micklethwait, John, and Adrian Wooldridge. 2003. *The Company: A Short History of a Revolutionary Idea*. New York, New York: The Modern Library.
Mikkelson, Barbara. 2011. "The Reich Stuff?" *Urban Legends Reference Pages*. April 29. Published on Snopes.Com, the home of the *Urban Legends Reference Pages*. San Fernando Valley, California: San Fernando Valley Folklore Society. http://www.snopes.com/cokelore/fanta.asp.
Mikkelson, Barbara, and David P. Mikkelson. 2007. "Loco Motive." *Urban Legends Reference Pages*. September 29. Published on Snopes.Com, the home of the *Urban Legends Reference Pages*. San Fernando Valley, California: San Fernando Valley Folklore Society. http://www.snopes.com/history/govern/trains.asp.
Mill, John Stuart. 1887 8th ed. *A System of Logic*. New York, New York: Harper Brothers.
Miller, Charles A. 2000. "The Rich Fields of Nature." In S. Day et al. eds. 2000: 57–61.
Mills, C. Wright. 1942. "Locating the Enemy: The Nazi Behemoth Dissected." *Partisan Review* vol. 9 (September-October, no. 4): 432–37.
Milward, Alan S. 1976. "Fascism and Economy." In W. Laquer ed.: 379–414.
———. 1977. *War, Economy and Society, 1939–1945*. Berkeley, California: University of California Press.
Mises, Ludwig, von. 1966 3d revised ed. [1949.] *Human Action: A Treatise on Economics*. Chicago, Illinois: Henry Regnery Company.
———. 1981. [1922.] *Socialism: An Economic and Sociological Analysis*. J. Kahane, trans. Indianapolis, Indiana: Liberty Fund. Republished in the *Online Library of Liberty*. Indianapolis, Indiana: Liberty Fund. http://oll.libertyfund.org/index.php?option=com_staticxt&staticfile=show. php%3Ftitle=1060&Itemid=99999999.
Moalem, Sharon, and Jonathan Prince. 2007. *Survival of the Sickest: A Medical Maverick Discovers Why We Need Disease*. New York, New York: William Morrow.
Mommsen, Wolfgang. 1993. "Government Without Parties." In L. E. Jones and Retallack eds. 1993: 347–374.
Monks, Robert A. G., and Nell Minow. 1991. *Power and Accountability*. New York, New York: HarperBusiness.
Montagu, Lord, and David Burgess-Wise. 1995. *The Daimler Century: The First History of Britain's Oldest Car Maker*. Sparkford, United Kingdom: Patrick Stephens Limited.
Moon, John F. 1974. *Rudolf Diesel and the Diesel Engine*. East Sussex, United Kingdom: Wayland Publishers.
Moore, Michael. 2003. *Dude, Where's My Country?* New York, New York: Warner Books.
Moore, R. Laurence. 1974. "Directions of Thought in Progressive America." In L. Gould ed. 1974: 35–54.
Mosse, George L. 1964. *The Crisis of German Ideology: Intellectual Origins of the Third Reich*. New York, New York: Grosset & Dunlap.
———. 1975. *The Nationalization of the Masses: Political Symbolism and Mass Movements in Germany from the Napoleonic Wars Though the Third Reich*. New York, New York: Howard Fertig.
———. 1985, trade paperback. [1978.] *Toward the Final Solution: A History of European Racism*. Madison, Wisconsin: The University of Wisconsin Press.
Mosse, George L., ed. 1966. *Nazi Culture: Intellectual, Cultural and Social Life in the Third Reich*. Salvator Attanasio, trans. New York, New York: Grosset & Dunlap.
Mowry, George E. 1958. *The Era of Theodore Roosevelt, 1900–1912*. New York, New York: Harper & Row.

Msarovic, Mihajlo, and Eduard Pestel. 1974. *Mankind at the Turning Point: The Second Report to the Club of Rome*. New York, New York: Dutton.

Murray, Nicholas. 2002. *Aldous Huxley: A Biography*. New York, New York: Thomas Dunne Books.

Mussolini, Benito. 1933. "Roosevelt e il Sistema." *Bolletino del R. Ministero degli Affari Esteri*, vol. 40 (Luglio): 715–17.

———. 1937. *Gerarchia: Rassegna Mensile della Rivoluzione Fascista* [*Hierarchy: The Monthly Review of the Fascist Revolution*], vol. 17. Retrieved on Google Books.

Naam, Ramez. 2005. *More Than Human: Embracing the Promise of Biological Enhancement*. New York, New York: Broadway Books.

Nazi Party. 1937. *Richtlinien für die Leibeserziehung in Jungenschulen* (*Guidelines for Physical Education in Boys' Schools*). Berlin, Germany: Weidmann'sche Verlagbuchhandlung (Weidmann's Publishing House Bookstore).

———. 1966. [1937.] "Physical Education and National Socialism." In G. L. Mosse ed. 1966: 281–82.

Neisser, Arden. 1983. *The Other Side of Silence*. New York, New York: Alfred A. Knopf.

Nesse, Randolph M., and George C. Williams. 1995. *Why We Get Sick: The New Science of Darwinian Medicine*. New York, New York: Times Books.

Nisbett, Richard E. 2003. *The Geography of Thought: How Asians and Westerners Think Differently . . . And Why*. New York, New York: The Free Press.

Novick, Sheldon M. 1992. "Justice Holmes Philosophy." *Washington University Law Quarterly*, vol. 70: 703–53.

Nozick, Robert. 1974. *Anarchy, State, and Utopia*. New York, New York: Basic Books.

Obama, Barack. 2006. "The American Dream Is Possible in Today's Economy." In K. Burns ed. 2006: 35–44.

———. 2012. "Transcript: Read the Full Text of Obama's Speech on the GOP Budget." *Los Angeles Times*. Tuesday, April 3. Republished on the *Los Angeles Times* website. Los Angeles, California: Los Angeles Times Company. http://articles.latimes.com/print/2012/apr/03/news/la-pn-obama-speech-transcript-20120403.

Olasky, Marvin. 1992. *The Tragedy of American Compassion*. Washington, D.C.: Regnery Gateway.

*Olsen, Jonathan. 1999. *Nature and Nationalism: Right-Wing Ecology and the Politics of Identity in Contemporary Germany*. New York, New York: Palgrave Macmillan.

Olson, Steve. 1989. *Shaping the Future: Biology and Human Values*. Washington, D.C.: National Academy [of Sciences] Press.

———. 2002. *Mapping Human History: Discovering the Past Through Our Genes*. Boston, Massachusetts: Houghton Mifflin Company.

Ormerod, Paul. 2000, trade paperback. [1998.] *Butterfly Economics: A New General Theory of Social and Economic Behavior*. New York, New York: Pantheon Books.

———. 2005. *Why Most Things Fail: Extinction and Economics*. New York, New York: Pantheon Books.

O'Rourke, P. J. 2003. [1991.] *Parliament of Whores: A Lone Humorist Attempts to Explain the Entire U.S. Government*. New York, New York: Grove Press.

———. 2007. *On "The Wealth of Nations."* New York, New York: Atlantic Monthly. 2007.

Orwell, George. 1941. *The Lion and the Unicorn*. New York, New York: Penguin Books.

———. 1968a. *The Collected Essays, Journalism and Letters of George Orwell* (4 vols.). London, United Kingdom: Martin Secker & Warburg.

———. 1968b. [1945.] "Freedom of the Park." In Orwell 1968a vol. 4: 37–40.

———. 1968c. [1940.] "Review: *Mein Kampf* by Adolf Hitler." In Orwell 1968a vol. 2: 12–14.

———. 1968f. [1944.] "Review: *The Road to Serfdom* by F. A. Hayek; *The Mirror of the Past* by K. Zilliacus." In Orwell 1968a vol. 3: 117–19.

Osborn, Frederick. 1968. *The Future of Human Heredity: An Introduction to Eugenics in Modern Society*. New York, New York: Weybright and Talley.

Osborn, Henry Fairfield. 1932. "Birth Selection Versus Birth Control." *Science*, vol. 76 (no. 1965, August 26): 173–79.

———. 1934. [1932.] "Birth Selection Versus Birth Control." In H. F. Perkins ed. 1934: 29–41.

Oshinsky, David. 2005. *Polio: An American Story*. New York, New York: Oxford University Press.

———. "Preventive Medicine." *New York Times*. February 4. New York, New York: The New York Times Company. http://www.nytimes.com/2007/02/04/books/review/Oshinsky.t. html?pagewanted=print&_r=0.

Outman, James L., and Elisabeth M. Outman. 2003a. *Industrial Revolution Almanac*. Detroit, Michigan: Thomson Gale.

———. 2003b. *Industrial Revolution Biographies*. Detroit, Michigan: Thomson Gale.

*Paarlberg, Robert. 2010. *Food Politics: What Everyone Needs to Know*. New York, New York: Oxford University Press.

Padgen, Anthony. 2013. *The Enlightenment: And Why It Still Matters*. New York, New York: Random House.

Paglia, Camille. 1990. *Sexual Personae: Art and Decadence from Nefertiti to Emily Dickinson*. New Haven, Connecticut: Yale University Press.

Parenti, Michael. 1986. *Inventing Reality: The Politics of Mass Media*. New York, New York: St. Martin's Press.

———. 1989. *The Sword and the Dollar: Imperialism, Revolution, and the Arms Race*. New York, New York: St. Martin's Press.

———. 1995 6th ed. *Democracy for the Few*. New York, New York: St. Martin's Press.

———. 2004. *Superpatriotism*. San Francisco, California: City Lights Books.

Parsons, Talcott. 1932. "Economics and Sociology: Marshall in Relation to the Thought of His Time." *Quarterly Journal of Economics*, vol. 46 (no. 2, February): 316–347.

Patterson, James T. 1972. *Mr. Republican: A Biography of Robert A. Taft*. Boston, Massachusetts: Houghton Mifflin.

Paulovicova, Nina. 2007. "The Immoral Moral Scientist: John Maynard Keynes." *Past Imperfect*, vol. 13: 24–55. Republished as a PDF on the University of Alberta website. Alberta, Canada: University of Alberta. http://ejournals.library.ualberta.ca/index.php/pi/article/viewFile/1591/1117.

Payne, Stanley G. 1980. *Fascism: Comparison and Definition*. Madison, Wisconsin: The University of Wisconsin Press.

**Peikoff, Leonard. 1993a. [1982.] *The Ominous Parallels: The End of Freedom in America*. New York, New York: Meridian.

**———. 1993b. [1991.] *Objectivism: The Philosophy of Ayn Rand*. New York, New York: Meridian.

**Peikoff, Leonard, ed. 1990c paperback. *Voice of Reason: Essays in Objectivist Thought*. New York, New York: Meridian.

Pence, Gregory E. 2002. *Designer Food: Mutant Harvest of Breadbasket to the World?* Lanham, Maryland: Rowman & Littlefield.

Pendergrast, Mark. 1993. *For God, Country and Coca-Cola: The Unauthorized History of the Great American Soft Drink and the Company That Makes It*. New York, New York: Charles Scribner's Sons.

Perkins, Harry F., ed. 1934. *A Decade of Progress in Eugenics: Scientific Papers of the Third International Congress of Eugenics*. Baltimore, Maryland: Williams and Wilkins Company.

Perlstein, Rick. 2001. *Before the Storm: Barry Goldwater and the Unmaking of the American Consensus*. New York, New York: Hill and Wang.

Petersen, William. 1979. *Malthus*. Cambridge, Massachusetts: Harvard University Press.

Petrucci, Francesca, ed. 2002. *The Welfare State: Past, Present, and Future*. Pisa, Italy: Edizioni Plus.

Pflanze, Otto. 1990. *Bismarck and the Development of Germany* vol. 3: *The Period of Fortification, 1880–1898*. Princeton, New Jersey: Princeton University Press.

Phillips-Fein, Kim. 2009. *Invisible Hands: The Making of the Conservative Movement from the New Deal to Reagan*. New York, New York: W. W. Norton & Company.

Pinker, Steven. 1997. *How the Mind Works*. New York, New York: W. W. Norton.

———. 2000, paperback. [1994.] *The Language Instinct: How the Mind Creates Language*. New York, New York: Harper Perennial.

*———. 2011. *The Better Angels of Our Nature: Why Violence Has Declined*. New York, New York: Viking.

Plato. 1892a 3d revised ed. *The Dialogues of Plato* translated into English with Analyses and Introductions by B. Jowett, M.A. in Five Volumes (5 vols.). Benjamin Jowett, ed. Oxford, United Kingdom: Oxford University Press. Republished in the *Online Library of Liberty*. Indianapolis, Indiana: Liberty Fund. http://tinyurl.com/6ywt2a.

———. 1892b vol. 3. *The Dialogues of Plato* vol. 3: *Republic, Timaeus, Critias*. Benjamin Jowett, ed. Oxford, United Kingdom: Oxford University Press. Republished in the *Online Library of Liberty*. Indianapolis, Indiana: Liberty Fund. http://oll.libertyfund.org/index. php?option=com_staticxt&staticfile=show.php%3Ftitle=767&Itemid=99999999.

Plotz, David. 2005. *The Genius Factory: The Curious History of the Nobel Prize Sperm Bank*. New York, New York: Random House.

Poliakov, Leon. 1974 U.S. version. [1971.] *The Aryan Myth: A History of Racist and Nationalist Ideas in Europe*. Edmund Howard, trans. New York, New York: Basic Books.

Ponte, Lowell. 2002. "Political Scientist." *Front Page Magazine*. May 22. David Horowitz, editor-in-chief. Los Angeles, California: The David Horowitz Freedom Center. http://frontpagemag.com/Articles/ReadArticle.asp?ID=4093.

Popper, Karl. 1971b vol. 2. *The Open Society and Its Enemies* vol. 2: *Hegel, Marx, and the Aftermath*. Princeton, New Jersey: Princeton University Press.

———. 1996a. [1954.] "Immanuel Kant: Philosopher of the Enlightenment (A Lecture to Commemorate the 150th Anniversary of Kant's Death." In K. Popper 1996b paperback: 126–136.

———. 1996b paperback. *In Search of a Better World: Lectures and Essays from Thirty Years*. New York, New York: Routledge.

Porter, Roy. 2004 U.S. version. [2003.] *Flesh in the Age of Reason*. New York, New York: W. W. Norton & Company.

Porteus, Stanley D. 1945. *Calabashes and Kings: An Introduction to Hawaii*. Palo Alto, California: Pacific Books.

Posner, Richard A., ed. 1992. *The Essential Holmes: Selections from the Letters, Speeches, Judicial Opinions, and Other Writings of Oliver Wendell Holmes, Jr.* Chicago, Illinois: University of Chicago Press.

Potter, Steven. 2010. *Designer Genes: A New Era in the Evolution of Man*. New York, New York: Random House.

*Pounds, Norman J.G. 1989. *Hearth & Home: A History of Material Culture*. Bloomington, Indiana: Indiana University Press.

Powell, Jim. 2000. *The Triumph of Liberty: A 2,000-Year History, Told Through the Lives of Freedom's Greatest Champions*. New York, New York: The Free Press.

———. 2008. *Great Emancipations: How the West Abolished Slavery*. New York, New York: Palgrave Macmillan.

Proctor, Robert N. 1999. *The Nazi War on Cancer*. Princeton, New Jersey: Princeton University Press.

Proudhon, Pierre-Joseph. 1876. [1840.] *What Is Property?: An Inquiry Into the Principle of Right and of Government*. Benjamin Ricketson Tucker, trans. Princeton, Massachusetts: Benjamin R. Tucker. Republished on the website of the University of Virginia Library. Charlottesville, Virginia: University of Virginia. http://etext.virginia.edu/toc/modeng/public/ProProp.html.

Rader, Benjamin G. 1966. *The Academic Mind and Reform: The Influence of Richard T. Ely in American Life*. Lexington, Kentucky: University Press of Kentucky.

Rall, Ted. 2004. *Wake Up, You're Liberal: How We Can Take America Back from the Right*. Brooklyn, New York.

**Ralston, Richard E., ed. 1999. *Why Businessmen Need Philosophy*. Marina del Rey, California: Ayn Rand Institute Press.

**Rand, Ayn. 1961b. *For the New Intellectual*. New York, New York: Signet.

**———. 1961c. "For the New Intellectual." In Rand 1961b: 10–57.

**———. 1963a. "The Ethics of Emergencies." *The Objectivist Newsletter*, vol. 2 (no. 2, February): 5–6.

**———. 1963d. "The Nature of Government." *The Objectivist Newsletter*, vol. 2 (no. 12, December): 45–46, 49–50.

**———. 1963e. "Review of *Aristotle* By John Herman Randall, Jr." *The Objectivist Newsletter*, vol. 2 (no. 5, May): 18–19.

**———. 1970 paperback. [1964.] *The Virtue of Selfishness: A New Concept of Egoism*. Ayn Rand, ed. New York, New York: Signet.

**———. 1984c. [1960.] "Faith and Force : The Destroyers of the Modern World." In Rand 1984e: 58–76.

**———. 1984e. *Philosophy: Who Needs It*. Leonard Peikoff, ed. New York, New York: Signet.

**———. 1984f. [1974.] "Philosophy: Who Needs It." In Rand 1984e: 1–11.

**———. 1985, mass market paperback. [1957.] *Atlas Shrugged*. New York, New York: Signet.

**———. 1990a. [1961.] "The Intellectual Bankruptcy of Our Age." In Peikoff ed. 1990c paperback, 85-99.

**———. 1990b expanded 2d ed. [1979.] *Introduction to Objectivist Epistemology*. Harry Binswanger and Leonard Peikoff, eds. New York, New York: Meridian.

**———. 1990c. [1963.] "Review of *Aristotle* By John Herman Randall, Jr." In Peikoff ed. 1990c paperback: 6–12.

**———. 1992a, mass market paperback. [1943.] *The Fountainhead*. New York, New York: Signet.

**———. 1992b, hardcover. [1957.] *Atlas Shrugged*. New York, New York: Dutton.

**———. 1997, paperback. [1995.] *Letters of Ayn Rand*. Michael S. Berliner, ed. New York, New York: Plume.

**———. 1999c. *Return of the Primitive: The Anti-Industrial Revolution*. Peter Schwartz, ed. New York, New York: Meridian.

———. 2000. [1964.] *The "Playboy" Interview With Philosopher Ayn Rand: A Wide-Ranging Discussion of the Basic Ideas of Objectivism*. Interview conducted by Alvin Toffler. Gaylordsville, Connecticut: Second Renaissance Books.

**———. 2007, paperback. [1957.] *Atlas Shrugged*. New York, New York: Signet.

**Rand, Ayn, ed. 1982a. [Originally published in separate issues from 1966 to 1971.] *The Objectivist: Volumes 5–10, 1966–1971*. Palo Alto, California: Palo Alto Book Service.

Randi, James. 1982a. *Flim-Flam!: Psychics, ESP, Unicorns and Other Delusions*. Amherst, New York: Prometheus Books.

Random House Unabridged Dictionary. 2014. New York, New York: Random House.

Ravitch, Diane. 2000. *Left Back: A Century of Failed School Reforms*. New York, New York: Simon & Schuster.

Reagan, Ronald. 2004a. [1981.] "Remarks at the Conservative Political Action Conference: Washington, D.C., March 20, 1981." In R. Reagan 2004b paperback: 93–101.

———. 2004b, paperback. [1989.] *Speaking My Mind: Selected Speeches*. New York, New York: Simon & Schuster.

Redbeard, Ragnar. 2005. [1890.] *Might Is Right, or Survival of the Fittest*. Springfield, Missouri: Dil Pickle Press.

Redmond, Lynn, prod. 2003. "The Pursuit of Happyness." *20/20*. Friday, January 17. Bob Brown, reporter. Television broadcast. American Broadcasting Company (ABC).

Rees, Robert. 2003. "Right Thinking." *Honolulu Advertiser* (Sunday, June 29): B1, B4. Republished on the Honolulu Advertiser's website. Honolulu, Hawaii: The Honolulu Advertiser. http://tinyurl.com/6vc5u

———. 2004. "Ahem! I Beg to Differ With You, Mr. Slater." *Honolulu Advertiser*. (Monday, April 26): A6. Republished on the Honolulu Advertiser's website. Honolulu, Hawaii: The Honolulu Advertiser. http://tinyurl.com/4vjuw.

Regis, Ed. 2003. *The Info Mesa: Science, Business, and New Age Alchemy on the Santa Fe Plateau*. New York, New York: W. W. Norton & Company.

Reid, James S., ed. 1874. *The Academica of Cicero*. London, United Kingdom: Macmillan and Co.

Renehan, Edward J., Jr. *Commodore: The Life of Cornelius Vanderbilt*. New York, New York: Basic Books.

Ridley, Matt. 1997. *The Origins of Virtue: Human Instincts and the Evolution of Cooperation*. New York, New York: Viking.

———. 1999. *Genome: The Autobiography of a Species in 23 Chapters*. New York, New York: HarperCollins.

———. 2000. "The New Eugenics: Better Than the Old." *National Review*. July 31. Republished online in *FindArticles.Com*. http://tinyurl.com/2x8av9.

———. 2010. *The Rational Optimist: How Prosperity Evolves*. New York, New York: Harper.

Riess, Curt. 1948. *Joseph Goebbels*. Garden City, New York: Doubleday & Company, Inc.

Riis, Jacob. 1997. [1890.] *How the Other Half Lives: Studies Among the Tenements of New York*. New York, New York: Penguin Books.

Roach, Marion. 2005. *The Roots of Desire: The Myth, Meaning, and Power of Red Hair*. New York, New York: Bloomsbury.

Roback, Jennifer. 1986. "The Political Economy of Segregation: The Case of Segregated Streetcars." *Journal of Economic History* vol. 56 (no. 4, December): 893–917.

Robbins, Lionel. 1998. *A History of Economic Thought: The LSE Lectures*. Steven G. Medema and Warren J. Samuels, eds. Princeton, New Jersey: Princeton University Press.

Robbins, Tim, et al., prods. 1999. *Cradle Will Rock*. Tim Robbins, dir. DVD. Touchstone Pictures.

Robinson, Greg. 2001. *By Order of the President: FDR and the Internment of Japanese Americans*. Cambridge, Massachusetts: Harvard University Press.

Roland, Paul. 2008. *The Nazis and the Occult: The Dark Forces Unleashed By the Third Reich*. Edison, New Jersey: Chartwell Books.

Roosevelt, Theodore. 1907. "A Letter from President Roosevelt on Race Suicide." *American Monthly Review* vol. 35 (no. 5): 550–51.

———. 1913b. "Foundations of the Nineteenth Century." In T. Roosevelt 1913c: 233–243.

———. 1913c. *History as Literature*. New York, New York: Charles Scriber's Sons.

———. 1913d. *Progressive Principles: Selections from Addresses Made During the Presidential Campaign of 1912*. Elmer H. Youngman, ed. New York, New York: Progressive National Service.

Root, Damon W. 2007. "The Party of Jefferson." *Reason*, vol. 39 (no. 7, December). Republished on *Reason Online*. Washington, D.C.: Reason Foundation. http://reason.com/archives/2007/11/27/the-party-of-jefferson/print.

———. 2010. "Progressive History 101 (Minus All the Uncomfortable Racism, Sexism, and Support for Eugenics." *Hit & Run* blog. April 19. Washington, D.C.: Reason Foundation. http://reason.com/blog/2010/04/19/progressive-history-101-minus.

———. 2012. "Frederick Douglass, Classical Liberal." *Reason*, vol. 44 (no. 4, August/September): 56. Republished on *Reason Online*. Washington, D.C.: Reason Foundation. http://reason.com/archives/2012/07/17/frederick-douglass-classical-liberal/print.

Root, Jesse. 1898–99. [1798.] *Reports of Cases Adjudged in the Superior Court and Supreme Court of Errors from July, A.D. 1789, to June, A.D., 1793* (2 vols.). Waterbury, Connecticut: The Dissell Publishing Company.

Ross, Dorothy. 1991. *The Origins of American Social Science*. New York, New York: Cambridge University Press.

Ross, Edward A. 1906. *Social Control*. New York, New York: Macmillan.

Rothbard, Murray N. 1973b. *For a New Liberty: The Libertarian Manifesto*. New York, New York: The Macmillan Company.

———. 2004 scholars' ed. [1962.] *"Man, Economy, and State," with "Power and Market."* Auburn, Alabama: Ludwig von Mises Institute. Republished on the website of the Ludwig von Mises Institute. Auburn, Alabama: Ludwig von Mises Institute. http://www.mises.org/rothbard/mes.asp.

Rothman, Sheila M., and David J. Rothman. 2003. *The Pursuit of Perfection: The Promise and Perils of Medical Enhancement*. New York, New York: Pantheon Books.

*Rothman, Stanley, and S. Robert Lichter. 1996. "Is Environmental Cancer a Political Disease?" In P. Gross et al. eds. 1996, trade paperback: 231–245.

Rothschild, Michael. 1990. *Bionomics: The Inevitability of Capitalism* (later titled *Bionomics: Economy as Ecosystem*). New York, New York: Henry Holt and Company.

Rousseau, Jean-Jacques. 1909. [1762.] *Émile; or Treatise on Education*. William H. Payne, trans. and ed. New York, New York: D. Appleton and Company.

———. 1921. [1762.] *Émile; or Education*. Barbara Foxley, trans. London, United Kingdom: J. M. Den and Sons. Republished in the *Online Library of Liberty*. Liberty Fund: Indianapolis, Indiana. http://oll.libertyfund.org/index.php?option=com_staticxt&staticfile=show.php%3Ftitle=2256&Itemid=28.

———. 1923. [1761.] *The Social Contract and Discourses*. G. D. H. Cole, trans. London, United Kingdom: J. M. Dent and Sons. Republished in the *Online Library of Liberty*. Liberty Fund: Indianapolis, Indiana. http://tinyurl.com/36oone.

*Rubin, Barry, and Judith Colp Rubin. 2004. *Hating America: A History*. New York, New York: Oxford University Press.

Rubin, Barry, and Judith Colp Rubin, eds. 2002. *Anti-American Terrorism and the Middle East: A Documentary Reader*. New York, New York: Oxford University Press.

Rubin, John, prod. 2008. "Ape Genius." *Nova*. John Rubin, dir. Aired Tuesday, February 19. Television broadcast. Public Broadcasting Service (PBS) and the Corporation for Public Broadcasting (CPB). Transcript online at http://www.pbs.org/wgbh/nova/transcripts/3504_apegeniu.html.

Rubin, Paul H. 2002. *Darwinian Politics: The Evolutionary Origin of Freedom*. New Brunswick, New Jersey: Rutgers University Press.

Ruse, Michael. 1979. *The Darwinian Revolution: Science Red in Tooth and Claw*. Chicago, Illinois: The University of Chicago Press.

Russell, Thaddeus. 2010. *A Renegade History of the United States*. New York, New York: Free Press.

Sade, Marquis (Donatien), de. 1965a. *Justine, Philosophy in the Bedroom, & Other Writings*. New York, New York: Grove Press.

———. 1965b. *Philosophy in the Bedroom*. In Sade 1965a: 179–446.

———. 1988. [1797.] *Juliette*. New York, New York: Grove Press.

Sagan, Carl. 1995. *The Demon-Haunted World: Science as a Candle in the Dark*. New York, New York: Random House.

Samuelson, Paul A. 1980 11th ed. *Economics*. New York, New York: McGraw-Hill.

Samuelsson, Per-Olof. 2012. "Ludwig von Mises on Egoism Vs. Altruism." *The House at POS Corner*. April 29. Stockholm, Sweden: Per-Olof Samuelsson. http://perolofsamuelsson1.wordpress. com/2012/04/29/ludwig-von-mises-on-egoism-vs-altruism/.

Saul, John Ralston. 1993, paperback. [1992.] *Voltaire's Bastards: The Dictatorship of Reason in the West*. New York, New York: Vintage Books.

Sax, Boria. 2000. *Animals in the Third Reich: Pets, Scapegoats, and the Holocaust*, New York, New York: Continuum International Publishing Group.

Say, Jean-Baptiste. 1823. "Letter from J. B. Say to the Author." In A. Hodgson 1823 2d ed.: 59–60.

Sayet, Evan. 2012. *KinderGarden of Eden* vol. 1: *How the Modern Liberal Thinks*. Los Angeles, California: Evan Sayet.

Schapiro, J. Salwyn. 1963. *Condorcet and the Rise of Liberalism*. New York, New York: Octagon Books.

Schemm, Hans. 1935. *Hans Schemm Spricht: Seine Reden und Sein Werk* (*Hans Schemm Speaks: His Speeches and His Work*). G. Kahl-Furthmann, ed. Gauleitung der Bayerischen Ostmark, Hauptamtsleitung des national-sozialistischen Lehrerbundes: Gauverlag Bayerische Ostmark.

———. 1966. [1935.] "Ten Calories More Character." In G. L. Mosse ed. 1966: 282–84.

Schivelbusch, Wolfgang. 2006. *Three New Deals: Reflections on Roosevelt's America, Mussolini's Italy, and Hitler's Germany, 1933–1939*. Jefferson Chase, trans. New York, New York: Metropolitan Books.

Schlesinger, Arthur M., Jr. 2003, trade paperback. [1960.] *The Age of Roosevelt: The Age of Upheaval, 1935–1936*. New York, New York: The Houghton Mifflin Company.

Schmidt, Oscar. 1879. "Science and Socialism." *Popular Science Monthly* vol. 14 (no. 5, March): 577–590. Republished on *Wikisource*. San Francisco, California: Wikimedia Foundation. http://en.wikisource.org/wiki/Popular_Science_Monthly/Volume_14/March_1879/Science_and_Socialism.

Schneider, Dorothy, and Carl J. Schneider. 2007. [2000.] *Slavery in America*. New York, New York: Facts on File.

*Schoenfeld, Gabriel. 2001. "The Punch-Card Conspiracy: A Journalist Explores Relations Between IBM and the Third Reich." *New York Times*. March 18. Republished online at http://tinyurl.com/9o7ys.

Schoenwald, Richard L. 1968. "Town Guano and Social Statics." *Victorian Studies*, vol. 9 (Supplement, Summer): 691–710.

Schoorl, Evert. 2012. *Jean-Baptiste Say: Revolutionary, Entrepreneur, Economist*. London, United Kingdom: Routledge.

Schramm, Peter W., et al. eds. 2006. *TeachingAmericanHistory.Org*. Ashland, Ohio: Ashbrook Center for Public Affairs at Ashland University. http://www.teachingamericanhistory.org/.

Schroeder, Alice. 2008. *The Snowball: Warren Buffett and the Business of Life*. New York, New York: Bantam Books.

Schüddekopf, Otto-Ernst. 1960. *Linke Leute von Rechts, Die Nationalrevolutionären Minderheiten und der Kommunismus in der Weimarer Republik* (*People Left from Right, The National Revolutionary Minorities and Communism in the Weimar Republic*). Stuttgart, Germany.

Schumpeter, Joseph A. 1950. [1942.] *Capitalism, Socialism, and Democracy*. New York, New York: Harper & Brothers.

———. 1996 revised ed. [1954.] *History of Economic Analysis*. Elizabeth Boody Schumpeter, ed. New York, New York: Oxford University Press.

Schurz, Carl. 1964. [1889.] "Forests, Prosperity, and Progress." In Commager ed. 1964 2d ed.: 88–91.

Schwartz, Barry. 1986. *The Battle for Human Nature: Science, Morality, and Modern Life*. New York, New York: W. W. Norton & Company.

Schwartz, Michael. 1992. "Sozialistiche Eugenik: Eugenische Sozialtechnologien in Diskurs und Politik der Deutschen Sozialdemokratie, 1890–1933." Diss. Munster, Germany: University of Munster.

**Schwartz, Peter. No date (n.d.). [1985.] *Libertarianism: The Perversion of Liberty*. Irvine, California: Ayn Rand Bookstore.

Schwarze Korps, Das. 1966. [July 27, 1937.] "Throttling the Retail Trade." In G. L. Mosse ed. 1966: 362–63.

Schwarzschild, Léopold. 1947. *The Red Prussian: The Life and Legend of Karl Marx*. New York, New York: Charles Scribner's Sons.

Schweizer, Peter. 2006. *Do As I Say (Not As I Do): Profiles in Liberal Hypocrisy*. New York, New York: Broadway Books.

Scott, John, and Gordon Marshall, ed. 2007. *A Dictionary of Sociology* 3d ed. New York, New York: Oxford University Press.

Scruton, Roger. 1982. *Kant*. New York, New York: Oxford University Press.

Scurr, Ruth. 2006. *Fatal Purity: Robespierre and the French Revolution*. New York, New York: Metropolitan Books.

Seifer, Marc J. 1998. *Wizard: The Life and Times of Nikola Tesla—Biography of a Genius*. Secaucus, New Jersey: Carol Publishing Group.

Seldes, George. 1960. *The Great Quotations*. New York, New York: Lyle Stuart.

Shafir, E., ed. 2013. *The Behavioral Foundations of Public Policy*. Princeton, New Jersey: Princeton University Press.

Shaler, William 1935. *Journal of a Voyage Between China and the Northwestern Coast of America, Made in 1804*. Claremont, California: Saunders Studio Press.

Shelley, Mary. 2011a. "Appendix C: Introduction to the 1831 Edition." In M. Shelley 2011b: 437–443.

———. 2011b. *The Original "Frankenstein."* New York, New York: Random House.

Shepard, Cyn, ed. 1999–2008. *4-20: A Columbine Site*. Peoria, Arizona: C. Shepard. See online at http://acolumbinesite.com/index.html.

Shepherd, A. P. 1983 U.S. version. [1954.] *Scientist of the Invisible—Spiritual Science: The Life and Work of Rudolf Steiner*. New York, New York: Inner Traditions International.

Sherman, Howard. 1972. *Radical Political Economy: Capitalism and Socialism from a Marxist-Humanist Perspective*. New York, New York: Basic Books.

Shermer, Michael. 2003b. "The Domesticated Savage: Science Reveals a Way to Rise Above Our Natures." *Scientific American* vol. 289 (no. 3, September). Republished on *Scientific American*'s website. New York, New York: Scientific American, Inc. http://tinyurl.com/2aok9f.

———. 2004. *The Science of Good and Evil: Why People Cheat, Gossip, Care, Share, and Follow the Golden Rule*. New York, New York: Times Books.

———. 2006. "Darwin on the Right: Why Christians and Conservatives Should Accept Evolution." *Scientific American*, vol. 292 (no. 4, October). Republished on *Scientific American*'s website. New York, New York: Scientific American, Inc. http://tinyurl.com/qbsh5.

———. 2009, paperback. [2008.] *The Mind of the Market: How Biology and Psychology Shape Our Economic Lives*. New York, New York: Henry Holt and Company.

Shirer, William L. 1990. [1959.] *The Rise and Fall of the Third Reich*. New York, New York: Simon & Schuster.

*Shone, Steve J. 2004. "Cultural Relativism and the Savage: The Alleged Inconsistency of William Graham Sumner." *The American Journal of Economics and Sociology*, vol. 63 (no. 3, July): 697–715. Republished online at *FindArticles.Com*. Second Street, San Francisco: LookSmart, Ltd. http://tinyurl.com/3dh22t.

Sidney, Algernon. 1750. [1698.] *Discourses Concerning Government* (2 vols.). Edinburgh, United Kingdom: G. Hamilton and J. Balfour.

———. 1996. [1698.] *Discourses Concerning Government*. Thomas G. West, ed. Indianapolis, Indiana: Liberty Fund. Republished in the *Online Library of Liberty*. Indianapolis, Indiana: Liberty Fund. http://oll.libertyfund.org/index.php?option=com_staticxt&staticfile=show.php%3Ftitle=223&Item-id=99999999.

Siebarth, Werner, ed. 1936. *Hitlers Wollen* (*Hitler's Wants*). Munich, Germany: Frz. Eher Verlag (Frz. Eher Publishing House).

Silber, Kenneth. 2005. "The First Eugenicist." *Reason*, vol. 37 (no. 2, July). Republished on *Reason Online*. Washington, D.C.: Reason Foundation. http://www.reason.com/news/show/32283.html.

Silver, Lee M. 1998, paperback. [1997.] *Remaking Eden: How Genetic Engineering and Cloning Will Transform the American Family*. New York, New York: Bard.

———. 2006. *Challenging Nature: The Clash of Science and Spirituality at the New Frontiers of Life*. New York, New York: HarperCollins.

*Simon, Julian L. 1980. *The Ultimate Resource*. Princeton, New Jersey: Princeton University Press.

*———. 1998, trade paperback. [1996 hardcover.] *The Ultimate Resource 2*. Princeton, New Jersey: Princeton University Press.

*Simon, Julian L., ed. 1995, trade paperback. *The State of Humanity*. Cambridge, Massachusetts: Blackwell Publishers.

Simon, Linda. 2004. *Dark Light: Electricity and Anxiety from the Telegraph to the X-Ray*. Orlando, Florida: Harcourt.

Singer, Peter. 1999. *A Darwinian Left*. New Haven, Connecticut: Yale University Press.

Sinclair, Upton, ed. 1915. *The Cry for Justice: An Anthology of the Literature of Social Protest*. Philadelphia, Pennsylvania: The John C. Winston Company.

Sinklair, George. 1971. *The Racial Attitudes of American Presidents*. New York, New York: Doubleday.

Sirota, David. 2007. *Hostile Takeover: How Big Money and Corruption Conquered Our Government—And How We Can Take It Back*. New York, New York: Broadway Books.

Skousen, Mark. 2001 trade paperback. [2001 hardcover.] *The Making of Modern Economics: The Lives and Ideas of the Great Thinkers*. Armonk, New York: M. E. Sharpe.

———. 2007. "*Atlas Shrugged*—50 Years Later." *Christian Science Monitor*. March 6. Republished on the website of Mark Skousen. New York, New York: Mark Skousen. http://www.mskousen.com/2007/03/atlas-shrugged-50-years-later/.

Smith, Adam. 1790 6th ed. [1759.] *The Theory of Moral Sentiments*. Edinburgh, United Kingdom: A. Millar, in the Strand, and A. Kincaid and J. Bell. Republished online in the *Library of Economics and Liberty*. Indianapolis, Indiana: Liberty Fund. http://www.econlib.org/Library/Smith/smMS.html.

———. 1904. [1776.] *An Inquiry Into the Nature and Causes of the Wealth of Nations* (2 vols.). Edwin Cannan ed. London, United Kingdom: Methuen and Co., Ltd. Republished online in the *Library of Economics and Liberty*. Indianapolis, Indiana: Liberty Fund. http://www.econlib.org/library/Smith/smWNtoc.html.

———. 1982. [1759.] *Glasgow Edition of the Works and Correspondence of Adam Smith* (7 vols.) David Daiches Raphael and Alec Lawrence Macfie, eds. Indianapolis, Indiana: Liberty Fund, Inc. Republished in the *Online Library of Liberty*. Indianapolis, Indiana: Liberty Fund, Inc. http://tinyurl.com/65m4qy.

Smith, Denis Mack. 1982. *Mussolini*. New York, New York: Alfred A. Knopf.

Smith, G. W., ed. 2002. *Liberalism: Rights, Property and Markets*. New York, New York: Routledge.

Smith, George H. 1978. "Will the Real Herbert Spencer Please Stand Up?" *The Libertarian Review*, vol. 7 (no. 11, December): 14–18. Roy A. Childs, Jr., editor-in-chief. San Francisco, California: Libertarian Review, Inc.

———. 1981. "Herbert Spencer's Theory of Causation." *The Journal of Libertarian Studies*, vol. 5 (no. 2, Spring): 113–152. Murray N. Rothbard, ed. New York, New York: The Center for Libertarian Studies.

*———. 1989, trade paperback. [1979.] *Atheism: The Case Against God*. Amherst, New York: Prometheus Books.

———. 1991-A. *Atheism, Ayn Rand, and Other Heresies*. Buffalo, New York: Prometheus Books.

———. 1991-H. "Justice Entrepreneurship in a Free Market." In G. H. Smith 1991-A: 295–324.

———. 1991-K. [1978.] "Will the Real Herbert Spencer Please Stand Up?" In G. H. Smith 1991-A: 239–250.

Smith, Richard Norton. 1997. *The Colonel: The Life and Legend of Robert R. McCormick*. Boston, Massachusetts: Houghton Mifflin Company.

Snyder, Jack. 2000. *From Voting to Violence: Democratization and Nationalist Conflict*. New York, New York: W. W. Norton & Company.

Soros, George. 1997. "The Capitalist Threat." *The Atlantic Monthly*, vol. 279 (no. 2, February): 45–58. Republished on the *Atlantic* website. New York, New York: The Atlantic Monthly Company. http://www.theatlantic.com/past/docs/issues/97feb/capital/capital.htm.

Specter, Michael. 2009. *Denialism: How Irrational Thinking Hinders Scientific Progress, Harms the Planet, and Threatens Our Lives*. New York, New York: Penguin Press.

*Spencer, Herbert. *Social Statics*. 1851. London, United Kingdom: John Chapman. Republished in the *Online Library of Liberty*. Indianapolis, Indiana: Liberty Fund. http://tinyurl.com/2fw89o.

———. 1870 2d ed. *First Principles: Of a New Philosophy*. New York, New York: D. Appleton and Company.

———. 1886. *The Study of Sociology*. New York, New York: D. Appleton and Company.

———. 1891a. *Essays: Scientific, Political, & Speculative* (3 vols.). London, United Kingdom: Williams and Norgate.

———. 1891b. [1872.] "Mr. Martineau on Evolution." In Spencer 1891a, vol. 1: 371–388.

———. 1891c. [12 March 1864.] "Reasons for Dissenting from the Philosophy of M. Comte." In Spencer 1891a, vol. 3: 59–80.

———. 1897. *The Principles of Sociology* vol. 3 of 3. New York, New York: D. Appleton and Company.

———. 1902a. "Exaggerations and Mis-Statements." In Spencer 1902b: 145–156.

———. 1902b. *Facts and Comments*. New York, New York: D. Appleton and Company.

———. 1902c. "Imperialism and Slavery." In Spencer 1902b: 157–171.

———. 1906. *The Principles of Sociology* vol. 2 of 3. New York, New York: D. Appleton and Company.

———. 1907a. "The Development Hypothesis." In Spencer 1907g: 35–37.

———. 1907b. "The Genesis of Science." In Spencer 1907g: 38–73.

———. 1907f. [1857.] "Progress: Its Law and Cause." In Spencer 1907g: 7–34.

———. 1907g. *Seven Essays Selected from the Works of Herbert Spencer*. London, United Kingdom: Watts & Co.

———. 1908a. *The Life and Letters of Herbert Spencer*. David Duncan, ed. London, United Kingdom: Methuen & Co.

———. 1908b. [1873.] "To J. E. Cairnes, 21 March 1873." In Spencer 1908a: 161.

———. 1910a. [1897.] *The Principles of Sociology* vol. 1 of 3. New York, New York: D. Appleton and Company.

———. 1910b. [1897.] *Various Fragments*. New York, New York: D. Appleton and Company.

———. 1978. [1897.] *The Principles of Ethics* (2 vols.). Indianapolis, Indiana: Liberty Classics. Republished in the *Online Library of Liberty*. Indianapolis, Indiana: Liberty Fund. http://tinyurl.com/yvo78j.

*———. 1981a. [1884.] "The Coming Slavery." In Spencer 1981c: 31–70. Republished in the *Online Library of Liberty*. Indianapolis, Indiana: Liberty Fund. http://oll.libertyfund.org/?option=com_staticxt&staticfile=show.php%3Ftitle=330&chapter=119773&layout=html#a_2420254.

*———. 1981b. [1884.] "The Great Political Superstition." In Spencer 1981c: 123-166. Republished in the *Online Library of Liberty*. Indianapolis, Indiana: Liberty Fund. http://oll.libertyfund.org/?option=com_staticxt&staticfile=show.php%3Ftitle=330&chapter=119745&layout=html&Itemid=27.

*———. 1981c. [Rpt. of 1892 ed.; 1st ed. published 1884.] *The Man Versus the State, With Six Essays on Government, Society and Freedom*. Indianapolis, Indiana: LibertyClassics. Also republished online in *The Library of Economics and Liberty*. Indianapolis, Indiana: Liberty Fund. http: //www.econlib.org/library/LFBooks/Spencer/spnMvStoc.html.

*———. 1981b. [1884.] "The New Toryism." In Spencer 1981c: 5–30. Republished in the *Online Library of Liberty*. Indianapolis, Indiana: Liberty Fund. http://oll.libertyfund.org/?option=com_staticxt&staticfile=show.php%3Ftitle=330&chapter=119773&layout=html#a_2420254.

*———. 1981e. [1884.] "Postscript." In Spencer 1981c: 167–177. Republished in the *Online Library of Liberty*. Indianapolis, Indiana: Liberty Fund. http://oll.libertyfund.org/?option=com_staticxt&staticfile=show.php%3Ftitle=330&chapter=119746&layout=html&Itemid=27.

*———. 1981f. [1842–43.] "The Proper Sphere of Government." In Spencer 1981c: 181–264. Republished in the *Online Library of Liberty*. Indianapolis, Indiana: Liberty Fund. http://oll.libertyfund.org/?option=com_staticxt&staticfile=show.php%3Ftitle=330&chapter=119773&layout=html#a_2420254.

*———. 1981f. [1884.] "Sins of Legislators." In Spencer 1981c: 71–122. Republished in the *Online Library of Liberty*. Indianapolis, Indiana: Liberty Fund. http://oll.libertyfund.org/?option=com_staticxt&staticfile=show.php%3Ftitle=330&chapter=119744&layout=html&Itemid=27.

———. 1981h. [1860.] "The Social Organism." In Spencer 1981c: 383–434 Republished in the *Online Library of Liberty*. Indianapolis, Indiana: Liberty Fund. http://oll.libertyfund.org/?option=com_staticxt&staticfile=show.php%3Ftitle=330&chapter=119773&layout=html#a_2420254.

Spiegel, Henry William. 2002 3d ed., trade paperback. [1971.] *The Growth of Economic Thought*. Durham, North Carolina: Duke University Press.

Spiro, Jonathan Peter. 2009. *Defending the Master Race: Conservation, Eugenics, and the Legacy of Madison Grant*. Burlington, Vermont: University of Vermont Press.

Sponsel, Leslie E. 2013. "Industrial-Growth Society Putting All Our Futures in Danger." *Honolulu Star-Advertiser* (Tuesday, October 15): A11. Honolulu, Hawaii: Honolulu Star-Advertiser.

Standage, Tom. 2009. *An Edible History of Humanity*. New York, New York: Wallace & Company.

Standiford, Lee. 2005. *Meet You in Hell: Andrew Carnegie, Henry Clay Frick, and the Bitter Partnership That Transformed America*. New York, New York: Crown Publishers.

Stanley, Tim, et al., eds. 2013. *Justia*. Mountain View, California: Justia Company. http://www.justia.com/.

Starr, Paul. 1982. *The Social Transformation of American Medicine*. New York, New York: Basic Books.

Staudenmaier, Peter. 2013. "Organic Farming in Nazi Germany: The Politics of Biodynamic Agriculture, 1933–1945." *Environmental History* no. 18: 383–411.

*Steigmann-Gall, Richard. 2003. *The Holy Reich: Nazi Conceptions of Christianity, 1919–1945*. New York, New York: Cambridge University Press.

Steinberg, Jonathan. 2011. *Bismarck: A Life*. New York, New York: Oxford University Press.

Steinberg, Neil. 2005. "Wal-Mart's Red Alert." *New York Daily News*. February 13. Republished on the website of the *New York Daily News*. New York, New York: New York Daily News. http://tinyurl.com/ysz5gb.

*Steinweis, Alan E. 1993. *Art, Ideology, & Economics in Nazi Germany: The Third Reich Chambers of Music, Theater, and the Visual Arts*. Chapel Hill, North Carolina: The University of North Carolina Press.

Stepelvich, Lawrence S., ed. 1983. *The Young Hegelians: An Anthology*. New York, New York: Cambridge University Press.

Stephen, James Fitzjames. 1883. *A History of Criminal Law in England* (3 vols.). London, United Kingdom: Macmillan and Co.

Stern, Fritz. 1961. *The Politics of Cultural Despair*. Berkeley, California: University of California Press.

———. 1977. *Gold and Iron: Bismarck, Bleichröder, and the Building of the German Empire*. New York, New York: Alfred A. Knopf.

———. 1987. *Dreams and Delusions: The Drama of German History*. New York, New York: Alfred A. Knopf.

———. 2006. *Five Germanys That I Have Known*. New York, New York: Farrar, Straus and Giroux.

Sternheimer, Karen. 2010. *Connecting Social Problems and Popular Culture: Why Media Is Not the Answer*. Boulder, Colorado: Westview Press.

Stieg, Margaret F. 1992. *Public Libraries in Nazi Germany*. Tuscaloosa, Alabama: University of Alabama Press.

Stokes, Gale. 1993. *The Walls Came Tumbling Down: The Collapse of Communism in Eastern Europe*. New York, New York: Oxford University Press.

Stossel, John. 2006a. *Myths, Lies, and Downright Stupidity: Get Out the Shovel—Why Everything You Know Is Wrong*. New York, New York: Hyperion.

Strathern, Paul. 2002. *Heidegger in 90 Minutes*. Chicago, Illinois: Ivan R. Dee.

Strossen, Nadine. 2000, paperback. [1995.] *Defending Pornography: Free Speech, Sex, and the Fight for Women's Rights*. New York, New York: New York University Press.

Sullum, Jacob. 1998. *For Your Own Good: The Anti-Smoking Crusade and the Tyranny of Public Health*. New York, New York: The Free Press.

———. 1999. "To Your Health!" Reason Foundation op-ed. *Reason Online*. September 13. Washington, D.C.: Reason Foundation. http://www.reason.com/news/show/34344.html.

*Sumner, William Graham. 1911. *What Social Classes Owe to Each Other*. New York, New York: Harper and Brothers. Republished in the *Online Library of Liberty*. Indianapolis, Indiana: Liberty Fund, Inc. http://tinyurl.com/34vafr.

———. 1940. [1906.] *Folkways: A Study of the Sociological Importance of Usages, Manners, Customs, Mores, and Morals*. Boston, Massachusetts: Gina and Company.

*———. 1969-A. [c. 1880s.] "The Challenge of Facts." In Keller and Davie eds. 1969 vol. 2: 97–122.

*———. 1969-B. "The Concentration of Wealth: Its Economic Justification." In Keller and Davie eds. 1969 vol. 2: 163–172.

———. 1969-C. [1899.] "The Conquest of the United States By Spain." In Keller and Davie eds. 1969 vol. 2: 266–303.

*———. 1969-D. "The Forgotten Man." In Keller and Davie eds. 1969 vol. 1: 466–496.

*———. 1969-E. "The Influence of Commercial Crises on Opinions About Economic Doctrines." In Keller and Davie eds. 1969 vol. 2: 44–66.

*———. 1969-F. "Laissez Faire." In Keller and Davie eds. 1969 vol. 2: 468–477.

*———. 1969-G. "Liberty and Responsibility, Part 1 of 9." In Keller and Davie eds. 1969 vol. 1: 310–314.

*———. 1969-I. "Liberty and Responsibility, Part 3 of 9: Liberty and Discipline." In Keller and Davie eds. 1969 vol. 1: 320–25.

———. 1969-J. "Purpose and Consequences." In Keller and Davie eds. 1969 vol. 1: 11–19.

———. 1969-K. "Rights." In Keller and Davie eds. 1969 vol. 1: 358–362.

———. 1969-L. "The Scientific Attitude of Mind." In Keller and Davie eds. 1969 vol. 1: 43–54.

———. 1969-M. "Some Natural Rights." In Keller and Davie eds. 1969 vol. 1: 363-68.

———. 1992. *On Liberty, Society, and Politics: The Essential Essays of William Graham Sumner*. Robert C. Bannister, ed. Indianapolis, Indiana: Liberty Fund. Republished in 2004 in the *Online Library of Liberty*. Indianapolis, Indiana: Liberty Fund, Inc. http://tinyurl.com/3c9lnu.

Sumner, William Graham, with Albert Galloway Keller. 1927. *The Science of Society* vol. 1. New Haven, Connecticut: Yale University Press.

Swade, Doron. 2001. *The Difference Engine: Charles Babbage and the Quest to Build the First Computer*. New York, New York: Viking.

Talmon, Jacob L. 1955. *The Origins of Totalitarian Democracy*. London, United Kingdom: Secker and Warburg.

Tame, Chris. 1998. "An Economic Misinterpretation of History: A Critique of J. K. Galbraith's Account of American Capitalism." *Historical Notes* vol. 1 (no. 6): 1–7. Republished on the website of the Libertarian Alliance. London, United Kingdom: Libertarian Alliance. PDF: http://tinyurl.com/24e2df.

Tarbell, Ida M. 1939. *All in a Day's Work: An Autobiography*. New York, New York: Macmillan.

Taylor, Joan Kennedy. 1974. "Slavery: America's Fatal Compromise." In T. R. Machan ed. 1974: 235–244.

Teixeira, Ruy, and John Halpin. 2010. *The Progressive Tradition Series* Pt. 2: *The Progressive Tradition in American Politics*. April. Washington, D.C.: Center for American Progress. http://www.americanprogress.org/issues/2010/04/pdf/progressive_traditions.pdf.

Thies, Clifford, and Ryan Daza. 2011. "Richard T. Ely: The Confederate Flag of the AEA?" *Econ Journal Watch*, vol. 8 (no. 2, May): 147–156. Republished on the website of *Econ Journal Watch*. Daniel T. Klein, ed. Fairfax, Virginia: George Mason University. http://econjwatch.org/articles/richard-t-ely-the-confederate-flag-of-the-aea.

Thomas, Norman. 1971. *A Socialist's Faith*. Port Washington, New York: Kennikat Press.

**Thompson, C. Bradley. 1998. *John Adams and the Spirit of Liberty*. Lawrence, Kansas: University Press of Kansas.

Thompson, J. M. 1939. *Robespierre*. Oxford, United Kingdom: Basil Blackwell.

Thompson, Larry. 1994. *Correcting the Code: Inventing the Genetic Cure for the Human Body*. New York, New York: Simon & Schuster.

Thüring, Bruno. 1936. *Deutsche Mathematik* (*German Mathematics*). Theodor Vahlen, ed. Leipzig, Germany: Kommissionsverlag von S. Hirzel (Von S. Hirzel Commission Publishing).

———. 1966. [1936.] "Nature Presupposes a Spiritual Disposition." In G. L. Mosse ed. 1966: 208–215.

Thurow, Lester. 1997. *The Future of Capitalism: How Today's Economic Forces Shape Tomorrow's World*. New York, New York: Penguin.

Tilney, Nicholas L. 2003. *Transplant: From Myth to Reality*. New Haven, Connecticut: Yale University Press.

Tismaneanu, Vladimir. 2012. *The Devil in History: Communism, Fascism, and Some Lessons of the Twentieth Century*. Berkeley, California: University of California Press.

Toffler, Alvin. 1981, paperback. [1980.] *The Third Wave*. New York, New York: Bantam Books.

Toland, John. 1976. *Adolf Hitler*. New York, New York: Doubleday & Company.

Tone, Andrea. 2001. *Devices and Desires: A History of Contraceptives in America*. New York, New York: Hill and Wang.

Toussenel, Alphonse. 1947. *Les Juifs, Rois de l'Époque, Histoire de la Féodalité Financière* (*Jewish Kings of the Era, A history of Financial Feudalism*) (2 vols.). Paris, France: École Sociétaire.

Townsend, Ebenezer. 1921. [1888.] *Extract from the Diary of Ebenezer Townsend, Jr., Supercargo of the Sealing Ship "Neptune" on Her Voyage to the South Pacific and Canton . . . Arranged and Indexed for the Hawaiian Historical Society By Bruce Cartwright*. Honolulu, Hawaii: Hawaiian Historical Society Reprints no . 4.

Trachtenberg, Alan. 1982. *The Incorporation of America: Culture and Society in the Gilded Age*. New York, New York: Hill and Wang.

**Tracinski, Robert. 2005. "Goodwill Versus Duty." *Capitalism Magazine* "Dollars & Crosses" Blog. January 7. Mark Da Cunha, ed. Freeport, The Bahamas: Bahamas 2000. http://www.capmag.com/news.asp?id=1335.

*Tracy, Antoine-Louis-Claude, Destutt de. 1817. "A Treatise on Political Economy: To Which is Prefixed a Supplement to a Preceding Work on the Understanding or Elements of Ideology; With an Analytical Table, and an Introduction on the Faculty of the Will." Thomas Jefferson, trans. Georgetown, Washington, D.C.: Joseph Milligan. Republished in the *Online Library of Liberty*. Indianapolis, Indiana: Liberty Fund, Inc. http://tinyurl.com/29bwqa.

Trefil, James. 1997. *Are We Unique?: A Scientist Explores the Unparalleled Intelligence of the Human Mind*. New York, New York: John Wiley & Sons.

Trivers, Robert. 1981. "Sociobiology and Politics." In E. White ed. 1981: 1–44.

Tuccille, Jerome. 1975. *Who's Afraid of 1984?* New Rochelle, New York: Arlington House.

Tuchman, Barbara. 1970. *The Proud Tower: A Portrait of the World Before the War, 1890–1914*. New York, New York: Macmillan.

Tudge, Colin. 1995. [1993.] *The Engineer in the Garden—Genes and Genetics: From the Idea of Heredity to the Creation of Life*. New York, New York: Hill and Wang.

———. 2000. *The Impact of the Gene: From Mendel's Peas to Designer Babies*. New York, New York: Hill and Wang.

*Turner, Henry Ashby. 1985. *German Big Business and the Rise of Hitler*. New York, New York: Oxford University Press.

Uglow, Jenny. 2003. *The Lunar Men: Five Friends Whose Curiosity Changed the World*. New York, New York: Farrar, Straus and Giroux.

University of Virginia Library. 2012. *Electronic Text Center*. Charlottesville, Virginia: University of Virginia. http://www2.lib.virginia.edu/scholarslab/.

Valeri, Valerio. 1985. *Kingship and Sacrifice: Ritual and Society in Ancient Hawaii*. Chicago, Illinois: The University of Chicago Press.

Valk, Anne M. 2003/2004. "Art for the People: The Rediscovery and Preservation of Progressive and WPA-Era Murals in the Chicago Public Schools, 1904–1943." *Journal of the Illinois State Historical Society* (no. 1, Winter). http://findarticles.com/p/articles/mi_qa3945/is_200301/ai_n9333977.

Vander Vliet, Amy, ed. 2013. *Resources on Faith, Ethics, and Public Life*. Washington, D.C.: Berkeley Center for Religion, Peace, & World Affairs at Georgetown University. http://berkleycenter.georgetown.edu/resources.

van Schaik, Carel. 2004. *Among Orangutans: Red Apes and the Rise of Human Culture*. Cambridge, Massachusetts: The Belknap Press of Harvard University Press.

Vedantam, Shankar. 2007. "If It Feels Good to Be Good, It Might Only Be Natural." *Washington Post*. Monday, May 28. Republished on the website of the *Washington Post*. Arlington, Virginia: The Washington Post Company. http://www.washingtonpost.com/wp-dyn/content/article/2007/05/27/AR2007052701056_pf.html.

Voltaire. 1792. *Œuvres de Voltaire: Facéties* [*Works of Voltaire: Facetiousness, Jokes*]. Paris, France: Stoupe.

Vorzimmer, Peter J. 1977. *Charles Darwin: The Years of Controversy—"The Origin of Species" and Its Critics, 1859–1882*. Philadelphia, Pennsylvania: Temple University Press.

Waal, Frans, de. 2009. *The Age of Empathy: Nature's Lessons for a Kinder Society*. New York, New York: Three Rivers Press.

Wagener, Otto. 1985. *Hitler: Memoirs of a Confidant*. Henry Ashby Turner, Jr., ed. Ruth Heim, trans. New Haven, Connecticut: Yale University Press.

Wainwright, Milton. 1990. *Miracle Cure: The Story of Penicillin and the Golden Age of Antibiotics*. Oxford, United Kingdom: Basil Blackwell.

Waldrop, M. Mitchell. 1993, trade paperback. [1992.] *Complexity: The Emerging Science at the Edge of Order and Chaos*. New York, New York: Simon & Schuster.

Ward, Joseph P., ed. 2003. *Britain and the American South: From Colonialism to Rock and Roll*. Oxford, Mississippi: University Press of Mississippi.

Ward, Lester Frank. 1891. "Neo-Darwinism and Neo-Lamarckism." *Proceedings of the Biological Society of Washington* 6: 71.

———. 1902. "Contemporary Sociology" (in three parts). *The American Journal of Sociology* vol. 7: 749–762.

———. 1903. *Pure Sociology*. New York, New York: Macmillan Co.

———. 1907a. [1883.] *Dynamic Sociology; or Applied Social Science* (2 vols.). New York, New York: Appleton.

———. 1907b. "Social and Biological Struggles." *American Journal of Sociology* vol. 13: 289–299.

———. 1913. "Eugenics, Euthenics, and Eudemics." *American Journal of Sociology* vol. 18 (no. 6, May): 737–754.

———. 1975. [1871–73.] "Education (1871–1873)." In Karier ed. 1975, 145–159.

Watson, James D. 2003. *DNA: The Secret of Life*. New York, New York: Alfred A. Knopf.

Watson, James D., ed. 2005. *Darwin: The Indelible Stamp—The Evolution of an Idea*. Philadelphia, Pennsylvania: Running Press.

Watson, Peter. 2010. *The German Genius: Europe's Third Renaissance, the Second Scientific Revolution, and the Twentieth Century*. New York, New York: HarperCollins.

Watts, Steve. 2005. *The People's Tycoon: Henry Ford and the American Century*. New York, New York: Alfred A. Knopf.

Webb, Beatrice. 1980, paperback. [1926.] *My Apprenticeship*. New York, New York: Cambridge University Press.

Webb, Sidney. 1910–1911. "Eugenics and the Poor Law: The Minority Report." *Eugenics Review* 2: 233–241.

———. 1912. "The Economic Theory of a Legal Minimum Wage." *Journal of Political Economy*, vol. 20 (no. 10, December): 973–998.

Webb, Sidney, and Beatrice Webb. 1920. [1897.] *Industrial Democracy*. London, United Kingdom: Longmans Green.

Webber, Carolyn, and Aaron Wildavsky. 1986. *A History of Taxation and Expenditure in the Western World*. New York, New York: Simon & Schuster.

Weingart, Peter, et al. 1988. *Rasse, Blut un Gene: Geschichte der Eugenik und Rassehygiene in Deutschland*, Frankfurt, Germany: Suhrkamp.

*Weissmann, Gerald. 1996a. [1995.] "Ecosentimentalism: The Summer Dream Beneath the Tamarind Tree." In P. R. Gross et al. eds. 1996, trade paperback: 483–89.

———. 1996b. [1995.] " 'Sucking With Vampires': The Medicine of Unreason." In P. R. Gross et al. eds. 1996, trade paperback: 179–187.

Welling, George M., ed. 2012. *American History from Revolution to Reconstruction and Beyond*. Website of the University of Groningen. Groningen, The Netherlands: University of Groningen. http://www.let.rug.nl/usa/.

Wells, H. G. 2005. [1949.] "'The Nineteenth Century' from *The Outline of History*." In J. Sakolsky ed. 2005: 9–16.

Werth, Barry. 2009. *Banquet at Delmonico's: Great Minds, the Gilded Age, and the Triumph of Evolution in America*. New York, New York: Random House.

Western, David, and Mary C. Pearl, eds. 1989. *Conservation for the Twenty-First Century*. New York, New York: Oxford University Press.

Whelan, Elizabeth M. 1995. "The Carcinogen or Toxin of the Week Phenomenon: The Facts Behind the Scares." In J. L. Simon ed. 1995, trade paperback: 595–607.

White, Elliot, ed. 1981. *Sociobiology and Human Politics*. Lexington, Massachusetts: Lexington Books.

White, Leslie A. 1968. "Nations as Sociocultural Systems." *Ingenor* no. 5 (Autumn): 5–19.

———. 1975. *The Concept of Cultural Systems: A Key to Understanding Tribes and Nations*. New York, New York: Columbia University Press.

Whittemore, Robert C. 1964. *Makers of the American Mind*. New York, New York: Morrow.

Wilson, Colin. 2000. *Rogue Messiahs: Tales of Self-Proclaimed Saviors*. Charlottesville, Virginia: Hampton Roads Publishing Company.

Wilson, David Sloan. 2003. *Darwin's Cathedral*. Chicago, Illinois: University of Chicago Press.

———. 2007. *Evolution for Everyone: How Darwin's Theory Can Change the Way We Think About Our Lives*. New York, New York: Delacorte Press.

Wilson, Edward O. 1978. *On Human Nature*. Cambridge, Massachusetts: Harvard University Press.

———. 1980 abr. ed. [1975.] *Sociobiology*. Cambridge, Massachusetts: The Belknap Press of Harvard University.

———. 1992. *The Diversity of Life*. London, United Kingdom: Allen Lane.

———. 1998. *Consilience: The Unity of Knowledge*. New York, New York: Alfred A. Knopf.

Wilson, Edward O., and Frances M. Peter, eds. 1988. *Biodiversity*. Washington, D.C.: National Academy Press.

Wilson, (Thomas) Woodrow. 1913. *The New Freedom: A Call for the Emancipation of the Generous Energies of a People*. New York, New York: Doubleday, Page & Company.

———. 1968a. *The Papers of Woodrow Wilson* (5 vols.). Arthur S. Link, ed. Princeton, New Jersey: Princeton University Press.

———. 1968b. [1887.] "The Study of Administration." In W. Wilson 1968a: 365–67.

———. 2005. [1913.] *The New Freedom: A Call for the Emancipation of the Generous Energies of a People*. Republished online by Project Gutenberg. http://www.gutenberg.org/files/14811/14811.txt.

Wiltshire, David. 1978. *The Social and Political Thought of Herbert Spencer*. New York, New York: Oxford University Press.

Winston, George P. 1972. *John Fiske*. New York, New York: Twayne Publishers.

Winter, J. Methuen. 1974. "The Webbs and the Non-White World: A Case of Socialist Racialism." *Journal of Contemporary History* vol. 9 (no. 1, January): 181–192.

Wood, Allen, ed. 2001. *Basic Writings on Kant*. New York, New York: Modern Library.

Wood, Neal. 1991, paperback. [1988.] *Cicero's Social and Political Thought*. Berkeley, California: University of California Press.

WordNet 3.0. 2006. Princeton, New Jersey: Princeton University Press.

Wright, Robert. 1994, hardcover. *The Moral Animal: Evolutionary Psychology and Everyday Life*. New York, New York: Pantheon Books.

———. 2000. *Nonzero: The Logic of Human Destiny*. New York, New York: Pantheon Books.

Yafa, Stephen. 2005. *Big Cotton: How a Humble Fiber Created Fortunes, Wrecked Civilizations, and Put America on the Map*. New York, New York: Viking.

Yoder, Jon A. 1975. *Upton Sinclair*. New York, New York: Frederick Ungar Publishing Co.

*Younkins, Edward W. 2002. *Capitalism and Commerce: Conceptual Foundations of Free Enterprise*. Lanham, Maryland: Lexington Books.

Zakaria, Fareed. 2003. *The Future of Freedom: Illiberal Democracy at Home and Abroad*. New York, New York: W. W. Norton & Company.

———. 2011. [2008.] *The Post-American World, Release 2.0*. New York, New York: W. W. Norton & Company.

Zinser, Lin, and Paul Hsieh. 2007–2008. "Moral Health Care Versus 'Universal Health Care.'" *The Objective Standard*, vol. 2 (no. 4, Winter): 9–41. Republished online on the *Objective Standard* website. Craig Biddle, ed. Glen Allen, Virginia: The Objective Standard. http://www.theobjectivestandard.com/issues/2007-winter/moral-vs-universal-health-care.asp.

Zubrin, Robert. 2012. *Merchants of Despair: Radical Environmentalists, Criminal Pseudo-Scientists, and the Fatal Cult of Antihumanism*. New York, New York: Encounter Books.

Zuk, Marlene. 2007. *Riddled With Life: Friendly Worms, Ladybug Sex, and the Parasites That Makes Us Who We Are*. Orlando, Florida: Harcourt.

Zupan, Marty, interviewer. 1984. "Interview With Henry Hazlitt." *Reason*, vol. 16 (no. 7, December): 36–40.

Index

About the Author

Prior to writing this book, **Stuart K. Hayashi** worked as an analyst and aide at the Hawaii State Capitol in both legislative houses. His writings have appeared in *Ideas in Action TV* (formerly *Tech Central Station*) and the *Honolulu Advertiser*. In 2014, the official Twitter account of national television journalist John Stossel quoted him by name, and one section of the official website of Thomas Jefferson's Monticello was revised and updated to incorporate information that Mr. Hayashi provided.

CPSIA information can be obtained at www.ICGtesting.com
Printed in the USA
BVOW09*0444090215

386768BV00003B/4/P

9 780739 186701